PRIMATE ECOLOGY:

PROBLEM-ORIENTED FIELD STUDIES

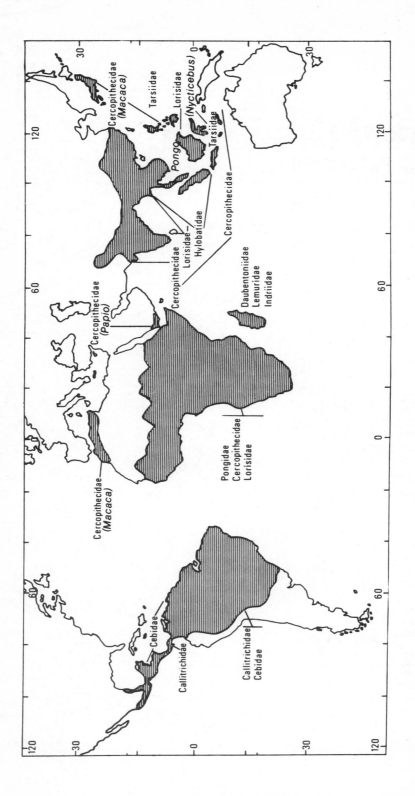

PRIMATE ECOLOGY:

PROBLEM-ORIENTED FIELD STUDIES

ROBERT W. SUSSMAN
Washington University

JOHN WILEY & SONS
New York
Chichester
Brisbane
Toronto

Library of Congress Cataloging in Publication Data:

Main entry under title: Primate ecology

Includes indexes.

1. Primates—Ecology.   2. Mammals—Ecology.
I. Sussman, Robert W., 1941-

QL737.P9P67236   599'.8'045   79-17828
ISBN 0-471-03823-7

Printed in the United States of America

10   9   8   7   6   5   4   3   2   1

PREFACE

Research on the ecology of naturally living primates has consisted of three major types of enquiry: general descriptive natural history; problem-oriented field research; and purely theoretical studies. Until recently most primate field studies have been of the first type. Natural history studies include detailed descriptions of the habits and habitats of the animals. This research provides us with basic knowledge of free-ranging primate populations, and is thus a necessary prerequisite for problem-oriented studies. Primatologists have collected data on a number of species living in their natural environment. However, for the most part, books presently available on primate behavior and ecology, either as textbooks or as readers, have concentrated solely on descriptive, natural history investigations, and usually on the most studied and easily accessible species.

An excellent brief history of field primatology up to the early 1960s can be found in Southwick (1963). Descriptive natural history of primates has continued up to the present but, as Southwick stated in 1963: "For most species of primates, our field knowledge is surprisingly sparse or non-existent" (p. 5). This is still true today, as will be discussed further in my introductory remarks to the various sections of this book.

Within the past 15 years the number of field investigations increased tremendously, and many primatologists have begun to do problem-oriented field studies. In these studies specific questions are posed. These questions are usually formulated either from general ecological theory or directly from data gathered during natural history research. Because of the problem orientation, the research is often, but not always, of shorter duration than that of more general natural history studies. It does not attempt to describe all aspects of the behavior of the species. This type of fieldwork is now becoming the most frequent and most important focus in primatology.

I consider the articles in this book to be a representative sample of this relatively new approach. These investigations are some of the earliest attempts at problem-oriented field research on primates. However, I have tried to choose articles that have somewhat similar themes and use relatively similar methods. In most of the articles, comparisons are made between certain aspects of the behavior and ecology of more than one species living in the same forest and utilizing potentially the same resources or between populations of one species living in different types of environment. In these studies, the investigator can relate specific habitat preferences to differences in the morphology and social behavior of the species or populations under study. Because of the necessity of comparison, the fieldworkers have developed methods enabling them to quantify much of their data. These, however, are some of the first attempts at quantification of data collected in the field on habitat utilization by primates. I suspect that the techniques of data collection and the methods of analysis will improve rapidly. In fact, many studies now in progress and some already in press have become more sophisticated in both data collection and analysis. However, the articles in this book will stand as examples of the beginning of a new direction in primatology.

The book is divided into four initial sections arranged along systematic lines—prosimians, New World monkeys, Old World monkeys, and apes; a fifth section contains theoretical selections that deal with the question of relationships between social structure and ecology among primates. In organizing the first four sections I have used the following classification:

Order: Primates
   Suborder: Prosimii (Prosimians)
      Infraorder: Lemuriformes (lemurs)
         Superfamily: Lemuroidea
            Family: 1.   Lemuridae
                     2.   Indriidae
         Superfamily: Daubentonioidea
            Family: Daubentonioidae (aye aye)
      Infraorder: Lorisiformes
         Family: Lorisidae
            Subfamily: 1.   Lorisinae (lorises)
                          2.   Galaginae (galagos)
      Infraorder: Tarsiiformes (tarsiers)
         Family: Tarsiidae
   Suborder: Anthropoidea
      Superfamily: Ceboidea (New World monkeys)
         Family: 1.   Cebidae
                    2.   Callitrichidae
      Superfamily: Cercopithecoidea (Old World monkeys)
         Family: Cercopithecidae
            Subfamily: Cercopithecinae
                        Colobinae
      Superfamily: Hominoidea (apes and man)
         Family: Hylobatidae (lesser apes)
         Family: Pongidae (greater apes)
         Family: Hominidae (man)

Although there are 16 selections in these four sections, because of the nature of the studies, certain aspects of the behavior and ecology of over 40 species of primate are presented.

The theoretical papers in the final section are presented in chronological order so that the reader can trace the stream of thought and the arguments as they are presented in the literature. These articles focus mainly on the following questions: What is the relationship between the environment or environments in which a species of primate is found and the social structure of that species? In other words, is there a relationship between ecology and social structure, and is this a necessary relationship? Given specific environmental conditions, how can a particular social structure increase the reproductive success of certain individuals? To what extent does genetic relationship (phylogeny) determine social structure regardless of environment? As will be seen as one reads the theoretical papers, although the questions remain the same, as more data are collected, the answers in many ways become more difficult to obtain. This is because we find that we need to make more subtle distinctions between the variables we are measuring. However, intensive, quantified, and comparative studies, such as those presented in this book, should allow us to collect the type of data needed to answer such questions and to formulate and test more sophisticated theories in primatology and in general ecology.

I would like to thank Pamela Ashmore and Robert Jamieson for their assistance in proofing and indexing the final manuscript and to dedicate this book to my parents, Louis and Helen Sussman.

Robert W. Sussman

References cited
Southwick, C.H. 1963. Introduction. *In* Primate Social Behavior, C.H. Southwick (ed.). Van Nostrand Reinhold, New York.

Further Bibliography
Clutton-Brock, T.H. (ed.). 1977. Primate Ecology: Field Studies on Feeding and Ranging Behavior of Lemurs, Monkeys, and Apes. Academic Press, London. (An excellently edited reader containing a number of problem-oriented studies on feeding behavior).
De Vore, I. (ed.). 1965. Primate Behavior: Field Studies of Monkeys and Apes. Holt, Rinehart and Winston, New York. (An early reader including natural history studies carried out in the late 1950s and early 1960s.)
Jay, P.C. (ed.). 1968. Primates: Studies in Adaptation and Variability. Holt, Rinehart and Winston, New York. (A reader dedicated to K.R.L. Hall, one of the first primatologists to do quantitative field studies on primates.)
Jolly, A. 1972. The Evolution of Primate Behavior. Macmillan, New York.

(A textbook and reference book on primate behavor. The relationships and distinctions between behavior in the field and in the laboratory are not always made clear. However, this is the first and best attempt to synthesize the extremely diverse literature in primatology.)

Kummer, H. 1971. Primate Societies: Group Techniques of Ecological Adaptation. Aldine, Chicago. (An excellent book on primate ecology that, however, concentrates on ground-dwelling forms.)

Napier, J.R. and P.H. Napier. 1967. A Handbook of Living Primates. Academic Press, London. (A catalog of the living primates, organized by genera.)

Southwick, C.H. 1963. Primate Social Behavior. Van Nostrand Reinhold, New York. (The first general reader concentrating on primate field studies. It contains a number of historically important papers.)

CONTENTS

PART 1 FIELD STUDIES

Prosimians

New World Monkeys

Old World Monkeys

Apes

PART 2 THEORETICAL PAPERS

FIELD STUDIES

PROSIMIANS

introduction:
field studies on prosimians

Prosimians (the so-called "lower primates") were the earliest primates to evolve, about 60 million years ago. They are often considered "primitive" in relation to the "more advanced" anthropoids (so-called "higher primates"), monkeys, apes, and humans. However, it must be kept in mind that the living prosimians are the end products of 60 million years of evolution. Although some may retain certain early primate features, many are very specialized, both morphologically and behaviorally.

As can be seen in the classification given in the Preface, the suborder Prosimii is divided into three major groups, or infraorders: Lorisiformes, Tarsiiformes, and Lemuriformes. Lorisiformes are represented by one family (Lorisidae) which, because of major differences in locomotor anatomy and behavior, is divided into two subfamilies: Lorisinae and Galaginae. All Lorisiformes are nocturnal (active at night and asleep during the day). The Lorisinae are slow-moving animals, two species of which are found in Asia and two in

tropical forests of Western Africa. Galagines are
fast-moving animals, all six species of which are
confined to Africa. Although a number of natural
history studies have been done on the African
forms of Lorisiformes, very little is known about the
Asian lorises.

Selection I by Charles-Dominique describes a
study of five species of lorisid in a tropical
evergreen forest in Gabon. This study began in
1965, and it is one of the longest continuous field
studies on primates in their natural habitats.
Charles-Dominique provides us with an excellent
understanding of the variability in behavior and
ecology found among nocturnal prosimians. He
illustrates how, by selecting different habitats within
the same forest, the five species avoid competing
with one another for essential resources.

Tarsiiformes also have nocturnal activity
cycles. Although there were a number of genera of
tarsiers during the early evolution of primates (in
the Eocene Epoch), there is only one living genus,
Tarsius. The three living species are found in
Southeast Asia on many islands of the Malay
Archipelago. There is practically nothing known
about naturally living populations of *Tarsius*. The
only study now available (Fogden 1974) came about
by accident when a number of tarsiers were trapped
in mist-nets during a study of birds in Sarawak,
Borneo.

The living Lemuriformes are all found on the
islands of Madagascar and the Comoros. These
primates have probably been isolated on
Madagascar since the Eocene and have radiated
into a number of diverse niches on this large island.
In the present classificatory scheme there are three
families (see preface) and 21 species. Among the
Lemuriformes are the only diurnally active
prosimians. In fact, approximately half of the
species are diurnal. Detailed studies have been
done on four of the diurnal species and one
nocturnal species and preliminary studies on six
others (see Richard and Sussman 1974 for
references). Practically nothing is known about the
natural behavior of 10 of the species.

The detailed studies of the Lorisiformes by Charles-Dominique, along with the studies of *Tarsius* (Fogden 1974) and the nocturnal lemuriform *Microcebus* (Martin 1974) indicate that these nocturnal prosimians share a generally similar social organization. In all of these forms, female ranges are small and overlap to a variable degree. Male ranges are usually larger and scarcely overlap with one another, although they overlap with one or more of the female ranges. This basic social pattern (referred to as "solitary but social" by Charles-Dominique) is found in a number of nocturnal mammals and may have been a primitive feature of the placental mammals.

As can be seen in the articles by myself and by Richard (Chapters 2 and 3), the diurnal lemurs have some behavioral and ecological adaptations that converge with those of anthropoids in a number of ways. Furthermore, all diurnal lemurs live in relatively permanent social groups, as do all anthropoids. The structure of these groups, however, is quite variable. This is true even within the same population for *Propithecus* (Selection 2). The study by Richard compares populations of *Propithecus* living in two vastly different environments. My study (Selection 3) of *Lemur catta* and *Lemur fulvus* compares these two species, both where they are sympatric and where they do not coexist. The three articles in this section illustrate three different methods utilized to quantify behavioral data collected in the fields.

Since the arrival of humans on Madagascar, approximately 2000 years ago, 14 species of lemur have become extinct. At present, all of the living species are gravely in danger of extinction. If conservation efforts are unsuccessful, we may never have the opportunity to learn more from these extremely unique and beautiful animals.

REFERENCES CITED

Fogden, M.P.L. 1974. A preliminary study of the western tarsier, *Tarsius bancanus* Horsefield. *In* Prosimian Biology, R.D. Martin, **G.A. Doyle, and A.C. Walker, eds.** Duckworth, London.

Martin, R.D. 1973. A review of the behavior and ecology of the lesser mouse lemur. *In* Comparative Ecology and Behaviour of Primates, R.P. Michael and J.H. Crook, eds. Academic Press, London.

Richard, A.F. and R.W. Sussman. 1975. Future of Malagasy lemurs: Conservation or extinction? *In* Lemur Biology, I. Tattersall and R.W. Sussman, eds. Plenum Press, New York.

FURTHER BIBLIOGRAPHY

Baldwin, L.A. and G. Teleki. 1977. Field Research on tree shrews and prosimians: An historical, geographical, and bibliographical listing. Primates, 18: 985-1007.

Charles-Dominique, P. 1977. Ecology and Behavior of Nocturnal Primates. Columbia University Press, New York. (A continuation and review of the study reported in Chapter 1 with a current review of our knowledge of nocturnal prosimians.)

Martin, R.D., G.A. Doyle, and A.C. Walker, eds. 1974. Prosimian Biology. Duckworth, London. (A large collection of articles on all aspects of prosimian biology.)

Richard, A.F. 1978. Behavioral Variation: Case study of a Malagasy lemur. Buckmell University Press, Lewisburg. (A more detailed account of data presented in Chapter 2 with excellent chapters on methodology and on relationships between ecology and social structure.)

Tattersall, I. and R.W. Sussman, eds. 1975. Lemur Biology. Plenum Press, New York. (A collection of articles summarizing current research on Malagasy lemurs.)

1

Ecology and Feeding Behaviour of Five Sympatric Lorisids in Gabon

1974

P. CHARLES-DOMINIQUE

INTRODUCTION

This study was carried out in Makokou (Gabon) in the period 1965-1969, on the basis of four study visits of 7, 8, 14 and 3 months' duration, respectively. Field-work was carried out from the CNRS Laboratory in Gabon (originally known as the Mission Biologique au Gabon, under the direction of Professor P.P. Grassé), which is now referred to as the Laboratoire de Primatologie et d'Écologie Equatoriale (Director: A. Brosset).

The study region is located in the heart of the Congolese block of rain-forest (Ogoue-Ivindo basin). It lies 550 km. from the Atlantic coast and 0.4° latitude north of the equator. Along the access routes (roads, certain water-courses, etc.), the forest has been degraded for cultivation which has been subsequently abandoned at various times, thus giving rise to areas of secondary forest at different stages of reconstitution. Primary forest is found a few km. away from these inhabited zones.

Five prosimian species live sympatrically in Gabon; two lorisines:

1. *Perodicticus potto edwardsi* (Bouvier, 1879). Body-weight 1100 gm. Head + body length 327 mm.; tail-length 52 mm.
2. *Arctocebus calabarensis aureus* (De Winton, 1902). Body-weight 200 gm. Head + body length 244 mm.; tail-length 15 mm.

and three galagines:

Reprinted from *Prosimian Biology* by Martin, Doyle, and Walker, Editors by permission of the University of Pittsburgh Press. Copyright © 1974 by Gerald Duckworth & Co., Ltd.

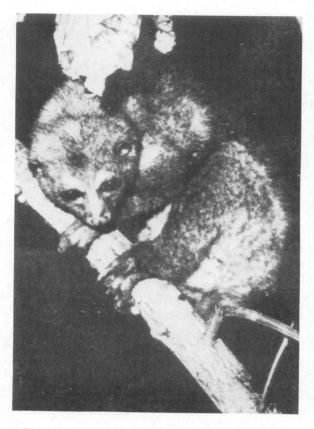

The potto, *Perodicticus potto edwardsi* Bouvier, 1879.

1. *Galago demidovii* (Fischer, 1808). Body-weight 61 gm. Head + body length 123 mm.; tail-length 172 mm.
2. *Galago alleni* (Waterhouse, 1837). Body-weight 260 gm. Head + body length 200 mm.; tail-length 255 mm.
3. *Euoticus elegantulus elegantulus* (Le Conte, 1857). Body-weight 300 gm. Head + body length 200 mm.; tail-length 290 mm.

(These figures represent averages based on 33, 30, 66, 17 and 39 specimens respectively.)

The Lorisinae and the Galaginae represent two quite distinct subfamilies: the former are exclusively slow climbers which never leap and always move around slowly and cautiously, whilst the latter are rapid and vigorous leapers.* These behavioural differences are paralleled by numerous anatomical adaptations, primarily affecting the limbs and the tail. In particular, the tarsus is extremely developed in the Galaginae, whilst the tail is very

Allen's bushbaby, *Galago alleni* Waterhouse 1837.

reduced in Lorisinae. By contrast, the skull, the dentition, the digestive tract and the reproductive organs exhibit only very slight differences which would not, in themselves, justify a separation into two subfamilies.

The leaping specialisation of the bushbabies is generally regarded as providing a means of rapid escape, at the same time permitting exploitation of an extensive home-range. On the other hand, the slow and deliberate gait of the lorises has been given different interpretations by different authors: " . . . high-grade specialisations for an arboreal existence" (Hill); " . . . [1] directly related to catching prey such as insects and roosting birds"

---

*Many authors have exaggerated the locomotor performance of the bushbabies. Sanderson,[5] in particular, writes of a leap of over 10 m. made by *E. elegantulus*, with the animal purportedly gaining in height! We have measured leaps of 2 m. for *G. demidovii*, and of 2.5 m. for *G. alleni* and *E. elegantulus*, with take-off and landing at the same level. The present record is a leap of 5.5 m. (orthogonal projection of horizontal displacement) made by *E. elegantulus*, with a loss in height of 3.1 m.

(Walker).[2] Some authors have, in fact, expressed astonishment that these animals, which are apparently so vulnerable, have not been decimated by predators. From the author's observations in the forest,[3] it seems very likely that some kind of cryptic mechanism is involved, though this would of course operate only in the natural habitat of these species. (The eye, particularly that of raptors, is more sensitive to rapid movement than to slow progression, and many nocturnal arboreal predators are guided primarily by the auditory sense in localising their prey.) A cryptic mechanism would not, in any case, be the sole means of defence. As a last resort, when an encounter occurs, the lorises utilise active defence mechanisms which vary from species to species. The behaviour of the potto and the angwantibo towards predators has already been described, along with the morphological adaptations involved.[3]

In sum, the two lorisid groups have developed two radically different methods of escaping from predators. Whereas the bushbabies flee rapidly once detected by a predator, the lorises avoid detection by means of an elaborate pattern of slow locomotion. It will be shown that such adaptation has been associated with extensive modification of the feeding behaviour and the diet of the lorisines.

ECOLOGY

The lorisids represent a tiny fraction of the forest mammalian fauna; there are 5 species among 120 mammal species living sympatrically at Makokou. Over and above this, these lorisid species live at relatively low population densities. Systematic counting along pathways and the results of trapping[3] indicate the following average densities per square km.:

Perodicticus potto	8 per square km.
Arctocebus calabarensis	2 per square km.
Galago demidovii	50 per square km.
Galago alleni	15 per square km.
Euoticus elegantulus	15 per square km.

Using the same techniques, we have calculated far higher densities for lemurs in Madagascar, where the prosimians represent the bulk of the mammalian fauna.[4] However, one must make allowance for heterogeneous distribution of populations, which occur as "nuclei" ("noyaux") which can be separated to varying degrees. The extreme case seems to be that of the angwantibo, which is abundant in certain parts of the forest (7 per square km.) and virtually lacking over large areas. Nevertheless, we have observed the sympatric occurrence of three, four and even five of these prosimian species in some areas. In fact, numerous interspecific encounters were observed in the forest, without any sign of attack or escape behaviour. In general, any two lorisids of different species which encounter one another exhibit a brief bout of mutual observation and then continue on their way. Thus, any hypothesis of direct interspecific competition based on aggression must be discarded.

The lorisids are all nocturnal. Certain authors, on the basis of observations made in captivity, have suggested that they are partially diurnal or crepuscular.[5,6,7] Such observations must arise from artefacts of captivity, since all of the individuals followed in the forest showed themselves to be strictly nocturnal. In fact, although the lorises only move around during the night, the bushbabies exhibit some initial activity in twilight. But night falls so rapidly at the equator that the bushbabies are only moving around for ten minutes or so before the twilight has faded. The following values for the luminosity of the sky were measured at the times when activity began:

Euoticus elegantulus	300 to 100 lux
Galago demidovii	150 to 20 lux
Galago alleni	50 to 20 ux

even then, it must be remembered that the light penetrating into the undergrowth of primary forest represents only one hundredth of the values measured above the canopy.

Quite erroneously — again as a result of observations in captivity — numerous authors have reached the conclusion that all lorisids use tree-holes as retreats. Of course, when placed in cages deprived of foliage, the animals do actually retreat into the only boxes placed at their disposition. However, of the five species studied, *Galago alleni* is the only one which sleeps in tree-holes under natural conditions. Usually, such tree cavities are in split hollow trunks, which can be entered from above. The bushbaby can spend the whole day clinging to the internal face of such a "chimney." Sometimes a rudimentary nest is built at the base with a few collected leaves. Interspecific competition for daytime retreats can be excluded; the observer has to examine a large number before finding one which is occupied. In addition, *Galago alleni* seems to be quite tolerant of other mammal species. We observed one individual sleeping a few metres away from an arboreal rodent *(Anomalurus erythronotus)* and a group of five bats in a hollow trunk of *Scyphocephalium ochocoa*. The two other bushbaby species sleep singly or in small groups on a small, leafy branch or in a tangle of lianes. *Galago demidovii* will also sleep in spherical nests constructed with green leaves.[3,8,9] The two lorisine species sleep singly on branches or lianes, protected by foliage.

The modern concept of the "ecological niche" involves a large number of factors. Here, we will only examine the spatial localisation of the animals and their respective diets. These two factors cannot be dissociated, since food-seeking occupies the major part of the activity period of these lorisid species.

SPATIAL LOCALISATION

In order to define in an objective manner the localisation and nature of the supports utilised, we considered the following criteria:

- height relative to the ground
- orientation of the support
- diameter of the support
- nature of the support (ground, small trunk, liane base, large trunk, large branch, foliage, foliage mixed with lianes, lianes)

Every time a lorisid was encountered in the forest, we noted the various characteristics of the support on which the animal was *first* seen. (The presence of the observer could modify the selection of any subsequent pathway taken by the animal.) A large number of such observations (642 sightings of the five species) permitted statistical consideration of the data, which are discussed in detail in a previous publication.[3] All that will be given here is a brief description of the forest biotopes most frequently visited by each lorisid species (see Fig.1):

1. Perodicticus potto

The potto inhabits the canopy (10-30 m. in primary forest; 5-15 m. in secondary forest). It is an exclusive climber, using supports with a wide range of sizes (1-30 cm. diameter). This permits the potto to pass from tree

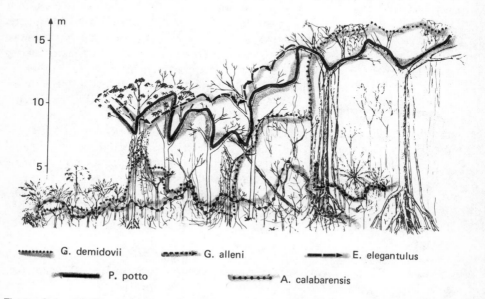

.... G. demidovii   ━•━•━► G. alleni   ━━━► E. elegantulus

━━━━ P. potto   ┉•┉•┉ A. calabarensis

Fig. 1a. Schematic illustration of the paths taken by the five prosimian species in young secondary forest, established secondary forest and flooded forest (passing from left to right).

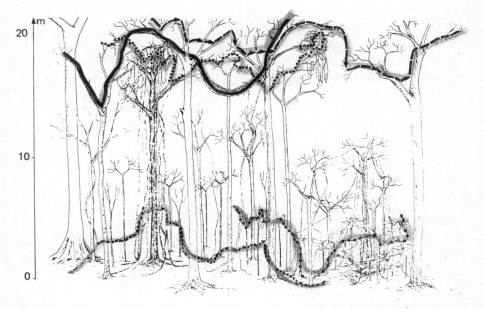

Fig. 1*b*. Similar illustration of the characteristic pathways taken by the five prosimian species in primary forest.

to tree by successively utilising large forks (20%) and small branches which interlock from one branch to another (21%). Lianes which permit short-cuts are also utilised (20%). The support orientations are: horizontals 39%, obliques 35%, verticals 26%. In exceptional cases, the potto descends to the ground (escape from a conspecific, crossing a deforested area, etc.). When this occurs, the potto is extremely cautious and heads directly for the nearest tree.

2. *Arctocebus calabarensis*

The angwantibo lives at 0-5 m., both in primary forest and in secondary forest, though it may rarely flee up to 10-12 m. when greatly alarmed. When progressing, this slow climber generally utilises small lianes passing between the small bushes in the undergrowth (43%). Quite frequently, the angwantibo is sighted in the foliage of these small bushes (43%), where it hunts for insects. This species will descend quite readily to the ground to eat fallen fruits or to follow the tracks made by terrestrial mammals (4%). 40% of the supports utilised have a diameter of less than 1 cm.; 52% are between 1 and 10 cm. in diameter. Orientations of supports are: horizontals 20%, obliques 30%, verticals 50%.

3. Galago demidovii

This bushbaby lives primarily in dense vegetation invaded by small lianes (35%) and in foliage (25%). Accordingly, it is found at a great height in primary forest (10-30 m.) and low down in secondary forest (0-10 m.); the vertical distribution follows the distribution of the biotopes utilised. 64% of the supports used are less than 1 cm. in diameter; 25% are between 1 and 5 cm. in diameter. The support orientations are: horizontals 22%, obliques 30%, verticals 48%.

4. Galago alleni

This bushbaby species is more or less restricted to primary forest, living at a height of 0-2 m. It hunts for animal prey at ground-level, at the same time collecting fallen fruits; but locomotion consists primarily of leaping from one vertical support to another (small tree-trunks; bases of large lianes). The supports utilised have the following diameters: 9% less than 1 cm., 60% between 1 and 5 cm., 19% between 5 and 10 cm. The support orientations are: horizontals 6%, obliques 13%, verticals 81%.

5. Euoticus elegantulus

This bushbaby, which scarcely ever descends to the ground, lives in the canopy up to 50 m. However, it may descend to a height of 3-4 m. on certain large lianes which provide droplets of gum. Much progression is along large branches (62%) or large lianes (30%), and along the larger tree-trunks. The special structure of the nails, which terminate in small "claws," permits movement head-downwards or head-upwards along large, smooth trunks where no other lorisid species would be able to maintain a grasp. It will be seen that this adaptation is related to the diet. The support orientations are: horizontals 22% obliques 51%, verticals 27%.

To summarise the above information, it can be said that within each sub-family the closest species are ecologically separated by the height of the forest stratum exploited. In the Lorisinae, the potto exploits the canopy and the angwantibo lives in the undergrowth. Among the Galaginae, *Euoticus elegantulus* principally exploits the canopy, whilst *Galago alleni* inhabits the undergrowth. *Galago demidovii* provides a special case: this species depends upon dense vegetation, in which it can move easily thanks to its small size, and it follows the distribution of such vegetation (high up in primary forest and low down in secondary forest).

This latter example provides a good illustration of the fact that the height of the stratum exploited is not due to a "height preference" exerted by each animal. It is an indirect consequence of choice of particular vegetation zones in each case. We were able to observe that the same applies to *Galago alleni*. After hand-rearing two young animals captured at the age of 1 week (at which stage they can only move a few centimetres), we found that, from the age of 1 month onwards, they sought out all supports with a vertical orientation (especially chair and table legs). Such selection, which has an apparent hereditary basis, is accompanied by a specific leaping "style" which is adapted to this kind of support.[3] In addition, *Galago alleni* captured as adults and placed in secondary forest (where they scarcely ever occur naturally) moved around at heights of 10-15 m. They utilised vertical or oblique branches at this height when progressing, whereas the average height utilised in primary forest is 1-2 m. With *Galago alleni*, it seems very likely that the vertical orientation of supports of a certain size contributes greatly to the localisation of this species in the undergrowth zone of primary forest.

It will be seen that spatial localisation of these lorisid species is no more than the bare framework of ecological separation permitting closely related species to avoid dietary competition. In parallel with such spatial localisation, there has been dietary specialisation, which has particularly affected feeding behaviour.

DIET

Direct observation of lorisids under natural conditions is difficult. When they are surprised by the light-beam of a headlamp they remain immobile for a few seconds and then disappear rapidly into the vegetation. Under these conditions, analysis of the diet could only be carried out through examination of stomach contents. In Gabon, the prosimians are virtually untouched by hunters, and it was therefore possible to collect a large number of specimens without endangering the local prosimian populations. The animals were collected with the aid of a rifle, samples being taken at different times of the night and in all months of the year. This permitted us to study alimentary rhythms during the night as well as annual rhythms (examination of 174 digestive tracts for the 5 species). The food is quite finely chewed before swallowing; but it is still quite easy to separate the different constituents, which are mixed only to a slight extent in the stomach. We were thus able to separate animal prey from fruit and gums and to obtain fresh weights for these constituent fractions for each stomach examined. The fruits can be identified when the kernels are ingested, but it is often impossible to obtain a precise identification for animal prey. The analysis was therefore restricted to the level of large taxonomic groups: Coleoptera, Lepidoptera (both caterpillars and moths), Orthoptera, Hymenoptera (ants), Isoptera (termites), Myriapoda (centipedes and millipedes), Arachnida, Gasteropoda (slugs), Batracia.

After calculating for each species the percentage of the three principal dietary categories, we obtained the following results (Table 1):

Table 1
Stomach Contents

	Animal Prey	Fruits	Gums
Perodicticus potto	10%	65% (+ some leaves and fungi)	21%
Arctocebus calabarensis	85%	14% (+ some wood fibre)	—
Galago demidovii	70%	19% (+ some leaves and buds)	10%
Galago alleni	25%	73% (+ some leaves, buds and wood fibre)	— (small amounts only)
Euoticus elegantulus	20%	5% (+ some buds)	75%

If this first result were taken at face value, one might conclude that the bushbabies include one insectivore, one frugivore, and one gummivore, whilst the lorises represent one frugivore and one insectivore. Since the most closely related animals are separated by strafitication of the vegetational zones utilised, it would seem that there can be no dietary competition. However, the problem is in reality more complex than this. The percentage figures do not take into account differences in body size of the five prosimian species concerned — that is, they do not reflect the real weight of food ingested. By calculating the average weight of each of the three dietary categories per stomach, the following results are obtained (Table 2).

Table 2
Weights of Dietary Components

	Animal prey	Fruits	Gums
Perodicticus potto	3.4 gm.	21 gm.	7 gm.
Arctocebus calabarensis	2.0 gm.	0.3 gm.	0 gm.
Galago demidovii	1.16 gm.	0.3 gm.	0.15 gm.
Galago alleni	2.2 gm.	9.2 gm.	(negligible)
Euoticus elegantulus	1.18 gm.	0.25 gm.	4.8 gm.

Although the differences for fruits and gums are very clear, this is not the case with animal prey, which are consumed in virtually equal quantities by all five species. Insects (which represent the bulk of the animal prey) are dispersed in the vegetation, and a hunting animal must explore large areas in order to capture large numbers of prey. A small bushbaby and a large bushbaby cover approximately the same distance in the course of the night, and thus each has approximately the same number of opportunities to encounter insects. The same applies to the potto and the angwantibo.

This mechanism has already been described by Hladik and Hladik[10] for the platyrrhines of Panama, which consume roughly the same absolute quantities of insects regardless of their own body-size. The smaller platyrrhines obtain almost all of their food by hunting animal prey, whilst the larger species have to supplement their diet with plant food.

Among the bushbabies, *Galago demidovii* (60 gm.) relies almost entirely on hunting, whereas the two larger species supplement their diets with fruits (*G. alleni*, 260 gm.) or with gums (*E. elegantulus*, 300 gm.). Among the lorises, the angwantibo (200 gm.) derives most of its food by hunting, whilst the potto (1100 gm.) augments its diet of insects with fruits and gums.

In captivity, lorisids are often seen to "ignore" fruits when they have a large quantity of insects available. The same applies in the forest, where pottos which begin the night with a "good hunting session" do not go on to eat fruits afterwards. All the individual pottos dissected which had 8-15 gm. of insect matter in the stomach had failed to eat fruits or gums. On the other hand, all individuals with less than 1 gm. of insect matter in the stomach had eaten 20-60 gm. of fruits and/or gums. In *Galago demidovii* the diet is observed to be slightly less insectivorous in the early part of the night, when the animals are still hungry, than in the second half of the night (35% fruits and gums devoured before midnight, and 20% thereafter). Accordingly, in the dry season (when insects are rarer) this species consumes 50% fruits and gums, as against 30% during the rest of the year.

Thus it seems probable that the availability of animal food is the major factor influencing the diet of these lorisid species. Naturally, secondary specialisations affect the feeding behaviour of the "large" species oriented towards gums or fruits.

1. Animal food

In this analysis, we have considered the lorisines and the galagines separately. However, it is conceivable that the members of the two subfamilies may compete with one another in hunting insects. An examination of the categories of prey taken by each species shows that this is not the case. Table 3 shows, in order of importance, the different prey categories found in the stomach contents. The most common categories are followed by percentage figures based on the relative weights of animal food types in the stomachs.

Among the galagines, 78% of the prey are beetles, nocturnal moths and grasshoppers, whereas in the lorisines caterpillars and ants represent 70% of the prey. Pottos particularly prey upon ants (*Crematogaster* sp.) which release large quantities of formic acid, on centipedes (*Spirostreptus* sp.) which release large quantities of iodine, and on "criquets puants" (malodorous orthopterans), which also emit repellent substances. The angwantibo feeds primarily on caterpillars, most of which are covered with stinging hairs. (The caterpillars eaten by the bushbabies never bear such hairs.) Thus, it would seem that the two lorisine species are specialised to tolerate "noxious" prey left untouched by the galagines.

Table 3
Animal prey in order of preference

	1	2	3	4	5
Perodicticus potto[a] (N = 41)	ants (65%)	large beetles (10%)	slugs (10%)	cater-pillars (10%)	orthopterans (malodorous), centipedes, spiders, termites
Arctocebus calabarensis (N = 14)	caterpill-ars (65%)	beetles (25%)	ortho-pterans	—	dipterans, ants
Galago demidovii (N = 55)	small beetles (45%)	small nocturnal moths	cater-pillars (10%)	hemi-pterans	orthopterans, millipedes, homopterans, pupae
Galago alleni (N = 12)	medium-sized beetles (25%)	slugs	noc-turnal moths	frogs (8%) ants (8%)	grasshoppers, termites, millipedes, pupae, caterpillars
Euoticus elegantulus[b] (N = 52)	grass-hoppers (40%)	medium-sized beetles (25%)	cater-pillars (20%)	nocturnal moths (12%)	ants, homopterans

[a] Examination of 41 stomachs from pottos did not reveal one case of vertebrate remains. However, whilst following a population of tame animals in the forest we were able to observe a female attempting to capture weaver-birds in a tree carrying a large number of nests. On another occasion, we suprised a female devouring a young frugivorous bat (*Epomops franquetti*). Capture of such prey must be relatively rare.
[b] No vertebrate remains were found in 52 *E. elegantulus* which were dissected; but we did once come across a tame individual eating a bird (*Camaroptera brevicauda*) in secondary forest.

This special tolerance of the lorisines for "noxious" prey permits the capture of a sufficient quantity with a minimum of movement from place to place. In fact, these efficiently protected "noxious" organisms almost always display certain forms, colours or (especially) odours, which normally signal their "unpalatable quality" to potential predators. They are easily found, and their immobility permits easy capture. (In the case of ants of the genus *Crematogaster*, the potto follows moving columns, licking them up.) Over and above this, these prey organisms — which are ignored by many other predators — are abundant.

In this case, the lorisines are actually exhibiting a *tolerance* rather than a *preference*. If, in captivity, pottos or angwantibos are presented with their habitual prey alongside grasshoppers or moths normally eaten by galagos, they will eat the latter insects. In actual fact, angwantibos exhibit a particular form of behaviour prior to eating caterpillars.These animals are gripped by the head in the angwantibo's teeth, and the two hands "massage" the prey for 10-20 seconds, with the result that many of the hairs are removed. Nevertheless, once it has eaten the caterpillar, the angwantibo spends some time wiping its snout and hands by rubbing them on a branch.

Despite their slow locomotion, which serves to protect them from predators, the lorisines succeed in capturing a sufficient quantity of animal prey to ensure a balanced diet. Proteins necessary for physiological equilibrium are found principally in animal prey or in the green parts of plants (leaves and buds). The primates are, in fact, either insectivorous/frugivorous (usually the more rapid species) or folivorous/frugivorous (usually the slower forms, with a few intermediate types).[10] Among the Lorisidae (insectivore-frugivores), the lorises exhibit an exceptional adaptation. These slow-moving animals have conserved the typical diet of the family Lorisidae by obtaining their proteins from a category of animal prey generally left untouched by insectivorous species.

One can therefore discount the idea of any competition between lorisines and galagines in predation on insects. A second contributory factor is important in this context: whereas the lorises capture resting, slow-moving prey, the bushbabies frequently capture rapid-moving prey, quite often when the latter are on the wing. In order to do this, the bushbaby projects its body forward with great rapidity, whilst maintaining a grasp on the branch with its hind-feet. The insect is trapped in flight with both hands, and the bushbaby returns to its starting position (immediately in the case of *Galago demidovii*, which returns like a spring, and only after a short time-lapse in the case of *Euoticus elegantulus*, which ends up suspended head-downwards after the capture).

On the one hand, the potto and the angwantibo — already separated by their height in the vegetational strata — hunt animal prey which are malodorous (potto) and irritating (angwantibo). On the other, the two large bushbaby species are separated by the heights at which they are active. *Galago demidovii* and *Euoticus elegantulus*, which both exploit the canopy in primary forest, are unlikely to compete with one another. The first species preys primarily upon small animals (beetles, moths), and the second feeds upon larger prey (grasshoppers, larger beetles).

2. Fruits

The dietary tables given above show that only the potto and Allen's bushbaby can be regarded as frugivorous. In fact, the other three species consume absolute quantities of fruits amounting to only 1/30th of that eaten by *Galago alleni* and 1/70th of that consumed by the potto. In general, all of the lorisids primarily select soft, sweet fruits (*Uapaca* sp., *Musanga ceroopioides*, *Ricinodendron africanus*). However, the potto can attack certain fruits with a hard exterior, and on two occasions we saw *Galago demidovii* profiting from the passage of a potto to eat the remains of a large fruit which the latter had opened. As far as *P. potto* and *G. alleni* are concerned, the former eats fruits in the canopy, whilst the second collects them on the ground. This excludes any competition between the two species. These field observations have been confirmed experimentally in captivity. In a cage containing a tree, the pottos preferred to eat fruits placed in the branches, whilst *G. alleni* preferentially ate those placed on the ground.

Trees which are in fruit not only attract animals which are seeking the fruit — they also attract predators. Frugivores are presented with an abundant source of food; but the more time they spend there, the more they expose themselves to predation. Thus, the optimal solution for them is to collect *rapidly* and *in large numbers* fruits which can be eaten in some protected place. These two necessities are met in different ways in different zoological groups: by the crop in birds, by the cheek-pouches in monkeys and rodents, by the rumen in ruminants, etc. The potto and Allen's bushbaby have particularly distensible stomachs which can contain up to 1/13th of the body-weight. With the other three species, by contrast, we have never found any individuals with more than 1/30th of their body-weights in the stomach. Whereas the species which are not specialised for a frugivorous diet primarily utilise their tooth-scraper for slow removal of small fruit morsels, the potto and Allen's bushbaby can swallow rapidly large pieces of fruit, often including the kernels. In 30-60 seconds, a potto can eat an entire banana, and Allen's bushbaby swallows cherry-sized pieces of fruit by pushing them into the mouth with both hands and keeping the head up. It is actually quite rare to find these two species in immediate proximity of trees in fruit. In general, they are found 30-50 m. away, digesting under cover.

In the forest, the sites of fruit production are diverse and continually fluctuating. Through systematic counts, we established that large trees with high productivity are roughly 50 times less numerous than small trees or lianes with medium or low productivity. This is of capital importance for the distribution of small territorial mammals with permanent, restricted home-ranges (in particular, murids and prosimians). When a large tree comes into fruit from time to time in such a home-range, it is exploited straight away; but for the rest of the year food is obtained from trees with low productivity. The latter suffice for mammals of small body-size, but they must be abundant enough to provide a permanent supply of ripe fruits in the existing home-ranges. At night, trees with high fruit productivity are rarely visited. Using traps, we never captured more than 2-3 individuals of any given mammal species in such a tree (i.e. just as many as around trees with low productivity). Conversely, trees with low fruit productivity are little exploited during the daytime — approximately ten times less, according to our counts. This is doubtless associated with the fact that diurnal animals detect fruits by sight (at long range, when large fruiting trees are involved), whereas most nocturnal mammals detect by smell fruit which is isolated and hidden in the vegetation.

In order to study the social life of the prosimians, we have conducted a great deal of trapping with banana as bait. The animals were marked and released, which permitted us — among other things — to investigate their natural feeding behaviour. A basket of lianes containing 10 bananas is discovered in 1-5 days by *Galago alleni* and in 1-10 days by *Perodicticus potto.* On subsequent nights, if the bananas are replenished as necessary, the animals will return regularly by direct routes, often immediately after waking. They spend a few hours close to the bait and then move on to other fruiting areas. If several baskets of bananas are placed in the home-range of one individual, they will all be discovered and 2-3 may be visited during one night. By dissecting numerous pottos and Allen's bushbabies which were collected at the end of the night, we were able to observe that they can eat 2-3 different types of fruit in one night, which must oblige them to visit several fruiting points every night.

From all of these observations, it would seem that frugivorous prosimians (and murids) exploit simultaneously several fruiting trees, and that they never cease to explore their home-ranges in search of new trees in fruit. This mechanism, which is based on memory and exploration, permits them to feed themselves even if habitually visited fruiting trees cease production, or if they have been visited and depleted by another animal. Through continuous exploration of the home-range, they can locate trees at the start of fructification which will replace those which are ceasing to bear fruit.

3. Gums

The tables show that the potto and the needle-clawed bushbaby *(E. elegantulus)* are the principal feeders on gums. *(G. demidovii* consumes only 1/50th of the quantity of gums eaten by *E. elegantulus,* whilst *G. alleni* and *Arctocebus calabarensis* scarcely eat gums at all.) Both the potto and *E. elegantulus* inhabit the canopy, so it might be expected that they compete for gums.

Gums form principally along trunks and large branches at the site of old wounds and holes made by the mouth-parts of homopterans, (Examination of stomach contents revealed the presence of certain Auchaenorhynch homopterans — Fulgorids, Membracids, Tibinicids, etc. — which were no doubt swallowed involuntarily along with the exudations of resins which they provoked.) In contrast to fruits, gums appear regularly at the same places throughout the year. However, their production — which is dependent upon the metabolism of the trees — may be diminished during the main dry season.

In equatorial West Africa, the main dry season (15 June-15 September) is characterised by almost complete absence of rain, continuous cloud cover during the daytime, and a reduction of 3-4° C in mean temperatures. During this period, we were able to identify a marked decrease in the biomass of insects and a reduction in fructification. In parallel, we have observed that the prosimians lose 1/10th to 1/14th of their body-weights during this critical period. The weight-loss is directly dependent upon food availability. For example, *G. demidovii* consumes an average of 0.65 gm. of insects per night during the main dry season, as against 1.28 gm. per night during the rest of the year (stomach contents taken between 20.00 hrs. and midnight). Under the same conditions, we found smaller quantities of insects and gums in the stomachs of *E. elegantulus* examined during the dry season than in those collected at other times of the year. Thus, any competition would be exaggerated during this critical period.

When the dietary habits of the lorisids are examined in greater detail, it emerges that the potto eats only fruits and insects during the dry season, whereas fruits, gums and insects are taken at other times of the year. Conversely, *E. elegantulus* examined during the dry season had not eaten any fruits, although they eat small quantities during the rest of the year. Throughout the critical period, apart from the animal prey consumed, the potto is thus strictly frugivorous and *E. elegantulus* is strictly gummivorous.

Observation of the feeding behaviour of these two species renders the mechanism of such competiton easily comprehensible. Whereas the activity of the potto is primarily oriented towards searching for and visiting trees which are in fruit. *E. elegantulus* spend most of their time visiting trees which are gum-producers. Needle-clawed bushbabies have an excellent memory for gum-production sites, and they follow veritable "rounds" which permit them to collect (with the aid of the tooth-scraper) tiny droplets of gum formed after their last visit. Visits are made almost every night, even though each animal must visit a very large number of production-sites (about 300) in order to collect sufficient quantities of gums. *E elegantulus* rapidly covers its "rounds" thanks to its powerful leaps, stopping only for a few minutes at each gum-exuding site. In addition, the "claws" at the ends of the nails permit access to sites which are inaccessible to the other prosimian species, right along the largest trunks. During the dry season, when the gums accumulate more slowly, *Euoticus* must visit a larger number of production-sites, whereas smaller "rounds" suffice at other times of the year. Under these latter conditions, large aggregations of gums collect on trees which are not often visited, and it is such gums which are eaten by the pottos. Indeed, the gums found in the stomachs of pottos are often harder and darker in colour than those found in stomachs of *E. elegantulus,* and they occur as large lumps.

Thus, one cannot really talk in terms of real competition for gums between these two species, since the gums eaten by the potto are those left untouched by *E. elegantulus*. Dietary specialisations are most evident during the critical period of the year, and it is doubtless during this time that natural selection for adaptive characters is most active.

CONCLUSIONS

In both the Lorisinae and the Galaginae, the different sympatric species are morphologically relatively similar, and their dietary requirements are quite comparable: insects, with a supplement of fruits and gums in the larger species. In captivity, the five lorisid species can be maintained easily without any provision of foods (e.g. gums, fruits or certain insects) of the kinds eaten under natural conditions. In our present captive colony, all five species have become perfectly adapted to the same diet: milk, banana, apple, pear and crickets. In the forest, it is primarily through exploitation of different vegatation strata that the various species avoid dietary competition. Their ecological delimitations (in particular, their stratification) follow from "preferences" which orient each species towards a particular type of support. This orientation is complemented by certain behavioural and anatomical specialisations associated with utilisation of supports. These anatomical adaptations are relatively minor; it is primarily behavioural differences which permit the separation of ecological niches.

ACKNOWLEDGMENTS

My thanks go to Professor P.P. Grassé and A. Brosset for their generous provision of facilities and assistance at the CNRS field laboratory in Makokou, Gabon, throughout this study. The photographs were taken by A.R. Devez.

I should also like to make a special note of thanks to my friend and colleague, R.D. Martin, who offered to translate the manuscript for this article, despite his heavy commitments with the organisation of the Research Seminar. Our discussions, which have been numerous and very rewarding, have permitted me to extract a number of valuable conclusions.

NOTES

1 **Hill, W.C.O.** (1953), *Primates: Comparative Anatomy and Taxonomy*, 1.*Strepsirhini*, Edinburgh.

2 **Walker, A.** (1969), "The locomotion of the lorises whith special reference to the potto," *E. Afr. Wild. J.* 7, 1-5.

3 **Charles-Dominique, P.** (1971), "Eco-éthologie des prosimiens du Gabon," *Biol. Gabon.* 7(2), 121-228.

4 **Charles-Dominique, P. and Hladik, C.M.** (1971), "Le *Lepilemur* du Sud de Madagascar: écologie, alimentation et vie sociale ," *Terre et Vie* 25, 3-66.

5 **Sanderson, I.T.** (1940), "The mammals of the North Cameroons forest area," *Trans. Zool. Soc. Lond.* 24, 623-725.

6 **Napier, J.R. and Napier, P.H.** (1967), *A Handbook of Living Primates*, New York, 258.

7 **Jones, C.** (1969), "Notes on ecological relationship of four species of lorisids in Rio Muni, West Africa," *Folia primat.* 11, 255-67.

8 Vincent, F. (1969), "Contribution à l'étude des prosimiens africains: le Galago de Demidoff," Doctoral thesis, CNRS, AO 3575.

9 Charles-Dominique, P. (1971), "Ecologie et vie sociale de *Galago demidovii*," *Z. f. Tierpsychol.*, Suppl. 9,7-41.

10 Hladik, A. and Hladik, C.M. (1969), "Rapports trophiques entre végétation et Primates dans la forêt de Barro Colorado (Panama)," *Terre et Vie* 23, 25-117.

2

Intra-Specific Variation in the Social Organization and Ecology of *Propithecus verreauxi*

1974

ALISON RICHARD

Abstract. This article presents information collected during an 18-month study of four groups of *Propithecus verreauxi* living in the north-west and south of Madagascar. Various aspects of the social organization and ecology of this species are discussed, including group size and composition, patterns of group dispersion, ranging and home-range utilization, diet and feeding behaviour, daily activity patterns and intra-individual relationships within each group. Particular attention is given to the nature and extent of regional, seasonal and local variation in these parameters, and to the possible underlying environmental correlates.

Key Words feeding, ranging, behavioural variation

INTRODUCTION

Sefaka

Propithecus verreauxi is found in the remaining deciduous forests of north, west and southern Madagascar, and also in the arid didierea forest that covers extensive areas of the south (Fig.1). Between April 1970 and September 1971, an intensive study was made of two groups of *P. verreauxi* living in mixed deciduous forest in the north-west of the island, and of two groups living in the didierea forest in the south.

The study had three primary objectives. The first was to collect detailed information on a prosimian species. Prior to 1970, Petter's (1962a,b,c,1965)

Reprinted from *Folia Primatologica, 22:*178-207 (1974) with permission from S. Karger AG, Basel.

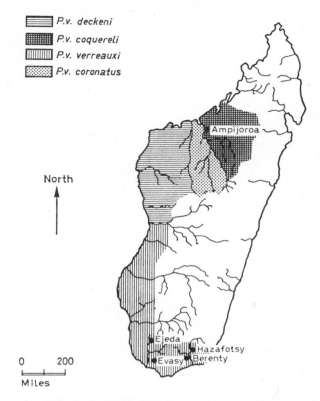

Fig. 1. The approximate distribution of the four sub-species, *P. v. coquereli, P. v. deckeni, P. v. coronatus* and *P. v. verreauxi,* locations at which surveys were made, and sites of two main study areas.

surveys and Jolly's (1966) field study were the only major published works on the behaviour and ecology of any prosimian species in the wild. In contrast, the literature on the Anthropoidea has proliferated during the past 10 years (e.g. Crook and Aldrich-Blake, 1968; DeVore, 1963, 1965; Goodall, 1963, 1965; Hall, 1962a,b; Jay,1965; Kummer and Kurt, 1963; Struhsaker, 1967,1969). There was a need to increase the amount of comparative material available on prosimians, in order to provide a broader overview of the whole spectrum of primate adaptations, and, possibly, some insight into ancestral primate patterns of behaviour.

The discovery of extensive regional variations in social organization has already demonstrated the limited applicability of the concept of "species-specific" behaviour to several Old World primate species (*Papio anubis,*Hall and DeVore, 1965; Rowell, 1966; *Presbytis entellus,*Jay, 1965; Ripley, 1967; Sugiyama, 1967; Yoshiba, 1968; *Cercopithecus aethiops,*Gartlan and Brain, 1968; Struhsaker, 1967). No such investigation of any prosimian species has been made, and the second aim of the study was to assess the flexibility of social organization in a prosimian species, by comparing groups from populations living in widely contrasting habitats.

Attempts have been made to produce a classification of primate social organizations, and to correlate variations in social organization with ecology (e.g. Crook and Gartlan, 1966). These efforts have not been altogether successful, partly through a lack of detailed information on the behaviour and ecology of many primate species. Since there are now extensive data on Old World and New World leaf-eating monkeys (Bernstein, 1968; Chivers, 1969; Clutton-Brock, 1972; Hladik and Hladik, 1969; Jay, 1965; Ripley, 1967), it seemed useful to provide comparative material on a prosimian species, a high percentage of whose diet consisted of leaves and shoots. The study's comparative approach was also used in this third context, in an effort to understand more clearly the processes by which ecology may influence social organization.

STUDY AREAS
The northern study area was located about 1 km from the forestry station at Ampijoroa, in the ecologically rich region of the Ankarafantsika; the sub-species *P.v.coquereli* is found in this region. The second study area was situated at Hazafotsy, about 80 km from the south coast in Reserve No.11 (Fig.1). This arid region, with its unique vegetation, supports the sub-species *P.v.verreauxi.*

Both deciduous and evergreen tree species were present in the northern forest. During the dry season, between April and September (which was also the cold season), deciduous trees lost their leaves. There was no well-defined canopy, most trees reaching a height of 12-16 m, with emergents of 30-35 m. Mean annual rainfall was 1,600 mm, and maximum/minimum temperatures of 38.5 and 14° C were recorded during the study. Although not so rich as some other parts of the Ankarafantsika, this particular locale was selected because there was no hunting there and animals were abundant and easily habituated, and also because the area was accessible by car at all times of year.

The vegetation in the southern study area was composed of xerophytic species rarely exceeding 13 m in height. The forest was dominated by species of the Didiereaceae family, particularly by *Alluaudia ascendens* and *A.procera.* Over 80% of the species in this forest were endemic, but overall species diversity was less in this region than in the northern area. Mean annual rainfall was 600 mm, and maximum/minimum temperatures of 44 and 8° C were recorded.

METHODS
15 months' quantitative observations were collected on the four study groups, of which twelve were subsequently analysed: data recorded early in the study were not sufficiently reliable to merit analysis. The data analysed

covered 3 months in the dry season (from April to September), and 3 months in the wet (from October to March), in each study area. 72 h of quantitative material was collected for each group, in each month, evenly distributed between 06.00 h and 18.00 h, and between the different age/sex classes. In three of the four groups, all animals were recognized individually throughout the study, and in all four groups it was possible to take one animal as the subject during the daily 12-hour observation period. Contact with the subject could usually be maintained continuously throughout the observation period. In this way, it was possible to establish a detailed record of one individual's activities for a whole day. A 72-hour, latitudinal sample made at half-hourly intervals on all group members showed close synchrony between the activities of members of a given group. Thus, the daily individual record is considered to be a fairly accurate index of group activities. At timed minute intervals during the observation period, the subject was described in terms of a number of categories, including its identity, height above the ground, activity, posture, the proximity and identity of its nearest neighbour.

In the northern study area, narrow trails were cut through the forest running north-south and east-west at measured 50-meter intervals. A minimum of vegetation was cleared in making these trails, and frequently only the paint-marks at 5 meter intervals along them betrayed their presence; the marking system also permitted immediate individual recognition of trails. This grid system covered the home-ranges of the two study groups, and was expanded where necessary when animals ranged into new areas. In the south, a similar system was used, although with the almost clear forest floor trails had only to be painted in and little initial clearance was required.

The trails facilitated rapid and quiet movement around the forest, particularly in the north where progress was sometimes hampered by the undergrowth. This was useful during habituation, but of little value subsequently. Habituated animals generally moved quite slowly and could be followed through the middle of grid-squares without difficulty. In both study areas, the main purpose of the grid system was to plot the movements and ranging patterns of each group and to determine the amount of time the group spent in different parts of its home-range.

The four groups were said to be habituated when all the members of each group would approach to within 2 m of me to feed. "Group" habituation is thus to be distinguished from the habituation of single animals within a group. Quantitative records were begun only when the whole group was habituated. The most striking aspect of the process of habituation was the animals' rapidly acquired, apparently almost total, indifference to my presence: no group took longer than 3 weeks to become completely habituated.

Concurrent with the study of the animals themselves, an analysis of the vegetation of each forest was made. The identity, height, spread and phenology of a total of 2,619 trees in the north and 3,136 trees in the south were recorded. These trees were sampled according to a "stratified random" technique (Southwood, 1971): two samples were made in each hectare square delineated by the grid system, and within each square these samples were randomly located. Each sample consisted of a circle with a measured radius of 5 m, and within these circles all trees with a trunk diameter greater than 3 cm at chest level were counted.

RESULTS *ecological data*

Social Organization and Behaviour

Three types of variation were found in the social organization and behaviour of the four groups studied. There were regional variations between the groups in each area; variation between seasons was observed in the behaviour of each group; thirdly, there was local variation, or variation between groups living in the same forest. The behavioral variation between seasons within a given group was generally found to be greater than that between any two groups at a given time. However, there was one important exception to this: a striking difference existed in the pattern of group dispersion between the two regions.

In the following discussion, the degree of regional, seasonal, and local variation observed in aspects of the social organization and behavior of *P. verreauxi* is described. The ecological parameters which may influence this variation are then considered.

Group Size and Composition. Surveys of group size and composition were made in five different forests. The location of these forests is shown in Fig. 1. Groups were counted ranging from 3 to 10 in size (Table 1), although groups of up to 13 have been reported in gallery forest along the Mangoky River (Sussman, personal commun.). In this study, no significant regional variation in group size was found.

Although the overall sex ratio for all groups censused was approximately unity, the sexual composition of individual groups varied to such an extent (Table 1) that no norm of group composition could be established. In 1963/64, Jolly (1966) found a significant excess of males in the gallery forest reserve at Berenty, but more recent censuses show that this imbalance has disappeared. In 1963, Jolly counted 23 males and 15 females in 10 groups of *P. verreauxi* living in the reserve; in 1964, she counted 24 males and 17 females in the same ten groups, and in 1970 the results were 23 and 22 (Jolly, 1972). The 1971 census carried out during this study, by Richard and Struhsaker, produced figures of 26 and 24. This is well within the limits of those fluctuations about a norm of 1:1 that are to be expected in any natural population (Rowell, 1972).

Table 1

Counts and partial analysis of the age and sex composition of all groups observed in each study area, and at Berenty, Evasy and Ejeda

Locality and date	Adult male	Adult female	Sub-adult	Juvenile	Infant	?	Total
Ampijoroa	2[a]	5					7
(northern study	1[a]	1	1(♂)	1(♂)	1		5
area), July 1970	3	1					4
	3	2	1		1		7
	1	2		1	1		5
	2	3					5
	3	5	1		1		10
	15	19	3	2	4		43
	adult						6
	composition						5
	unknown				1		4
							4
					1		4
Total number in 12 groups							66
Hazafotsy	2 [a]	2	1(♂)	1(♂)	2		8
(southern study	1 [a]	2	1(♂)		2		6
area),	2	1			1		4
September 1970	1	2		1	1		5
	2	4	1		1		8
	8	11	3	2	7		31
	adult						8
	composition						6
	unknown						5
							5
							3
Total number in 10 groups							58
Berenty,	3	2		1	2		8
September 1971	3	1		1	1	1	7
	2	1					3
	3	3				S-a: 1	7
	1	5			1	A: 1	8
	3	2					5
	2	2		1	2		7
	2	4		1	2		9
	3	3		1	2		9
	2	1		1		A: 1	5
	2 [b]						(2)
Totals for 10 groups	24	24		6	10	4	68

Table 1 (continued)

Locality and date	Adult male	Adult female	Sub- adult	Juvenile	Infant	?	Total
Evasy, May 1970	adult composition unknown						10 7 7 6 5 4 4 4
Total number in 8 groups							47
Ejeda, May 1970	adult composition unknown						7 7
Total number in 2 groups							14

[a] Study groups.
[b] These two males moved together in the area occupied by the ten groups listed.

Patterns of Group Dispersion. The area over which all four study groups ranged was divided into $50m^2$. If any member of a given group was seen in any one of these squares more than once, that square was said to constitute part of the home-range of that group. This probably led to an overestimate of the real area occupied by the group, but the method did provide a means of quantifying differential utilization of the home-range.

Home-range size varied between the groups from 6.75 to 8.50 ha, but this variation was not significant. However, in the south, each group had exclusive use of a much larger proportion of its home-range than in the north. Further, in the south this area of exclusive use, or "monopolized zone" (Jewell, 1966), comprised a large block of forest in the centre of each group's home-range; inter-group encounters took place only in the narrow zone of overlap around the periphery of the central monopolized zone. In contrast, areas of exclusive use within the home-ranges of the northern groups were scattered and may even have been an artifact of the observational techniques employed. Inter-group encounters occurred throughout the extensive areas of overlap with other groups. These differences are shown schematically in Fig.2.

Although the total number of inter-group encounters observed was higher in the north, the characteristic inter-group "battles" described by Jolly (1966) occurred proportionately more frequently in the south. In the north, encounters more often appeared to take the form of avoidance actions rather than confrontations: frequently, groups approached each other,

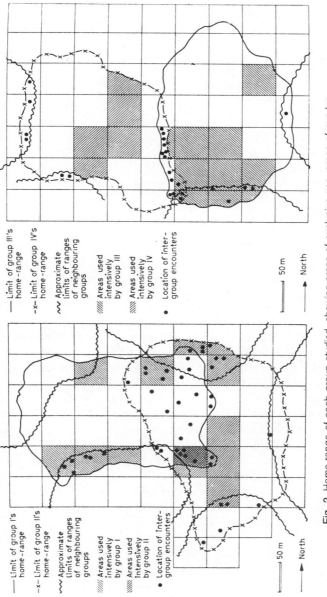

Fig. 2. Home-range of each group studied, showing areas of exclusive use, areas of overlap, areas of most intensive use and the locations of inter-group interactions.

— Limit of group I's home-range
-x— Limit of group II's home-range
∿ Approximate limits of ranges of neighbouring groups
▨ Areas used intensively by group I
▨ Areas used intensively by group II
● Location of inter-group encounters

50 m

↑ North

— Limit of group III's home-range
-x— Limit of group IV's home-range
∿ Approximate limits of ranges of neighbouring groups
▨ Areas used intensively by group III
▨ Areas used intensively by group IV
● Location of inter-group encounters

50 m

↑ North

stared and then moved apart. More active "battles" did occasionally occur, but they did not delimit any geographical area as they did in the south. Usually, they took place with reference to a preferred food source at which one of the groups was feeding at the beginning of the encounter.

In summary, dispersion in the north generally occurred through mutual avoidance between groups, and was associated with extensive home-range overlap, whereas in the south it was probably achieved through the maintenance of a defended territory by each group.

Ranging and Home-Range Utilization. Although the study groups each ranged over an area of approximately similar size, 1—2 ha of this area were used less than 5% of the time in all four instances. Further, a few parts of the home-range were used much more intensively by the group than others. The extent to which each group showed this tendency towards heavy use of paticular localities varied, but the variation was group-dependent rather than region-dependent. Areas of intensive use were arbitrarily defined as squares that, in 4 or more months, were among those in which a group spent 75% of its time. Their location is shown in Fig.2. Areas of intensive use were not necessarily clumped together (although in the south this was generally the case), nor was amount of use necessarily related to exclusivity of use: in neither study area were intensively used parts of the home-range always areas of exclusive use. No differentiation could be made between parts of the home-range according to the type of activity performed in them.

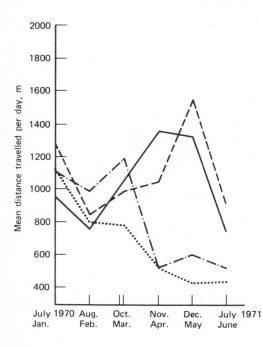

Fig. 3. Mean distance moved per day any individual, in each month and group.—= Group I;– – – =group II; . . . =group III; —·—· =group IV.

All four groups visited most parts of their home-range within 10-20 days, but although several general patterns of daily movement recurred, no evidence of a regular cycle of movement around the range was found. There was an overall difference in the mean daily distance moved between the northern and southern groups: the northern groups consistently moved further each day than the southern groups (Fig.3).

There was also significant, within-group seasonal variation in daily distance moved: in the wet season, groups sometimes moved more than twice as far in a day as they did during the dry season. Related to this increase in day ranges during the wet season, there was also a slight increase in the number of different grid-squares entered, and a larger increase in the total number of squares entered during each 6-day observation block (Table 2). This indicates that in the dry season groups tended to spend more time

Table 2
Number of squares entered, number of different squares entered, and the ratio between them, for each group for each month

Group	Month	Number of squares entered	Number of different squares entered	Ratio[a]
I	July	69	22	3:1
	August	62	20	3:1
	October	118	23	5:1
	November	144	25	5:8
	December	137	26	5:3
	July	69	17	4:0
II	July	90	24	3:7
	August	79	21	3:8
	October	122	23	5:3
	November	121	26	4:6
	December	167	27	6:2
	July	85	31	2:7
III	January	119	27	4:4
	February	88	23	3:8
	March	95	19	4:0
	April	69	20	3:4
	May	63	19	3:3
	June	55	17	3:2
IV	January	137	25	5:5
	February	107	24	4:4
	March	130	20	6:5
	April	69	21	3:3
	May	56	18	3:1
	June	64	17	3:8

[a] Ratio = total number of squares/number of different squares.

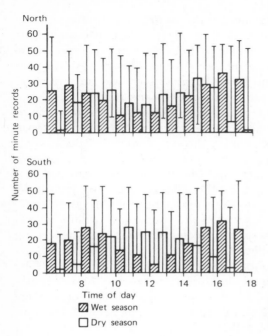

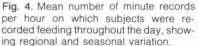

Fig. 4. Mean number of minute records per hour on which subjects were recorded feeding throughout the day, showing regional and seasonal variation.

in a smaller area of their home-range each day. However, although in the wet season animals tended to spread their time more evenly over a wider area, per unit time, the result was not an appreciable expansion of absolute home-range size but rather a shorter time-span within which the whole home-range was visited at least once.

Diet and Feeding Behaviour. The general characteristics of feeding behaviour were similar in both study areas. Animals adopted many feeding postures, enabling them to feed in most parts of any tree. In the wet season, all four groups tended to have two feeding bouts, one in the morning and one in the afternoon: in the dry season, there was one main feeding bout which reached a peak towards midday (Fig.4). Animals in the south fed for significantly shorter periods each day in the dry season than in the wet season. This difference was also seen in the north, but was not significant.

Regional, seasonal and local variation was found in the diet of *P. verreauxi.* Considering first regional variation, the species composition of the groups' diet was almost completely different in each study area. This was largely, but not uniquely, due to differences in the vegetational composition of the two forests. Four trees were identified which were common to both forests; these were *Commiphora pervilleana, Rothmannia decaryi, Cedrelopsis grevei* and *Baudouinia fluggeiformis. C. pervilleana* was eaten commonly by both northern groups, but in the south neither group spent more than 1% of total feeding time eating it. Contrarily, *R. decaryi* was eaten in quantity in the south, and rarely in the north; further, in the south, animals ate only the green and ripe fruit of this species whereas in the north animals ate only its large white flowers. The southern groups both spent over 1% of total feeding time eating *C.grevei* whereas neither northern group was ever seen to eat it. The reverse applied to *B.fluggeiformis,* which was an important dietary component in the north, and untouched in the south.

The diet of the southern groups contained fewer species than that of the northern groups. This does not, however, mean that the southern groups were more selective in their choice of foods: fewer tree species were present in the southern forest, and animals in fact fed on a proportionately wider range of the species that were available than did the animals in the north. Although the reasons underlying this difference are not known, by being proportionately less selective, the southern groups in effect maximized the diversity of their diet. Associated with this regional difference in diversity, the two southern groups spent a greater pecentage of total feeding time eating relatively few food species. (Fig.5).

Within each study area, a number of changes occurred seasonally. The direction of these changes was similar in both study areas, although generally more pronounced in the south. There was an almost complete change in the species composition of the diet of all four groups between seasons, as well as a significant difference in the food part being eaten. In the dry season, animals fed chiefly on adult leaves and dormant buds; in contrast,

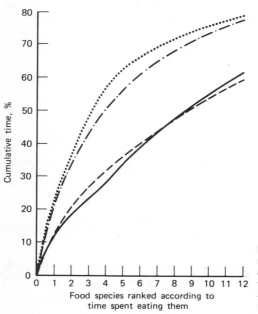

Fig. 5. Overall amount of time each group spent feeding on the twelve food species eaten more commonly by each group than any other food species.—=Group I; —=group II;-.-=group III; . . . =group IV.

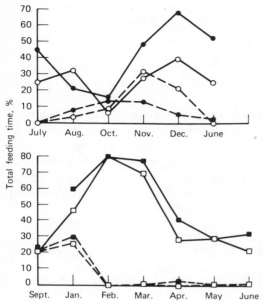

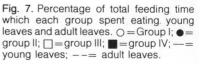

Fig. 6. Percentage of total feeding time which each group spent eating fruit, flowers and flower buds. ○ =Group I; ● = Group II; □ =group III; ■ =group IV; — = fruit; – –=flowers.

their diet in the wet season contained a high proportion of young leaves, flowers and fruit (Fig.6,7). It is also interesting to note that, during the dry season in the south, the bark and cambium of *Operculicarya decaryi* constituted

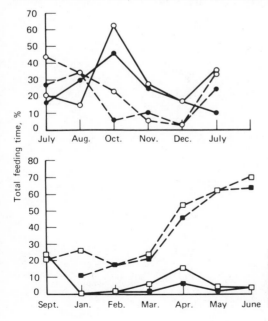

Fig. 7. Percentage of total feeding time which each group spent eating young leaves and adult leaves. ○ =Group I; ● = group II; □ =group III; ■ =group IV; — = young leaves; – –= adult leaves.

an important component in the animals' diet. This contained a high percentage by weight of water, and may have been critical for survival during this extremely arid period. In the north, animals ate bark to the exclusion of dead wood during the dry season, and dead wood almost to the exclusion of bark during the wet season. No evidence of the presence of insect life was visible to the human eye in this dead wood, and its significance for the diet of *P.v.coquereli* is not understood.

In addition to these seasonal differences in the species composition of the animals' diet and in the type of food being eaten, there was also a marked seasonal change in the diversity of diet of the two southern groups: both groups fed for longer periods on fewer species in the wet season and the overall diversity of their diet was reduced. There was no overall reduction in the diversity of the northern groups' diet, but if only those foods on which each group fed for more than 1% of total feeding time are considered, a seasonal effect can be seen: animals spent more time eating fewer species in the wet season than in the dry (Table 3).

Local variation, or variation between the two groups within each study area, was found in the species composition of their diet. In both the north and the south, only eight of the twelve species most commonly eaten by each group were the same (Table 4).

Daily Activity Patterns. In both study areas, the pattern of feeding, moving and resting was similar, and the nature of the changes in this pattern between seasons was also comparable (Fig.4,8). In the dry season, animals often did not move until 1 or 2 h after sunrise, at which point they would take up stations high in the trees, exposed to the sun. This "sunning" activity might last for over 1 h before the group finally moved off to feed. Animals fed more or less continuously until early afternoon, when they moved into the forks of trees and took up sleeping positions, in which they would remain until the following morning. The pattern in the wet season provided a striking contrast: animals were usually moving about and feeding

Table 3
Number of food species which each group ate for more than 1% of its total time spent feeding, in each season

Month	Number of species		Month	Number of species	
	group I	group II		group III	group IV
July and August	18	22	May and June	18	19
November and December	13	17	January and February	8	6

Table 4
Food species eaten by each group, ranked according to time spent feeding on
each (results are expressed as a percentage of total time spent feeding); species
eaten for less than 1% of total time spent feeding are not included

Food species	Time spent	Rank	Food species	Time spent	Rank
			Group I		
Drypetes sp. No. 18	12.4	1	*Cedrelopsis* sp. No. 123	1.5	15
Cedrelopsis sp. No. 471	5.4	2	*Baudouinia fluggeiformis*	1.5	16
Liana No. 215	5.3	3	*Erythroxylon* sp. No. 514	1.5	17
Dead wood	5.3	4	Sp. No. 433	1.5	18
Capurodendron microlobum	5.1	5	*Boscia* sp. No. 301	1.5	19
Rheedia arenicola Jerm and Perr	4.5	6	Sp. No. 211	1.4	20
Commiphora pervilleana	4.2	7	*Macphersonia gracilis*	1.1	21
Liana No. 312	4.0	8			
Liana No. 38	3.7	9	*Holmskioldia microcalyx*	1.1	22
Protorhus deflexa	3.4	10	Liana No. 452	1.1	23
Liana No. 36	3.3	11	*Mammea* sp. No. 5	1.1	24
Mundulea sp. No. 64	2.9	12	*Boscia* sp. No. 302	1.1	25
Liana sp. No. 13	2.2	13	*Boscia* sp. No. 603	1.0	26
Polyalthia sp. No. 116	2.1	14			
			Group II		
Drypetes sp. No. 18	11.7	1	Liana sp. No. 36	2.2	14
Cedrelopsis sp. No. 471	8.9	2	*Rheedia arenicola* Jerm and Perr	2.1	15
Liana No. 13	4.5	3	*Macphersonia gracilis*	1.9	16
Commiphora pervilleana	4.4	4	*Baudouinia fluggeiformis*	1.8	17
Bathiorhamnus louveli	3.9	5	*Boscia* sp. No. 301	1.6	18
dead wood	3.9	6			
Rhopalocarpus similis	3.7	7	*Mimusops* sp. No. 320	1.6	19
Liana No. 215	3.6	8	Liana sp. No. 296	1.6	20
Liana No. 38	3.3	9	*Holmskioldia microcalyx*	1.3	21
Boscia sp. No. 302	3.1	10	sp. No. 634	1.2	22
Protorhus deflexa	2.4	11	*Grewia* sp. No. 121	1.2	23
Capurodendron microlobum	2.3	12	*Malleastrum* sp. No. 240	1.0	24
Mammea sp. No. 5	2.2	13			

Table 4 (continued)

Food species	Time spent	Rank	Food species	Time spent	Rank
			Group III		
Terminalia			*Euphorbia plagiantha*	2.3	8
sp. No. 048	21.5	1	Liana sp. No. 054	2.1	9
Mimosa			Liana sp. No. 053	1.9	10
sp. No. 033	16.8	2	*Hagunta modesta*	1.8	11
Liana No. 042	12.3	3	Liana sp. No. 056	1.7	12
Grewia			Sp. No. 0113	1.4	13
sp. No. 089	7.7	4	Liana sp. No. 058	1.3	14
Terminalia			*Cedrelopsis grevei*	1.3	15
sp. No. 09	5.0	5	*Operculicarya decaryia*	1.2	16
Diospyros humbertii	4.1	6	Liana sp. No. 0146	1.1	17
Grewia					
sp. No. 059	3.2	7			
			Group IV		
Terminalia			*Rothmannia decaryi*	1.6	10
sp. No. 048	21.0	1	*Commiphora*		
Liana sp. No. 042	15.4	2	sp. No. 092	1.6	11
Mimosa			*Grewia*		
sp. No. 033	11.2	3	sp. No. 0126	1.5	12
Grewia			Liana sp. No. 0125	1.5	13
sp. No. 059	9.9	4	*Entada abyssinicus*	1.4	14
Hagunta modesta	4.8	5	Liana sp. No. 054	1.4	15
Terminalia			*Cedrelopsis grevei*	1.0	16
sp. No. 09	4.3	6	*Albizzia*		
Grewia			sp. No. 034	1.0	17
sp. No. 089	2.6	7	*Commiphora*		
Liana sp. No. 053	2.5	8	sp. No. 076	1.0	18
Diospyros humbertii	1.9	9			

before sunrise, and most feeding activity ceased by midmorning. The group then slept until the middle of the afternoon before resuming their feeding and foraging activities until after sunset. In both study areas, more time was spent in the shade during the wet season than during the cooler dry season. Animals, in both seasons, also spent more time in the sun in the early morning than later in the day (Fig.9).

Long distances were rarely travelled in one concerted movement: only when fleeing from dogs or man did a group move more than about 50 m without pausing. Although many postures were assumed while feeding, when travelling animals habitually used a "vertical clinging and leaping" mode of locomotion. In neither study area did they spend much time on the

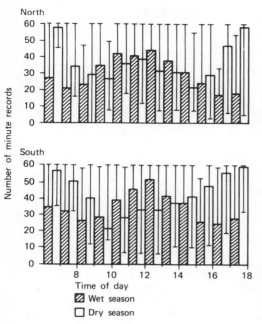

Fig. 8. Mean number of minute records per hour on which subjects were recorded resting throughout the day, showing regional and seasonal variation.

ground, but all other types of substrate were used extensively (Fig.10). The similarity of the results for substrate use (as well as for time spent in different postures — Richard, 1973) in each study area contrasts with the

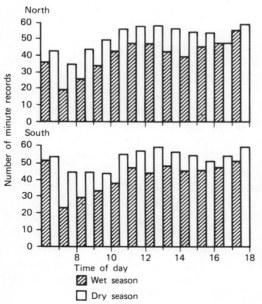

Fig. 9. Mean number of minute records per hour on which subjects were in the shade, showing regional and seasonal variation.

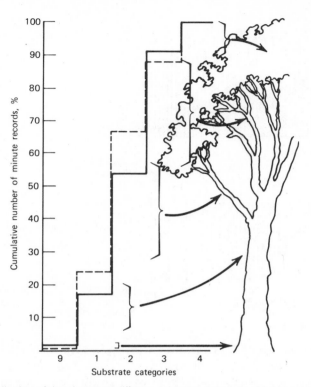

Fig. 10. Distribution of time between different substrate categories by groups in each region.— =Northern groups; − −=southern groups.

considerable differences apparent between the physical structure of the two forests. It would thus seem that animals were highly selective in their choice of substrate so that the differing physical parameters of the forests had little influence on the frequency with which different postures or modes of loco-motion were employed.

Social Structure and the Mating Season. The frequency of overt interaction of any kind between the members of each group studied was low: animals often moved, fed and rested for extensive periods in close proximity without any evident interaction. Grooming was the most common form of friendly interaction. The frequency of agonistic encounters varied between groups from 0.25 to 0.44 per animal per hour. In a total of 432 h of observation per group, the actual number of agonistic encounters observed in each group ranged from 107 to 191. This variation was group-dependent rather than region-dependent.

In all groups, most agonistic encounters outside the mating season occurred in a feeding situation, although aggression was also commonly seen when animals tried to handle a mother's infant: the mother would cuff and bite to prevent them gaining access to her infant. Using priority of access to food as the criterion of social dominance, a clear-cut hierarchy was evident: aggression in this context was uni-directional, outside the mating season, and the outcome of agonistic encounters could be predicted with a high degree of accuracy. However, this hierarchy was referred to only as the "feeding hierarchy" because there was no consistent correlation between the rank of individuals ordered according to the criterion of priority of access to food and their rank in hierarchies established according to the frequency of aggression, the direction and frequency of grooming, or preferential access to females during the mating season. Although the highest ranking animal in the three groups whose social structure was studied in depth was a female, dominance in the feeding hierarchy was not necessarily a function of sex: in one group, and adult male always had priority of access over an adult female.

Beyond the existence of this hierarchy, a number of factors tended to increase the complexity of relationships within each group, so that it would be an oversimplification to see the social structure of *P. verreauxi* in terms of a unitary theory of social dominance manifested in a simple, pervasive hierarchy. First, the presence of infants and the associated rise in their mothers' frequency of aggression probably played an important part in regulating social structure within the group. There was also limited evidence of what may be called idiosyncratic relationships between animals: these relationships consisted of interactions that occurred as manifestations of the characteristics of two individual animals rather than as relatively stereotyped encounters between members of age/sex classes. For example, in one group, a male dominant in the feeding hierarchy persistently directed aggression at a subordinate male, often apparently "gratuitously." It is suggested that this was due to individual "incompatibility," since adult males in other groups did not exhibit this high frequency of aggression. Finally, in two groups, there appeared to be a gradual change in process in the relationship between the sub-adult male and the juvenile throughout the period of the study: initially, these immature animals played together frequently and aggression was reciprocal. However, through time, the sub-adult became more and more assertive and the direction of aggression came close to being one-way.

In addition to local, and individual, differences in the frequency of aggressive behaviour, there was seasonal variation. In all four groups, agonistic encounters occurred more frequently in the wet season than in the dry; most animals contributed to this overall increase, and in no case was a decrease in frequency recorded (Table 5). The increase during the wet season may have been due to the fact that animals were generally more active and feeding for longer periods: since most aggression occurred in a feeding situation, an increase would be expected when animals were feeding more. It is also probable that the advent of the mating season resulted in heightened frequencies of aggression.

Table 5

Contribution of each animal to total group aggression in wet and dry seasons, and index of increased aggression

Group	Initiator	Contribution to group aggression				Index of increased aggression (wet n/dry n)
		wet season		dry season		
		n	%	n	%	
II	♀	73	71	60	83	1.2
	♂	16	15	8	11	2.0
	Y♂	10	10	4	6	2.5
	J	4	4	—	—	—
III	♀FD	13	17	13	48	1.0
	♀NFD	24	29	5	19	4.8
	♂F	38	46	7	26	5.4
	♂P	2	2	2	7	1.0
	Y♂	3	4	—	—	—
	J	2	2	—	—	—
IV	♀FI	85	56	24	62	3.5
	♀FNI	12	8	9	23	1.3
	♂R/INT	49	32	6	15	8.2
	SA♂♀	6	4	—	—	—
	Inf.	—	—	not present	not present	—

Grooming usually occurred during periods of inactivity. Adult females generally initiated less grooming and were more commonly groomed than other group members. Associated with this observation, there was a tendency for animals dominant in the feeding hierarchy to groom least and be groomed most, those subordinate in the hierarchy grooming most and being groomed least. However, the existence of frequent grooming as a function both of maternity and in response to persistent aggressive behaviour by one animal towards another complicated relationships and removed the simple linearity of the feeding hierarchy.

Copulation by 3 males with the 2 females in group IV was observed between March 3rd and 6th. This is referred to as the "copulatory period," and was the only time during the field study when male/male and male/-female mounting was observed. Flushing of the vulva of one of the females in group IV was noted on 2 days in the last week of January. The period between this observation and copulation in March is referred to as the "pre-copulatory period." Neither of the females in group III was seen to copulate, although they both produced infants the following August.

A number of behavioural changes were noted in group IV during the pre-copulatory period. These have been described in detail elsewhere

(Richard, in press), and only a brief summary of events is given here. During the pre-copulatory and copulatory periods, one or another female in group IV was usually the focus of male attention in that group, yet rejected the majority of male advances. Associated with this, there were significant increases in scent-marking and "roaming" by adult males. "Roaming" was said to occur when males, singly or in pairs, detached themselves from their own groups and made long forays into the home-ranges of other groups. 23 such excursions were observed during the mating season, involving males from either group III or group IV. At no other time of the year did males leave their groups for extended periods and travel far into the home-ranges of neighbouring groups. In addition to these changes in male behaviour, there was an increase in the frequency of inter-group aggression, involving all adult animals. In group IV, there was also an increase in the frequency of intra-group aggression. This latter change was not, however, noted in group III.

The increase in intra-group agggression in group IV was accompanied by some degree of breakdown in group structuring: previously subordinate animals began initiating aggression against dominant animals, thus removing the uni-directionality of the feeding hierarchy.

Copulation was only observed between members of different groups. However, it should be noted that the sample size was small, and the data available on that sample limited. I was not present during the mating season in the north, so no data are available for a regional comparison.

Ecological Correlates of Behavioural Variation

A comparison was made between the same groups in different seasons, and between populations of the same species, *P.verreauxi*, living in different habitat types. Data were collected on the two aspects of ecological variation that appeared to be important with regard to the populations studied, namely climate and vegetation. Certain correlations emerged between the observed behavioural and ecological variation. These are considered below. However, although such correlations may signify causal relationships, this can be neither assumed nor demonstrated in the present study.

Temperature and Day-Length. It is likely that daily exposure to extremes of high and low temperature (Fig. 11) has favoured the development of behavior patterns in *P.verreauxi* that help maintain a constant body temperature. These patterns include "sunning " behaviour, huddling together at night, and a tendency to spend more time exposed to direct sunlight during the cooler dry season, and also in the early morning in both seasons, when temperatures are still low after the night.

Although no controlled experiments have yet been carried out on the physiology and, specifically, thermoregulation of *P.verreauxi*, Bourlière and Petter-Rousseaux (1953) and Bourlière *et al.* (1956) have investigated the

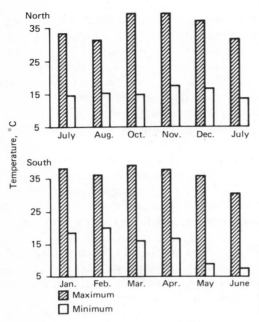

Fig. 11. Maximum and minimum temperatures recorded in each region, each month.

relation of rectal temperature to fluctuations in ambient temperature in some of the smaller prosimians. They found that animals in captivity exposed to wide temperature fluctuations did not maintain a constant body temperature. Rainey's (1970) work on the rock hyrax, *Heterohyrax brucei*, showed that this mammal is unable to maintain its body temperature at a constant level by physiological mechanisms alone, when exposed to the normal range of temperatures it encounters in its wild habitat: behavioural responses to changes in ambient temperature are also critical. I believe that, similarly, the behavioural responses of *P. verreauxi* were critical for thermoregulation, but until controlled experiments are performed on animals, exposed to variations in ambient temperature, which are (1) able to bring to bear both physiological and behavioural responses, and (2) permitted only to make a physiological response, the importance of the behavioural component in thermoregulation remains open to speculation.

It is possible that the change in activity pattern during the dry season was partly attributable to the drop in temperatures at that time: animals became active only when the ambient temperature had risen several degrees from its dawn level and they had sat in the sunshine for an hour or more. Even unhabituated animals were sluggish at that time and rarely moved off or responded strongly and immediately to my presence, although they always did so at other times of day. This late start to the day's activities in the dry season may have been due to animals' need first to raise their body temperature after being exposed to the relatively low night-time temperatures. It is also possible that the increased energy output required for thermoregulation in the dry season, due to exposure to greater extremes of temperature, may have reduced the amount of energy available for other activities, and thereby contributed to the overall reduction in activity and, particularly, the cessation of all play behaviour, during the dry season.

Stoltz and Saayman (1970) have shown that the daily distance covered by the chacma baboon in the northern Transvaal is inversely related to maximum daily temperature; animals tend to move further on cooler days. (No seasonal effect was found, but nor was there a significant seasonal difference in maximum daily temperatures.) Clutton-Brock (1972) reported that a group of *Colobus badius* ranged *further* each day during the cool dry season than it did during the warmer wet season. Thus, caution should be exercised·in inferring causal relationships from simple correlations observed between temperature and activity. It is very likely that in *P. verreauxi* factors such as the distribution and availability of food were also important in determining daily patterns of activity and the length of day ranges. On the available evidence, the relative importance of temperature and these other factors cannot be determined.

In both study areas, day length varied between seasons by about 3 h. This may have contributed to seasonal variation in activity patterns. Pariente (in press) has shown that light levels provide critical triggering mechanisms in the activity cycle of *Lepilemur mustelinus*. However, *P. verreauxi* was usually moving and feeding in the halflight before sunrise in the wet season, and continued to feed in the evening until it was too dark for observations to be made; in contrast, during the dry season, animals rarely moved until 2 or 3 h after sunrise, and were usually settled for the night about 3 h before sunset. Thus, light level did not play a crucial role in determining the onset and cessation of activity in the groups studied, although it may have been a contributing factor.

Vegetation. The pattern of group dispersion was strikingly different in each study area. It was not possible to demonstrate the ecological factors which were responsible for this variation. However, out of a number of possible explanations the following hypothesis, based on estimated differences in the distribution, size and availability of food resources, is put forward as being the most probable. In the north, it is argued, groups required their total home-range in the course of a whole year. Their diet contained a wide range of foods which were in many cases distributed in small, scattered resource units that occurred rarely in the forest and were often only seasonally available (Richard, 1973). As these different foods became available, so the area shifted over which animals had to range. Thus, although at any one time the availability of a few nutritively critical foods within the group's home-range probably just met the group's requirements, the total food available was likely to be in excess of its requirements. This permitted extensive overlap between the ranges of neighbouring groups. It is likely that it was more efficient for groups to range regularly throughout their total home-range rather than through a smaller, seasonally shifting area: the former, wider ranging pattern allowed them regularly to "monitor" the presence or absence of scattered, seasonally available food sources. In contrast, total food availability probably did become a limiting factor for the southern groups towards the end of the dry season, because of the effect on the vegetation of the extreme aridity at that time. It is suggested that, during these critical periods, the food available within each home-range could support only one group without its carrying capacity being exceeded. The round-the-year territoriality found in the south may have been an adaptive response to this minimum foraging area requirement which operated at times of greatly reduced food availability.

In the dry season in both study areas, animals moved short distances each day and fed for short periods on a wide variety of food species. In the wet season, they moved further each day, and fed for longer periods on fewer species. The possible contributing role of fluctuations in temperature to this seasonal variation has already been referred to. Changes in the availability and distribution of food may also have played a part in determining this seasonal variation. It was estimated that there was a higher density of vegetation in the northern study area, and it is likely that this reflected higher primary productivity in that area. It is also probable that in both study areas productivity was greater in the wet season than in the dry. Reduced productivity in the dry season may have been associated with a decrease in the availability of food, although it should be emphasized that there is no necessary correlation between overall productivity and the availability of food for any one animal species. If such a decrease is assumed, it can be postulated that this caused a reduction in the amount of time animals spent feeding. This would have led to a net reduction of available energy and hence a decrease in the length of day ranges. However, the underlying assumption cannot be demonstrated and may well be invalid.

An alternative hypothesis is that seasonal changes in ranging patterns may have been determined in part by changes in the distribution and availability of foods and associated changes in the degree of selectivity exercised by the animals. The vegetational analysis indicated that many tree species in both forests tended to be widely scattered (Richard, 1973). Further, during both seasons, over 50% of the species eaten by the study groups were quite rarely occurring in both forests. In the wet season, animals mainly ate the fruit of a few species. In order to seek out adequate supplies of these few species, they had to range widely each day. In the dry season, little or no fruit was available and animals were much less selective in their choice of food. They ate the leaves of many species and thus did not have to travel far each day in order to find adequate, if unappetizing, food. Thus, it is argued, these closely related changes in food availability and selectivity effectively altered the distribution of the main components of each group's diet and, hence, the groups ranging pattern.

DISCUSSION AND CONCLUSIONS

In *Propithecus verreauxi,* the group is the basic unit of social organization. However, the group is more appropriately considered as a foraging party of mutually familiar animals than as a reproductive unit of predictable composition; groups were found, in the course of the study, that fitted the description "one-male group," "family group," "age-graded male group," as well as "multi-male group." While ecological factors may play a part in limiting group size, there was no evidence of pressures operating to bring about a norm of sexual compositon within the group.

Jolly's observation of an excess of males in the population of *P. verreauxi* at Berenty in 1963/64 appears to have been a local, temporary phenomenon. Censuses made in 1970 and 1971 indicated a sex ratio of approximately 1:1. The reasons for the imbalance 10 years previously can only be speculative: it is possible that a disease differentially affecting females reduced the number of females in groups at that time, or that there was a higher rate of immigration into the reserve by males than females. Some support for the latter theory is given by the fact that males may leave their group and join a new one both in and out of the mating season (Richard, 1973, in press). In contrast, no female was seen to change groups in the course of the study. This pattern of female stability and male mobility is in keeping with the evidence from many primate societies (see Rowell, 1972; Altmann and Altmann, 1970).

Male mobility was highest during the mating season: males often left their groups at this time and mated with females belonging to other groups. This pattern of mating ensured some degree of outbreeding, and is probably important in a species which habitually lives in small groups. However, the extensive fighting between males which was associated with the intrusion of "outsider" males into groups suggests that the mating season has functions additional to that of promoting outbreeding. It also apparently operates to produce intra-sexual selection between adult males. It is possible that the social upheaval and fights of the mating season test the "fitness" of the males in terms of their ability to survive relatively long periods of high energy output. As an integral part of the social structure, the mating season should be viewed not as a cohesive social force but as a catalyst, permitting an extensive breakdown and reshuffle both between and within groups that are, for the rest of the year, more or less isolated.

The influence of climate and vegetation on the social organization of *P. verreauxi* have been discussed. It should be emphasized that other parameters may have played a critical role, too. Chivers (1969), for example, has suggested that changes in the population density of *Alouatta palliata,* the howler monkey, on Barro Colorado Island, is leading to changes in social organization. However, it is difficult to determine whether the density factor is in fact operating directly or through its effect on food resources. Again, the influence of predation pressures on social structure has been discussed extensively (e.g. Crook and Gartlan, 1966). While the paucity of data concerning predation on *P.verreauxi* suggests that the influence of predators is minimal, this inference may be misleading. It is possible, if unlikely, that there are significant pressures due to predation, or that the social organization seen today was formed at a time in the past when predation was a much more critical factor than it is today.

While some of the behavioural variation seen may have been associated with parameters of which no account was taken in this study, the author believes that some of it, particularly the "local" variation, was random and of no adaptive significance. For example, while ecological factors may determine the upper limit of group size and may favour social rather than "solitary" living, within these limits group size and sex composition may be arbitrary. Similarly, local differences in the species composition of diet, and in the amount of time spent feeding on particular foods, may reflect arbitrary differences of "tradition" between groups rather than a real difference in food availability. Regional differences in the amount of time spent feeding on the four tree species common to both forests may also be due to tradition, although a number of other explanations are possible: environmental factors such as soil type may have modified the nutritional value of these species in each region, making them important dietary components in the one and not in the other. In cases where a species was eaten in one area and not in the second, it is possible that other species may have been preferred alternatives, providing equal, or greater, nutritional value.

From this survey of various aspects of what is here loosely termed the "social organization" of *P.verreauxi*, perhaps the most important feature which emerges is the variability of this animal. Although the variation found was not as striking as that reported in, for example, the langur (Yoshiba, 1968), it is nonetheless clear that an accurate picture of the behavioural capabilities of *P.verreauxi* cannot be reflected by a brief study of any one group living in one particular forest. Within broadly defined limits, a general pattern does exist, but within these limits the variation is such as to suggest that in at least this prosimian species behavioural adaptability is not significantly less than that reported for many of the Anthropoidea.

SUMMARY

During an 18-month field study of *Propithecus verreauxi*, quantitative data were collected on four groups of animals all of whom were habituated to the observer's presence. Two groups lived in the rich, semi-deciduous forest of north-west Madagascar. The other two groups ranged in a second study area, located in the arid Didierea forest in the extreme south of the island. Data were also collected on the structure, diversity, distribution and phenology of vegetation in each study area; records were obtained of daily maximum/minimum temperatures and rainfall in each study area.

Group size ranged from 3 to 10; no significant regional variation in group size was found. In contrast, patterns of group dispersion were strikingly different in the two study areas. Regional, seasonal and local variation was found in ranging patterns and home-range utilization, diet, and daily activity patterns. Variation in ecological parameters, specifically vegetation and climate, were found to be associated with the observed behavioural variation. A causal relationship between ecological and behavioural factors cannot be demonstrated, however, and it is postulated that at least some of the observed behavioural variation was "random," and of no adaptive significance.

ACKNOWLEDGMENTS

I would like to thank Prof.J.R. Napier for his continual support and advice during this research, and Dr. Alison Jolly, Dr. R.D. Martin, Dr. T. Clutton-Brock, and Dr. P. Lattin for their help and comments. I am appreciative of the cooperation given me by the Malagasy authorities, without which the study would not have been possible.

The field work was supported by a Royal Society Leverhulme Award, the Explorers' Club of America, the Boise Fund, the Society of the Sigma Xi, a NATO Overseas Studentship, the John Spedan Lewis Trust Fund for the Advancement of Science, and the Central Research Fund of London University.

REFERENCES

Altmann, S.A. and Altmann, J.: Baboon ecology (Chicago Univ. Press, 1970).

Bernstein, I.S.: The lutong of Kuala Selangor. Behaviour *32:*1-16 (1968).

Bourlière, F. et Petter-Rousseaux, A.: L'homéothermie imparfaite de certains prosimiens. C.R.Soc. Biol. *147:*1594 (1953).

Bourlière, F.; Petter, J.-J. et Petter-Rousseaux, A.: Variabilité de la temperature centrale chez les lémuriens. Mém. Inst. Sci. Madagascar *10:*303-304 (1956).

Chivers, D.J.: On the daily behaviour and spacing of howling monkey groups. Folia primat. *10:*48-102 (1969).

Clutton-Brock, T.H.: Feeding and ranging behaviour of the red colobus monkey; PhD thesis Cambridge (1972).

Crook, J.H. and Aldrich-Blake, P.: Ecological and behavioural contrasts between sumpatric ground-dwelling primates in Ethiopia. Folia primat. *8:*192-227 (1968).

Crook, J.H. and Gartlan, J.S.: Evoluton of primate societies. Nature, Lond. *210:*1200-1203 (1966).

DeVore, I.: A comparison of the ecology and behavior of monkeys and apes; in Washburn Classification and human evolution (Aldine, Chicago 1963).

DeVore, I.: Changes in the population structure of Nairobi Park baboons 1959-1963; in Vegteborg The baboon in medical research (Texas Univ. Press, Austin 1965).

Gartlan, J.S. and Brain, C.K.: Ecology and social variability in *Cercopithecus aethiops* and *C.mitis;* in Jay Primates: studies in adaptation and variability (Holt, Rinehart & Winston, New York 1968).

Goodall, J. (van Lawick-): Feeding behaviour of wild chimpanzees: a preliminary report. Symp. zool. Soc. Lond.*10:* 39-47 (1963).

Goodall, J. (van Lawick-): Chimpanzees of the Gombe Stream Reserve; in DeVore Primate behaviour: field studies of monkeys and apes (Holt, Rinehart & Winston, New York 1965).

Hall, K.R.L.: Numerical data, maintenance activities, and locomotion of the wild chacma baboon, *Papio ursinus.* Proc. zool. Soc. Lond. *139:* 181-220 (1962a).

Hall, K.R.L.: The sexual, agonistic and derived social behaviour patterns of the wild chacma baboon, *Papio ursinus.* Proc. zool. Soc. Lond. *139:* 283-327(1962b).

Hall, K.R.L. and DeVore,I.: Baboon social behaviour; in DeVore Primate behaviour: Field studies of monkeys and apes (Holt, Rinehart & Winston, New York 1965).

Hladik, A. and Hladik, C.M.: Rapports trophiques entre végétation et primates dans la forêt de Barro Colorado (Panama). Terre Vie *1:* 25-117 (1969).

Jay, P.: The common langur of North India; in DeVore Primate behaviour: field studies of monkeys and apes (Holt, Rinehart & Winston, New York 1965).

Jewell, P.A.: The concept of home range in mammals. Symp. zool. Soc. Lond. *18:* 85-110 (1966).

Jolly, A.: Lemur behavior. (Chicago Univ. Press, Chicago 1966).

Jolly, A.: Troop continuity and troop spacing in *Propithecus verreauxi* and *Lemur catta* at Berenty (Madagascar). *Folia primat. 17:* 321-334 (1972).

Kummer, H. and Kurt, F.: Social units of a free-living population of hamadryas baboons. Folia primat.*1:* 1-19 (1963).

Pariente, G.: Influence of light on the activity rythms of two Malagasy lemurs: *Phaner furcifer* and*Lepilemur mustelinus leucopus;* in Martin, Doyle and Walker Prosimian biology (Duckworth, London, in press).

Petter, J.-J.: Recherches sur l'écologie et l'éthologie des Lémuriens malgaches. Mém. Mus. nat. Hist. nat., Sér. A *27:* 1-146 (1962a).

Petter, J.-J.: Ecological and behavioural studies of Madagascar lemurs in the field. Ann. N.Y. Acad. Sci.*102:* 267-281 (1962b).

Petter, J.-J.: Ecologie et éthologie comparées des Lemuriens Malgaches. La Terre et la Vie *109:* 394-416 (1962c).

Petter, J.-J.: The lemurs of Madagascar; in DeVore Primate behavior: field studies of monkeys and apes (Holt, Rinehart & Winston, New York 1965).

Rainey, M.: Aspects of physiological and behavioural temperature regulation in the rock hyrax, *Heterophyrax brucei;* MA thesis Nairobi (1970).

Richard, A.: Ecology and social organization of *Propithecus verreauxi;* PhD thesis London (1973)

Richard, A.: Patterns of mating in *Propithecus verreauxi;* in Martin, Walker and Doyle Prosimian biology (Duckworth, London, in press).

Ripley, S.: Intertroop encounters among Ceylon gray langurs *(Presbytis entellus)* ; in Altmann Social communication among primates (Chicago Univ. Press. Chicago 1967).

Rowell, T.E.: Forest-living baboons in Uganda. J.Zool., Lond. *149:* 344-364 (1966).

Rowell, T.E.: Social behavior of monkeys (Penguin Books, 1972).

Southwood, T.R.E.: Ecological methods: with particular reference to the study of insect populations (Chapman & Hall, London 1971).

Stoltz, L.P. and Saayman, G.S.: Ecology and behaviour of baboons in the Northern Transvaal. Ann. Transv. Mus. *26:* 5 (1970).

Struhsaker, T.T.: Ecology of vervet monkeys *(Cercopithecus aethiops)* in the Masai-Amboseli Game Reserve, Kenya. Ecology *48:* 891-904 (1967).

Struhsaker, T.T.: Correlates of ecology and social organization among African Cercopithecines. Folia primat. *11:* 80-118 (1969).

Sugiyama, Y.: Social organization of hanuman langurs; in Altmann Social communication among primates (Chicago Univ. Press, Chicago 1967).

Yoshiba, K.: Local and intertroop variability in ecology and social hebavior of common Indian langurs; in Jay Primates: studies in adaptation and variability (Holt, Rinehart and Winston, New York 1968).

3

Ecological Distinction in Sympatric Species of *Lemur*

1974

ROBERT W. SUSSMAN

INTRODUCTION

In forests throughout Madagascar, there are many sympatric species of Lemuriformes. The dynamics of the interaction between these species are, however, as yet poorly understood. Generally, coexistence of related species is made possible by differential exploitation of the environment, which minimises competition. Differential exploitation of the environment, in most cases, is the result of habitat selection in which the species have particular preferences for different portions of the shared environment. Where species choose different habitats, the particular preferences may be the result of adaptations to different environmental conditions existing in places where the species are allopatric, or the result of divergence caused directly by interaction between the sympatric populations. With the latter phenomenon, referred to as *character displacement*,[1] sympatric populations of two species will tend to differ in one or more characteristics in which allopatric populations of the same two species may be similar.

A study of the ecological relations between two sympatric forms, therefore, may give indications as to the nature of the forces causing differentiation. Furthermore, once differences in ecological preferences are recognised, it is possible to formulate and test hypotheses concerning the

def.

Reprinted from Prosimian Biology by Martin, Doyle, and Walker, Editors by permission of the University of Pittsburgh Press. © 1974 by Gerald Duckworth & Co., Ltd.

relationship between ecology, morphophysiology, and social behaviour of the species in question.

Lemur fulvus rufus male.

Lemur fulvus rufus female.

Lemur catta.

Propithecus verreauxi verreauxi.

In this article, I will describe some of the results of a study on two species of *Lemur: Lemur fulvus rufus* and *Lemur catta* .The study is discussed in more detail elsewhere.[2] Populations of *L. f. rufus* range from the north-west to the south-west of Madagascar. Populations of *L. catta* are found from the south-west to the more arid south. In many forests of the south-west, the two species are sympatric. (Fig.1).

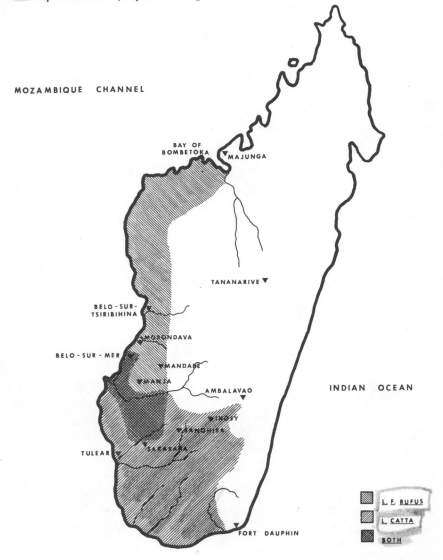

Fig. 1. Geographical distribution of *Lemur fulvus rufus* and *Lemur catta*. Populations are not continuous within these areas, but are only found where suitable primary vegetation exists.

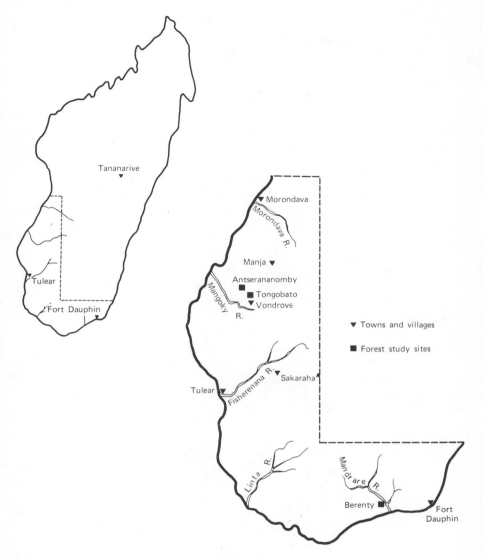

Fig. 2. Study sites for intensive observations of behavior.

Intensive studies were carried out in three forests: Antserananomby, Tongobato, and Berenty (Fig.2). At Antserananomby, *L. f. rufus* and *L. catta* are sympatric, and in the other two forests they are allopatric. Data on the utilisation of space and time, diet and social structure were collected on both the allopatric and sympatric populations of *L. f. rufus* and *L. catta*. The study was carried out between September 1969 and November 1970, during which period the animals were observed for a total of 830 hours.

GENERAL ECOLOGICAL DISTRIBUTION OF THE TWO SPECIES IN THE SOUTH-WEST OF MADAGASCAR

Populations of *Lemur fulvus rufus* and *Lemur catta* coexist in the region between 20° 44' and 23° 12' south latitude. These two species, along with *Propithecus verreauxi verreauxi,* are the only diurnal lemur species found in the south-west of Madagascar. The south-west is a region of transition between the moist, deciduous forests of the north-west and the desert-like vegetation of the south. In the south-west, the diurnal lemur species are found mainly in two types of primary forest: (1) closed canopy, deciduous forests and (2) brush and scrub forests. The deciduous forests are dominated by *Tamarindus indica* trees, which make up a continuous, closed canopy about 7 to 15 m. in height. These forests have been classified by Perrier de la Bathie as "bois des terrains silicieux."[3] Forests of this type present essentially the same characters from the north-west to the south-west, and similar forests are also found along the large rivers in the south of Madagascar. The brush and scrub forests reach the height of only about 3 to 7 m. and the under-brush is very dense. Perrier de la Bathie classified brush and scrub forests as "bois des terrains calcaires." The vegetation of Madagascar is described in detail by Perrier de la Bathie, Humbert, and Humbert and Darne.[3]

An initial survey was conducted in the south-west within the area in which populations of *L. f. rufus* and *L. catta* are sympatric. Data from the survey indicate that the ecological distribution of the two species differs within this area. *L. f. rufus* is found without *L. catta* only in small, circumscribed, continuous canopy forests. *L. catta* exists alone only in brush and scrub forests. *L. f. rufus* and *L. catta* coexist only in areas in which a continuous canopy forest merges directly into a primary brush and scrub forest. I have called these "mixed" forests. Within the mixed forests, *L. f. rufus* was seen only in those portions with a continuous canopy, whereas *L. catta* could be found in all parts of these forests. *P. v. verreauxi* inhabits both continuous canopy and mixed forests, but was never seen in brush and scrub forests. This is probably due to the fact that the density and stratigraphy of these latter forests necessitate terrestrial locomotion.

The survey data, therefore, suggest that *L. f. rufus* and *L. catta* have preferences for different habitats. These preferences seem to be related to different locomotor adaptations in the two species. While *L. f. rufus* is limited to areas with a continuous, closed canopy, *L. catta,* because much of its travel is done on the ground, can exploit a number of regions which differ in ecological structure.

INTENSIVE STUDY

Three forests were chosen for intensive study: Antserananomby and Tongobato in the south-west, and Berenty in the south (Fig.2). The basic physiognomy of the three forests is similar. Kily trees *(Tamarindus indica)* form a closed, continuous canopy in all cases. However, Antserananomby is a mixed forest in which only seven of the total ten hectares studied contain a closed canopy.

Lemur fulvus rufus and *Lemur catta* coexist at Antserananomby, whilst only *L. f. rufus* is found at Tongobato, and only *L. catta* is found at Berenty. *Propithecus verreauxi* inhabits all three forests. The nocturnal lemur species found at Antserananomby are: *Lepilemur mustelinus ruficaudatus, Phaner furcifer, Cheirogaleus medius,* and *Microcebus murinus. L. m. ruficaudatus, P. furcifer,* and *M. murinus* were observed at Tongobato. *L. m. leucopus, C. medius,* and *M. murinus* are found at Berenty. Many of the same genera of mammals, birds, and reptiles are found in all three forests.

The study was conducted in December 1969 and in March and April 1970 at Tongobato, from July through September 1970 at Antserananomby, and in November 1970 at Berenty. The temperatures during the day were similar in all three forests. The evening temperatures, however, were higher at Tongobato and Berenty than at Antserananomby (Table 1). The days were also longer at Tongobato and Berenty (5.00 hrs. to 18.30 hrs.) than they were at Antserananomby (6.00 hrs. to 18.00 hrs.).

Table 1

Average maximum and minimum temperatures for the months of study at each forest[a]

Month	T_x	T_n
Tongobato		
December 1969	36.1	19.2
March 1970	36.2	19.9
April 1970	35.3	16.3
Antserananomby		
July 1970	32.1	11.8
August 1970	35.5	13.9
September 1970	35.5	15.5
Berenty		
November 1970	34.7	19.3

T_x = average maximum daily temperature in °C.
T_n = average minimum daily temperature in °C.

[a] The temperatures are taken from those recorded at the meteorological station closest to each forest.

1. Utilisation of the forest strata

Data were collected at five-minute intervals on the number of individuals engaged in each of six activities— feeding, grooming, resting, moving, travel, and other— and the levels of the forest at which the activities were performed.

Five forest levels were recognised: Level 1 is the ground layer of the forest, which includes herb and grass vegetation. Level 2 is the shrub layer, 1-3 m. above the ground. This layer is usually found in patches throughout closed canopy forests, but is much more dense and is the dominant layer in brush and scrub regions. Level 3 consists of small trees, the lower branches of larger trees, and saplings of the larger species of trees. This layer ranges from 3-7 m. above the ground. Level 4 is the continuous or closed canopy layer. It usually ranges from 5 to 15 m. in height. The dominant tree of the closed canopy, at all three forests, is the kily. Level 5 is the emergent layer and consists of the crowns of those trees which rise above the closed canopy and are higher than 15 m. All three forests in which I made intensive studies are primary forests; the tree layers are quite distinct. With observations in which the forest level could not be clearly distinguished, the level was not recorded.

Each observation of an animal constituted an individual activity record (IAR) collected in a given five-minute time sample.[4] The number of individual activity records for *L. f. rufus* was 7,084 at Antserananomby and 2,896 at Tongobato. The number of IARs for *L. catta* was 5,383 at Antserananomby and 6,861 at Berenty. The IARs were summed for each half-

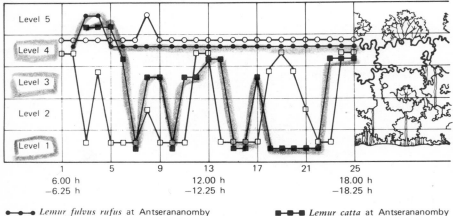

●—●—● *Lemur fulvus rufus* at Antserananomby   ■—■—■ *Lemur catta* at Antserananomby
○—○—○ *Lemur fulvus rufus* at Tongobato   □—□—□ *Lemur catta* at Berenty

Fig. 3. The forest level at which the highest percentage of animals was observed during each of the 25 half-hour periods for the species studied at each of the forests.

hour from 6.00 hrs. to 18.25 hrs. The day was thus divided into 25 half-hour periods. Comparisons were made between the quantitative data collected on the two species and between those collected on allopatric populations of the same species.

Fig.3 shows the level at which the highest percentage of animals was observed during each of the 25 half-hour periods. Data on both species (from all three forests) are represented in this figure. Both at Tongobato and at Antserananomby, *L. f. rufus* was found in the continuous canopy for almost all of the half-hour periods. The vertical displacement of *L. catta* followed a pattern throughout the day which was related to the foraging and movement patterns of this species. The pattern for *L. catta* was similar in both forests.

Table 2 includes the mean percentage of animals observed at each level for each activity and at each level overall, regardless of activity, for the 25 half-hour periods. Again, the intraspecific comparisons show the populations of each species to be similar in their use of vertical habitat, regardless of the forest or the presence or absence of the other species. *L. f. rufus* was very specific in its choice of vertical habitat. A high percentage (over 90%) of the activities of *L. f. rufus* took place in the top layers of the forest (levels 3, 4, and 5). In all six activities, *L. f. rufus* spent the highest percentage of time in level 4, the continuous canopy of the forest. *L. catta*, on the other hand, was found in all forest levels. For the most part, *L. catta* moved and travelled on the ground, rested during the day in the low trees (levels 2 and 3), and rested at night in level 4. *L. catta* fed in all of the levels of the forest.

At both forests in which *L. catta* was studied, it spent more time on the ground than in any one of the other four levels (36% at Berenty and 30% at Antserananomby). Perhaps the most significant finding is that *L. catta* travelled on the ground 65% (Antserananomby) and 71% (Berenty) of the time. A high percentage of the individual movement of *L. catta* was also carried out on the ground. *L. f. rufus* travelled in level 4 88% of the time at Antserananomby and 83% of the time at Tongobato. *L. f. rufus* was seen on the ground in less than 2% of the observations overall.

When *L. catta* travelled in the trees, its locomotor behaviour differed from that of *L. f. rufus*. When groups of *L. f. rufus* moved horizontally through the trees, the animals ran along the fine terminal branches of the large trees. These branches were generally horizontal in relation to the ground and formed a continuous series of pathways throughout the closed canopy. *L. catta*, rather than moving along the fine, horizontal branches, would usually climb the large, oblique branches and leap from one of these branches to another. Thus, they moved through the trees by performing a series of runs and leaps, in each case running up an oblique branch and leaping down to another.

Table 2
Mean percentage of animals observed at each level for each activity, and
regardless of activity

a. *Lemur fulvus rufus* at Antserananomby

Activity	Level				
	1	2	3	4	5
Feeding	0.00	2.58	15.60	59.66	22.16
Grooming	0.00	0.54	9.34	78.24	11.88
Resting	0.00	1.20	14.05	79.60	5.15
Moving	2.61	0.60	18.14	66.16	12.48
Travel	0.00	0.00	2.60	87.84	9.56
Other	0.00	8.33	5.95	54.70	31.02
All Activities	0.44	2.21	10.95	71.03	15.38

b. *Lemur fulvus rufus* at Tongobato

Activity	Level				
	1	2	3	4	5
Feeding	2.17	11.40	5.01	44.66	36.76
Grooming	0.00	0.60	9.93	79.99	9.49
Resting	0.34	0.83	8.03	79.91	10.89
Moving	6.02	1.27	4.14	78.64	9.93
Travel	0.95	3.69	5.92	82.94	6.50
Other	0.00	5.00	15.00	56.18	23.82
All Activities	1.58	3.80	8.00	70.39	16.23

c. *Lemur catta* at Antserananomby

Activity	Level				
	1	2	3	4	5
Feeding	27.58	12.78	25.23	19.01	15.40
Grooming	4.37	15.90	20.42	46.00	13.31
Resting	9.54	16.74	25.62	35.02	13.08
Moving	44.73	5.88	21.62	17.58	10.20
Travel	64.70	1.81	13.47	17.46	2.57
Other	28.70	5.02	27.31	22.50	16.47
All Activities	29.93	9.69	22.28	26.26	11.84

Table 2 (continued)

d. *Lemur catta* at Berenty

Activity	Level				
	1	2	3	4	5
Feeding	30.84	15.10	41.56	12.32	0.18
Grooming	10.16	23.35	27.97	38.38	0.14
Resting	14.01	26.14	28.18	31.68	0.00
Moving	55.32	13.51	15.82	15.35	0.00
Travel	70.93	6.44	5.48	17.15	0.00
Other	35.71	17.27	20.61	26.41	0.00
All Activities	36.16	16.97	23.27	23.55	0.05

Morphological differences have been found between the foot of *L. f. rufus* and that of *L. catta*. In *L. f. rufus*, the heel is covered with hair, whereas in *L. catta* the heel is naked. In fact, *L. catta* is the only species of prosimian in which the plantar pad extends proximally to the heel. In this feature, the foot of *L. catta* closely resembles that of monkeys. An examination of the skeletons of the feet of *L. f. rufus* and *L. catta* has shown that the mid-tarsal joints (calcaneo-cuboid and talo-navovicular) are less mobile in *L. catta* and that the foot of *L. catta* may be adapted for what Morton has termed metatarsi-fulcrumation.[5] Further studies on the morphology and analyses of films of the locomotion of *L. f. rufus* and *L. catta* may allow more precise relationships to be found between the anatomy, locomotor behaviour, and vertical pattern of habitat preferences of the two species.

2 Utilisation of the horizontal habitat

Groups of *L. f. rufus* were identified and their locations were recorded throughout the study. Three groups of *L. f. rufus* were studied intensively at Tongobato and six groups at Antserananomby. At Antserananomby, there were twelve groups within the seven hectares of closed canopy forest. These groups were identified and their locations were noted on prepared maps whenever they were sighted. Groups of *L. f. rufus* usually could not be followed throughout the whole day because of the difficulty in remaining in continuous contact with this arboreal species. One group of *L. catta* at Antserananomby and two groups at Berenty were studied intensively. The groups of *L. catta* were usually followed throughout the day, their movements being recorded on prepared maps.

Figs. 4 and 5 illustrate the extreme differences in the size of the home-

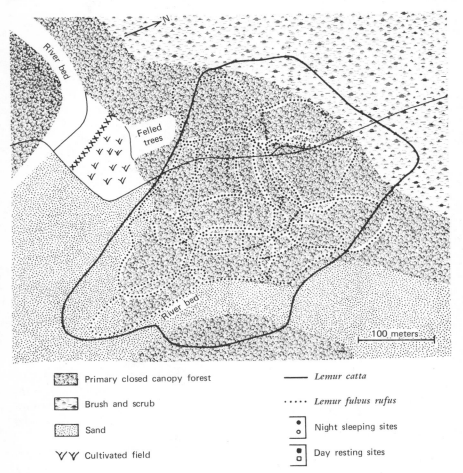

Fig. 4. Antserananomby. Home-ranges of *Lemur catta* and *Lemur fulvus rufus*. This map includes the home-range of one group of *L. catta* (19 animals) and twelve groups of *L. f. rufus* (112 animals in all).

ranges and day-ranges between *L. f. rufus* and *L. catta* at Antserananomby. In this forest, the home-ranges of twelve groups of *L. f. rufus* (112 animals in all) were included within the home-range of one group of 19 *L. catta* (Fig.4).

The day-ranges and home-ranges of *L. f. rufus* were very small, both at Antserananomby and at Tongobato. Although it was generally difficult to remain in contact with groups of *L. f. rufus*, at Antserananomby I was able to follow three groups for two days each (Fig.5). The average day-range for the six days was between 125 and 150 m. The home-ranges of the groups

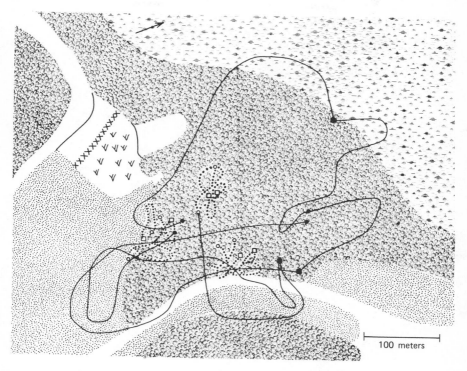

Fig. 5. Antserananomby. Comparative day-ranges of *Lemur catta* and *Lemur fulvus rufus*. This map includes a sample of three day-ranges of one group of *L. catta* (group AC-1) and two day-ranges of three groups of *L. f. rufus*. (Key as for Fig. 4.)

studied at Antserananomby averaged 0.75 hectares (Fig.4), while those of three groups at Tongobato averaged 1.0 hectares. The home-ranges of *L. f. rufus* were not rigidly defended, and the boundaries overlapped extensively.

L. catta had much larger day- and home-ranges than *L. f. rufus*. The average day-range for *L. catta* at Antserananomby was approximately 920 m. (Fig.6). At Berenty, in a sample of four days, the average day-range was 965 m. The home-range of the group of *L. catta* at Antserananomby (group AC-1) was 8.8 hectares (Fig.4). The home-range of group BC-1 at Berenty was 6.0 hectares. Jolly reported a home-range of 5.7 hectares for the group she studied at Berenty.[6] At Antserananomby, *L. catta* spent approximately 58% of its time in zones outside of the closed canopy area, even though these zones represented only 30% of its total home-range. By contrast, all of the home-ranges of the groups of *L. f. rufus* at this forest were located within the 7 hectares in which there is a closed canopy.

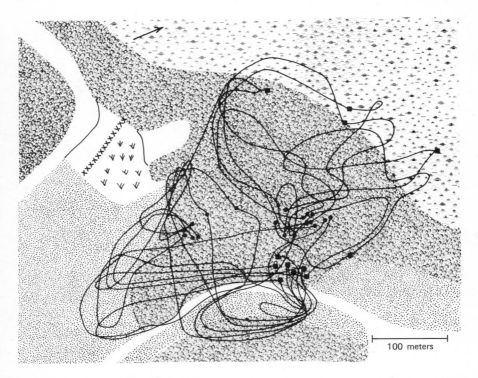

Fig. 6. Antserananomby. Day-ranges of *Lemur catta* (group AC-1), including all of the ranges for which the group was followed throughout the day. (Key as for Fig. 4.)

The population density of *L. f. rufus* was much higher than that of *L. catta*. The population density of each species was calculated from the average group size (see Tables 10 and 11) and the average area of the home-range. The population density of *L. f. rufus* at Tongobato was calculated only from data on groups TF2, TF3, and TF4.

Lemur catta

Antserananomby	215 per square km.
Berenty*	250 per square km.
Average	233 per square km.

*Jolly estimated the density of *L. catta* at Berenty to be 320 per square km.[6] This would give an average density of 261 per square km. for populations of *L. catta* studied to date.

	Lemur fulvus rufus
Antserananomby	1222 per squre km.
Tongobato	900 per square km.
Average	1061 per square km.

3. Diet

The number of animals feeding and the level of the forest in which they were feeding was recorded during the collection of five-minute activity records. The plant and the part being eaten were recorded directly on the data sheet. Plant specimens were collected whenever possible and were later identified by Armand Rakotozafy of the Laboratoire Botanique of ORSTOM, Tananarive. No stomach content samples were obtained, because it was not feasible to shoot any of the animals during the field study. The dry weights of the plants and the nutritional values of the specimens have not yet been analysed. Therefore, the dietary schedules of the two species have been determined from direct observation only. Samples of faeces were collected, but have not yet been analysed.

The differences between the diet of *L. catta* and *L. f. rufus* are related to the differences in habitat selection in the two species. *L. f. rufus* stayed in the continuous canopy of the forest and moved horizontally from tree to tree. It was rarely seen in areas that required movement on the ground. It fed over 90% of the time in the upper three forest levels (see Table 2). *L. catta*, on the other hand, was frequently observed feeding in all of the available forest layers. It spent about 58% of its time outside of the portion of the forest with a continuous canopy. *L. catta* fed over 65% of the time in levels 1, 2, and 3.

The restricted use of space by *L. f. rufus* is related to a less varied diet in this species than that found in *L. catta*. The plants which *L. f. rufus* was observed eating during the study are listed in Tables 3 and 4. I recorded *L. f. rufus* eating only 8 plant species at Tongobato and 11 at Antserananomby. There was a total of 13 different plant species eaten by *L. f. rufus* in the two forests. *L. catta*, on the other hand, had a much more varied diet in both of the forests in which it was studied. Both at Antserananomby and Berenty, *L. catta* was observed to feed on 24 different plant species (Tables 5 and 6). At Antserananomby. *L. catta* ate 12 plant species which were not eaten by *L. f. rufus*. Ten of these were ground plants or bushes, or plants found only outside the closed canopy portion of the forest.

L. f. rufus appears to be specialised in its choice of diet. A few species of plants made up a large proportion of its diet, and kily leaves were the main staple (Table 7). During the period of observation at Tongobato, three species of plant constituted more than 80% of the diet of *L. f. rufus*: *Flacourtia ramontchi*, *Tamarindus indica*, and *Terminalia mantaly*. At Antserananomby,

Table 3
Plant species eaten by *Lemur fulvus rufus* at Tongobato

Acacia rovumae Olia[a]
Acacia sp.[a]
Alchornea sp.
Flacourtia ramontchi[a]
Lawsonia alba Lamk
Tamarindus indica[a]
Terminalia mantaly[a]
Vitex beravensis[a]

[a] Plant species eaten by *Lemur fulvus rufus* at both Tongobato and Antserananomby.

Table 4
Plant species eaten by *Lemur fulvus rufus* at Antserananomby

Acacia rovumae Olia[a]	*Rinorea greveana*
Acacia sp.[a]	*Tamarindus indica*[a]
Ficus soroceoides Bak	*Terminalia mantaly*[a]
Flacourtia ramontchi[a]	*Tisomia* sp.
Papilionaceae	*Vitex beravensis*[a]
Quisivianthe papinae	

[a] Plant species eaten by *Lemur fulvus rufus* at both Tongobato and Antserananomby.

Acacia rovumae, *Ficus soroceoides*, and *Tamarindus indica* accounted for over 85% of the diet of *L. f. rufus*. Kily *(Tamarindus indica)* leaves made up 42% of the diet of *L. f. rufus* at Tongobato and about 75% of its diet at Antserana-

Table 5
Plant species eaten by *Lemur catta* at Antserananomby

Acacia rovumae Olia	*Mimilopsis* sp.
Acacia sp.	*Paederia* sp.
Acalypha sp.[a]	*Paederia* sp.
Achyranthes aspera L.[a]	Papilionaceae
Adenia sp.	*Poupartia caffra* H. Perr
Alchornea sp.	*Quisivianthe papinae*
Commicarpus commersonii	*Tamarindus indica*[a]
Ficus cocculifolia	*Terminalia mantaly*
Ficus soroceoides	*Vitex beravensis*
Ficus sp.	
Flacourtia ramontchi	
Grevia sp.	
Three species unidentified (small trees)	

[a] Plant species eaten by *Lemur catta* at both Antserananomby and Berenty.

Table 6

Plant species eaten by *Lemur catta* at berenty

Acalypha sp.[a]	*Ehretia* sp.
Aizoaceae	*Hzima tetracantha* Lamk
Albizzia polyphylla Forven	*Mangifera indica* L.
Annona sp.	*Melia azedarach* L.
Achyranthes aspera L.[a]	*Opuntia vulgaris* Mill
Boerhaavia diffusa L.	*Phyllanthus* sp.
Cardiaspermum halicacabum	*Pithecelobium dulce* Benth
Cassia sp.	*Rinorea greveana* H. Bn.
Celtis philippensis Blanco	*Tamarindus indica*[a]
Cissompelos sp.	*Zehneria* sp.
Combretum sp.	*Zizyphus jujuba* Lamk
Crateva excelsa Boj	One species unidentified (vine)

[a] Plant species eaten by *Lemur catta* at both Berenty and Antserananomby.

nomby. This degree of specialisation is rare among those primate species that have been studied. It has been reported to occur, however, in *Lepilemur mustelinus leucopus*.[7]

The diet of *L. catta* was much less restricted than that of *L. f. rufus* (Table 7). At Antserananomby, the following species made up over 70% of the diet:

Achyranthes aspera	*Grevia* sp.
Alchornea sp.	*Mimilopsis* sp.
Ficus soroceoides	*Poupartia caffra*
Flacourtia ramontchi	*Tamarindus indica*

The kily tree provided 24% of the diet of *L. catta* in this forest. However, less than half of this (11%) consisted of leaves. Kily pods made up 12% of the diet, and ground plants (*Achyranthes aspera* and *Mimilopsis* sp.) 15%.

The diets of *L. catta* at Berenty and Antserananomby consisted of different plant species, but were quite similar in design. At Berenty, the following species accounted for over 80% of the observed plants eaten by *L. catta:*

Achyranthes aspera	*Phyllanthus* sp.
Boerhaavia diffusa	*Pithecelobium dulce*
Cassia sp.	*Rinorea greveana*
Melia azedarach	*Tamarindus indica*

The kily tree made up 23% of the diet. Kily leaves were eaten in 12% of the observations and kily pods in 10%. The fruit of two other trees (*Rinorea greveana* and *Pithecelobium dulce*) together accounted for 40% of the observed feeding activity.

Both *L. catta* and *L. f. rufus* ate the fruit, leaves, flowers, bark, and sap of various species of plants (Table 8). The amount of time that they spent feeding on various parts of the plants depended upon the fruiting or blossoming seasons of the plants. As with differences in the number of plant

Table 7

The number and percentage of individual activity records (IARs) for feeding on identified plant species

a. Lemur fulvus rufus at Antserananomby

Plant Species	Number of IARs	Percentage of IARs
Tamarindus indica	1802	75.68 %
⎡ Leaves ⎤	⎡1793⎤	⎡75.30⎤
⎣ Fruit ⎦	⎣ 9 ⎦	⎣ 0.38⎦
Acacia sp.	156	6.55
Ficus soroceoides Bak	141	5.92
Acacia rovumae Olia	74	3.10
Terminalia mantaly	21	0.88
Quisivianthe papinae	18	0.75
Other	169	7.06
Total	2381	99.94 %

b. Lemur fulvus rufus at Tongobato

Plant Species	Number of IARs	Percentage of IARs
Tamarindus indica	276	48.85
⎡ Leaves ⎤	⎡237⎤	⎡41.95⎤
Flowers	26	4.60
Fruit	10	1.77
⎣ Bark ⎦	⎣ 3⎦	⎣ 0.53⎦
Terminalia mantaly	127	22.47
Flacourtia ramontchi	69	12.21
Acacia rovumae Olia	38	6.72
Vitex beravensis	18	3.18
Other	37	6.54
Total	565	99.97 %

Table 7 (continued)

c. *Lemur catta* at Antserananomby

Plant Species	Number of IARs	Percentage of IARs
Tamarindus indica	374	24.36 %
⌈ Fruit ⌉	⌈ 183 ⌉	⌈ 11.92 ⌉
| Leaves |	| 174 |	| 11.33 |
⌊ Flowers ⌋	⌊ 17 ⌋	⌊ 1.11 ⌋
Small trees:	320	20.84
⌈ *Alchornea* sp. ⌉		
| *Flacourtia ramontchi* |		
| *Grevia* sp. |		
⌊ *Poupartia caffra* H. Perr ⌋		
Ground plants (all species)	225	14.65
Ficus soroceoides Bak	194	12.63
Vines (all species)	140	9.12
Quisivianthe papinae	94	6.12
Vitex beravensis	82	5.34
Ficus cocculifolia	40	2.60
Acacia rovumae Olia	18	1.17
Other	48	3.12
Total	1535	99.95 %

d. *Lemur catta* at Berenty

Plant Species	Number of IARs	Percentage of IARs
Tamarindus indica	519	22.97 %
⌈ Leaves ⌉	⌈ 274 ⌉	⌈ 12.13 ⌉
| Fruit |	| 225 |	| 9.96 |
⌊ Bark ⌋	⌊ 20 ⌋	⌊ .88 ⌋
Rinorea greveana H. Bn.	474	20.98
Pithecelobium dulce Benth	433	19.16
Phyllanthus sp.	137	6.06
Melia azedarach L.	132	5.84
Ehritia sp.	130	5.74
Ground plants (all species)	124	5.48
Opuntia vulgaris Mill	84	3.71
Annona sp.	38	1.68
Vines (all species)	27	1.19
Other	161	7.10
Total	2259	99.91 %

Table 8

Number and percentage of individual activity records for feeding on identified parts of plants

a. Lemur fulvus rufus at Antserananomby

Part of plant eaten	Number of IARs	Percentage of IARS
Leaves	2123	89.16 %
Fruit	161	6.76
Flowers	90	3.77
Bark	7	.29
Total	2381	99.98 %

b. Lemur fulvus rufus at Tongobato

Part of plant eaten	Number of IARs	Percentage of IARs
Leaves	275	52.08 %
Fruit	224	42.43
Flowers	26	4.92
Bark	3	.56
Total	528	99.99 %

c. Lemur catta at Antserananomby

Part of plant eaten	Number of IARs	Percentage of IARs
Leaves	670	43.64 %
Fruit	516	33.61
Herbs	225	14.65
Flowers	124	8.07
Total	1535	99.97 %

d. Lemur catta at Berenty

Part of plant eaten	Number of IARs	Percentage of IARs
Fruit	1335	59.30 %
Leaves	550	24.43
Flowers	137	6.08
Herbs	124	5.50
Bark, sap, cactus	105	4.66
Total	2251	9.97 %

species eaten by *L. catta* and *L. f. rufus*, the proportion of fruit or flowers fed upon by the two species is related to differences in the use of vertical and horizontal space. For example, a small tree (*Ficus cocculifolia*) produced fruits the size of large apples. I saw *L. f. rufus* feed on the fruit of this tree in a forest on the bank of the Mangoky River. There was only one of these trees in the forest at Antserananomby, and it was located on the side of a dry river bed opposite to the continuous canopy portion of the forest. *L. f. rufus* rarely crossed the river, and was never seen in this tree. *L. catta* on the other hand, regularly crossed the river and foraged in this tree for a number of days while the fruit was ripe.

4. Utilisation of time

Once a group of animals was located, counts were made at five-minute intervals of the number of individuals engaged in each of six activities: feeding, grooming, resting, movement (i.e. movement of an individual), travel (i.e. movement of the group), and other (i.e. sunning, play, fighting, etc.). Each animal observed during each five-minute interval constituted an individual activity record (IAR). The IARs for each five minutes were combined for each half-hour period from 6.00 hrs to 18.25 hrs., and the percentage of IARs for each activity was calculated for each half-hour. The day was then divided into five phases as shown in Fig.7, which includes the means of the percentages calculated for the half-hour periods within each of the phases.

The most striking differences between *L. f. rufus* and L. catta are seen in the data from Antserananomby (Figs. 7*a* and 7*b*). In this forest, *L. f. rufus* rested throughout the afternoon, whereas *L. catta* rested only during the midday phase (phase 3). Data from all three forests indicate that *L. f. rufus* rests more than *L. catta* during the day. The activity/rest ratios for the hours from 6.00 hrs. to 18.25 hrs. were calculated from the data in Table 9. The ratios are as follows:

	Lemur fulvus rufus
Tongobato	50/50 = 1.00
Antserananomby	44/56 = 0.79

	Lemur catta
Antserananomby	59/41 = 1.44
Berenty	61/39 = 1.56

L. f. rufus fed very early in the morning and late in the afternoon and travelled little to obtain its food. *L. catta* began to feed later in the morning and stopped earlier in the evening than *L. f. rufus*. Groups of *L. cattta*

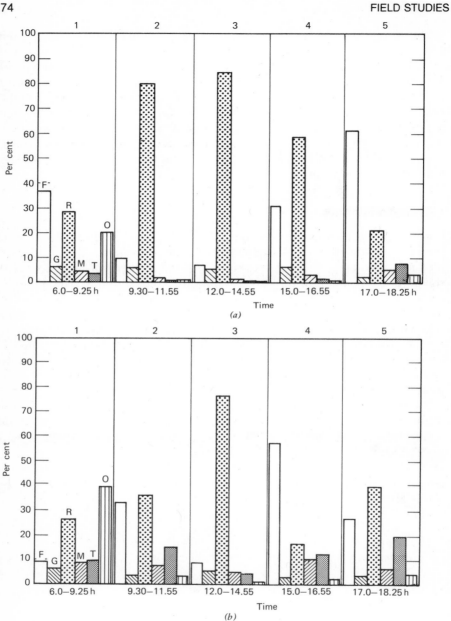

Fig. 7. Mean percentage of individual activity records for each of six activities during five phases of the day. F = Feeding; G = Grooming; R = Resting; M = Moving; T = Travel; O = Other. (a) *Lemur fulvus rufus* at Antserananomby. (b) *Lemur catta* at Antserananomby. (c) *Lemur fulvus rufus* at Tongobato. (d) *Lemur catta* at Berenty.

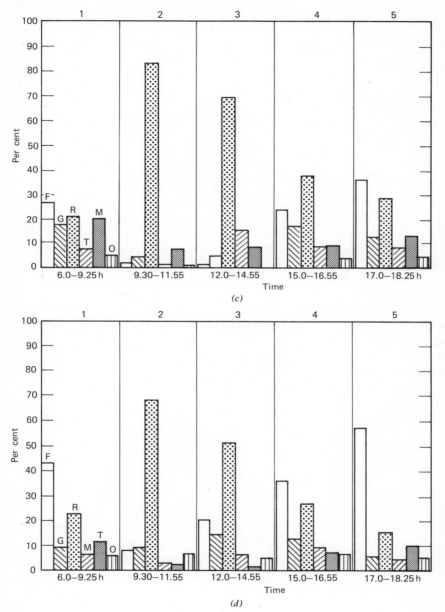

(c)

(d)

travelled greater distances to obtain food, foraging throughout the afternoon. At Antserananomby. *L. f. rufus* fed mainly during the first and fifth phases and *L. catta* during the second and fourth. In all phases but the third

Table 9
Mean percentages of individual activity records for each activity each half hour

Species and forest	Activity					
	Feeding	Grooming	Resting	Moving	Travel	Other
Lemur fulvus rufus						
Tongobato	16.59	11.30	49.73	7.77	12.21	2.41
Antserananomby	26.22	5.25	56.58	3.09	2.47	6.40
Lemur catta						
Antserananomby	24.94	4.67	41.42	7.45	11.34	10.19
Berenty	31.12	10.97	38.63	6.30	6.91	6.07

(12.00 to 14.55 hrs), when both species were resting, the patterns of activity were different.

At Berenty, groups of *L. catta* began to feed earlier in the morning and the peak of their resting activity was earlier than at Antserananomby. They also began to feed earlier in the afternoon and stretched out their afternoon feeding activity over a longer period of time. These differences, which can be related to the early sunrise and high early morning temperatures at Berenty, make the behaviour pattern of *L. catta* at this forest superficially similar to that of *L. f. rufus* in the other study areas. Even though the

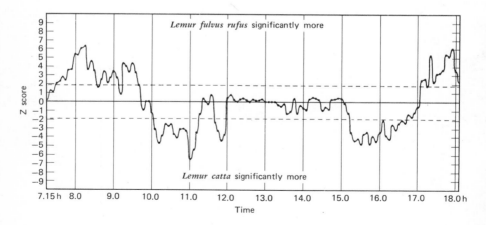

Fig. 8. Z scores for the feeding activity of *Lemur fulvus rufus* and *Lemur catta* at Antserananomby computed for each five-minute observation period. A score equal to or more than 1.96 means that *L. f. rufus* feeds significantly more than *L. catta* (p. ≤.05). A score equal to or less than −1.96 means that *L. catta* feeds significantly more than *L. f. rufus* (p ≤.05).

temperature and time of sunrise at Tongobato were more similar to those at Berenty than to those at Antserananomby, the pattern of behaviour of *L. f. rufus* was similar in both of the forests in which it was studied. The high percentage of movement and travel of *L. f. rufus* at Tongobato resulted from the difficulty in habituating the animals to the observer, because local inhabitants frequently hunted the lemurs in this forest. Seasonal differences (in this case, time of sunrise and sunset and time of the maximum temperature during the day) seemed to affect the daily activity cycle of *L. catta* more than that of *L. f. rufus*.

L. f. rufus and *L. catta* both exhibited "sunning" behaviour early in the morning at Antserananomby. This is reflected in the high percentage of "other" activities during the first phase of the day. *L. catta* sunned twice as much as *L. f. rufus* during this phase, and sunning may be an important thermo-regulating mechanism in this species. Very little sunning was observed for either species in the other two forests. The mean minimum temperatures at night at Berenty and Tongobato were much higher than those at Antserananomby during the months of observation in these forests.

Significance tests were run on the resting and feeding activities of *L. f. rufus* and *L. catta* at Antserananomby. Since the samples for different five-minute periods are not independent, Z scores were computed for each five-minute interval (Figs. 8 and 9). The methods used for these calculations are described by Sussman [2] and Sussman, O'Fallon and Buettner-Janusch (in prep.).

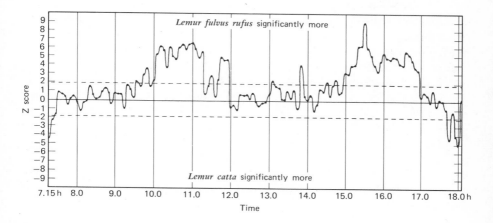

Fig. 9. Z scores for the resting activity of *Lemur fulvus* and *Lemur catta* at Antserananomby computed for each five-minute observation period. A score equal to or more than 1.96 means that *L. f. rufus* rests significantly more than *L. catta* (p ≤.05). A score equal to or less than −1.96 means that *L. catta* rests significantly more than *L. f. rufus* (p ≤.05).

Fig. 8 illustrates the fact that *L. f. rufus* fed significantly more in the early morning and late evening, and *L. catta* fed significantly more in the late morning and late afternoon. Feeding behaviour in the midday phase was not significantly different; at this time, both species were mainly resting. Throughout the day there are very few times (a total of 25 minutes) when *L. catta* rested significantly more than *L. f. rufus* (Fig. 9). There are only 26 five-minute periods in which a higher percentage of *L. catta* than *L. f. rufus* were observed resting.

Statistical analyses comparing data from different forests were difficult to interpret because of the differing forest conditions. Significance tests on the behaviour of populations in different locations only indicate whether or not the behaviour is significantly different. They do not give us any insight into whether differences are due to characteristics of the populations or to environmental conditions.

5. Group size and composition

Groups were counted throughout the study period at each forest, and census data on each group were checked and rechecked continually. In all three forests, census data were collected by more than one observer. In most cases, two observers worked together.[*]

The following criteria were used to distinguish between sex and age classes. There are conspicuous differences between the sexes in *Lemur fulvus rufus*. The male is grey and has a pronounced white face mask with a bright red-orange tuft of hair on the top of the head. The female is red-orange, with white patches over her eyes and a black bar across the forehead and between the eyes. These distinctions are obvious even in the juvenile animals. The coloration of infants, however, is often misleading, and sex determination of these young animals is not possible. In *Lemur catta* there are no differences in pelage between the sexes. In order to sex individuals of this species, it was necessary to get into strategic position to observe the genitalia of the animals. It was not possible to determine the sex of juvenile or infant *L. catta*.

Each individual was placed into one of three age categories—infant, juvenile or adult. Infants were those animals that were still being carried by their mothers. This included animals up to the age of about four to five months. Juveniles were those animals that were no longer being carried, but had not yet reached full size. However, in the field it is often difficult to

[*] Linda Sussman, Alain Schilling, and field guides Folo Emmanuel and Bernard Tsiefatao assisted me in collecting census data.

distinguish older juveniles from adults. Adult size is attained at about two years of age. In both *L. catta* and *L. f. rufus*, births occur once a year within a short period of time (usually a two-week period). The young all mature at approximately the same rate and the achievement of different stages of maturity is synchronous.

Tables 10 and 11 include the census data collected on *L. f. rufus* and *L. catta*. The 17 groups of *L. f. rufus* which I censused ranged in size from 4 to 17 animals. The average size of the groups was 9.5 animals. The average size of the three groups of *L. catta* which I censused was 18.0 animals. Including the counts of Jolly, and Klopfer and Jolly[6,8] all of the groups of *L. catta* which have been censused consist of 12 to 24 individuals and the average size of the groups is 18.8 animals. Thus, the sizes of groups of *L. catta*, on the average, are approximately twice those of groups of *L. f. rufus*.

Adult sex ratios were similar in groups of both *L. catta* and *L. f. rufus*. There were slightly more adult females than adult males in groups of both species. It is possible that groups of *L. catta* contain proportionately more infants per year than do those of *L. f. rufus*. The ratio of infants to non-infants for *L. f. rufus* at Tongobato was 1/4.66, while that for *L. catta* at Berenty was 1/3.38. Since there were no infants during the time of the census at Antserananomby, the ratio of juveniles (mainly one-year olds) to adults was calculated. This ratio was 1/5.11 for *L. f. rufus* and 1/3.75 for *L. catta*.

A characteristic which is not revealed by the census data is the tendency towards peripheralisation of subordinate males in groups of *L. catta*. This characteristic was associated with a well-defined dominance hierarchy. There was no noticeable hierarchy in groups of *L. f. rufus*. Jolly reported that peripheralisation of subordinate males occurred in the group of *L. catta* she studied intensively at Berenty.[6] The groups I studied also showed a tendency towards the peripheralisation of subordinate males. In one of the groups at Berenty, two adult males were actively kept away from the centre of the group. These two peripheral males were always together and sometimes intermingled with juveniles of the group, but were chased whenever they approached the centre of the group too closely.

CONCLUSIONS

Differences were found in the utilisation of space and time, the diet and the social structure of *Lemur fulvus rufus* and *Lemur catta*. These differences were related to the habitat preferences of the two species, and in each case were independent of the presence or absence of the other species. This indicates that the differences are not caused by the interaction between the populations of *L. f. rufus* and *L. catta*, but by adaptations to basically different environmental conditions which occur where the species are allopatric.

Table 10

Composition of groups of *Lemur fulvus rufus*

Name of group	Adult Male	Adult Female	Juvenile Male	Juvenile Female	Infant	Total	Adult Sex ratio M:F
Antserananomby[a]							
AF-1	4	6	1	1	0	12	1:1.50
AF-2	4	5	1	0	0	10	1:1.25
AF-3	4	3	0	1	0	8	1.33:1
AF-4	2	5	2	1	0	10	1:2.50
AF-5	4	6	0	2	0	12	1:1.50
AF-6	3	2	0	0	0	5	1.50:1
AF-7	3	4	1	1	0	9	1:1.33
AF-8	2	2	0	0	0	4	1:1.00
AF-9	2	2	1	0	0	5	1:1.00
AF-10	4	5	2	0	0	11	1:1.25
AF-11	5	7	1	2	0	15	1:1.40
AF-12	4	4	1	0	0	9	1:1.00
Totals	41	51	10	8	0	110[a]	41:51 = 1:1.24
Means	3.42	4.25	.83	.66	0	9.17	
Tongobato							
TF-1	5	8	0	0	4	17	1:1.60
TF-2	2	3	1	0	1	7	1:1.50
TF-3	4	5	0	1	2	12	1:1.25
TF-4	3	4	0	0	1	8	1:1.33
TF-5	3	2	0	1	1	7	1.50:1
Totals	17	22	1	2	9	51	17:22 = 1:1.29
Means	3.40	4.40	.20	.40	1.80	10.20	
Overall totals	58	73	11	10	9	161	58:73 = 1:1.26
Overall means	3.41	4.29	.64	.59	.53	9.47	

[a] There was also an extra male group consisting of two individuals. Therefore, the total number of animals censused at Antserananomby was 112.

The quantitative data on the utilisation of the vertical habitat indicate that *L. f. rufus* and *L. catta* have distinctive preferences for different forest strata. *L. f. rufus* is very specific in its choice of vertical habitat; *L. catta,* on the other hand, is frequently found in all forest layers. In all three forests studied, vertical habitat preferences do not seem to be altered by the presence or absence of the other species. This indicates that the two species do not compete directly for the use of a particular vertical forest stratum. It also suggests that the differences in the choice of vertical habitat are not the result of interaction between the two species.

The day-ranges and home-ranges of *L. f. rufus* are much smaller than those of *L. catta,* and the population density is much higher. The home-ranges of groups of *L. f. rufus* at Antserananomby and Tongobato are restricted to portions of the forests with a continuous canopy. The home-ranges are not rigidly defended and have overlapping boundaries. The high density of the population and the small, overlapping ranges of the groups allow *L. f. rufus* to completely fill these restricted forest areas. This pattern of group distribution may be adaptive for the exploitation of small areas which are relatively rich in food supply and in which the food is evenly distributed.[9]

Populations of *L. catta* are found in the west and south of Madagascar in many areas in which there are no closed canopy forests but only brush and scrub forests. Many of these scrub forests are very dry and the vegetation is patchy and likely to be sparse for at least part of the year. The large home-ranges and day-ranges of groups of *L. catta* may be genetically fixed adaptive responses to these arid environments. Trends in population density and range size similar to those found for *L. f. rufus* and *L. catta* have been reported for forest monkeys and savannah or grassland monkeys, respectively.[9, 10, 11, 12, 13]

Table 11

Composition of groups of *Lemur catta*

Name of Group	Adult Male	Adult Female	Juvenile	Infant	Total	Adult Sex ratio M:F
Antserananomby						
AC-1	7	8	4	0	19	1:1.14
Berenty						
BC-1	4	5	2	4	15	1:1.25
BC-2	7	6	3	4	20	1.17:1
Totals	11	11	5	8	35	11:11 = 1:1.00
Means	5.50	5.50	2.50	4.00	17.50	
Overall totals	18	19	9	8	54	18:19 = 1:1.05
Overall means	6.00	6.30	3.00	2.70	18.00	

Differences in the diet of *L. f. rufus* and *L. catta* can also be related to differences in the habitat preferences of the two species. Although both seem to be opportunistic in their choice of food, the less restricted use of the vertical and horizontal habitat by *L. catta* allows this species a greater opportunity to obtain a more varied food supply. *L. f. rufus,* having a more limited vertical and horizontal range, has a more specialised relationship to the environment.

The foraging habits of *L. f. rufus* and *L. catta* seem to be adaptations to different econiches. *L. f. rufus* inhabits closed canopy forests, which in most cases are found in small, relatively uniform, circumscribed areas. In these forests, the population density of *L. f. rufus* is extremely high. The ability to exist almost exclusively on a diet of kily leaves, supplemented with the fruit of those trees which happen to be within the range of any one group, allows *L. f rufus* to exploit efficiently this type of environment. A group of *L. catta* continuously forages within its range and exploits a number of plants in a large area. Within a period of seven to ten days, the group visits most of its total range. It is likely that this continuous foraging behaviour may be adaptive to arid environments in which the food supply is sparse. The constant surveillance of a large (or relatively large) range allows groups of *L. catta* to exploit a number of different plants which may have a patchy distribution over a wide area.

The utilisation of time by the two species further indicates that behavioural adaptations of *L. f. rufus* and *L. catta* have developed in response to different environmental conditions. *L. f. rufus* feeds early in the morning and late in the afternoon; it rests throughout the day and spends almost all of the warmest portion of the day in the closed canopy. The temperature of the micro-environments in which *L. catta* is found throughout the day is more likely to vary. At night, *L. catta* sleeps in the closed canopy portion of the forest. However, during the day it spends much of the afternoon in unshaded portions of the forest. In these areas, the ambient temperature is usually quite high. At Antserananomby, where the evening temperatures were low, *L. catta* sunned for long periods during the morning.

The distribution of *L. catta* within the forests in which it was studied, and the fact that it is found in the hot, arid regions of southern Madagascar, indicate that this species is well adapted to warm and dry climates. *L. f. rufus* is not found in the south of Madagascar, and it remains in the shaded portion of the forest throughout the day. Furthermore, *L. f. rufus* is active only during the coolest hours of the day. It seems likely that *L. catta* uses behavioural thermo-regulation (e.g. sunning) to help maintain its body temperature and in this way is able to utilise habitats with large ranges of ambient temperature.

The sizes of groups of *L. catta* are, on average, approximately twice those of groups of *L. f. rufus.* Terrestrially adapted species have generally been considered to form larger groups than arboreally adapted spe-cies.[9, 10, 11] There are, however, several exceptions, and phylogenetic rela-tionships also play a role in the determination of group size.[12, 14] Before relationships can be established between ecology and group size or social structure, further studies must be made on arboreally adapted species and their use of the environment.

There is a tendency in groups of *L. catta* towards peripheralisation of subordinate males. In a very dry environment, where food is sparse, the exclusion of excess males could be very adaptive for the survival of the group.[15, 16] A study of *L. catta* in arid regions may reveal groups containing many more adult females than males.

In summary, *L. f. rufus* seems to be a very specialised, arboreally adapted species. *L. catta* is able to exploit a number of different habitats. However, the behavioural adaptations which have been found in *L. catta* seem to have developed in response to the more arid, warm environments in which this species is found (where resources are likely to be limited). *L. f. rufus* shows a configuration of behavioural adaptations which conform to those found in many arboreal species of New and Old World monkeys, whereas some behavioural adaptations of *L. catta* parallel those of many terrestrial primates living in savannah or grassland econiches.

ACKNOWLEDGMENTS

I would like to thank Dr Brygoo of l'Institut Pasteur de Madagascar, Président M. Manambelona, Secrétaire Général du Comité National de la Recherche Scientifique, M. Ramanantsoavina, Directeur des Eaux et Forêts, and the staff of ORSTOM, Tananarive for their assistance and cooperation while I was in the field. I am also indebted to many people of the villages of Manja and Vondrove, whose kind hospi-tality is deeply appreciated. I am especially grateful for the continued assistance (both academic and otherwise) offered by Dr J. Buettner-Janusch throughout the period of this study. My wife, Linda, aided me in all aspects of the research and I am, of course, especially indebted to her. The study was supported in part by Research Fellowship MH46268-01 of the National Institute of Mental Health, United States Public Health Service and by a Duke University Graduate Fellowship.

NOTES

[1] **Brown, W.L. and Wilson, E.O.** (1956), "Character displacement," *Sys-tematic Zool.* 5, 49-64.

[2] **Sussman, R.W.** (1972), "An ecological study of two Madagascan primates: *Lemur fulvus rufus* and *Lemur catta*," PhD thesis, Duke University.

[3] **Perrier de la Bathie, H.** (1921), "La végétation malgache," *Ann. Mus. Colon. marseille*, 3rd ser., 9, 1-268; Humbert, H. (1954), "Les territoires phytogéographiques de Madagascar: leur cartographie," *LIX/ Colloque International du Centre National de la Recherche Scientifique*, Paris, pp. 439-

48; Humbert, H. and Darne, G.C. (1965), *Notice de la Carte de Madagascar,* Toulouse.

4 **Crook, J.H. and Aldrich-Blake, P.** (1968), "Ecological and behavioural contrasts between sympatric ground-dwelling primates in Ethiopia," *Folia primat.* 8, 192-227.

5 **Morton, D.J.** (1924), "Evolution of the human foot," *Amer. J. Phys. Anthrop.* 7, 1-52.

6 **Jolly, A.** (1966), *Lemur Behavior,* Chicago.

7 **Charles-Dominique, P. and Hladik, C.M.** (1971), "Le *Lepilemur* du sud de Madagascar; écologie, alimentation et vie sociale," *Terre et Vie* 25, 3-66.

8 **Klopfer, P.H. and Jolly, A.** (1970), "The stability of territorial boundaries in a lemur troop," *Folia primat.* 12, 199-208.

9 **Crook, J.H. and Gartlan, J.S.** (1966), "Evolution of primate societies," *Nature* 210, 1200-3.

10 **DeVore, I.** (1953), "A comparison of the ecology and behaviour of monkeys and apes," in Washburn, S.L. (ed.), *Classification and Human Evolution,* Chicago, 301-19.

11 **Aldrich-Blake, F.P.G.** (1970), "Problems of social structure in forest monkeys," in Crook, J.H. (ed), *Social Behaviour in Birds and Mammals,* New York, 79-101.

12 **Crook, J.H.** (1970), "The socio-ecology of primates," in Crook, J.H. (ed), *Social Behavior in Birds and Mammals,* New York, 103-66.

13 **Denham, W.W.** (1971), "Enery relations and some basic properties of primate social organisation," *Amer. Anthrop.* 73, 77-95.

14 **Struhsaker, T.T.** (1969), "Correlates of ecology and social organisation among African cercopithecines," *Folia primat.,* 11, 80-118.

15 **Kummer, H.** (1967), "Dimensions of a comparative biology of primate groups," *Amer. J. Phys. Anthrop.* 27, 357-66.

16 **Kummer, H.** (1971), *Primate Societies,* Chicago.

NEW WORLD MONKEYS

introduction
field studies on new world monkeys

The suborder Anthropoidea probably evolved from
a tarsierlike ancestor sometime during the
Oligocene (about 35 million years ago). It is
divided into three superfamilies: Ceboidea (New
World monkeys); Cercopithecoidea (Old World
monkeys); and Hominoidea (apes and humans). All
anthropoids have diurnal activity cycles except one
monospecific genus of New World monkey, *Aotus*.
Like the diurnal prosimians, all anthropoids live in
permanent social groups. As will be seen in the
following chapters, however, the structure and
organization of these groups are extremely variable.
Furthermore, group structure is not necessarily
related to the taxonomy of the animal. Some closely
related animals have very different social structures,
while the social structure of some species in
different superfamilies can be quite similar.

The Ceboidea are divided into two families,
Callitrichidae and Cebidae. All Ceboids are diurnal,
except *Aotus*, and highly or exclusively arboreal.
They are found from southern Mexico and Central

America throughout the Amazon as far south as 30° S. Longitude in South America (see frontispiece). Unlike prosimians and monkeys in Africa and Asia, most species within the same genus of South American monkey displace each other geographically (see Selection 4). This suggests some extensive differences in the evolutionary history of primates of these areas.

The Callitrichids are small primates, weighing about 100 to 450 g. They differ from other New World monkeys in that they have claws instead of nails on their fingers and toes (except the large toe or hallux), and they have lost both the upper and lower third molar. Only one of the 35 species of Callitrichidae, *Sanguinus geoffroyi*, has been studied in any detail (see Baldwin et al. 1977). From what is known of the social behavior of the family, all species seem to live in family groups (mother, father, and offspring), and females normally give birth to twins. Males carry infants during the normal progression of the group and generally take a dominant role in caring for the young. Family groups and male participation in caring for infants are also found in the two genera of the Cebidae subfamily Aotinae, *Aotus* and *Callicebus.*

There are 11 genera of Cebidae, ranging in size from about 600 g (*Callicebus*) to 11 kg (spider monkeys). As stated above, all of these primates are forest-dwelling and highly arboreal. It might seem surprising, but long-term detailed studies of the New World monkeys are very rare. Furthermore, most of these have been done at the Smithsonian Institute research station, Barro Colorado, a small island in the Panama Canal. The most studied genera are *Ateles* (spider monkeys), *Alouatta* (howler monkeys), and *Cebus* (capuchins) but only certain species of even these genera are well known. In a recent survey of the literature on the superfamily Ceboidea, Baldwin et al. (1977) found that of the 64 species, only 17% (11 species) were represented by short-term or long-term projects, and 23% (15 species) have only been surveyed or incidentally observed. Sixty percent (38 species) remain completely unstudied.

The chapters included in this section all describe behavioral and ecological contrasts in sympatric species of Cebidae. In Chapter 4 Kinzey and Gentry provide preliminary data to support the hypothesis that two species of *Callicebus* (*C. torquatus* and *C. moloch*) live in contiguous but not overlapping types of forest and are, thus, sympatric habitat isolates. They suggest that this pattern of distribution corresponds to different habitat preferences in the species that in turn are related to differences in diet and dental morphology. Sympatric habitat isolation, although relatively common in other taxa, has not been reported previously among primates. Thorington (Selection 5) shows how the feeding and foraging behavior of *Cebus* and *Saimiri* relate to other general behavioral characteristics. Without an understanding of the interrelationships between things such as concentration span and insect foraging, interpretation of the behavior of these two species in laboratory and especially psychological experiments might be nonsensical. The Kleins (Selection 6) describe the "fission fusion" social structure of *Ateles belzebuth*, a social structure in some ways similar to that found in some cercopithecines (see especially Selection 21) and in chimpanzees (Selection 16). By contrasting habitat utilization of four sympatric taxa, the Kleins illustrate how social structure might be related to resource distribution. They emphasize that the general dietary categories of frugivore, folivore, or insectivore might not be meaningful, since there are a number of dietary patterns that crosscut these traditional categories. Relationships between ecology and social structure, therefore, might be more subtle than usually assumed. This possibility is discussed further in some of the more recently published chapters in Part II.

REFERENCES CITED

Baldwin, L.A., T.L. Patterson, and G. Teleki, 1977. Field research on Callitrichid and Cebid monkeys: An historical, geographical, and bibliographical listing. Primates, 18: 485-507.

FURTHER BIBLIOGRAPHY
Hershkovitz, P. 1977. Living New World Monkeys (Platyrrhini). Volume 1. University of Chicago Press, Chicago. (An 1117 page reference book, detailing all aspects of the biology of the Callitrichidae. Volume 2 will deal with the Cebidae.)
Moynihan, M. 1976. The New World Primates. Princeton University Press, Princeton, N.J. (A somewhat popularized overview and review of our knowledge of the New World primates. The author includes many of his own previously unpublished observations.)

4

Habitat Utilization in Two Species of *Callicebus*

1974

WARREN G. KINZEY AND ALWYN H. GENTRY

Abstract. In the Neotropics most congeneric species of primates are allopatric; however, the titi monkeys, *Callicebus moloch* and *C. torquatus,* appear to be sympatric habitat isolates. At a study site in Peru, where *C. torquatus* has been studied since 1974, *C. t. torquatus* and *C. m. discolor* are sympatric and each is found in a different type of vegetation. *C. torquatus* is found almost entirely in vegetation that grows on white sand soils. The known distribution of white sand in the Neotropics is coincident with the distribution of *C. torquatus.* Dietary and dental morphological differences between the two species are correlated with the differences in habitat. Each subspecies of *C. moloch* and *C. torquatus* is distributed around a previously postulated Pleistocene refuge area. We suggest that the present-day distribution of these two species is the result of adaptation to different habitats, followed by Pleistocene separation of local populations in mesic refugia, and that the present-day overlap of the two species has been made possible by the difference in habitat preferences between the two species.

Habitat isolation (Mayr, 1963) refers to the segregation of closely related sympatric (or potentially sympatric) species resulting from their living in contiguous (or overlapping) habitats that differ in some biotic, edaphic, or climatic factor or factors. Among nonprimates habitat isolation has been well known since the time of Darwin. Among primates, however, habitat isolation of congeneric species is rare and does not appear to have been reported previously in the literature.

89

In the Old World, Congeneric sympatrics often divide their habitat in part through forest stratification, *e.g.*, *Lemur* (Sussman, 1974), *Galago* (Charles-Dominique, 1974), *Colobus* (Booth, 1956; Clutton-Brock, 1973), and *Cercopithecus* (Booth, 1956; Struhsaker, 1969). While vertical stratification of Neotropical primates is known (*e.g.*, Tokuda, 1968), it has not been reported for congeneric primate species in the Neotropics.

Among Neotropical primates only capuchin monkeys (*Cebus apella* together with *C. albifrons* or *C. nigrivittatus*), several pairs of tamarins (*e.g.*, *Saguinus imperator* and *S. fuscicollis* in Peru; *S. nigricollis* and *S. fuscicollis* in Colombia), and titi monkeys (*Callicebus moloch* and *C. torquatus*) are congeneric sympatrics. Species of *Alouatta* and *Ateles* are known to overlap at their geographic boundaries (Napier, 1976; Rossan and Baerg, 1977), but otherwise these congeneric species are allopatric. Marmosets (*Callithrix*) have been regarded by some as comprising sympatric species, but following the extensive studies of Hershkovitz (*e.g.*, see Hershkovitz, 1975) the three species of *Callithrix* are regarded as allopatric. Pairs of *Cebus* and *Saguinus* occur in similar habitats and are even found, at least occasionally, in mixed species groups. Only two (of the three) species of titi monkey, *C. moloch* and *C. torquatus*, appear as habitat isolates among the Neotropical primates.

The existence of habitat isolation between the two sympatric species of *Callicebus* has not previously been recognized. They have both been collected in Peru, and in western Brazil (Hershkovitz, 1963). (Fig. 1). *Callicebus torquatus torquatus* is sympatric both with *Callicebus moloch discolor* and with *C. m. cupreus*. *C. moloch* and *C. torquatus lugens* have also been reported as sympatric in southern Colombia by Hernandez-Camacho and Cooper (1976) and by Klein and Klein (1976). Soini (1972) was the first to recognize a possible habitat difference between the two species when he stated (1972:30) that *C. t. torquatus* "does not mix with sympatric *C. m. discolor*. It prefers higher terrain away from the rivers while the latter keeps to the river banks." Similarly, Moynihan was aware of differences in habitat of *C. torquatus:* "Its geographical range, in the broad or loose sense, overlaps that of *moloch*. Still the term 'overlap' may convey a misleading impression, for the two species have some different habits." (1976:81)

We have observed *C. torquatus torquatus* and *C. moloch discolor* at *Estación Biológica Callicebus* (EBC) about 4 km south of the Río Nanay in northeastern Peru where *C. torquatus* has been the object of extended study since 1974 (Kinzey, 1977a, 1977b; Kinzey *et al.*, 1977). In addition ecological differences between the two species have been noted (Kinzey, 1978). It is the purpose of this article to present further data on their ecological differences and to offer a hypothesis to explain some of the observed habits and distributional features of the titi monkey, *Callicebus torquatus*. The hypothesis can be tested in as yet unstudied areas of its distribution.

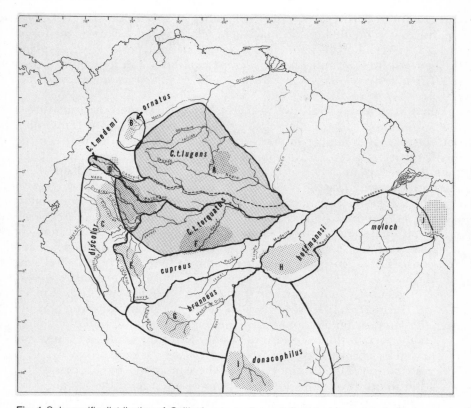

Fig. 1 Subspecific distribution of *Callicebus torquatus* and *C. moloch,* correlated with Amazonian Pleistocene refugia proposed by Brown et al. (1974) based on butterfly biogeography. Division of Napo refugium into two, modified after Brown (1976). The approximate area of distribution of each subspecies is outlined. The distribution of *C. torquatus* is shaded (light) with a dashed line between the subspecies, *C.t.lugens* and *C.t.torquatus.* The refuge areas (shaded heavy) are: A = Imerí refuge, B = Macarena refuge (Villavicencio of Brown), C = Napo refuge, D = Norte-Napo refuge (Putumayo of Brown), E = Eastern Peru refuges, F = Juruá refuge, G = Eastern Peru-Madre de Dios refuge (Inambari of Brown), H = Rondônia-Aripuanã refuge, I = Yungas refuge, and J = Belém refuge.

HABITAT ISOLATION AT *ESTACIÓN BIOLÓGICA CALLICEBUS*

At EBC *C. torquatus* has been observed almost entirely in areas where the substrate consists of white sand overlain by thin topsoil of varying thickness. Such substrates support vegetations referred to locally as *varillal,* which is physiognomically similar to campina forest (*campinarama*) of the Central Amazonian region in Brazil as described by Anderson *et al.* (1975). There are several types of *varillal,* with varying degrees of drainage and thickness of topsoil, but they all have an abundance of vertical trunks of small diameter, a relatively closed canopy, reduced ground cover, a paucity of lianas, and few (or at least greatly reduced number of species of) palms. *Varillals* are distributed on the tops of hills and in relatively level areas of the forest which are not seasonally inundated. The various subtypes of *varillal* (upland leached-soil forests) utilized by *Callicebus torquatus* are described in Kinzey (1977b) and the vegetation at EBC is described in further detail by Revilla (1974).

In upland regions well away from the Nanay River there are broad areas of distribution of *C. torquatus,* abundant white sand soils, and streams at the bottoms of slopes appear in many cases to form boundaries between the highly territorial groups. The monkeys enter vegetation immediately adjacent to streams (locally called *irapayal*), but primarily to feed upon ungurahui palm fruit (*Jessenia bataua* = *J. polycarpa*) which grows there. *Irapayal* is named after the irapay palm, *Lepidocaryum tessmanii,* which is the most abundant of several palms to grow there. This streamside, or *irapayal,* vegetation has dense undergrowth, abundant lianas, a discontinuous canopy of irregular height, and no upper storey emergent trees. The ground is damp and consists of a dense mat of intertwined tree roots and humus overlying brownish sand. In 1974, when the ungurahui fruit was ripe, only 3% of *C. torquatus* group I's active day was spent in the *irapayal* vegetation (Kinzey, 1977b). During the 1976 study season, when the ungurahui fruit was not ripe, the same group did not venture into *irapayal* at all (unpublished data).

As the terrain slopes toward the Nanay River the white sand substrate is replaced by a brown clayey latosol and *C. torquatus* is replaced by *C. moloch.* The distribution of the two species is no where more remarkable than at one small hill where two groups of *C. torquatus* on the top of the hill are completely surrounded by territories of groups of *C. moloch.* On the top of the hill the substrate is white sand and in the surrounding territory it is clay. Thus, within the area of the *Callicebus* study site at EBC there is a strong correlation between clay soils and the distribution of *C. moloch,* and between white sand soils and the distribution of *C. torquatus.* We propose that the latter species of titi monkey is especially adapted to living in vegetation that grows on white sand soils.

DIETARY PATTERNS

There are several important features of tropical vegetation (Janzen, 1974) on white sands that relate to the ecological niche of *Callicebus torquatus.* White sand vegetation has a reduced productivity and lower rate of turnover. In such a nutrient poor environment leaf and fruit loss must be minimized. The adoption of an evergreen habit is universal among plant species and in turn requires leaves able to minimize loss to predation. Leaves and fruits on white sand soils are notably more scleromorphic. Furthermore, Janzen (1974) maintains that leaves in such habitats have relatively large amounts of toxic secondary compounds. Plant species diversity is low and thus the variety of fruits available is reduced. Moreover, the incidence of insects on white sand substrates, at least in Amazonia, is relatively higher than in other habitats (Adis, 1977 and pers. comm.).

The insect diversity pattern has been documented only for seasonally inundated habitats, but is probably extrapolatable to non-inundated sites (Adis, pers. comm.) and contrasts with Janzen's (1974) suggestion of low insect diversity in Asian white sand communities. Adis suggests that low rates of litter decomposition may favor a build up in size and diversity of the white sand arthropod community. Relevant to the diversity in arthropod fauna may be the evolutionary strategies favored by many insect groups which stress co-evolutionary specializations to detoxify specific kinds of plant secondary compounds. Such specialist strategies should render insects relatively capable of coping evolutionarily with the toxic compounds characteristic of the white sand vegetation as compared to more generalized mammalian herbivores. For a white sand dwelling species like *C. torquatus* the insects themselves might provide a better nutrient source than leaves, the more readily available alternative protein source.

Each of the above factors is directly relevant to the habits of *C. torquatus*, especially when compared with those of *C. moloch*. Both species of titi monkeys are basically frugivorous, spending roughly 70% of their feeding time on fruits (Kinzey, 1978). The diet of fruits is supplemented almost entirely with leaves by *C. moloch*. *Callicebus torquatus* eats few leaves and supplements its frugivorous diet with a variety of arthropods including particularly ants, beetles, and spiders (Kinzey and Schuh, in prep.). Since searching for insects and spiders requires large amounts of foraging time (compared with feeding on leaves), *C. torquatus* tends to feed continuously throughout the day, rather than in peaks with long rest intervals as do *C. moloch* (and other largely folivorous primates.) As predicted by Hladik (1975) for insectivores, the territory over which *C. torquatus* ranges is relatively large compared with that of *C. moloch*. This appears to correlate both with generally lower habitat productivity and with the utilization by *C. torquatus* of a dispersed, low biomass protein-rich supplementary food resource like insects, rather than a higher biomass food resource like leaves.

Relative to that of *C. moloch* the frugivorous part of the diet of *C. torquatus* is composed of fewer fruits with only 2-3 fruit species comprising 50% of its diet during any given season. This low diversity of the primary food resource of *C. torquatus* correlates with the low taxonomic diversity of white sand vegetation. For example figs (*Ficus*) are important components of non-white sand forests and are important fruits in the diet of *C. moloch; C. torquatus* has not been observed to eat figs and we have not observed *Ficus* growing on white sands at EBC. Concentration on a few fruiting species at a time may also be related to the mast-fruiting phenomenon (Janzen, 1974) characteristic of some white sand plant species.

DENTAL MORPHOLOGY

Most of the fruits fed upon by *C. torquatus* (*e.g.*, those of *Clarisia racemosa, Jessenia bataua, Tovomita* sp., and *Virola* sp.) are relatively hard and sclero-morphic. It has been previously suggested (Kinzey, 1978) that certain mor-phological features of the molar teeth of *C. torquatus* are particularly suited to reducing hard food items and chitinous insects, and it appears that in *C. torquatus* dental features have evolved which are especially advantageous for dealing with such food items. For example crushing and grinding areas such as the talonid basin are larger and deeper on the lower molars of *C. torquatus* than on those of *C. moloch.* Further, since the leaves in its upland forest habitat are especially high in secondary coumpounds, *C. torquatus* has very little leafy material in its diet, presumably not having developed mech-anisms for detoxifying such substances (Freeland and Janzen, 1974) in con-trast to the folivorous howler monkey (Glander, 1975). Thus, *C. torquatus* shows behavioral, morphological, and probably physiological characteris-tics which appear to adapt it to impoverished white sand habitats.

In the case of *C. moloch* the basically frugivorous diet is supplemented with substantial amounts of leaves. Each of the first two lower molars of *C. moloch* has a longer shearing crest (the cristid obliqua) than does the corre-sponding tooth of *C. torquatus,* even though the overall length of the teeth in the latter is greater (Kinzey, 1978). Increasing length of molar shearing crests has been equated with folivory in primates of *Callicebus* body size (Kay, 1975). In species of *Callicebus* the longer shearing crest has also been as-sociated with a more folivorous diet (Kinzey, 1977c).

GEOGRAPHIC DISTRIBUTION

If *C. torquatus* has become selectively adapted to living in habitats on white sand soils, coincident distributions of these monkeys and white sand soils in the Neotropics might be predicted. White sand soils (podzols) have long been recognized in tropical lowlands (e.g., Richards, 1941); unfortunately, there is still no definitive description of their precise locations in the Neo-tropics. The white sand distributions noted below are based mostly on personal observations of A.G. In general, the northern tributaries of the Amazon from the Rio Negro west are black water rivers draining predomi-nantly white sand soils, while the Amazon's southern tributaries are not. The Rio Negro, for example, is named for its dark humic water that drains the predominantly white sand soils along its western and northern tributar-ies in southern Colombia, southern Venezuela and northwestern Brazil. Most of the major tributaries of the upper Orinoco in southern Venezuela and adjacent Colombia also drain white sand regions. In Amazonian Peru white sand soils and black water are characteristic of the region north of the Marañon and Amazonas, but not of the southern regions drained by the

Ucayali and Huallaga. White sands also occur in the Guianas outside the distributional area of *Callicebus*.

With two possible exceptions the area of distribution of *C. torquatus* is exactly coincident with the Amazonian region of the white sand soils, as predicted (Fig. 1). There are three subspecies of *C. torquatus*. *C. t. torquatus* is found in the Amazon River drainage, as far east as the Rio Negro on the left bank and as far east as the Rio Purús on the right; *C. t. medemi* is known only from the upper reaches of the Putumayo and Caquetá Rivers in southeastern Colombia; *C. t. lugens* is found in the upper drainage on both sides of the Rio Negro in Brazil and Colombia and the southern tributaries of the Orinoco River in southern Venezuela.

The northermost collecting locality of *C. torquatus* at Maipures is near the northern limit of white sand south of Puerto Ayacucho. The eastern distributional limit of both *C. torquatus* and more or less continuous white sand are near the Rio Negro. East of Manaus in Amazonian Brazil white sand is limited to small isolated patches called campinas. White water (*i.e*, nonwhite sand) of the Rio Branco separates the Rio Negro area from the disjunct area of white sand in the Guianas. The presumed southern limit of predominantly white sand substrate, near the course of the Amazon, is also the southern distributional limit of *C. torquatus,* with the exception of two collecting localities, one from Jaburú on the Rio Purús and one from 7° S on the Rio Juruá. White sand may well occur in this poorly known region as it does along the major black water tributaries of the Purús, the Ituxi, and Ipixuna and along parts of the white-water Rio Javari further west (Prance, pers. com.). The western limit of white sand is even more poorly established but is probably near the eastern frontier of Ecuador. White sand substrates are apparently absent from the upper Ríos Tigre and Corrientes near the Ecuadorian border in Amazonian Peru and we know of no white sand areas in Equador. Further north along the Río Putmayo white sand must occur west at least to the Río Zubineta (Rio Yavineto; near the westernmost locality of *C. torquatus torquatus,* Fig. 1) to judge from the occurrence there of *Jacaranda obtusifolia* spp. *obtusifolia* H&B, a plant restricted to white sand substrates in Amazonian Peru.

Only the range of the subspecies, *C. t. medemi,* in the headwaters of the Caquetá and Putumayo in southern Colombia is perhaps outside the distribution of white sand substrates, although Cuatrecasas (pers. comm.) reports white sand beaches along the Putumayo at Puerto Asis. Interestingly this subspecies is in some respects (*e.g.,* black hands in adults, and lack of throat patch in some specimens) the most generalized form of *C. torquatus,* and somewhat intermediate in pelage color with *C. moloch.* In any event, it is distinct enough from *C. t. torquatus* and *C. t. lugens* that its possible noncorrelation with white sand habitats may be largely irrelevant to our thesis.

C. moloch, in contrast, has a much wider geographic distribution and does not occur in the Rio Negro drainage. Much of the area of its distribution, where *C. torquatus* does not occur in southern Peru, Bolivia and Paraguay, has not been reported to contain white sand soils. Major rivers in several cases (the Tapajóz, Madeira, and Ucayali) form boundaries between subspecies. In southern Peru at the junction of the Ríos Inuya, Urubamba, and Ucayali, the distributions of *C. m. discolor, C. m. cupreus,* and *C. m. brunneus* may be contiguous if collecting localities are correct.

SUGGESTED EVOLUTIONARY DIVERSIFICATION

Callicebus torquatus is morphologically the more specialized of the two species of titi monkey, and probably evolved from *C. moloch,* (Hershkovitz, 1963), in the area between the Napo and Guaviare Rivers. This is exactly the area in which white sand substrates predominate. We differ, however, with Hershkovitz's view (as interpreted by Haffer, 1974) of the probable mode of evolutionary diversification in Amazonian mammals like *Callicebus.* An emerging concensus (*e.g.,* Brown, 1976; Haffer, 1969, 1974; Prance, 1973; Vanzolini, 1970, 1973; Wetterburg, 1976) holds that present distributions of species in Amazonia are largely products of Pleistocene climatic fluctuations. According to this theory, the ranges of forest species have been repeatedly disrupted by glacial-epoch dry periods with the species surviving only in isolated refugia, pockets of relatively mesic conditions mostly scattered around the periphery of Amazonia. Alternation of periods of isolation with periods of secondary contact during wetter interglacial epochs would have provided ideal conditions for speciation and rapid evolution. *Callicebus* fits neatly into this picture.

Differentiation of *C. torquatus* from *C. moloch* may have taken place in an earlier dry cycle, probably in the northern Amazonian white-sand area Imerí refugium near the Rio Negro (Fig. 1). A third species, *C. personatus,* concomitantly differentiated in the Serra do Mar refugium of coastal southeast Brazil. Subspecific differentiation within the two Amazonian species took place when the populations were again fragmented into refugia during a more recent dry period. In fact there is a subspecies of *Callicebus* for each of the major Amazonian-area refugia postulated by Brown (1976) from his studies of butterflies.

The striking correlation of the pattern of geographic variation shown by *Callicebus* with that of such birds as the *Pteroglossus aracari* and *P. bitorquatus* superspecies (Haffer, 1974) and the remarkable coincidence of *Callicebus*'s specific and subspecific distributional limits with Prance's (1977) phytogeographical divisions of Amazonia are strong evidence that evolution in *Callicebus* was similarly influenced by Pleistocene climatic fluctuations. Today the various conspecific *Callicebus* subspecies occupy mutually exclusive allopatric ranges while overlap between the species is made possible by habitat isolation.

SUMMARY

We suggest that the evolution of *C. torquatus* is related specifically to the ecology of white sand substrates and that this adaptation has resulted in the observed habitat isolation of *C. torquatus* from the more generalized *C. moloch* in areas of sympatry.

In summary, our hypothesis is based upon the following evidence: (1) observation of differential distribution of the two species of titi monkey at EBC; (2) dietary preference of *C. torquatus* for items more readily obtainable from, and less toxic in, white sand habitats; (3) dental morphology particularly suited to reducing such dietary items; and, (4) correlation of the geographical distributions of white sand substrates and *C. torquatus*. We hope that presentation of the hypothesis at this time will stimulate investigation of *C. torquatus* in areas where it has not previously been studied, in order to test the hypothesis.

ACKNOWLEDGMENTS

The field work on which this investigation was based was supported by grants RF-10067 and RF-11147 from the City University of New York, a grant from the Explorers Club, and grants from Educational Expeditions International (Earthwatch), to WK, and by NSF grants DEB 75-20325A02 and OIP 75-18202 to AG. Financial support for technical assistance and for preparation of the manuscript was provided by grant RF-11356 to WK. We are grateful to the Department of Mammalogy, American Museum of Natural History, New York, for use of its facilities and permission to study specimens in its collections. We thank Barbara L. Sleeper for assistance and Audrey Stanley for preparation of the illustration. For providing information and/or commenting on the manuscript we thank J. Adis, A. Anderson, J. Cuatrecasas, Karl Koopman, J. Linderman, Prue Napier, G. Prance, J. Revilla C., R. W. Sussman, and Suzanne Ripley.

LITERATURE CITED

Adis, J. 1977. Terrestrial arthropods in Amazonian inundation forests. *In* Henk Wolda (ed.), Resumenes para el IV Symposium Internacional de Ecologia Tropical. Panama City, Panama.

Anderson, A.B., G.T. Prance, and B. W. P. de Albuquerque 1975. Estudos sobre a vegetação das Campinas Amazonicas--III. Acta Amazonica 5: 225-246.

Booth, A.H. 1956. The distribution of primates in the Gold Coast. J. West African Sci. Assoc. 2: 122-133.

Brown, K.S., Jr. 1976 Geographical patterns of Evolution in Neotropical Forest Lepidoptera (Nymphalidae: Ithomiinae and Nymphalinae-Heliconiini).*In* H. Descimon, ed. Biogeographie et Evolution en Amerique Tropicale. Laboratoire de Zoologie de l'Ecole Normale Superieure, Paris.

Brown, K.S., Jr., P. M. Sheppard, and J.R.G. Turner 1974 Quaternary Refugia in Tropical America: Evidence from Race Formation in *Heliconius* Butterflies. Proc. Roy. Soc. London 187B: 369-378.

Charles-Dominique, P. 1974. Ecology and feeding behaviour of five sympatric lorisids in Gabon. *In* R.D. Martin, G. A. Doyle, and A. C. Walker (eds.), Prosimian Biology. Duckworth, London.

Clutton-Brock, T.H. 1973. Feeding levels and feeding sites of red colobus *(Colobus badius tephrosceles)*in the Gombe National Park. Folia primat. 19: 368-379.   '

Freeland, W.J., and D.H. Janzen. 1974. Strategies in herbivory by mammals: the role of plant secondary compounds. Amer. Nat. 108: 269-289.

Glander, K.E. 1975. Habitat description and resource utilization: a preliminary report on mantled howling monkey ecology. *In* R.H. Tuttle (ed.), Socioecology and psychology of primates. Mouton, The Hague.

Haffer, J. J. 1969. Speciation in Amazonian forest birds. Science 165: 131-137.

———. 1974. Avian speciation in tropical South America. Nuttall Ornithological Club. Museum of Comparative Zoology. Harvard Univ., Cambridge, Mass. 390 pp.

Hernández-Camacho, J., and R.W. Cooper. 1976. The nonhuman primates of Colombia. *In* R.W. Thorington, Jr., and P.G. Heltne (eds.), Neotropical primates, field studies and conservation. National Academy of Sciences, Washington, D.C.

Hershkovitz,P. 1963. A systematic and zoogeographic account of the monkeys of the genus *Callicebus* (Cebidae) of the Amazons and Orinoco River Basins. Mammalia 27: 1-79.

———. 1975. Comments on the taxonomy of Brazilian marmosets (*Callithrix*, Callitrichidae). Folia primatol. 24: 137-172.

Hladik, C.M. 1975. Ecology, diet and social patterning in Old and New World primates. *In* R.H. Tuttle (ed.), Socioecology and psychology of primates. Mouton, The Hague.

Janzen, D.H. 1974. Tropical blackwater rivers, animals, and mast fruiting by the Dipterocarpaceae. Biotropica 6: 69-103.

Kay, R.F. 1975. The functional adaptations of primate molar teeth.Amer. J. Phys. Anthrop. 43: 195-216.

Kinzey, W.G. 1977*a.* Diet and feeding behaviour of *Callicebus torquatus. In* T.H. Clutton-Brock (ed.), Primate ecology: studies of feeding and ranging behaviour in lemurs, monkeys and apes. Academic Press, London.

———. 1977*b.* Positional behavior and ecology in *Callicebus torquatus.*Yearbook of Physical Anthropology 20:468-480.

———. 1977*c.* Dietary correlates of molar morphology in *Callicebus* and *Aotus.* Amer. J. Phys. Anthrop. 46:142.

———. 1978. Feeding behavior and molar features in two species of titi monkey. *In* D. J. Chivers and J. Herbert (eds.) Academic Press, London.

Kinzey, W.G, A.L. Rosenberger, P. S. Heisler, D.L. Prowse, and J.S. Trilling. 1977. A preliminary field investigation of the yellow-handed titi monkey, *Callicebus torquatus torquatus*, in northern Peru. Primates 18: 159-181.

Klein, L.L., and D.J. Klein. 1976. Neotropical primates: aspects of habitat usage, population density, and regional distribution in La Macarena, Colombia. *In* R.W. Thorington, Jr., and P.G. Heltne (eds.), Neotropical primates, field studies and conservation. National Academy of Sciences, Washington, D.C.

Mayr, E. 1963. Animal species and evolution. Harvard University Press.

Moynihan, M. 1976. The New World primates, adaptive radiation and the evolution of social behavior, languages, and intelligence. Princeton University Press.

Napier, P.H. 1976. Catalogue of primates in the British Museum (Natural History), Part I: Families Callithrichidae and Cebidae. British Museum (Natural History), London.

Prance, G.T. 1973. Phytogeographic support for the theory of Pleistocene forest refuges in the Amazon Basin, based on evidence from distribution patterns in Caryocaraceae, Chrysobalanaceae, Dichapetalaceae and Lecythidaceae. Acta Amazonica 3: 5-28.

————. 1977. The phytogeographic subdivisions of Amazonia and their influence on the selection of natural reserves. *In* G. Prance and T. Elias (eds.), Extinction is Forever. New York Botanical Garden, New York.

Revilla, J. 1974. Descripción de los tipos de vegetación en Mishana, Río Nanay, Loreto, Perú. PAHO project AMRO-0719 report. Pan American Health Organization, Washington, D.C.

Richards, P.W. 1941. Lowland tropical podsols and their vegetation. Nature 148: 129-131.

Rossan, R.N., and D.C. Baerg. 1977. Laboratory and feral hybridization of *Ateles qeoffroyi panamensis* Kellogg and Goldman 1944 and *A. fusciceps robustus* Allen 1914 in Panama. Primates 18: 235-237.

Soini, P. 1972. The capture and commerce of live monkeys in the Amazonian region of Peru. International Zoo Yearbook 12: 26-35.

Struhsaker, T.T. 1969. Correlates of ecology and social organization among African cercopithecines. Folia primatol. 11: 80-118.

Sussman, R.W. 1974. Ecological distinctions in sympatric species of *Lemur*. *In* R.D. Martin, G.A. Doyle, and A.C. Walker (eds), Prosimian Biology. Duckworth, London.

Tokuda, K. 1968. Group size and vertical distribution of New World monkeys in the basin of the Rio Putumayo, the Upper Amazon. Proc. 8th int. Congr. Anthrop. Ethnol. Sci., vol. 1, Science Council of Japan, Tokyo.

Vanzolini, P.E. 1970. Zoologia sistematica, geografia e a origem das espécies. Inst. Geografia, Univ. Sao Paulo. Teses e Monogr. 3. 56 pp. (as interpreted by Haffer, 1974, and Prance, 1973).

————. 1973. Paleoclimates, relief, and species multiplication in equatorial forests. *In* Meggers *et al.* (eds.), Tropical Forest Ecosystems in Africa and South America: a comparative review. Smithsonian Institution, Washington, D.C.

Wetterberg, G.B. 1976. Uma Análise de Prioridades em Conservacão da Natureza na Amazonia. PNUD/FAO/IBDF/BRA-45, Série Técnica no. 8.

5

Feeding and Activity of *Cebus* and *Saimiri* in a Colombian Forest

1966

R. W. THORINGTON, JR.

INTRODUCTION

Field studies of New World primates have been conducted by a variety of different investigators who have brought different talents to bear on the subject (e.g. Altmann 1959; Bernstein 1964; Carpenter 1934, 1935; Kühlhorn 1939; Mason 1966). However, the most basic facts about the ecology of most species of platyrrhines are unknown. We can not detail the ecological differences between the different species nor state which differences are the most significant. For example, five to seven different species of monkeys cohabit a number of small forests in Colombia. These animals eat many of the same foods in the forests and appear to make similar demands on their habitat. It is not evident to what degree these monkeys compete with each other for food or other essentials nor what is limiting population size among these species.

The adaptation of platyrrhines to different ecological roles in the forests of South and Central America has obviously entailed differentiation at many levels of organization. These morphological, physiological, and behavioral differences are being documented in increasing detail. These species characteristics have evolved in response to, and hence can ultimately be understood only in the light of, the ecological adaptations of these animals to their environment. It is the purpose of this article to present a few facts about the ecological differences between two species of platyrr-

From Stark, Schneider, Kuhn (eds.), *Progress in Primatology*, Stuttgart (G. Fischer) 1966. With permission of the publisher.

hines and to explore a few of the implications of these differences and other aspects of the biology of these species.

STUDY SITE

This study was carried out at Hacienda Barbascal, east of San Martin, Colombia. The area has been described by Mason, 1966. This part of the Colombian llanos is typified by gallery forests growing along water courses and escarpments. These forests are surrounded by grassland. An abrupt edge between the forests and grasslands is presently maintained by the grazing of cattle and by fire. Troops of monkeys in the different forests seem quite isolated from each other by this intervening grassland. There are few observations and only scanty circumstantial evidence to document the frequency with which these monkeys cross extensive tracts of grass.

My observations were confined to some of the monkeys in a forest called "Monte Secoe." The northern part of the forest is approximately 750 yards long and 250 yards wide. It is connected by an isthmus of palms to a more extensive tract of forest, but the monkeys I was observing intensively seldom, if ever, crossed this isthmus. The forest was more open on the eastern edge, closer and more festooned with lianas on the western side. It was bordered to the north by a road and by some recently cleared and burned forest, to the east and west by grassland. The forest has been diminished in recent years. Within this forest, there were five species of monkeys:

Aotus trivirgatus, Callicebus moloch, Alouatta seniculus, Saimiri sciureus, and *Cebus apella. Saimiri* and *Cebus* were the most common species.

Saimiri sciureus

The troop of squirrel monkeys which inhabited the northern part of Monte Seco consisted of eighteen animals. Four infants were born during the course of the study (26 January to 28 March 1965). (This troop was variously estimated by passers-by to consist of 50 to 100 animals.) A larger troop of 30 to 40 animals occasionally entered the southern end of the forest.

At night and during the warmest hours of the day, the squirrel monkeys remained in the dense western side of the forest. These were not necessarily the coolest parts of the forest at mid-day — in fact, one place favored by the monkeys at mid-day was the hottest part of the forest at that time, due to the wind coming across the grassland — but solar radiation was low. At night the animals slept at altitudes of 30-60 feet, whereas they frequently rested at lower levels during the day. They became active in the morning before it was light on the forest floor. They usually moved directly to some nearby fruiting trees and began to feed. After feeding a short while they would move quickly to another tree with fruit. This progression from one fruiting tree to another would continue but at a more leisurely pace as the morning wore on. The same sort of activity occurred during the afternoon, after mid-day siesta, until the troop congregated again at one of the several areas in which they would spend the night.

All animals of the troop seemed to spend the night in a common area, though this was difficult to document. During the day, the whole troop was together very little of the time. Instead the animals traveled through the forest in small groups. I was able to obtain good counts of group size on 88 different occasions. On almost half of those occasions, the group size ranged from 5 to 8 animals. The mode was 7 animals per group. These groups were frequently unisexual. Commonly the dominant males of the troop traveled together. Another common group comprised the pregnant females, subsequently the females with infants. Young animals of the troop more frequently associated with the females than with the adult males.

Individual *Saimiri* did not spend long in any one fruit tree. Feeding times averaged three minutes per tree. The tree might be full of *Saimiri* for a much longer time as one or more groups of *Saimiri* passed through it, but any particular animal would remain in a tree for only two to four minutes, before moving to another. Three facts suggest that this behavior did not result from a depletion of the fruit. 1. After one group of monkeys had moved out of a tree, another group might move into it, almost immediately. 2. A group of *Saimiri* might return to the same tree several times during a day. 3. The wasteful manner in which the monkeys eat fruit suggests that the supply is superabundant. Neither did the brief feeding time per tree seem to reflect a satiation with particular fruits. The animals would sometimes move from one tree to another tree of the same species. Rather, the reason for this behavior is to be found in the foraging that the animals do between fruiting trees.

Between different fruiting trees, the individual animals followed different routes. Only at certain difficult crossings did animals follow one another. As they moved through the foliage, they foraged among the leaves for insects and arachnids. When not feeding on fruits or sleeping, the *Saimiri* were constantly foraging — unrolling dead leaves, investigating bunches of leaves, peering here and there. I observed them to catch a number of insects, and the stomach contents that I have examined all contained insect fragments. Small hawks sometimes followed, above the canopy, and caught insects which the monkeys stirred up in their foraging activities. The kind of restless inquisitiveness typical of their foraging behavior seems to be a pervasive characteristic of the species.

It is difficult to observe social interactions in the forest, but it is possible to obtain a general idea of the degree and kinds of interactions between the animals by their vocalizations, correlated with occasional observations in which all the participating animals can be seen. The social activity of the *Saimiri* is influenced by their pattern of feeding. When the animals are in fruiting trees, there is some jockeying for position and much more interaction than there is when the animals are foraging. While foraging, the *Saimiri* are generally quiet. Amid the foliage, there is little chance for overt dominance behavior in feeding. In the early morning and late afternoon there may be intense interactions between the animals, related to selection of favorable roosting places or to feeding when more animals are congregated in an area. At mid-day, when larger numbers of animals congregate in one area, the adults tend to be quiet and somnolent, while the juvenile animals play with each other.

Cebus apella

A troop of 5 capuchins regularly inhabited the northern end of Monte Seco. Other animals would enter from the south, apparently a troop of 7 *Cebus*. A few *Cebus* had been removed from Monte Seco for another project before my arrival. Some may have been shot by poachers in the forest. Thus these troops had been larger.

The *Cebus* ecology seemed to be similar to that of the *Saimiri*. *Cebus* frequented the same fruiting trees, commonly at the same time as the *Saimiri*, and frequently could be found with the *Saimiri* at other times. They spent the night and the middle of the day in situations similar to those favored by the *Saimiri*. There were occasional aggressive interactions between the young *Cebus* and some of the *Saimiri*, but the adult *Cebus* seemed to pay no attention to the *Saimiri*. Pregnant female *Saimiri* tended to remain in the vicinity of the *Cebus* troop, which traveled through the forest at a much slower rate than did the other groups of *Saimiri*.

Early in the day, the *Cebus* moved into fruiting trees and proceeded to feed. At this and other times during the day, they fed in fruit trees for much longer periods than did the *Saimiri*, averaging 20 minutes before moving out of the tree. They did not visit as many fruiting trees nor as frequently as did the *Saimiri*.

The foraging of the *Cebus* is quite different from that of the *Saimiri:* it is carried out in a different manner and is directed at a different part of the arachnid and insect fauna of the forest. The monkeys move into dead trees and investigate dead limbs and lianas. They spend hours examining these and extracting insects from them. To do this they pull off bark, break open branches with their teeth, and probe with their hands and fingers into holes and crannies. Very typical behavior involves breaking off a part of a vine or branch, biting into it at one point, if there are no insects there biting into it at another point, 9 or 12 inches further down the vine, and proceeding thusly until the whole vine has been investigated. When a *Cebus* finds some beetle grubs or an ant nest in the vine or branch, he will break open the whole of it in the process of extracting the insects.

The characteristics and significance of this kind of foraging are several.

First, this is a solitary pastime, requiring much diligence and concentration. The *Cebus* are separated from each other for hours at a time, with only intermittent vocal and visual contact. The troop may be spread out so that the animals are 50 to 100 yards apart. The young animals may stay close together but quite distant from the adult male or females. Second, this type of foraging requires far more manipulation of the environment than does that of the *Saimiri*. In the fruiting trees, this same manipulation of objects in the environment was obvious in the carrying of fruits or nuts from one place to another and the pounding or rubbing of them on branches. *Cebus* use their tails dexterously while foraging and feeding. In comparison, they make little use of it when they are moving rapidly through the forest, suggesting that this striking character of some platyrrhines may be basically a feeding adaptation. Third, the nutritional significance is that these *Cebus* are supplementing their frugivorous diets with many insects, which are presumably providing the animals with most of their proteins. The bulk of these are insects that the *Saimiri* can not obtain from the same environment. There are only a few foraging techniques which *Saimiri* and *Cebus* share: both species investigate dead palms and unroll the curled fronds where they find insects and arachnids.

The use of urine in grooming, which has been discussed by Nolte (1958) in captive animals, was very evident on some occasions. Several of these were feeding situations when the *Cebus* were actively and excitedly extracting ants from dead branches. On another occason, the *Cebus* appeared to be bothered by mosquitos or flies. In both occasions, the animals seemed to use the urine to prevent or alleviate insect bites, probably the latter. This possible functional explanation of the behavior will not serve as an explanation of its ritualization observed in captivity, however. Common to the situations I observed and those described by Nolte is the degree of excitement exhibited by the animals which are gooming with urine. Also I observed behavior akin to anting, but I can not asseverate that it was homologous to that described by Nolte.

SUMMARY

1. One of the major ecological differences between *Saimiri* and *Cebus* is the manner in which each forages for insects. This insectivorous behavior probably provides each species with its major source of protein. The two species do not compete with each other for this part of their diets.

2. These different foraging procedures involve different patterns of activity and movement through the forest and hence they also influence the social interactions between the animals.

3. Other behavioral characteristics of these species, such as manipulative ability, concentration span, and problem solving, are probably determined by the different adaptations of these animals for extracting food from their environment.

ACKNOWLEDGEMENTS
This study was supported by N. I. H. Grant FR 00168-03 to Harvard University. I am grateful to Dr. J.C.S. Paterson, Director of the ICMRT program in Cali, Colombia and to Dr. William A. Mason, who made this study possible. The hospitality of the Botero family, owners of Hacienda Barbascal is gratefully acknowledged.

REFERENCES
Altmann, Stuart, 1959: Field observations on a howling monkey society. Journal of Mammalogy *40:* 317-330.
Bernstein, I.S., 1964: A field study of the activities of howler monkeys. Animal Behavior 12: 92-97.
Carpenter, C.R., 1934: A field study of the behavior and social relations of howling monkeys. Comparative Psychology Monographs *10:* 1-168.
Carpenter, C.R., 1935: Behavior of red spider monkeys in Panama. Journal of Mammalogy *16:* 171-180.
Kühlhorn, F., 1939: Beobachtungen über das Verhalten von Kapuzineraffen in freier Wildbahn. Zeitschrift für Tierpsychologie *3:* 147-151.
Mason, W.A., 1966: Social organization of the South American Monkey, *Callicebus moloch:* a preliminary report. Tulane Studies in Zoology *13:* 23-28.
Nolte, A., 1958: Beobachtungen über das Instinktverhalten von Kapuzineraffen (*Cebus apella* L.) in der Gefangenschaft. Behaviour *12:* 183-207.

6

Social and Ecological Contrasts Between Four Taxa of Neotropical Primates

1975

LEWIS L. KLEIN AND DOROTHY J. KLEIN

From October, 1967, through November 1968, observations were made of four neotropical primates (*Ateles belzebuth, Alouatta seniculus, Cebus apella*, and *Saimiri sciureus*) sympatric in the floodplain forest of the Colombian national park, La Macarena. The park is situated approximately 3° North latitude between 73 and 74° West longitude, and comprises about 11,000 square kilometers (4,250 square miles). It consists of three very general types of terrain: a small mountain range approximately parallel to the eastern cordillera of the Andes, foothills, and floodplains. Two-thirds of the park's area consists of floodplains, primarily of three rivers, the Guejar, Ariari, and Guayabero, and several of their smaller tributaries. These rivers ultimately drain into the Orinoco. The study site on the north bank of the Guayabero and most of the areas surveyed were located on or near floodplains, terrain which was inundated either yearly or at least once every several years. Forests found at the study site were not, however, floristically uniform, despite their extensive continuity, taxonomic heterogeneity, freedom from slash-and-burn agricultural practices, and annual inundation. At the approximately three-square-mile study site, a sector of a much larger continuous forest, eight different types of vegetational community were identified,

Reprinted by permission of Mouton Publishers from "Social and Ecological Contrasts Between Four Taxa of Neotropical Primates" in *Socioecology and Psychology of Primates* (R.H. Tuttle, ed.). Pages 59-85. World Anthropology. Mouton, The Hague. 1975.

using gross criteria of presence and abundance of several easily identifiable trees or readily apparent differences in forest structure.

CENSUS METHODS

Ateles belzebuth was the primary subject of these observations. Experience during the survey period consistently revealed that the number of animals encountered at any one location, even on the same day, was extremely variable. In making a census of the populations, therefore, we adopted the following methodological criteria and standards for determining the number and composition of animals moving and in contact with one another at any given time.

1. Census counts were tallied only when animals were kept under observation for a minimum period of fifteen minutes.
2. Tallies were considered complete only if no additional spider monkeys were seen or heard moving nearby for the same minimum period of fifteen minutes and if no spider monkey vocalizations were heard from areas within 200 yards not traceable to one of the animals under observation or already counted.

It was assumed that these counts included all those individuals potentially in constant or at least frequent intermittent and audible contact with one another. Spider monkeys so considered were also usually simultaneously engaged in similar or complementary activities, e.g. feeding, resting, grooming, and travelling.

We made 498 counts meeting the above criteria in an eleven-month period between December 4, 1967, and November 22, 1968. Visual contact with these units, hereafter referred to as subgroups, was maintained by one or both of the observers for sixty-six minutes on the average. Contact with any single subgroup was considered terminated if all members of the subgroup were lost to sight or if a new subgroup formed either by merger or division and persisted for a minimum of fifteen minutes.

SUBGROUP SIZES

The median subgroup encountered over the eleven-month period was composed of 3.5 independently locomoting spider monkeys. Infants under one year were not considered independent. Subgroups of two comprised 21 percent of the total number of subgroups; this was the modal size subgroup. Subgroups of four individuals represented 16 percent of the total, and isolated animals, 15 percent of the total. Subgroups of eight or more comprised 15 percent of the subgroups. (See Table 1.)

Table 1.
Percentage of subgroups consisting of one to eight or more independently
locomoting *A. belzebuth*

Subgroup size	Frequencies	Percent of total subgroups
1	75	15
2	105	21
3	72	14
4	78	16
5	36	7
6	36	7
7	20	4
8–22	76	15
Total:	498	99

SUBGROUP SPACING

Distances as great as one-half mile between different subgroups did not
appear to be unusual. Estimates of this kind were made possible by the
observers' ability to recognize individual animals and the ability of one
observer to follow an animal or animals as it left a subgroup it had been with
while the other observer stayed and moved with the remaining animal or
animals. Spacing between members of what were considered the same subgroup varied with the ongoing activity but was rarely more than 200 yards.
In general, interanimal spacing was greater between feeding than resting
spider monkeys; spacing also tended to be uniform within large and medium-sized fruiting trees. (For further detail see Klein 1972.)

Subgroups of the most commonly observed sizes as well as isolates were
observed when the animals were engaged in all types of major activities.
There was no indication that subgroups assembled into larger units or
reassembled at specific "lodge" trees before nightfall for sleeping (cf. Carpenter 1935). On twenty-six separate occasions subgroups were followed
for several hours until after dusk when all activities had terminated. The
median size subgroup on these occasions was three, the mode two, and the
range one to eleven, which was not too different from the subgroup statistics
of all observations. On several of these occasions we were able to recontact
and follow the same animals the next morning until they made renewed
contact with other conspecifics. It was quite clear, then, that at least occasionally, and probably more often, they had been sleeping at distances at
least as great as one-quarter to one-half mile from members of the same
social group.

ISOLATES

Seventy-five instances of isolated animals were recorded, constituting 15
percent of the total number of subgroups observed. Observer contact was

maintained on the average for fifty-five minutes before the animal either encountered others (22 percent of cases) or was lost to sight (78 percent of cases). Adult females without infants constituted 66.7 percent of the total number of isolates; females with infants, 10.7 percent of cases; adult males, 22.7 percent of cases. No juvenile or infant was ever observed isolated for longer than a few minutes. The maximum periods that animals were actually observed continuously and known to remain isolated were as follows: adult female, eight and a half daylight hours and then overnight; adult female with infant, six hours; adult male, one and a half hours. It was concluded that temporary isolation of most adult spider monkeys for periods as long as one to three days was a regular occurrence at the study site. (See Table 2.)

SUBGROUP COMPOSITION

The composition of all subgroups was variable. Subgroups composed entirely of females, entirely of males, and bisexual subgroups containing one to five adult males were observed on occasion.

Subgroups of females without males comprised approximately 45 percent of observed subgroups. These maleless subgroups ranged in size from two to eleven females. They were observed for periods ranging from fifteen minutes to almost twelve hours. In the latter case, the same group was recontacted and followed most of the next day, during which time they still had not made contact with a male. There was no indication that subgroups without males were any more or less stable than subgroups with males.

Subgroups composed entirely of adult males comprised approximately 4 percent of all observed subgroups. They ranged in size from two to four

Table 2
Frequency and biological status of isolated *A. belzebuth* at study site

	Number of observed isolated A. belzebuth of O group	Number of observed isolated A. belzebuth of O, S, and H groups combined	Percent composition of O group isolates[a]	Percent composition of isolates[a] of O, S, and H groups combined
Isolated male	13	17	20	23
Isolated independent female	45	50	69	67
Isolated female with infant	7	8	11	11
Isolated juvenile	0	0	0	0
Total frequency:	65	75		

[a] Isolated animals comprised 15 percent of all scored subgroups.

animals. These all-male subgroups were observed to remain isolated from other conspecifics for periods ranging from fifteen minutes to three and a half hours. Because males were more difficult to follow than females, subgroup stability was assumed to be approximately similar to that of exclusively female subgroups.

The remaining subgroups censused were composed of adults of both sexes. The number of males present ranged from one to five, and some subgroups were observed in which males outnumbered females. (See Tables 3, 4, 5.)

SUBGROUP ASSOCIATIONS — AN EXAMPLE

As a consequence of our ability to identify individually most spider monkeys at the study site, it became clear that the same animals were at times both isolated and members of subgroups of all sizes and compositions. Subgroup membership over a period of days was usually unstable. However, their composition was drawn from a larger, mutually exclusive network of animals that usually intereacted peacefully with one another.

Table 3

Composition of subgroups of four or more independently locomoting *A. belzebuth*

	Frequency		Range of subgroup sizes
Bisexual subgroups	152 (66 percent)		4–22
Bisexual with two or more adult males		76 (33 percent)	4–22
Bisexual with one adult male		76 (33 percent)	4–10
Subtotal:		152 (66 percent)	
Single-sexed groups	77 (34percent)		4–11
Entirely male		2 (1 percent)	4
Entirely female with no dependent young		13 (6 percent)	4–5
Entirely female with dependent infants or juveniles		13 (6 percent)	4–8
Entirely female with and without dependent young		49 (21 percent)	4–11
Subtotal:		77 (34 percent)	
Total:	229 (100 percent)		

Table 4
Composition of subgroups of three independently locomoting *A. belzebuth*

	Frequency	Percent
♂ ♂ ♂	3	4
♂ ♂ ♀	2	3
♂ ♂ ♀inf	2	3
♂ ♀ ♀	10	14
♂ ♀ ♀inf	1	1
♂ ♀ Juv	11	16
♀inf ♀inf ♀inf	1	1
♀inf ♀inf ♀	0	0
♀inf ♀inf Juv	7	10
♀inf ♀ ♀	1	1
♀inf Juv Juv	0	0
♀inf ♀ Juv	9	13
♀ ♀ ♀	11	16
♀ ♀ Juv	12	17
♀ Juv Juv	1	1
Juv Juv Juv	0	0
Total:	71	100

Table 5
Composition of subgroups of two independently locomoting *A. belzebuth*

	Frequency	Percent
♂ ♂	11	10
♂ ♀	26	25
♂ ♀inf	1	1
♂ Juv	1	1
♀ ♀	23	22
♀ Juv	23	22
♀ ♀inf	8	8
♀inf ♀inf	3	3
♀inf Juv	8	8
Juv Juv	1	1
Total:	105	101

The subgroup membership protocols of one known animal should serve to illustrate the degree of stability and unstability occuring. Mo, an adult female who gave birth to a female infant on or around January 29, 1968, was encountered on 54 percent of those days at the study site on which spider monkeys were continuously observed for fifteen minutes or longer.

Mo was observed as an isolate on two occasions, September 17 and September 20, 1968. On September 17 she was initially encountered in a tree with an adult male at 9:00 A.M. He left her at 9:20 and from that time until 2:20 P.M., when she and her infant were lost to sight, she made no contact with other conspecifics. She was isolated for five hours. Mo was relocated again on September 20, still accompanied only by her infant. Contact was maintained with her at that time for two hours. The following day she was seen in a subgroup of ten to twelve animals.

Besides being observed as an isolate, Mo was also seen in subgroups of most sizes. She was a member of subgroups of two to four animals in 38 percent of all our encounters with her; of subgroups of five to seven animals on 26 percent of encounters; and of subgroups of eight or more animals at the remaining 35 percent of encounters. (See Tables 6 and 7.)

Table 6

Frequency with which the adult female Mo was observed in subgroups of one to eight or more independently locomoting spider monkeys and in association with one or more of the following males —Crestless, E–D, Two–Dot, Scar–Face; females —H–F, B–F, W–C, H–M

	Number of occasions	Number of observed associations with:							
		Males Crestless	E–D	Two–Dot	Scar–Face	B–F	W–C	H–F	H–M
Isolated:	2	—	—	—	—	—	—	—	—
Subgroups of:									
2	14	1	1	0	0	0	0	3	0
3	16	0	0	1	0	2	1	7	1
4	26	4	2	1	0	3	5	7	1
5	12	2	2	2	0	2	3	3	1
Subgroups of:									
2–5	68	7	5	4	0	7	9	20	3
Subgroups of:									
6	18	5	6	4	0	8	7	6	3
7	8	1	2	2	0	3	4	4	1
8 or more	52	25	32	27	0	38	35	38	16
Total:	148	38	45	37	0	56	55	68	23

Table 7
Percentage of subgroups containing Mo and AT LEAST one male, and/or one
additional female with infant, one female with juvenile, one unattached juvenile,
and one independent female

Subgroups of:	*One male*	*One additional female with infant*	*One female with juvenile*	*One unattached juvenile*	*One independent female*
			Percentage of subgroups containing at least:		
2	14	21	0	42	21
3	6	44	19	69	44
4	23	31	35	62	62
5	42	33	50	92	58
2-5	21	32	18	68	57
6	56	33	67	72	72
7	75	62	50	100	88
8 +	94	73	100	98	98

Her most frequent associates, aside from her infant, who was seen with
her at all times, were a specific female juvenile between two and three years
of age on 68 percent of all encounters; an adult female who gave birth
around April 5, 1968, on 32 percent of all encounters; and a single adult
female who gave birth to probably her first infant in the first week of November, 1968, on 57 percent of all encounters. Mo was, at various times, also
seen in subgroups with three different and clearly distinct adult males. She
was seen in the company of all three of these males, any combination of two,
or any one of these males singly. These associations were observed to last
for periods as long as five hours or more. Although Mo's association patterns were extensive, there were certain well-known spider monkeys at the
study site with which she was never observed, e.g. a specific adult male with
extensive scarring around his mouth (Scar-Face). Not only was she never
observed in pacific proximity to this monkey, she was also never observed
with any of that male's indentifiable female and male associates.

GROUP AFFILIATION
On the basis of such patterns of individual animal association, it was discovered that at the study site there were three mutually exclusive social units
of spider monkeys whose members did not interact peacefully with one
another, although spatial proximity between them occasionally occurred
(see below). The members of each social unit appeared to share a common

home range. The size of the home range of the social network studied most intensively was approximately one to one and a half square miles; those of the other social groups were either the same or larger. Home ranges of these mutually exclusive networks of groups partially overlapped, approximately 20 to 30 percent. The frequency with which different parts of the home range were utilized appeared to be a function of the preference of specific animals and of such factors as tree density, canopy height, vegetative associations, amount of time elapsed since the last intergroup agonistic interaction, and seasonal changes in the availability of ripe fruit.

Relations between individuals of the different social networks were usually agonistic, conspicuously so in the case of adult males. For example, although E-D, one of the males we observed most often, frequently associated in both small and large subgroups with other males more often than with any particular females, he was never observed moving in a subgroup either with the male Scar-Face or with any of Scar-Face's identifiable male or female associates. On several occasions E-D and Scar-Face or E-D and some of Scar-Face's associates were observed within 100 yards of one another. Proximity of this degree was usually associated with aggressive displays and vocalizations, chasing, or active avoidance. On one such occasion E-D and Scar-Face were observed within 100 to 200 yards of one another, whooping and growling in each other's direction. Throughout this observed interaction E-D was either in physical contact with or within twenty feet of one of his two frequent adult male associates and a male juvenile who also growled, whooped, and charged in Scar-Face's direction. After about an hour of displays, these three males moved away, followed by several female associates who had remained inconspicuously nearby during the confrontation. On another occasion, E-D, who had been traveling in a subgroup with four adult females and one juvenile female, was observed marking branches with saliva and sternal gland secretions and charging toward two male associates of Scar-Face after having moved to within seventy-five feet of them. Although Scar-Face was not with them at that time, he was observed in the company of these two males about thirty minutes subsequent to the agonistic interaction with E-D.

ATELES FEEDING BEHAVIOR

Methods of Study

Data on *A. belzebuth* diet were obtained by observations of feeding and checks of fecal material. Relative percentage importance of specific food items was determined from data on the durations of all individual animal feeding bouts — animal feeding minutes. Each feeding bout was considered to have

frugtovores

terminated when the individual spider monkey (1) had clearly ceased eating for a period of three minutes or more; (2) was observed engaging in an activity incompatible with feeding, e.g. resting; (3) moved into another tree; or (4) began eating a different substance. Termination times could be most accurately recorded when animals moved out of the tree they had been feeding in; fortunately, this was the way most feeding actually ended. Adult *A. belzebuth* were observed only three times over the year eating more than one type of substance successively within the same tree, e.g. leaves and fruit. From our experience with captive *A. belzebuth*, we were able to ascertain that ingested solids, i.e. seeds and rinds, pass through the intestinal tract in solid feces from four to twelve hours after consumption. This allowed us roughly to gauge the representativeness of the dietary data obtained by observations of feeding animals. The correspondence was usually good. (For further detail see Klein 1972.)

Results

The feeding duration data revealed that over the year ripe fruit comprised the overwhelming part of *Ateles'* diet. About 83 percent of all feeding was on ripe fruit and included more than fifty different taxa. The remaining feeding time was spent on tree leaves and buds (5 percent), epiphytic leaves and stems (2 percent), and dead or decaying wood (10 percent). Flowers were positively observed to have been eaten on only two occasions, in both cases by a juvenile. Nonripe fruit constituted less than 1 percent of the diet. No clearcut instances of *Ateles* eating either invertebrates, mammals, birds, reptiles, or amphibians, or the eggs of insects, amphibians, or birds were observed, although some termites, fig insects, or other invertebrates may have been eaten inadvertently. All published reports of *Ateles* eating vertebrates or invertebrates (e.g. Richard 1970; Carpenter 1935; Wagner 1956) were carefully scrutinized, and none appear to be based upon actual observations, in contrast to inferences. (For detail see Klein 1972.) (See Table 8.)

Table 8
Relative amounts of time spent by *A. belzebuth* between February 1, 1968, and November 22, 1968, eating fruits, leaves and stems, flowers, and wood

Types of edible substance consumed	Percentage of total feeding time observed
Fruit	83
Tree leaves and buds	5
Epiphytic leaves and stems	2
Dead wood	10
Flowers	> .1
Insects, animals, and eggs	0

Characteristics of Trees from Which A. belzebuth *Ate Ripe Fruit.* Trees from which *A. belzebuth* ate ripe fruit ranged in height from fifty to a hundred and fifty feet and in crown width from approximately twenty-five to more than two hundred feet in diameter. Fruit was borne in almost all cases at least forty feet from the ground. In many of the important fruit-bearing trees under ninety feet tall, with the exception of palms, fruit appeared to be most abundant either around the crown or on a side with relatively greater lateral exposure.

There was no single or general spatial pattern of fruit tree dispersion or of fruit concentrations within single trees and no temporal pattern of ripening to which all, or even most, of the commonest type of large trees at our study site conformed. It was convenient to distinguish, therefore, the following categories of fruiting trees at the study site.

Type I trees were widely dispersed throughout the forest and individually bore ripe fruit over a relatively brief period from a few days to a week. Type I trees were subdivided on the basis of crown diameter and/or the degree to which the ripe fruit they bore was concentrated within them. Type Ia trees were those whose crowns were greater than sixty feet across. Most of the figs used by the primates at the study site were of this category. All their fruit ripened in a very short period and most were widely dispersed. Type Ib trees bore ripe fruit briefly, their crowns were moderately broad (between twenty-five and sixty feet across), and the distribution of fruit within the tree scattered. No important *Ateles* food tree was included in this category. Type Ic trees were those whose fruit ripened in a short period and whose crowns were narrow, under twenty-five feet across (e.g. a species of *Pseudolmedia*).

Type II trees generally grew in close proximity to others of the same kin but bore ripe fruit over slightly longer periods than did Type I, usually for one or two weeks. On the basis of crown diameter, they were subdivided into Type IIa, very broad crowned trees (e.g. *Brosimum*) and Type IIb, moderately broad corwned trees (e.g. second species of *Pseudolmedia*). No Type IIc trees, those with narrow crowns, were an important food source for *Ateles*.

Type III trees were widely dispersed, as were Type I, but in contrast to Type I, they bore ripe fruit over a considerably longer period of time, two to four weeks. Type IIIa trees had crown diameters greater than sixty feet (e.g. *Simarouba*) and Type IIIb trees had crown diameters of about twenty-five to sixty feet (e.g. *Virola* sp., *Heisteria* sp., *Protium* sp., and *Rheedia* sp.).

Type IV trees occurred, as did Type II, in close proximity to others of the same taxon, but they bore ripe fruit for longer periods, from two to four weeks. Type IVa trees with crown diameter greater than sixty feet were *Chrysophyllum* sp. and *Callophylum* sp. Type IVb trees had crown diameters less than sixty feet but more than twenty-five feet (e.g. *Hyeronima* and *Pouteria*).

Type V trees were widely dispersed, as were Types I and III. However, Type V trees bore at least some ripe fruit over very long periods of time, from one to three or four months. They all had small crown diameters, under twenty-five feet, and their fruits were circumscribed within even narrower boundaries. Palms of the genera *Euterpe* and *Iriartea* fell into this category and constituted an important food source of *A. belzebuth*.

Mature and Immature Fruit. Almost all the fruits *A. Belzebuth* were observed to eat appeared either from color and/or taste to be fully mature. The ratio of observed animal feeding time devoted to ripe versus unripe fruit was a hundred to one. Green fruit was eaten in significant amounts only during periods in which ripe fruit was scarce or nonexistent. Thus, utilization of specific taxa of trees for food was usually limited to those periods in which some specimens bore ripe fruit.

Fruit Size and Feeding Competitors. Ripe fruits observed to have been eaten frequently by *A. belzebuth* ranged in size from those less than an eighth of an inch in diameter to some more than two inches, including certain *Inga* sp. whose pods frequently exceeded twelve inches in length, although in this case only the seeds and some of the pod flesh were consumed. This wide range in fruit sizes meant that *A. belzebuth* shared and competed with other arboreal-feeding vertebrates of disparate sizes, including birds as small as *Forpus* parrots (five to six inches long) and mammals as large as coatimundis (two feet from head to base of tail). Those animals observed to have most frequently eaten fruits also eaten by *A. belzebuth* included *Alouatta seniculus*, observed to have utilized thirteen of the taxa used by *Ateles; Cebus apella*, twelve taxa; *Ramphastos tucanus* (white-throated toucan), *Pipile cumanensis* (white-headed piping guan), and *Saimiri sciureus*, six taxa.

Fruit Structure. Almost all the ripe fruits observed to have been eaten frequently by *A. belzebuth*, with the exception of *Ficus* sp., contained a single seed or seeds clearly distinguishable from and harder than the surrounding flesh, pulp, juice, or aril. Additionally, about 50 percent of the fruits had a leathery or hard outer covering (pericarp). Adults swallowed most of the seeds whole, and intact and recognizable seeds comprised the major volume of their feces in most cases.

Daily Variety. Individual *A. belzebuth* were usually observed to have eaten two or more varieties of ripe fruit within any one day. On only one of the sixty-seven days on which we were able to observe spider monkeys for four or more hours, were they observed to have eaten just one variety of mature fruit. On the remaining sixty-six days of four or more hours of observations, the number of different taxa of mature fruit eaten ranged from two to nine, with a median of 3.3 varieties per day. Over the entire year, an average daily rate of one taxon of ripe fruit per ninety-five minutes of daily observation time was recorded, but the variety of ripe fruits available and eaten varied from period to period. The diet of *A. belzebuth* at its most monotonous level consisted of one new variety of ripe fruit per 195 minutes of daily observa-

Table 9

Average number of different taxa of fruit eaten daily during successive periods
of the year

Inclusive dates	Number of different fruits eaten daily	Minutes of observation	Rate at which different fruits are eaten daily in minutes of observation time	
February 1–14	46	3,252	1 fruit per	71 minutes
February 15–29	45	4,252	1	94
March 1–12	35	3,679	1	105
April 1–13	45	3,822	1	85
April 18–30	27	2,410	1	89
May 1–14	34	2,480	1	73
May 16–27	24	1,267	1	53
June 18–24	5	245	1	50
September 5–13	13	1,818	1	140
September 16–29	24	2,329	1	97
October 2–14	12	2,314	1	193
October 15–22	14	2,175	1	155
November 11–22	22	2,360	1	107

tion time; their diet at its most diverse level consisted of one new variety of
ripe fruit approximately every hour. (See Table 9.)

CORRESPONDENCE BETWEEN DIETARY CHANGES, SUBGROUP SIZE, AND SUBGROUP STABILITY

Median and modal subgroup size, subgroup stability, and the frequency of
isolates varied over the year. If just those months in which observation
conditions were at least moderately good are considered (i.e. ignoring the
period of extensive forest flooding), intermonth variations in median subgroup size ranged from 2.2 to 5.6 animals. Subgoups of eight or more spider
monkeys accounted for over 20 percent of the subgroups counted during
the months of May and October and represented less than 10 percent of all
subgroups encountered during the months of December, 1967, and January, September, and November, 1968. Isolated but individually recognizable spider monkeys were encountered every month of the year, but they
made up more than 10 percent of the total number of subgroups observed
in December, 1967, and January, April, September, October, and November, 1968. The frequency with which isolated independent spider monkeys
were encountered does not appear to be a simple negative reflection of the
frequency with which large subgroups are formed, because in October,
1968, the relative frequencies of encounters both with isolated animals and

with large subgroups of eight or more animals were considerably higher than usual. (See Table 10.)

Some of these grouping tendencies appeared to correspond with the changing conditions of abundance, dispersion, and variety of ripe fruit available. The general picture of correspondence was the following. At those times of year (September, December, and January) when small, widely dispersed trees bearing ripe fruit for lengthy periods of time (Type V trees — palms) were the most important and almost sole source of ripe fruit for spider monkeys (in September palm fruits accounted for 53 percent of observed feeding time), *Ateles* social organization reflected maximal group fission and subgroup stability. Isolated animals were observed frequently during September (28 percent of all subgroup encounters) and appeared to remain isolated for periods ranging from several hours to several days, and almost all subgroups were composed of four or fewer animals (80 percent of those observed for more than fifteen minutes). The median subgroup size was 2.2, the mode 2. Subgroups appeared to be relatively stable; the mean duration of observation time per subgroup was ninety minutes during September, when *Ateles* were eating mainly palm fruit. The situation was probably the same for the month of December, 1969, but fewer data were collected.

In contrast, in October, when large trees bearing ripe fruit for short to moderately lengthy periods (Type IIa trees) constituted the most important source of food for spider monkeys (over 60 percent of their diet was composed of a single taxon of this type, *Brosimum*), large, relatively stable sub-

Table 10

Intermonth variations in subgroup size of all *A. belzebuth* observed at study site

Months	Median	Mode	Range	Number of encounters	Percent of encounters with isolated individuals	Percent of encounters with subgroups of eight or more
January	2.9	3	1–17	64	17	2
February	4.0	2	1–18	103	5	16
March	4.3	4	1–10	48	4	12
April	3.7	2	1–11	82	12	15
May	5.6	3	1–20	43	7	40
June–August	2.7	1	1–6	19	32	0
September	2.2	2	1–11	47	28	6
October	5.2	1	1–22	44	21	43
November	2.2	2	1–11	36	25	3
December	1.5	1	1–4	12	50	0
Total over year:	3.5	2	1–22	498	15	15

groups were more frequent. Median subgroup size was almost five animals; subgroups smaller than five accounted for less than 50 percent of all observations, and groups of eight or more animals comprised more than 40 percent of our encounters. Mean observation time per subgroup was ninety-five minutes.

Correspondence between fruiting patterns and subgroup sizes and dispersion was considerably more complicated when a greater variety of trees with differing patterns of dispersion and ripening characteristics were bearing mature fruit in quantity. For example, in February a wider than average variety of fruits was available and frequently used by spider monkeys. These included trees of Types Ia, Ic, IIb, IIIb, and IVb. The size and composition of subgroups during this period reflected the annual statistics, although subgroups appeared to be less stable than usual. Median subgroup size during February was 4.0, subgroups of eight or more individuals accounted for 16 percent of subgroups censused, while the number of isolates was infrequent, 5 percent. Mean duration of observation time per subgroup was 70.1 minutes, relatively unstable in contrast to September and October.

Some additional kinds of observations and arguments reinforce the view that size and stability of spider monkey subgroups were affected by the types and amounts of ripe fruit available at any one time. First, although dominance interactions and overt aggression were observed relatively infrequently between spider monkeys of the same social group (thirteen supplantations and sixty more intense episodes of intragroup agonistic interactions observed over 627 hours of observation), a large percentage of those observed occurred when at least one of the participants was feeding (40 percent of the supplantations and 23 percent of the agonistic interactions). All but two of these episodes occurred when animals were feeding on trees with crowns smaller than sixty feet on substances restricted to an area less than twenty-five feet in diameter (Types Ic and V). Second, the number of spider monkeys that would simultaneously feed in any single fruiting tree was relatively limited, e.g. compared with *Saimiri* at the study site. The number of simultaneously feeding spider monkeys observed within the same tree, in fact, was also frequently smaller than the number of individuals included in the same subgroup. Ten *Ateles* was the maximum number of animals ever observed feeding in the same tree at the same time, and numbers greater than eight were observed in exceptionally large trees on only a few occasions, and then for only a few minutes. These maximum figures were smaller than the maximum number observed resting in similarly sized trees and even the very same trees when they were not bearing fruit. For example, fifteen independently locomoting animals were observed resting within a single *Brosimum* tree for thirty minutes. Trees preferred as daytime resting sites were large trees not bearing ripe fruit, such as *Brosimum*, *Chrysophyllum*, and *Ficus* (Types IIa, IVa, and Ia).

The following were probably also related: (1) adult spider monkeys that were feeding generally appeared to be evenly spread out within fruiting trees, and frequently a change of location by one animal was soon followed by location changes re-establishing a uniform spread; (2) spider monkeys were frequently observed waiting for periods of at least several minutes on the periphery of a fruiting tree, entering only after some other individuals had exited; and (3) all spider monkeys appeared reluctant to rest in the same tree they had just been feeding in, although in some cases the same trees when not bearing ripe fruit were observed to be preferred resting sites.

The features reviewed so far concerning the characteristics of important food sources (i.e. tree dispersion, seasonal fruiting, and fruit concentrations), seasonal correspondences between these characteristics and subgroup sizes and stability, and feeding space and agonistic interactions combine in a way that appears to make palm fruit a central factor in understanding the social organization of spider monkeys. Palm fruits appeared to have been eaten by spider monkeys most frequently when other fruits were scarce and during those periods when very small subgroups were most frequent. At the study site they represented a relatively unique source of food for several reasons. (1) Many palm trees of certain species bore ripe fruit over exceptionally long periods, perhaps as great as three to four months, although (2) frequently, only a very small proportion of the fruit was ripe on any given day. Thus, from the standpoint of a spider monkey, palms may constitute a very reliable, if sparse, source of food. (3) The fruits of each palm tree all grow within a rather compact cluster, which makes it difficult for more than one animal to feed simultaneously without making actual physical contact with one another. (4) At the study site, the trees of the two most important species of palms in the diet of *A. belzebuth,* i.e. *Iriartea ventricosa* and *Euterpe* sp., were found over large sectors of appropriate terrain either uniformly or randomly dispersed, rather than clumped. As a consequence of these characteristics of dispersion, and in combination with the other characteristics of palms noted above, feeding from a large number of palms by any single individual spider monkey would have probably increased the possibility of becoming a participant in agonistic interactions and would have required a considerable amount of travel for minimal amounts of mature fruit at a time when nonpalm sources of food were usually in short supply. Maximal dispersion and minimal association with conspecifics would appear to be an adaptive response of animals with as specialized a feeding base as spider monkeys appear to have.

COMPARISONS WITH SYMPATRIC TAXA

It was quite clear that the degree of daily variability and seasonal change in *Aletes* subgrouping was not characteristic of the other primate taxa sympatric with *Ateles* at the study site, which shared many of the same fruit sources and inhabited the same trees. (See Table 11.)

Table 11
Estimated study site population densities, group size, and number of groups present

	Estimated population density per square mile	Range of group size[a]	Number of groups utilizing 3-square-mile study site
S. sciureus	50–80	25–35	3–6
C. apella	15–25	6–12	4–6
A. seniculus	30–75	3–6	6–15
A. belzebuth[b]	30–40	17–22	3

[a] Independently locomoting animals
[b] See text for methods of determining total group size from subgroup data

Alouatta seniculus - howler

Alouatta seniculus were observed at the study site for approximately seventy hours. They were observed on fifty-two occasions for minimal periods of fifteen minutes, and on twenty-one of these occasions for an hour or more. Groups were considerably smaller than those reported for Barro Colorado Island (see, e.g. Carpenter 1965; Chivers 1969) and ranged in size from three to six animals. All groups encountered contained at least one adult male. Most groups consisted of one adult male and two to three females with associated immatures. The few groups observed with two mature males were uncommon, and the second male in all observed cases appeared to be either very old or subadult. In contrast to *Ateles*, *A. seniculus* females were not observed without males nearby, although instances of one or two isolated adult male howlers were seen.

Relative to *Ateles* subgroups, the composition and membership of these bisexual groups of howlers was both stable and cohesive. Distances between members of each bisexual group rarely exceeded fifty to a hundred feet, the crown diameter of one or two large trees. Occasionally several howler groups were contacted on successive days in the same general area, but at certain seasons groups of howlers concentrated at distances at least as far as one-quarter mile from where similar concentrations were observed at other times (see Klein and Klein 1973). Group size and composition did not appear to be affected by seasonal shifts of habitat use.

Our unsystematic observations of howler feeding habits were in general quite consistent with what has been described for *A. palliata* on Barro Colorado Island (Hladik and Hladik 1969). Major proportions of their diet consisted of mature and immature leaves and mature and immature fig fruit. Also included were quantities of ripe fruit, decaying wood, and leaf stems.

They were never observed to eat palm fruit at our study site, nor are palms listed in the specific food catalogues for *A. palliata* of Carpenter (1934) or Hladik and Hladik (1969) (cf. Neville 1972).

In contrast to palm fruit, the fruit of fig trees (*Ficus* spp.) appears to play a more important role in the diet of *Alouatta* than of *Ateles*. Several unidentified species of *Ficus* were observed to have been eaten more frequently than by *Ateles*. In addition, a greater number of *Ficus* species were observed to have been eaten more frequently by *Alouatta* than by *Ateles*. These disparities in favor of *Alouatta* were noted despite the lesser amounts of time spent in visual contact with them than with *Ateles*. Moreover, on three separate occasions spider monkeys were observed to have ignored the fruit of large fig trees through which they passed, although howlers were feeding in them just before or after. In contrast, ripe fruit of large trees other than *Ficus* were rarely ignored by spider monkeys, unless they had been feeding in the same type of tree immediately beforehand.

A similar feeding difference appears to be characteristic of the populations of the two genera on Barro Colorado Island. Altmann (1959), for example, notes: "During the first two weeks of the present study, about 95 per cent of the diet of the howlers of the laboratory group consisted of figs. After that, the diet shifted to about 50 per cent figs." Hladik and Hladik (1969:63) note:

> The different species of *Ficus,* particularly *F. insipida,* played a primary role
> in the diet of howlers (more than 50% of the biomass of all quantitative
> samples; immature fig fruit composed 100% of the contents of one
> specimen's stomach). The genus *Ficus* furnished mature fruit several times in
> a year at complementary periods. (Translated from the French.)

Of additional interest are the radical differences between the genus *Ficus* and the palms with respect to (1) tree size, (2) distribution of fruit, and (3) length of fruiting period. In general, those figs that by growth pattern are or become trees are usually among the largest trees in the evergreen forest, bear large quantities of fruit throughout their crowns or at least on many of their branches, and bear ripe fruit over an extremely brief period, generally on the order of three to four days, falling into our category Type Ia (Hladik and Hladik 1969; Condit 1969; L. Klein, personal observation).

Although no data are available on the subject, it is thought that differences between *Ateles* and *Alouatta* with respect to the differential utilization of figs and palms may have a physiological basis. Several types of figs are known to be very high in nitrogen (Hladik and Hladik 1969), latex, and a proteolytic enzyme, ficin (Condit 1969). On the other hand, the flesh of many palms contains exceptionally large amounts of vegetable oils (Corner 1966).

The comparative dietary difference between *Ateles* and *Alouatta* with respect to figs and palm fruits illustrates one of the dangers of grossly labeling primates in terms of dietary categories such as frugivorous, insectivorous, etc. Furthermore, it also illustrates one of the problems with using a more refined approach based on percentage differences as long as categories as gross as "fruits," "leaves," "insects," etc., continue to be used. Such categories may be overly simple and even misleading when more specific dietary substances are the actual differentiating factors. Although howler monkeys are definitely more herbivorous (folivorous) and less frugivorous than spider monkeys, they also eat more immature fruit and fig fruit than do spider monkeys.

It is suggested that the howlers' extensive use of such food as mature leaves, Type Ia figs, and green fruit mitigated the effects of seasonal change and seasonal scarcity on feeding competition and social cohesion.

Saimiri sciureus – squirrel

Saimiri sciureus were observed at the study site on sixty-eight occasions for fifteen or more continuous minutes, for a total of approximately sixty hours. On eighteen of these occasions they were observed for an hour or more. The longest continuous observation of a group of *Saimiri* lasted about two hours. Those group counts which were made under the best conditions of visibility noted between twenty-five and thirty-five independently locomoting animals and included at least several adult males, with the exception of two clear-cut encounters with isolated adult male *Saimiri*.

These bisexual groups of *Saimiri* were considerably more cohesive than comparably sized *A. belzebuth* groups. Except for the temporarily isolated males, the distances between individual squirrel monkeys of the same group rarely, if ever, exceeded fifty to a hundred yards.

The formation of subgroups as defined in this article was not observed to be characteristic of squirrel monkeys. The groups did not seem to be restricted to any particular part of sector of the study area at any particular time of the year.

The major portion of the diet of *Saimiri* as measured by feeding time appeared to consist of flushed, agile, and cryptically camouflaged insects. (For greater detail see Klein 1972; Thorington 1967.) Vegetable products such as leaves and ripe fruits did not appear to be exceptionally attractive to squirrel monkeys even if they were available in abundance, and *Saimiri* appeared to satiate themselves rapidly on these substances. Insects were frequently sought even when ripe fruit was immediately at hand.

The fruits observed to have been used most frequently by squirrel monkeys were those that occurred either abundantly scattered throughout the very large trees (Type Ia) or those that were borne on trees growing in close proximity to others of the same taxon and bearing fruit at about the same time (Type II and IV trees). Squirrel monkeys, like howler monkeys, were not observed eating palm fruits. They, too, may have been unable to digest these highly dependable but concentrated sources of food.

Major dietary differences between *Saimiri* and *Ateles* may be related to their respective patterns of social spacing. *Saimiri sciureus* collect much of their food by foraging, i.e. feeding as they move between and through fruiting or nonfruiting trees. Movement is usually continuous, and stationary feeding at a single location for more than ten-second intervals is rare. *Saimiri* were frequently observed investigating numerous places where insects were likely to be found and catching, killing, and ingesting insects such as katydids, grasshoppers, caterpillars, arachnids, and cicadas. Within limits, the number of mobile insects discovered and eaten by each individual squirrel monkey may be positively affected by the cumulative amount of branch and leaf disturbance, which in turn would be at least a partial function of the number of animals moving in relative proximity to one another when feeding. Cohesive social groups of small primates foraging extensively on cryptic but mobile insect prey may thus increase the foraging efficiency of most individual animals.

Spider monkeys do relatively little, if any, foraging, i.e. feeding while moving. It is difficult to imagine how foraging would be advantageous to spider monkeys, which are predominantly frugivorous, unless they were capable of supplementing their diet with insects and significant quantities of stems and leaves. This they were not seen to do, and they do not seem capable of effectively doing so even in captivity (Klein and Klein, personal observation).

Additionally, predation pressures were almost certainly higher on *S. sciureus* than on any of the other diurnal primates at the study site. The formation of cohesive groups of these rather active and, for their size, relatively far-ranging primates (home range estimated to be one-quarter to one-half square mile) may have had significant adaptive advantages in this respect as well.

Cebus appella - cupuchins

The data on *Cebus apella* grouping patterns were not nearly as adequate as the information on *A. belzebuth*. Nevertheless, certain features did become evident and were consistent with more complete reports of another species of the same genus (Oppenheimer 1968).

C. apella were observed at the study site for approximately fifty-five hours. They were encountered on approximately 200 occasions: sixty-three of these contacts were maintained for fifteen minutes or longer, and several periods of observation persisted for about two hours. The number of independently locomoting *C. apella* observed at any one time ranged from one to twelve; the median was approximately eight. Groups of this size always included adult animals of both sexes. Contacts with what appeared to be isolated *C. apella* occurred twice; both encounters persisted only a little longer than fifteen minutes. On one additional occasion, a group of one adult female and a juvenile were observed feeding in a tree for about one hour. Four to twelve animals comprised the remaining group counts, and six or more accounted for 75 percent of the total.

Distances between individual *C. apella* considered to be part of the same group ranged from a few feet to 150 to 200 yards. When foraging, the progression was usually considerably more spread out than *Saimiri*. More than half our encounters with *C. apella* occurred when they were in close proximity to — usually intermixed with — larger groups of foraging *Saimiri* (Klein and Klein 1973).

The diet of *C. apella* was considerably more diversified than that of *A. belzebuth*. Besides fruit, they were also observed eating flowers, insects, small vertebrates, leaf buds, and leaf stems. In comparing the diets of *Cebus capucinus, Alouatta villosa,* and *Ateles geoffroyi*, Hladik and Hladik (1969:82) came to a similar conclusion concerning the capuchins' catholicity, calling them "the most 'opportunistic' of all the Barro Colorado Island primates." A considerable portion of capuchin feeding time appeared to be devoted to searching for insects. Hladik and Hladik (1969) estimated that by weight the overall diet of *C. capucinus* consisted of 20 percent prey animals. Time devoted to foraging for insects probably comprises a considerably greater percentage of feeding time. Prey occasionally included small vertebrates, e.g. frogs and squirrels (Klein 1972; Oppenheimer 1968). Although some overlap with *Saimiri* insect and arachnid prey occurred (see Klein and Klein 1973), an important portion appeared to be different (Thorington 1967). The brown capuchins were frequently observed investigating, prying open or apart dead pieces of bark, branches, vines, palm sheaths, and certain types of insect nests.

Many of the fruits eaten by *C. apella* were among those most important in the diet of *A. belzebuth*. Significantly, these also included two of the most important palms, *Iriartea exorrhiza* and *Euterpe* sp. Moreover, the flowers of at least one of these palms (*Euterpe*) and the leaf stems of the other (*Iriartea*) were also eaten by capuchins, to the probable detriment of the future production of fruit.

Several social aspects of feeding which were noted above in the course of explaining seasonal changes in spider monkey subgroup size and dispersion were also observed to be characteristic of capuchins. For example, rigid spacing between individual adult *C. apella* simultaneously feeding in the same tree was sometimes apparent if they remained for a long enough period. Competition over certain feeding locations at times appeared to be even more conspicuous between capuchins than between spider monkeys. For example, more than one adult capuchin feeding simultaneously from a single palm tree was never observed, whereas two adult *Ateles* were observed on several occasions feeding simultaneously in a single palm, and once two adult *Ateles* fed simultaneously alongside one another while a third occasionally reached in from an adjoining tree. Obvious supplantations and agonistic interactions in or around a palm tree were also observed to occur relatively more frequently between capuchins than between spider monkeys.

Why, then, were capuchin groups not affected in the same way or at least to the same degree as *Ateles* by periodic fluctuations in fruit availability and concentration?

We suggest that there are two major dietary factors, although one is judged to be probably more important than the other. First, many of the insects most frequently utilized by *C. apella,* appeared to encourage relatively constant movement and avoidance of repetitive daily travel paths. For example, among the locations in which *C. apella* were most frequently observed searching for and catching insects were arboreal snags of dead branches and palm sheaths caught on lianas or small trees. As a consequence of foraging through these structures, much of the detritus fell to the ground. Intermittent rather than daily foraging over the same areas probably increased the overall supply by allowing new accumulations of decaying vegetation to develop where important varieties of insects could subsequently be easily accessible to the manipulative capuchins. However, continual movement in pursuit of better and more accessible sources of insects meant that there was little feeding advantage to be gained by remaining near exclusive or restricted sources of fruit even if they represented a highly stable source of food. This appeared, it will be recalled, to be one of the major causes of wide dispersion between extremely small subgroups of *Ateles.*

Perhaps even more important in stabilizing group size and cohesion in *C. apella* relative to *A. belzebuth* was the considerably more diversified nature of the capuchins' diet. The degree of daily eclecticism precluded the possibility that any individual capuchin, in contrast to individual spider monkeys, could at any time entirely inhibit the feeding behavior of other group members, despite the fact that it could, while in possession or proximity, prevent some conspecifics from eating a specific item or from approaching a specific source. In fact, *C. apella* of the same group were frequently observed feeding simultaneously on different substances while they remained in visual contact. It was their most common practice and contrasted radically with spider monkeys, whose diet was considerably more restricted; the latter were at times wholly dependent upon just a few scattered but dependable sources of food and were rarely observed feeding on different substances in sight of one another. Spider monkeys usually either ate, waited, or departed. They rarely appeared to have some less desirable food substance close at hand.

Also related to this dietary difference may be the apparently considerably greater propensity of *C. apella* to organize significant aspects of their daily social behavior around status differences and conspicuous signaling of submission and appeasement. Their communicatory repertoire appears to include visual signals, which at least in form, duration, discreteness, complexity, and function are most similar to the Old World monkeys of the

genera *Papio* and *Macaca*. The fact that many of the better-studied Old
World taxa also appear to be characterized by relatively constant and frequent intragroup visual contact and feed upon a rather similarly broad
spectrum of different substances, usually in sight of one another, may be of
adaptive significance in this respect.

CONCLUDING IMPLICATIONS

1.   Variations in the social organization of four diurnal neotropical
primates sympatric at a river-plain forest sector of approximately three
square miles were described. The observations reaffirm what has become
increasingly clear from work on Old World primates (e.g. Hall 1965; Struhsaker 1969; Aldrich-Blake 1968), i.e. that a wide range of different but
taxonomicaly correlated social organizations can occur within any single   *1.*
type of forest habitat despite the fact that the sympatric primates are all
essentially arboreal and share to varying extents specific dietary items.

2.   Traditional classification of primate diets into fruit, leaves, and
insects obscures important differences clearly relatable to feeding competi-   *2.*
tion and social behavior. There are a host of possible differences, both
crosscutting and subdividing these traditional categories, some of which
have been noted by others (e.g. Hladik and Hladik 1969). For our work,
within the category of ripe fruit, specific differences between palm and fig
fruit turned out to be an important discriminatory feature between the
feeding behavior of howler and spider monkeys. This difference, in turn,
was argued to be relatable to differences in social cohesion via competitive
feeding. In addition, ripening, dispersion, and crown-diameter characteristics of different fruiting trees were seen to be an important factor regulating
A. belzebuth subgroup size, subgroup stability, feeding and resting patterns,
and the frequency of agonistic interactions. Similarly, varying dietary em-
phases on either cryptic and mobile insects or concealed and sedentary ones
may explain some of the described social differences between *Saimiri* and
Cebus.

3.   It was found impossible, however, to relate characteristics of food
supply to social organization in any abstract manner without taking into   *3.*
consideration differences in taxa-specific feeding choices and strategies —
some of which appear to be both physiologically and anatomically based.
For example, differences in the ability and propensity to obtain and ingest
large quantities of mature leaves, insects hidden beneath bark and accumulation of leaf detritus, insects protected by spines or stingers, and rapidly
leaping but camouflaged leaf-eating insects may depend upon characteristics of gastrointestinal anatomy and physiology, manual dexterity, rapidity
of movement, and motor-visual coordination. A dietary factor that appeared
to be of major importance, but one that is generally overlooked in discus-   *4.*
sion of primate socioecology, was the degree to which specific primates are
able to utilize in an expeditious manner varied substances in any single day.

To different extents this modified the competitive effects of fruit tree dispersion and the abundance and concentration of ripe fruit. It has been suggested that rigid spacing, appeasement gestures, and the frequency of supplantations are related to the relatively great diversity of the diet of capuchin monkeys. In contrast, the flexibility and individual "fission-fusion" social grouping patterns of spider monkeys appear to be based upon a considerably more restricted range of potential food substances available each day. The considerable importance of sparse, widely dispersed ripe palm fruit, along with extreme concentration of the ripe fruit on any one tree, was correlated with minimal group size and maximal occurrence of isolated adult spider monkeys of both sexes.

4.   Implied in some of the material reviewed, but not discussed, was that simple positive correlations between vegetation or fruit density and single primate taxon population densities cannot be supported without considerable additional information on intertaxon dietary differences and the nature and extent of feeding competition both between primate species and between primates and sympatric mammalian, avian, and perhaps invertebrate taxa.

5.   The outlined descriptive material on social grouping in *A. belzebuth* suggests that "fission-fusion" social grouping as the outcome of association preferences of individual adult animals of both sexes within definable social networks has evolved independently in at least two taxonomically distinct primate families: Cebidae and Pongidae (Nishida 1968). This suggests that the factors responsible are more likely to be ecological than phylogenetic (cf. Kummer 1971).

REFERENCES

Aldrich-Blake, F. P. G. 1968 A fertile hybrid between two *Cercopithecus* spp. in the Budongo Forest, Uganda. *Folia Primatologica* 9:15–21.

Altmann, S. A. 1959 Field observations on a howling monkey society. *Journal of Mammalogy* 40:317–330.

Carpenter, C. R. 1934 A field study of the behavior and social relations of howling monkeys. *Comparative Psychology Monographs* 10(2).
1935 Behavior of red spider monkeys in Panama. *Journal of Mammalogy* 16:171–180.
1965 "The howlers of Barro Colorado Island," in *Primate behavior: field studies of monkeys and apes.* Edited by I. DeVore, 250–291. New York: Holt, Rinehart and Winston.

Chivers, D. J. 1969 On the daily behaviour and spacing of howling monkey groups. *Folia Primatologica* 10:48–102.

Condit, I. J. 1969 *Ficus: the exotic species.* University of California, Division of Agricultural Sciences.

Corner, E. J. H. 1966 *The natural history of palms.* Berkeley: University of California Press.

Hall, K. R. L. 1965 "Ecology and behavior of baboons, patas, and vervet monkeys in Uganda," in *The baboon in medical research*. Edited by H. Vagtbord, 43–61. San Antonio: University of Texas Press.

Hladik, A., C. M. Hladik 1969 Rapports tropiques entre végétation et primates dans la forêt de Barro Colorado (Panama). *La Terre et la Vie* 1:25–117.

Klein, L. 1972 "The ecology and social organization of the spider monkey, *Ateles belzebuth*." Unpublished doctoral dissertation. University of California, Berkeley.

Klein, L., D. Klein 1973 Observations on two types of neotropical primate intertaxa associations. *American Journal of Physical Anthropology* 38:649–653.

Kummer, H. 1971 *Primate societies: group techniques of ecological adaptation.* Chicago: Aldine-Atherton.

Neville, M. K. 1972 The population structure of red howler monkeys (*Alouatta seniculus*) in Trinidad and Venezuela. *Folia Primatologica* 17:56–86.

Nishida, T. 1968 The social group of wild chimpanzees in the Mahali mountains. *Primates* 9:167–224.

Oppenheimer, J. R. 1968 "Behavior and ecology of the white-faced monkey, *Cebus capucinus*, on Barro Colorado Island." Unpublished doctoral dissertation, University of Illinois, Urbana.

Richard, A. 1970 A comparative study of the activity patterns and behavior of *Alouatta villosa* and *Ateles geoffroyi*. *Folia Primatologica* 12: 241–263.

Struhsaker, T. T. 1969 Correlates of ecology and social organization among African cercopithecines. *Folia Primatologica* 11:80–118.

Thorington, R. W., Jr. 1967 "Feeding and activity of *Cebus* and *Saimiri* in a Columbian forest," in *Progress in primatology*. Edited by D. Starck, R. Schneider, and H. J. Kuhn, 180–184. Stuttgart: Fischer.

Wagner, H. O. 1956 Freilandbeobachtungen an Klammeraffen. *Zeitschrift für Tierpsychologie* 13:302–313.

OLD WORLD MONKEYS

introduction
field studies on old world monkeys

The superfamily Cercopithecoidea is composed of
one family, Cercopithecidae, which is divided into
two subfamilies: Cercopithecinae and Colobinae.
The Cercopithecinae include the macaques (*Macaca*), of which there are 12 species widely distributed
in northern Africa and throughout southern and
eastern Asia as far as Japan. Besides having a wide
geographical distribution, macaques occupy a great
variety of habitats; some species are mainly arboreal
while others are highly terrestrial. All of the other
genera of cercopithecines are found in Africa, south
of the Sahara. These include the mainly arboreal
mangabeys (*Cercocebus*; 5 species) and guenons
(*Cercopithecus*; 23 species); and the mainly terrestrial
patas monkey (*Erythrocebus*), baboons (including 5
species of *Papio* and one of *Theropithecus*), and
mandrills (*Mandrillus*; 2 species).

The Colobinae are distinguished from the
Cercopithecinae by the absence of cheek pouches
and the development of an elaborate and sacculated
stomach. This latter adaptation allows efficient

digestion of cellulose, and the colobines are thus
often referred to as leaf-eating monkeys. The genus
Colobus (of which there are three major groups:
olive colobus, red colobus, and black and white
colobus) are all arboreal and are the only African
colobines. There are five genera of colobines in
Asia. The most widely distributed genus *Presbytis*
(14 species) is adapted to a wide range of habitats.
The other four genera (*Pygathrix, Rhinopithecus,
Simias,* and *Nasalis*) are quite localized in
distribution.

Dividing the cercopithecoids geographically,
the ecology of the African forms has been studied
much more extensively than the Asian forms. The
most studied of the Asian monkeys are *Presbytis
entellus* and *Macaca mulatta.* Over half of the species
of *Presbytis,* however, have either barely been
documented or remain totally undocumented. The
four localized genera of Asian Colobinae (*Pygathrix,
Rhinopithecus, Simias,* and *Nasalis*) have been almost
totally ignored (see Baldwin et al. 1975). Most of
our knowledge of *Macaca* comes from the excellent
long-term studies of the Japanese macaques and of
the rhesus macaques on the island of Cayo
Santiago, Puerto Rico. However, both of these
populations are provisioned and the latter
population was, of course, introduced. Considering
the widespread use of macaques in behavioral and
medical laboratory research, it is surprising that the
general ecology of these Asian monkeys is relatively
unknown. In this book, I have not included any
field studies of the Asian monkeys; this was mainly
a problem of space. However, some discussion of
the Asian forms will be found in the last section of
the book (especially Selections 23 and 26).

Of the African cercopithecoids, the more
terrestrial savannah forms have been the most
extensively and intensively studied of any
nonhuman primates (see Baldwin and Teleki 1972).
The arboreal forest-dwelling African monkeys
(*Colobus, Cercocebus,* and most *Cercopithecus*) are much
less well documented (see Baldwin et al. 1976).
Most of these genera have been studied only in
eastern Africa. Populations in western and in

central Africa have been relatively and almost
totally ignored. The articles included in the section
entitled "Arboreal Old World Monkeys" focus on
multiple species or synecological interactions of
arboreal African monkeys. The study by Booth
(Selection 7) was carried out in western Africa.
Although it is qualitative in nature, it is the earliest
attempt to document differential habitat utilization
in a number of sympatric primate species. Gartlan
and Strushaker (Selection 8) discuss polyspecific
associations among several species of African
forest-dwelling primates, especially *Cercopithecus.*
Mixed-species groups are common in *Cercopithecus*
and in some species of birds, and have been
observed in some South American monkeys.
However, these groups are rarely permanent, and
their function is not yet fully understood. In
Selection 9, Strushaker and Oates describe the ways
in which two species of African leaf-eating monkey
(*Colobus guereza* and *Colobus badius*) utilize different
resources in the same forest. Habitat utilization in
these two species is discussed further in Selection
25.

The articles in the section entitled "Terrestrial
Old World Monkeys" are on terrestrial African
monkeys. The Dunbars (Selection 10) compare the
habitat preferences in three sympatric Ethiopian
species, *Cercopithecus aethiops, Papio anubis,* and
Theropithecus gelada. The chapters by Rowell and
Harding are both on *Papio anubis.* Rowell (Selection
11) emphasizes that baboons living in forests do
not behave the same as those living in savannah
environments; Harding (Selection 12) points out
that even savannah baboons may not behave in the
way that had been previously suggested. These
articles illustrate the variability in baboon behavior,
as well as that of the methods used by
primatologists. In the final article in this section
(Selection 13), Altmann discusses how the social
structure of different species of baboon can be
related to specific ecological determinants.

REFERENCES CITED

Baldwin, L.A., M. Kavanagh, and G. Teleki. 1975. Field research on langur and proboscis monkeys: An historical, geographical, and bibliographical listing. Primates, 16: 351–363.

Baldwin, L.A. and G. Teleki. 1972. Field research on baboons, drills, and geladas: An historical, geographical, and bibliographical listing. Primates, 13: 427–432.

Baldwin, L.A., G. Teleki, and M. Kavanagh. 1976. Field research on colobus, guenon, mangabey, and patas monkeys: An historical, geographical, and bibliographical listing. Primates, 17: 233–252.

FURTHER REFERENCES

Freeland, W.J. 1977. Blood-sucking flies and primate polyspecific associations. Nature, 269:801-802. (This short paper suggests a unique explanation of the significance of primate polyspecific associations).

Napier, J.R. and P.H. Napier (eds.) 1970. Old World Monkeys: Evolution, Systematics and Behavior. Academic Press, London. (A reader of original papers focusing on a number of aspects and species of Old World primates.)

Roonwal, M.L. and S.M. Mohnot. 1977. Primates of South Asia: Ecology, Sociobiology and Behavior. Harvard University Press, Cambridge. (A review of work done on the primates of South Asia. The reference section is useful, but contains very few sources later than 1972!)

Arboreal
Old World
Monkeys

7

The Distribution of Primates in the Gold Coast

1956

A. H. BOOTH

Summary. Sixteen forms of Primate occur in the Gold Coast. Five may be considered savannah, and eleven forest forms. In certain areas of the High Forest, ten of these eleven forms are found together. Even after initial separation into leaf-eating and fruit-eating, diurnal and nocturnal forms, there remains between certain species a degree of ecological overlap. Competition appears to be reduced not so much by selective feeding as by stratification in a vertical sense, species having their typical levels for feeding, travelling, and sleeping.

INTRODUCTION

Ingoldby (1929) and Cansdale (1948) list fourteen species of Primate occurring in the Gold Coast. To these lists there have since been added one form deserving specific distinction, and one further race of a species already recorded. Several alterations in nomenclature have been found necessary as a result of studies in systematics and distribution recorded elsewhere (Booth, 1954, 1955, and *in press*).

The list of forms reads as follows: (Hill's classification, 1953)

Suborder Lorisoidea
Family Lorisidae
Perodicticus potto potto (P. L. S. Müller), 1766. Bosman's Potto.
Family Galagidae
Galago senegalensis senegalensis E. Geoffroy, 1796. Senegal Galago.

Reprinted from *Journal of the West African Scientific Association*, 2:122-133 (1956).

Galagoides demidovii demidovii (G. Fischer), 1808. Demidoff's Dwarf Galago.

Suborder Pithecoidea
Family Cercopithecidae
Cercopithecus aethiops sabaeus (Linnaeus), 1766. Western Green Monkey.
Cercopithecus aethiops tantalus Ogilby, 1841. Nigerian Green Monkey.
Cercopithecus diana roloway (Schreber), 1774. Gold Coast Diana Monkey.
Cercopithecus mona (Schreber), 1774. Mona Monkey.
Cercopithecus campbelli lowei Thomas, 1923. Lowe's Mona Monkey.
Cercopithecus petaurista petaurista (Schreber), 1774. Spot-nosed Monkey.
Erythrocebus patas patas (Schreber), 1774. Red Patas Monkey.
Cercocebus atys lunulatus (Temminck), 1853. White-Crowned Mangabey.
Papio anubis choras (Ogilby), 1843. Guinea Baboon.
Family Colobidae
Colobus polykomos vellerosus (I. Geoffroy), 1834. White-thighed Colobus.
Procolobus verus (van Beneden), 1838. Olive Colobus.
Procolobus badius waldroni (Hayman), 1936. Ashanti Red Colobus.
Family Pongidae
Pan troglodytes verus Schwarz, 1934. Western Chimpanzee.

The primary division of the Gold Coast into High Forest and Guinea Savannah vegetation zones is broadly reflected in the distribution of the Primate species. This broad correspondence has already been noted by Cansdale (1946). *Galago senegalensis, Cercopithecus aethiops, Erythrocebus patas,* and *Papio anubis* are savannah forms, the remainder occurring in High Forest. Fig. 1 shows the approximate distribution of High Forest in the Gold Coast, after Taylor (1952). The remainder of the country is covered with a Guinea Savannah type of vegetation (more or less degraded by heavy farming in some areas) except for the Accra Plain, a grass-scrub area. The Accra Plain is very different from Guinea Savannah: occasional clumps of low, very dense bush alternate with wide spaces of open grassland. Forest of a very dry type, which may formerly have been typical of the whole area, is found in scattered (mainly "juju") spots, such as Senya Beraku, Abokobi, Krobo Hill, and the Shai Hills. Typical Guinea Savannah, despite its grass ground-cover, is well-wooded, with low trees every few yards. Treeless areas, where they occur, are mainly on swampy ground or on soil exhausted by cultivation.

One further remark must be made concerning the division into main vegetation zones. Along river banks far into the Guinea Savannah we find a formation known as "fringing forest." There are also within the Guinea Savannah isolated patches of "relic" High Forest of a very dry type. Both these types of forest are liable to harbour a High Forest fauna. Typical High Forest mammals are been found along the Black Volta as far north as Wa.

We have seen that "forest," or vegetation dominated by woody growth and lacking a grass under-storey, may exist outside the High Forest zone

Fig. 1. Map of the Gold Coast, showing High Forest zone (after Taylor, 1952) stippled, and localities mentioned in the text, also the writer's main collecting centres. Fringing forest and forest outliers are not marked.

proper. But within the High Forest zone, subdivisions have been made. These subdivisions may be made, as by Aubréville (1949) on the basis of climate, or as by Taylor (1952) on the basis of an "association" having typical dominant emergent tree species. Other authors refer to the extent of the deciduous habit among emergent trees. The last is perhaps the most easily comprehended by the zoologist. "Rain forest," in which the emergent trees are, owing to a low saturation deficit throughout the year, largely evergreen, exists only in the south-western corner of the Gold Coast, approximately in the Ahanta-Nzima-Aowin districts. The remainder of the High Forest is "semi-deciduous," subject to a distinct dry season, in which a proportion of the emergent trees shed their leaves. Semi-deciduous High Forest may be further subdivided into 'moist' and "dry" types, having the implied differences in climate, and slightly different floristic composition.

These types of High Forest interdigitate and intergrade, so that no definite boundary can be marked on a map, despite continuous and distinguished botanical effort, either climatically or floristically.

Both Cansdale (1946) and Rosevear (1953) have drawn attention to the fact that mammals as a whole do not show any apparent discrimination between the various subdivisions of High Forest recognized by botanists. Cansdale's pronouncement on the subject carries more weight here, since he deals with the Gold Coast, and actually mentions two Primates in this connexion, the Red Colobus and the Chimpanzee. Both these species are found only to the west of Kumasi, yet neither is restricted to the rain forest or even to the moist semi-deciduous forest: both range freely throughout the High Forest zone, but only to the west of Kumasi.

OBSERVATIONS ON AUTECOLOGY

Perodicticus potto

Bosman's Potto occurs throughout the High Forest zone, in fringing forest almost to the north-western tip of the Gold Coast, and in forest outliers in the savannah. It is also found in coastal scrub, being reported frequently within the municipal boundary of Accra. Cansdale remarks that it is, in his opinion, one of the animals which is commoner in farmland and secondary forest than in mature High Forest. This may be regarded as a possibility. It is nocturnal and omnivorous (stomach contents have included fruits, insects, and a gecko). Its distribution is shown in Fig. 2A. Further biological information is available in Cansdale's publication (1946).

Galago senegalensis

This species is typical of "orchard bush," that is, of grass-covered country with numerous evenly-spaced, low-growing (fire-affected) trees; such areas

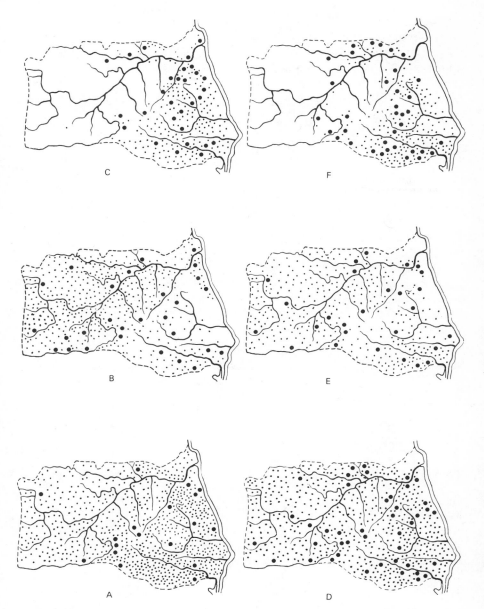

Fig. 2. Primate distribution maps. The large dots mark points where the species have been collected or definitely identified. The smaller dots indicate the probable range in the Gold Coast.
A. *Perodicticus polla* (black), *Galago senegalensis* (red), *Galagoides demidovii* (green).
B. *Cercopithecus diana* (black), *C, aethiops tantalus* (red), *C. a. sabaeus* (green).
C. *C. campbelli* (black), *C. mona* (red).
D. *C. petaurista* (black), *Erythrocebus patas* (red).
E. *Cercocebus atys* (black), *Papio anubis* (red).
F. *Colobus polykomos* (black), *Procolobus badius* (red), *Procolobus verus* (green).

are found in Guinea Savannah in the Gold Coast (Fig. 2A). The species also occurs in thorn-scrub to the north of the country in Sudan Savannah. Within the range of the species, there also occur forest outliers and fringing forest. This type of environment is avoided by the Senegal Galago (but see above notes on the Potto). The species does not occur in the Accra Plains grassland. It is nocturnal and omnivorous. Nests and daytime roosts are found, where available, in hollow trees, but the species also sleeps in leaves when such places are not available.

Galagoides demidovii

The range of Demidoff's Dwarf Galago is similar to that of the Potto, but does not include the drier forest outliers and fringing forest (Fig. 2A). These areas (by no means extensive) are probably the only timbered localities in the Gold Coast which are not inhabited by a Galago. Biological information on this species may be found in Cansdale (1944), but this article is concerned mainly with the species in captivity, where it clearly is quite catholic in its tastes. The few stomachs of wild-killed specimens examined by the writer suggest that this species is almost entirely insectivorous in the wild state. Sanderson's account (1937) of its habits, quoted with due caution by Osman Hill (1953) is utterly absurd, and can be disregarded. Low, tangled bush, whether in High Forest, in secondary forest or coastal scrub, is the normal habitat. Its brilliant eyes are rarely seen above 30 feet from the ground, and it leaps as do other Galagos.

Cercopithecus aethiops

As a species, the Green Monkeys are found throughout the Guinea Savannah zone, and on the fringes of the Accra Plain. The mutual boundary between the two races *sabaeus* and *tantalus* is approximately the line of the White Volta and the Volta (Fig. 2B). The form found on the Accra Plain is, however, the eastern race *C. a. tantalus,* which has thus made the crossing from Togoland. Within the broad area included in its range, the Green Monkey displays decided habitat preferences. Forest outliers, fringing forest along rivers and streams, and denser cover are preferred. Although they range freely into neighbouring orchard bush, farmland, or even village wells, where food may be dropped, the troops return at night to more congenial shelter. The species is frequently seen on the ground when travelling between feeding areas, but normally feeds among the branches, though grass seeds have once been found among the stomach contents. It is mainly frugivorous. T. S. Jones (private communication) reports that in Sierra Leone the Green Monkey (*C. a. sabaeus*) regularly invades the High

Forest zone wherever substantial agricultural clearings occur. In the Gold Coast, this occurs only on the edges of the High Forest.

Cercopithecus diana

Undoubtedly it is the rarest monkey in the Gold Coast, though not in captivity. The species is restricted entirely to the High Forest zone proper, where it is found only in mature forest, mainly in the west and south. Its range extends to western Akim (Fig. 2B). It is typical of the middle and upper storeys of the forest, only very rarely descending to lower levels. The Diana Monkey sleeps in, escapes to, and generally travels in the uppermost layers of the forest, using the emergent trees whenever possible. In this respect it is comparable with *Procolobus badius*. The species is believed to be entirely frugivorous (14 stomachs examined).

Cercopithecus campbelli and C. mona

Representatives of the *mona*-group are found throughout the High Forest zone in both primary and secondary forest, and in forest outliers and fringing forest at least as far north as Bamboi, though they tend to become scarce and local in that area (Fig. 2C). Lowe's Mona is limited in its range to the area west and south of the Volta-Black Volta, the true Mona Monkey being found to the east of that line. Along the Afram, however, and penetrating the Gold Coast to western Kwahu, *C. mona* overlaps with *C. campbelli* (Booth, 1955).

The species are almost entirely frugivorous (28 stomachs examined), and are typical of the lower and middle layers of the forest. They sleep in the middle storey, rarely ascending the tall emergent trees.

In the area of overlap, a slight degree of ecological differentiation has been achieved, *C. mona* tending to favour the river banks, while *C. campbelli* occupies the forest behind. Mixed parties have, however, been observed feeding together at sources of abundant food. Cansdale (1946) reports a Mona Monkey (doubtless *C. mona*) from the Volta estuary, in mangrove swamps.

Cercopithecus petaurista

The Spot-nosed Monkey (often incorrectly designated "Putty-nosed," a name which properly belongs to *C. nictitans*, a totally different species), ranges throughout the High Forest zone in the Gold Coast and Togoland, freely entering forest outliers, fringing forest and coastal scrub (Fig. 2D). It is abundant on Krobo Mountain and the Shai Hills. Essentially a dweller in thickets, it is found in High Forest mainly where a fallen emergent leaves a gap in the canopy, permitting more light to pierce through to ground level and a thick underbrush to grow. It is also common in swampy areas where emergents are few. It sleeps in the middle and lower layers.

The wide range over which the species is common is due to the fact that leaves as well as fruit are taken. Haddow (1952) has given a list of foods of a closely-related species in Uganda. Leaves, shoots, and flowers occur 114 times, fruits 107 times. As a result of its catholic tastes, the Spot-nosed Monkey can flourish in relatively small areas of forest, hence the frequency of its occurrence in areas which can only be described as "marginal" for forest species.

Erythrocebus patas

The range of the Red Patas Monkey is similar to that of the Senegal Galago (Fig. 2D). Like the latter, it is typical of orchard bush, rarely if ever entering forest outliers or fringing forest, except, in this case, to drink. Being largely terrestrial, and having great speed and considerable endurance over the ground, a troop may range widely in a single day. Little is known of its movements or feeding habits, and it is justly reputed one of the most difficult animals in the world to hunt. The majority of specimens are obtained either by trapping near farms, or as a result of chance encounters.

Cercocebus atys

The White-crowned Mangabey has a range similar to that of the Diana Monkey, but slightly wider, occurring in Kwahu, where the Diana is rare or absent (Fig. 2E). It is commonest, however, in the south and west of the Gold Coast. Cansdale is of the opinion that it favours rocky areas in the forest. My own impression is quite the reverse, in both the Gold Coast and the Ivory Coast: the areas in which the species is most likely to be found are those with abundant Rhaphia palm swamps and rice farms. In support of this impression, it may be pointed out that undoubtedly Rhaphia, rice, and Mangabeys decrease in their frequency of occurrence from the west to east, while rocky areas increase.

The Mangabey is the only Primate of the High Forest zone, other than the Chimpanzee, which is largely terrestrial. It rarely ascends above the lower storey in the forest and is probably entirely frugivorous, though a little animal food may also be taken. Unfortunately, the few stomach contents examined have been from animals taken near cultivation. In the more mature forests the species is extremely difficult to hunt. Secondary forest and farmland is readily invaded by this species.

Papio anubis

The Baboon in the Gold Coast occurs over approximately the same range

as the Green Monkey (Fig. 2E). Though typically found in the Guinea
Savannah zone, it also occurs on the eastern Accra Plain, making its head-
quarters among the rocks of Krobo and the Shai Hills. In the area from the
Afram Plains northward, almost every rocky eminence has its resident troop
of Baboons, which range widely during the day, returning generally at night.
Like the Green Monkey, the species may feed in thicker cover, or in more
open country, but it is more terrestrial in its habits than the Green Monkey.
Like the Red Patas, it ascends trees only to feed or to rest. Little is known
of the proportions of the various foods taken in the wild, but animal food
(insects, lizards, scorpions) is occasionally taken, the remains having been
found in the faeces. *Ficus* spp., *Adansonia, Kigelia,* and *Parkia* fruits figure
largely in the diet, the Baboon being one of the few mammals which can deal
effectively with the Baobab fruits. Roots, both wild and cultivated, are also
eaten, and forest farms are raided where they border on grassland.

Colobus polykomos

The "Black Colobus" is more affected than any other Gold Coast Primate
(with the possible exception of the Chimpanzee) by the predatory activities
of man. In demand both for meat and for its beautiful fur, it has been
exterminated over large areas of the Gold Coast. Probably the commonest
monkey where the human population is low, it is one of the rarest in intensively cultivated areas. In the High Forest zone it is now confined to primary
and mature secondary forest, but in sparsely inhabited districts it has a wide
range into forest outliers and fringing forest; it is found at least as far north
as Damongo, and along the Oti River between the latitudes of Wulehe and
Yendi (Fig. 2F).

  Within the High Forest zone it feeds at any level, from the topmost
branches of the emergents to low bushes round palm swamps. It is most
often observed in the upper layers probably because it is less cautious of the
hunter when thus protected. It sleeps in the middle and upper storeys. The
food of this species, as of all Gold Coast Colobidae, is normally exclusively
leaves. It may here be stated that 41 Black, 33 Red, and 17 Olive Colobus
stomachs have been examined, and not one contained any fragments identifiable as fruit or animal matter.

Procolobus verus

Like the Black Colobus, the Olive Colobus is a form typical of the High
Forest zone, which extends its range into the fringing forest, though not so
far as the former species (Fig. 2F). The ecological inter-relationships of the
two species in these marginal habitats will be discussed later in this communication.

  The Olive Colobus is comparable in its habitat preferences with the

Spot-nosed Monkey, already commented upon, but it tends to be concentrated around water, for instance, river banks, palm swamps. It enters secondary forest and low farm-bush, and is by no means the rarity that some workers have claimed it to be.

Procolobus badius

Confined to mature forests, the Red Colobus occurs no further east than Bekwai (Fig. 2F). It sleeps in, escapes to, and largely feeds in the tall emergent trees, rarely visiting the lower levels of the forest. Its extinction in the Gold Coast in the near future must be regarded as a probability, unless effective legislation to protect both the animal and its environment is forthcoming. It is not only the most specialized, but also the most unwary of all Gold Coast monkeys. At the present time, it numbers are still large in certain areas, but development of these areas will rapidly bring about its destruction.

Pan troglodytes

The range of the Chimpanzee is approximately co-extensive with that of the Red Colobus. Its total numbers in the territory certainly do not exceed 1,000, and are probably far less, but Cansdale's estimate of 50 animals is far too low. No legislation exists for the protection of the species, which is doomed to extinction in the Gold Coast within a very short time.

No ecological data is available for this species in the Gold Coast, other than that given by Cansdale (1946).

SYNECOLOGY AND COMPARATIVE DATA

Compared with such a region as, for instance, the Cameroons, the Primate fauna of the Gold Coast is rather poor in species. This makes the task of disentangling ecological relationships somewhat simpler. It is not here claimed that the picture is complete, since very little has been done in the way of identifying specific food plants. But the general ecological status of the various species has in most cases been assessed.

The Lorisoids have little in common with the Pithecoids on account of their small size and nocturnal, more or less solitary, feeding habits. The only possible ecological overlap here is between the Potto and the Dwarf Galago. But the Potto is by far the larger animal, hunts by stealth, and lies up during the day in any cluster of thick foliage. Dietetically it is probably less specialized than the Dwarf Galago, which would account for its wider range. The Dwarf Galago is an extremely active hunter, and lies up by day in a nest of leaves. It is doubtful if there is any appreciable competition between the two species.

The monkeys have a great deal in common in their way of life. Except for the Baboon, they overlap largely in size and weight. All are diurnal, gregarious, tend to be socially organized, and are to a greater or lesser extent arboreal. It is, therefore, inevitable that there should be some degree of ecological overlap. One indication of the extent of overlap is the occurrence of mixed feeding parties. The examples here given are of parties definitely observed to be taking identical food in the same spot. Aggregations, frequently observed, between leaf- and fruit-eating species are discounted, though they may have significance in protection against predators. The following mixed feeding-parties have been observed:—

Cercopithecus aethiops: with *Papio anubis, C. petaurista.*
Cercopithecus diana: with *C. campbelli.*
Cercopithecus campbelli: with *C. diana, C. mona, C. petaurista.*
Cercopithecus mona: with *C. campbelli, C. petaurista.*
Cercopithecus petaurista: with *C. aethiops, C. campbelli, C. mona.*
Erythrocebus patas: no mixed parties observed.
Cercocebus atys: observation of identical feeding uncertain, but mixed parties with *C. campbelli* and *C. petaurista* seen.
Papio anubis: with *Cercopithecus aethiops.*
Colobus polykomos: with *Procolobus verus, P. badius.*
Procolobus verus: with *Colobus polykomos.*
Procolobus badius: with *Colobus polykomos.*

The above observations fit generally into the ecological picture presented for the forest monkeys in Table 1.

Table 1
Vertical Range of Monkeys in Forests of South-West Gold Coast

		Sleeping			Travelling				Feeding				Food	
		Upper	Middle	Lower	Upper	Middle	Lower	Ground	Upper	Middle	Lower	Ground	Fruit	Leaves
Procolobus badius	..	+	−	−	++	+	−	−	++	+	(+)	−	−	+
Colobus polykomos	..	+	+	−	+	+	−	−	(+)	+	+	−	−	+
Procolobus verus	..	−	+	+	−	+	+	−	−	(+)	+	−	−	+
Cercopithecus diana	..	+	+	−	+	+	−	−	+	+	(+)	−	+	−
Cercopithecus campbelli	..	−	+	(+)	(+)	+	+	(+)	−	+	+	−	+	+
Cercocebus atys	..	−	+	+	−	+	+	++	−	−	+	+	+	−
Cercopithecus petaursita	..	−	+	+	−	+	++	(+)	−	(+)	++	−	+	+

The question of ecological overlap in Guinea Savannah species may first be dealt with, as being less complex. In the Guinea Savannah there can be no question of vertical stratification of habitat. The autecological notes already given illustrate the main points, being as follows:—

1.   The Baboon, by reason of its large size, is able to deal with food which is not available to the smaller monkeys. Its partiality for rocky areas is noted.
2.   The Green Monkey and the Red Patas favour contrasting habitats within the Guinea Savannah zone, the former haunting, the latter avoiding, denser cover.

There is, in fringing forest and forest outliers, a tendency towards ecological overlap between the Green Monkey on the one hand and the Mona and Spot-nosed Monkeys on the other. Here the following observations bear on their comparatively slight ecological competition:—

1.   The High Forest species (Mona and Spot-nose) rarely if ever stray from the forest strip into the surrounding orchard bush. The Green Monkey regularly does so.
2.   Where fringing forest is occupied by High Forest species, the Green Monkey remains on the very edge. Only where High Forest species are absent does the Green Monkey range freely throughout. A specific observation is of interest in this respect. At Dorimon, on the Black Volta near Wa, only Green Monkeys are commonly found in the fringing forest. Their reaction when hunted was to hide themselves in the denser cover near the water. On the other hand, at Bamboi, and near Wulehe, where the High Forest species also occur in the fringing forest, the Green Monkeys were observed normally to move out into the Guinea Savannah, either under cover of the grass or across bare ground if grass cover was not available.

The synecology of the High Forest monkeys is a more complex matter. It soon became clear, however, that ecologists' elucidations of habitat preferences in English woodland birds (e.g. Colquhoun and Morley, 1943) had a distinct bearing on the present problem. Whereas the ornithological problem resolves itself into the functions of feeding, sleeping, and nesting, in the case of monkeys we are dealing with feeding, sleeping, and travelling, the young being invariably carried by the mother. Of these three functions, clearly the most important from the point of view of competition is feeding. Sleeping quarters are abundantly available and, as indicated in Table I, they are frequently higher than the feeding level. In travelling, 'traffic jams' have occasionally been observed, particularly in the case of the Red Colobus, Diana, and Black Colobus Monkeys when disturbed in the emergent trees. They are of some importance in that they facilitate the attacks of the

Crowned Hawk-Eagle, *Stephanoaetus coronatus,* the only important non-human enemy of Gold Coast monkeys in the High Forest.

Competition for food is non-existent between the purely leaf-eating and the fruit-eating forms. The Leaf-eating Colobidae are very simply stratified: the Red Colobus occupies the uppermost layers, the Black Colobus the middle, and the Olive Colobus the lowest layers. Where the Red Colobus is absent, the Black Colobus moves higher, but there is no apparent change in the status of the Olive Colobus. Under all conditions of competition, it appears to be a thicket-haunter. At Mankrong, along the Afram River, the Black Colobus has been locally exterminated by hunters, while the Olive Colobus remains very common. Nevertheless, it does not ascend the higher trees to feed, despite the fact that the succulent *Triplochiton,* whose leaves are relished by monkeys, is available.

Among fruit-eating species the situation is similar. The upper and middle layers are the feeding-ground of the Diana Monkey, the middle and lower that of the Mona Monkey and Campbell's Mona, the lower and ground layers being occupied by the Mangabey in suitable localities.

The Spot-nosed Monkey, which regularly eats both fruit and leaves, comes into competition in the middle and particularly the lower layers with, mainly, the Olive Colobus, the two forms of Mona Monkey, and the Mangabey. That it holds this intermediate position successfully cannot be doubted. It is my impression, despite Collins' (1956) lack of observations, that it is the commonest High Forest monkey in the Gold Coast.

CONCLUSIONS ON ECOLOGICAL RANGE

To a certain extent, the comparative data help to explain the width of ecological range of the several species. It is to be expected that the Red Colobus and the Diana Monkey, which frequent the highest levels, should be absent from the eastern part of the territory, which is densely populated and much exploited for timber and agriculture. The restriction of the range of the Mangabey is, in my opinion, correlated with an observed preference for swampy areas. The low-level Mona and Spot-nosed Monkeys are, as might be anticipated, tolerant both of secondary forest and of fringing forest. The greater abundance of the latter is almost certainly due to its more varied diet. A small area will provide food for a troop of Spot-nosed Monkeys, while the more specialized Mona Monkeys require a greater area for foraging.

The only unexpected feature is the greater range of the Black as compared with the Olive Colobus. Here we must seek the answer in the behaviour of the two species. It has been emphasized that the Olive Colobus is a thicket-haunter, even where excellent food is available, without competition, at a higher level. The reluctance to come out into the open is shown not only in the vertical, but also in the horizontal sense. Whereas the Black Colobus may be observed sunning itself in the treetops, or equally crossing open roads, the Olive Colobus has never been observed to stray from dense cover. The marginal habitats occupied by the Black Colobus in the Northern Territories frequently necessitate the ability to traverse open ground, especially during the dry season. This the Black Colobus is psychologically adapted to do, while the Olive Colobus is not so adapted.

In conclusion, it must be stressed that the synecological picture outlined here is the result of only a few months' observations at scattered points in the Gold Coast, and at different times of year. Monkeys are highly adaptable animals, and their feeding habits and their food vary according to the district. Scattered identifications of stomach contents are of little use under the circumstances. To go more deeply into the matter of food requirements would involve extended periods of residence in a series of areas, and the slaughter of large numbers of animals. The only extant record of such an investigation is that of Haddow (1952), a contribution likely to remain unique for a long time.

ACKNOWLEDGEMENTS

The investigation, of which the above records form a part, was financed by a generous grant from the Academic Board of the University College of the Gold Coast. Much of the hunting and observation involved was done by Mr. Osumanie Moshi and Mrs. C. P. Booth. Grateful thanks are also due to Professor E. E. Edwards for facilities, advice, and encouragement.

REFERENCES

Aubréville, A. (1949). *Climats, forêts et désertification de l'Afrique tropicale.* Paris, Société d'Éditions Geographiques, Maritimes et Coloniales.

Booth, A. H. (1954). A note on the Colobus monkeys of the Gold and Ivory Coasts. *Ann. Mag. nat. Hist.* (12), 7, 857.

(1955). Speciation in Mona monkeys. *J. Mammal.* 36, 434.

(*in press*). The Cercopithecidae of the Gold and Ivory Coasts: geographic and systematic observations. *Ann. Mag. nat. Hist.*

Cansdale, G. S. (1944). The Lesser Bush Baby (*Galago demidovii demidovii* G. Fich.). *J. Soc. Preserv. Faun. Empire London* 50, 7.

(1946). *Animals of West Africa.* London, Longmans, Green and Co.

(1948). *Provisional checklist of Gold Coast mammals.* Accra, Government Printing Department.

Collins, W. B. (1956). The tropical forest: an animal and plant association. *Nigerian Field* 21, 4.

Colquhoun, M. K., and Morley, A. (1943). Vertical zonation in woodland bird communities. *J. Anim. Ecol.* 12, 75.

Haddow, A. J. (1952). Field and laboratory studies on an African monkey, *Cercopithecus ascanius schmidti* Matschie. *Proc. zool. Soc. London* 122, 297.

Hill, W. C. Osman (1953). *Primates:* I. *Strepsirhini.* Edinburgh, Univ. Press.

Ingoldby, C. M. (1929). On the mammals of the Gold Coast. *Ann. Mag. nat. Hist.* (10) 3, 511.

Rosevear, D. R. (1953). *Checklist and atlas of Nigerian mammals.* Lagos, Government Printer.

Sanderson, I. T. (1937). *Animal treasure.* London, Macmillan.

Taylor, C. J. (1952). *The vegetation zones of the Gold Coast.* Accra, Government Printing Department.

8

Polyspecific Associations and Niche 1972 Separation of Rain-Forest Anthropoids in Cameroon, West Africa

J. STEPHEN GARTLAN AND THOMAS T. STRUHSAKER

INTRODUCTION

Most studies of the ecology of rain-forest primates in West Africa have been qualitative (Malbrant & Maclatchy, 1949; Booth, 1956; Rahm, 1961), but Gautier & Gautier-Hion (1969) have made a detailed quantitative study of polyspecific associations of cercopithecids in Gabon, Jones & Sabater Pi (1968) have published notes on two species of *Cercocebus*, and Jones (1970) compares general food habits and gastro-intestinal anatomy of eight monkey species from Rio Muni.

The present article records the qualitative and quantitative results of two field studies of primates in West African rain-forest emphasizing polyspecific associations and ecological niche separation. Cameroon with its extremely rich primate fauna provides an excellent area for the study of these problems. The first study (by Thomas T. Struhsaker) extended from November 1966 to May 1968, comprising 17 months at a variety of sites in Cameroon, and one month in Rio Muni. The locality receiving the greatest attention was the mature forest near Idenau, West Cameroon (Fig. 1). The following primates were seen in or near the major study area; *Cercopithecus nictitans martini* Waterhouse 1838, *C. pogonias pogonias* Bennett 1833, *C. mona* Schreber 1775, *C. erythrotis camerunensis* Hayman 1940, *C. lhoesti preussi* Mats-

Reprinted and abridged from *Journal of Zoology, London, 168*:221-266 (1972) by permission of The Zoological Society of London.

chie 1898, *Cercocebus torquatus* Kerr 1792, *Mandrillus leucophaeus* Ritgen 1824, *Pan t. troglodytes* Blumenbach 1799, and *Galagoïdes demidovii* Fischer 1806. The second study (by J. Stephen Gartlan) was carried out between October 1967 and December 1968. Two weeks were spent in Rio Muni and the remainder of the time in Cameroon. The major study location was the Southern Bakundu Forest Reserve, West Cameroon. The following non-human primates were seen in or near the major study area: *Cercopithecus nictitans martini, C. mona, C. p. pogonias, C. erythrotis camerunensis, Mandrillus leucophaeus, Pan t. Troglodytes, Galagoïdes demidovii, Galago alleni* Waterhouse 1838, and *Euoticus elegantulus* (Le Comte) 1857. The two studies were complementary in that Struhsaker surveyed many areas, whereas Gartlan made an intensive study of the forest.

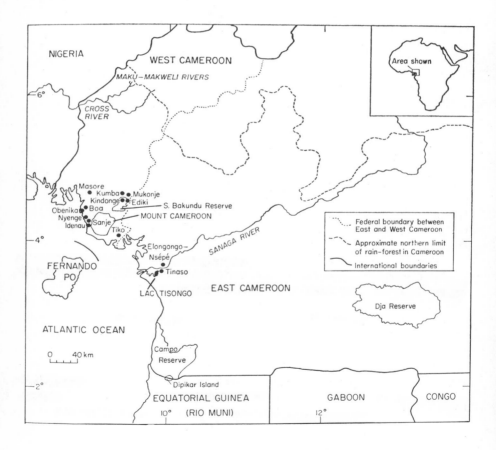

Fig. 1. Map of the study area, showing distribution of vegetation types.

METHODS AND STUDY AREAS

Methods of observation

Details of field techniques have been published elsewhere (Struhsaker, 1969, 1970; Gartlan, 1970). During the investigation J.S.G. spent a total of 2073 hours in the forest and was in visual contact with primates for 198 hours. T.T.S. spent 2803 hours in the forest with 1095 hours in contact with primates or within 30 yards of them.

RESULTS AND DISCUSSION

Polyspecific associations—their formation and disbandment

Many species of *Cercopithecus* in Cameroon are sympatric and, as has been observed elsewhere in Africa (Haddow, 1952; Booth, 1956; Jones & Sabater Pi, 1968; Gautier & Gautier-Hion, 1969; Aldrich-Blake, 1970), polyspecific associations are common. In this article two or more monkey species are said to be "in association" when members of the different species are in such proximity that members of one species are often interposed between conspecifics of the other. The term thus implies nothing about the duration of this proximity nor of its social or ecological implications.

In our studies the number of species in association ranged from two to six and on at least two occasions five species of *Cercopithecus* were seen together in one association; *C. mona, C. nictitans martini, C. erythrotis, C. p. pogonias* and *C. lhoesti.* The function of polyspecific associations is not clearly understood but feeding, foraging, resting and sleeping are the predominant activities occurring in them. Furthermore, although they are temporary, and vary in species composition from one area to another, certain combinations of species occur more frequently than others.

The impression was gained during field observations that polyspecific associations were frequently formed and disbanded. The following tentative generalizations stem from 50 clear-cut observations (T.T.S., 23; J.S.G., 27), 19 involving the formation and 31 the disbandment of a polyspecific association. Of those incidents observed by T.T.S., all but two occurred near Idenau, and all those observed by J.S.G. occurred in the Southern Bakundu Reserve.

Associations including *C. pogonias, C. erythrotis, C. nictitans* and *C. mona* were of longer duration than associations between any of them and *Pan troglodytes* or *Mandrillus leucophaeus.* No association including Chimpanzees was observed at Southern Bakundu and associations including Drills were both infrequent and transient. *C. nictitans* and *C. erythrotis* formed associations of longer duration and greater stability with each other than either of them did with *C. pogonias*. This was probably because *C. pogonias* groups had a much greater annual home range and a more extensive pattern of daily movement than did the other species. *C. lhoesti* (which did not occur at Southern Bakundu) associated less with other species at Idenau than did the other four *Cercopithecus* species in the area. No special behaviour was noted during the formation and disbandment of a polyspecific association, but many of the disbandments appeared to be the result of monkeys fleeing from the observer. As the detection of the observer by the monkeys generally resulted in this disruption of the association, our analysis of associations includes only those associations of completely known composition, excluding solitary animals and associations that either split or formed during observation; those which may, in other words, have been influenced by the presence of the observer.

Comparison with studies in Gabon. The study by Gautier & Gautier-Hion (1969) of polyspecific associations of cercopithecids in Gabon permits detailed comparison with our study as similar field techniques, methods of analysis and species are involved.

Stability of polyspecific associations. In the forest of Belinga, Gabon, it was found that interspecific associations were stable from day to day and from year to year. This was especially so with associations of *C. n. nictitans* and *C. cephus*. However, Gautier & Gautier-Hion also found daily changes in these associations at Belinga. They were most commonly formed in the early morning and late afternoon.

In contrast, our data clearly indicate the instability of polyspecific associations; they disbanded readily during the day and varied considerably with the seasons. It is possible, however, that the same monospecific groups did form polyspecific associations other than on the basis of chance alone. Definite proof of this is contingent upon individual recognition of several monkeys, which is sorely lacking from our study. One specific case does, however, at least indicate that among species showing seasonal variation in their interspecific associations the same monospecific groups may reunite in successive years. In May 1967 during the seasonal influx of *C. p. pogonias* into the major study area at Idenau a monkey with pelage colouration intermediate between *C. mona* and *C. p. pogonias* was seen in association with *C. nictitans*, *C. erythrotis* and *C. p. pogonias*. What appeared to be the same monkey was next seen one year later (April 1968) in the same area and again with *C. nictitans*, *C. erythrotis* and *C. p. pogonias* and again coincident with the seasonal influx of *C. p. pogonias*. However, this does not invalidate our conclusion that particular groups of one species form polyspecific associations with several other individual groups of any other given species. For example, *C. p. pogonias* at both Idenau and Southern Bakundu have been observed to move over an area encompassing most of the home range of two or three groups of *C. nictitans* within half a day. Furthermore, on two different occasions at

Idenau, one group of *C. p. pogonias* associated with two different groups of *C. nictitans* on the same day, with intervals between these associations of about 30 and 90 minutes.

Gautier & Gautier-Hion (1969) suggest that in the absence of hunting by man, polyspecific associations are stable, and that with increasing hunting pressure this stability breaks down. However, our data from Idenau do not support this suggestion. Hunting by humans did not occur during this study and yet the polyspecific associations were unstable.

Role of habitat on the composition and frequency of polyspecific associations. As in our study, Gautier & Gautier-Hion (1969) found that differences in habitat can have profound effects on polyspecific associations. For example, in the riverine forest at Ndjaddié, Gabon, they found no significant interspecific affinities.

Formation and disbandment. No specific behaviour patterns were employed in the formation or disbandment of polyspecific associations either in Cameroon or Gabon.

Relative tendencies of species to form polyspecific associations. In both studies most of the cercopithecids were observed in polyspecific associations significantly more often than in monospecific groups. Analysis of the aggregated data from Gabon showed two notable exceptions (Gautier & Gautier-Hion, 1969). *Cercopithecus neglectus* never associated with other species and *Miopithecus talapoin* did so only about half the time. A similar analysis of aggregated data from four samples in Cameroon also revealed two exceptional species; *Cercopithecus lhoesti* occurred more frequently in monospecific groups than with other species. *Cercopithecus mona* associated with other species but not with significant frequency with single species. At Southern Bakundu, however, although not forming significant bispecific associations with either *C. nictitans* or *C. erythrotis,* it showed a strong tendency to associate with both species simultaneously.

Social behaviour between species of Cercopithecus. No direct evidence of interspecific copulation was obtained, but at Idenau at least two monkeys with pelage and patterning intermediate between *C. mona* and *C. p. pogonias* were seen on several occasions and which may have been true hybrids (Struhsaker, 1970). Apart from responding to one another's alarm calls and forming mixed associations, only ten cases were observed of social interaction between *Cercopithecus* species.

Other interspecific relations

Predators. In Cameroon the most serious predator of anthropoid primates is man. Humans shoot and trap all the species mentioned in this article for food. In some localities certain species have undoubtedly been exterminated by this activity, e.g. the red colobus, *Colobus badius preussi* (Matschie) from the vicinity of Kumba.

The Crowned hawk-eagle *Stephanoaëtus coronatus* (Linnaeus) is probably the second most important predator of *Cercopithecus* species. At Southern Bakundu a *S. coronatus* was observed to catch and partly eat an adult female *C. nictitans martini*. During the attack the adult male of the *C. n. martini* group stood at the top of the canopy and gave high-intensity alarm barks in the direction of the predator, which was about 50 yards (46 m) distant; other group members hid. The barks continued until after the predator flew away. In another interaction with this species at Idenau both *C. nictitans* and *C. pogonias* gave alarm calls towards the bird whilst it was perched. The intensity of this calling appeared to increase as the eagle flew off, but terminated when it was out of view.

Ecological niche separation

The abundance and variety of sympatric primate species in the Cameroon forests provide an excellent opportunity for the study of niche separation and the applicability of Gause's (1934) concept.

Conclusions on niche separation. Ecological niche separation among the rain-forest anthropoids of Cameroon, especially in regard to food, is apparently effected in several different ways. Differences in body size and feeding apparatus probably separate *Mandrillus leucophaeus, Cercocebus* and *Pan* from the much smaller *Cercopithecus* species. But among the latter a unique example of congeneric sympatry exists. Five *Cercopithecus* species (*C. mona, C. lhoesti, C. nictitans, C. erythrotis* and *C. pogonias*) were found in the same forests, and on rare occasions all five banded together in the same temporary association. Because they are of similar body size, possess virtually the same anatomical devices for feeding, and differ primarily in pelage colour and pattern, the ecological segregation of the five species in regard to their food supply poses an especially interesting problem.

C. mona is most successful in mangrove swamp and *C. lhoesti* in montane forests, so that habitat preferences would appear to separate the bulk of their populations from one another and from the other three *Cercopithecus* species, which typically inhabit old secondary and mature lowland rain-forest. Vertical stratification in rain-forest helps to reduce competition for food between *C. erythrotis* and *C. mona* on the one hand, and *C. pogonias* and *C. nictitans* on the other, for the former pair tend to feed at lower levels than do the latter. Between *C. nictitans* and *C. pogonias* which feed at similar heights, competition for food is reduced by the much greater seasonal movements of *C. pogonias,* which bring the two species together only when there is a superabundance of food. Groups of *C. pogonias* have been observed to associate successively with different groups of *C. nictitans* on the same day, moving across large portions of the home range of the *C. nictitans* groups as they did so. *C. nictitans* are much more local and have considerably smaller home ranges than do *C. pogonias*.

Significance of polyspecific associations. In contrast to *Mandrillus leucophaeus*, *Cercocebus* spp. and *Miopithecus talapoin*, which live in large heterosexual groups containing more than one adult male (Struhsaker, 1969; Gartlan, 1970), all rain-forest *Cercopithecus* species so far studied show a strong tendency to live in small groups which contain a single adult male. Solitary adult and subadult males are common. It is not known whether different selective pressures have affected the evolution of *Cercopithecus* species or whether they have responded differently to similar evolutionary pressures, but the formation of polyspecific associations effectively serves to increase the size of their groups. The observations that some species-pairs form associations more often than others implies that they have some adaptive advantage and are more than chance phenomena. One advantage may be that when the size of a group is increased by association its ability to locate food is also augmented, without increasing intraspecific competition for food or copulation. This probably holds true with the associations between *C. cephus* and *C. n. nictitans* in Rio Muni and Gabon (Gautier & Gautier-Hion, 1969) and between *C. nictitans martini* and *C. erythrotis camerunensis* in Cameroon. Food competition between these pairs is minimized because *C. nictitans* feeds at higher levels than do *C. erythrotis* and *C. cephus*.

Our data do not readily suggest that polyspecific associations provide an advantage in food locating because the largest and most frequent associations were seen during the season when there was a superabundance of food. However, the situation may be analogous to that of migrating passerine birds from temperate regions which commonly form mixed flocks with neo-tropical resident passerines. It is presumed that the migrants are living on food supplies located by the resident supplies. A similar relationship may exist between *C. pogonias* which ranges widely throughout the year, and *C. nictitans martini* and *C. erythrotis,* which are much more local. The *C. pogonias* may make use of the food located by the resident groups of *C. nictitans* and *C. erythrotis.*

Some associations are probably the result of the breakdown of ecological barriers which previously separated two or more species. The association between *C. mona* and other *Cercopithecus* species would appear to be of this type. Data on the abundance of different *Cercopithecus* species in different habitats demonstrate that *C. mona* is most successful in mangrove swamp, less successful in secondary vegetation and least successful in mature lowland forest. The human agricultural invasion of African rain-forests is of relatively recent origin, beginning about 1 500 years ago (Murdock, 1959). Before that time secondary forests were probably restricted to river edges, to the interfaces of rain-forests with other kinds of habitat, and to small wind-fallen areas within the forest. Contact and hence association between *C. mona* and other *Cercopithecus* species would have been much less common than under present day conditions and would have been confined to the secondary vegetation along rivers and at the edge of mature forest. *C. mona* now becomes involved in more complex associations such as those seen at Southern Bakundu. There the species was rarely found in monospecific groups, sometimes in bispecific groups, but most frequently was seen in polyspecific associations with both *C. nictitans* and *C. erythrotis* simultaneously.

Because they are similar in body size, it is presumed that sympatric forest *Cercopithecus* species have similar predators. Indeed, all the species observed in this study reacted violently towards the most common predators, namely man and the Crowned hawk-eagle, *Stephanoaëtus coronatus*.

Consistent polyspecific associations, such as the common one between *C. nictitans* and *C. erythrotis* increase the number of monkeys present and, because the species characteristically move and feed at different heights, increase their vertical range in the forest. The chances of locating a predator before it attacks are thus improved (see also Gautier & Gautier-Hion, 1969) while the probability that any particular monkey will be attacked is reduced.

Despite these apparent advantages many polyspecific associations appear to be chance phenomena which do not result from natural selection. This is supported by the analysis of quantitative data using Fager's index of affinity, which showed that most pair combinations did not occur with significant frequency. Extreme examples of such insignificant associations are those formed by *Mandrillus leucophaeus* when passing through a group of *Cercopithecus* species or by *Cercopithecus* monkeys passing through a group of feeding *Pan troglodytes*.

These studies were made possible in Cameroon by kind permission of M. le Directeur des Eaux et Forêts et des Chasses, Yaoundé, and of Mr S. C. Tamanjong, Director of Forestry Services, Buea.

We acknowledge with thanks much help, technical assistance, information, facilities and hospitality from Dr. B. O. L. Duke (Director), Helminthiasis Research Unit, Kumba; T. S. Jones, Esq., former General Manager, Cameroon Development Corporation, Bota; the staff of the West Cameroon Forestry Department; Dr B. E. Leake, Department of Geology, University of Bristol; Dr F. N. Hepper and the staff of the Kew Herbarium. Dr B. O. L. Duke, Mr R. H. L. Disney and Mr J. F. Oates are thanked for comments on the manuscript.

The project would have accomplished much less in the time available without the tireless and cheerful assistance of our field assistant, Mr Ferdinand Namata.

Several drafts of this manuscripts were typed and corrected by Sue Gartlan.

The studies were financed by N. S. F. grants GB 5792 and GB 8476 and sponsored by the New York Zoological Society and the Rockefeller University, New York.

During the studies J. S. G. was on leave of absence from the Medical Research Council, London.

REFERENCES

Aldrich-Blake, F. P. G. (1970). *The ecology and behaviour of the blue monkey, Cercopithecus mitis stuhlmanni.* Ph.D. thesis. University of Bristol.

Booth, A. H. (1956). The distribution of primates in the Gold Coast. *Jl W. Afr. Sci. Ass.* 2:122-133.

Gartlan, J. S. (1970). Preliminary notes on the ecology and behaviour of the drill, *Mandrillus leucophaeus* Ritgen 1824. In *The old world monkeys:* 445-480. Napier, J. R. and Napier P. H. (Eds.) New York: Academic Press.

Gause, G. F. (1934). *The struggle for existence.* Baltimore: Williams & Wilkins Co.

Gautier, J. P. & Gautier-Hion, A. (1969). Les associations polyspecifiques chez les Cercopithecidae du Gabon. *Terre Vie* 116: 164-201.

Haddow, A. J. (1952). Field and laboratory studies on an African monkey, *Cercopithecus ascanius schmidti* Matschie. *Proc. zool. Soc. Lond.* 122: 297-394.

Jones, C. (1970). Stomach contents and gastro-intestinal relationships of monkeys collected in Rio Muni, West Africa. *Mammalia* 34: 107-117.

Jones, C. & Sabater Pi, J. (1968). Comparative ecology of *Cercocebus albigena* (Gray) and *Cercocebus torquatus* (Kerr) in Rio Muni, West Africa. *Folia primatol.* 9: 99-113.

Malbrant, R. & Maclatchy, A. (1949). *Faune de l'equateur africane français. 2. Mammifères.* Paris: Paul Lechevalier.

Murdock, G. P. (1959). *Africa, its peoples and their cultural history.* New York: McGraw-Hill Book Co., Inc.

Rahm, U. (1961). Esquisses mammalogiques de basse Côte d'Ivoire. *Bull. Inst. fr. Afr. noire.* 23A: 1229-1265.

Struhsaker, T. T. (1969). Correlates of ecology and social organization among African Cercopithecines. *Folia primatol.* 11: 80-118.

Struhsaker, T. T. (1970). Phylogenetic implications of some vocalizations of *Cercopithecus* monkeys. In *The old work monkeys:* 365-444. Napier, J. R. & Napier, P. H. (Eds.) New York: Academic Press.

9

Comparison of the Behavior and Ecology of Red Colobus and Black-and-White Colobus Monkeys in Uganda: A Summary

1975

THOMAS T. STRUHSAKER AND JOHN F. OATES

INTRODUCTION

The leaf-eating monkeys of the Old World constitute the subfamily Colobinae of the family Cercopithecidae. The majority of colobines occur in India and Southeast Asia; only three occur in Africa: the olive, black-and-white, and red colobus. The alpha taxonomy of the African Colobinae is in a confused state. The various forms of red colobus are placed either in *Procolobus*, *Piliocolobus*, or *Colobus*. The olive colobus is usually placed in *Procolobus*

Reprinted by permission of Mouton Publishers from "Comparison of the Behaviour and Ecology of Red Colobus and Black-and-White Colobus Monkeys in Uganda: A Summary" by T.T. Struhsaker and J. Oates in *Socioecology and Psychology of Primates* (R.H. Tuttle, ed.). Pages 103-123. World Anthropology. Mouton, The Hague. 1975.

We wish to thank the National Research Council and Mr. M. L. S. B. Rukuba, Chief Conservator of Forests, Forest Department of Uganda, for permission to study in Uganda and in the Kibale Forest. The staff of the Uganda Forest Department at Fort Portal and the Kanyawara Station were extremely helpful throughout our studies. Plant material was identified for us by Mr. A. B. Katende, assistant curator of the herbarium at Makerere University of Kampala, and Dr. Alan Hamilton, formerly of the Department of Botany, Makerere. We are grateful for many fruitful discussions on the subject of this article with Drs. Tim Clutton-Brock, Peter Marler, and Steven Green. Financial support was provided by the New York Zoological Society, Rockefeller University, United States National Science Foundation grant number GB 15147, and United States National Institutes of Mental Health grant number 1 R01 MH 23008-01, for which we are grateful.

and the black-and-white colobus in *Colobus* (Verheyen 1962; Dandelot 1968; Kuhn 1967, 1972; Rahm 1970; Groves 1970). The olive colobus is monotypic, whereas there are fourteen recognized forms of red colobus placed in one or six species, and as many as sixteen to twenty-two forms of black-and-white colobus included in two to four species (Napier and Napier 1967; Dandelot 1968; Rahm 1970).

This report is based primarily on two studies in the Kibale Forest of western Uganda, East Africa. Both studies were conducted in compartment 30 of the Kibale Forest, which is located adjacent to the Kanyawara Forest Station (0° 34' North, 30° 21' East, elevation 1,530 meters). The study area was comprised of mature rain forest, transitional in form between montane and lowland rain forest. The data considered here come from Struhsaker's study of the red colobus (*Colobs badius tephrosceles*) from May, 1970, through March, 1972, and Oates's study of the black-and-white colobus (*Colobus guereza occidentalis*) from October, 1970, through March, 1972. The emphasis of both studies was behavioral and ecological. Each study concentrated on one social group of the respective species.

This article concentrates on the differences and similarities in the social behavior and ecology of these two species and offers hypotheses to explain them. The reader is referred to our separate and detailed reports (Struhsaker i.p.; Oates 1974) for the basic data and information on methodology.

ECOLOGICAL COMPARISON OF THE TWO HOME RANGES

Although both studies were conducted in the same forest block, the home range of the main study group of red colobus (CW group) did not overlap the range of the main guereza group (group 4). The areas most heavily used by the two groups were separated by 900 to 950 meters. However, the northeast part of CW group's range was almost contiguous with the southwest corner of group 4's range.*

Botanically, the two home ranges were very similar. Strip enumerations of all trees that were at least ten meters tall were made in the home ranges of the two groups.† Based on these samples, the density of trees of this size class was estimated at 328 per hectare in the range of the badius group and 274 per hectare in the guereza home range. Much, if not all, of this differ-

---

*On the map grid system, which is explained in our detailed studies, the northeasternmost quadrat occupied by CW group was +4 +1 and the southwesternmost quadrat used by group 4 was +5 −1. This means that the extreme ends of the ranges of the two groups came within less than 100 meters of each other. The two quadrats most heavily used by the badius group were −2 −13 and −5 −13; the most heavily used quadrat of the guereza group was +11 +1.

†All trees within 2.5 meters of the enumeration trail were tallied. A total area of 1.43 hectares was enumerated by this method in the range of the badius group and 0.81 hectares in the home range of the guereza group.

ence in tree density is explained by the fact that 41 percent of the area enumerated within the guereza range had been selectively timbered in 1969. The guereza group, however, used this area little compared with the undisturbed mature forest, where the tree density was higher. Based on the same samples, the estimated number of tree species per unit area in the badius home range was lower (36 per hectare) than in the guereza range (47 per hectare). This difference is probably attributable to the difference in the size of the area sampled in the two ranges. The number of new species tallied in enumeration increases very rapidly in the initial several hundred meters, but as more area is covered, the number of new species encountered increases at a much lower rate. Consequently, enumerations covering relatively small areas have proportionally more species than do enumerations covering larger areas.

The tree-species composition of the two ranges was very similar. Only the ten most common species were considered, because these ten species comprised the majority of trees enumerated; 79.4 percent in the badius home range and 80.3 percent in the guereza range. Comparing the densities of the ten most common tree species in the two ranges revealed no significant difference and confirmed our impression that the two areas were very similar botanically (Table 1).*

In contrast to the general similarity in vegetation in the two areas, there was at least one pronounced difference in the primate faunas. There were definitely more groups of guereza in the vicinity of the guereza study area than in the badius study area (Table 2). The reason(s) for this pronounced difference in guereza density is not known. However, one obvious implication of this difference is that in the guereza study area, with its higher density of guereza groups, the probability of intergroup encounters is much higher than for groups of guereza living in the vicinity of the main study group of badius. In addition, there may also have been slightly fewer badius groups in the guereza area than in the badius range, but the reality of this difference is less certain. The other differences is relative abundance of primate species in the two areas are probably not significant (Table 2).

COMPARISON OF SOCIAL STRUCTURE AND SOCIAL BEHAVIOR

Group Size and Composition

The badius of the Kibale Forest lived in considerably larger social groups than did the guereza. Badius groups numbered about 19 to 80, with a central

---

*Wilcoxon's signed-ranks test for two groups arranged as paired observations (Sokal and Rohlf 1969) was used to compare the estimated densities in Table 1. Comparing the densities of the ten most common trees in the badius range with their densities in the guereza range revealed no significant difference (p = 0.084, two-tailed). Similarly, comparison of densities of the ten most common trees in the guereza range with their densities in the badius range revealed no difference (p > 0.10, two-tailed).

Table 1
Comparison of densities of ten most common tree species ($\geq$ 10 meters tall) in two study areas at Kanyawara, Kibale Forest, Uganda

Density (number/hectare) of ten most common trees in badius study area and density of same species in guereza study area

Species	Badius area	Guereza area
Diospyros abyssinica	65.7	27.2
Markhamia platycalyx	58.0	56.8
Celtis durandii	34.3	49.4
Uvariopsis congensis	25.2	19.7
Teclea nobilis	21.0	7.4
Funtumia latifolia	14.7	11.1
Strombosia scheffleri	14.7	11.1
Parinari excelsa	10.5	0
Chaetacme aristata	8.4	9.9
Millettia dura	7.6	0

Density (number/hectare) of ten most common trees in guereza area and density of same species in badius area

Species	Guereza area	Badius area
Markhamia platycalyx	56.8	58.0
Celtis durandii	49.4	34.3
Diospyros abyssinica	27.2	65.7
Uvariopsis congensis	19.7	25.2
Bosqueia phoberus	18.5	2.8
Strombosia scheffleri	11.1	14.7
Funtumia latifolia	11.1	14.7
Chaetacme aristata	9.9	8.4
Celtis africana	8.6	2.1
Teclea nobilis	7.4	21.0

tendency of approximately 50, whereas guereza groups ranged in size from 8 to 15, averaging 10.5. The main study group of badius averaged 20 in number and that of guereza 12. There was always more than one fully adult male in each badius social group, with the ratio of adult females to adult males varying from about 1.5 to 3.0. In marked contrast, there was usually only one fully adult male and three to four adult females in groups of guereza. The adult membership of social groups was relatively stable for both species. Solitary subadult and adult males of both species were quite common.

Table 2

Relative abundance of primate species in two study areas at Kanyawara, Kibae Forest, Uganda
$\bar{x}$ number of groups per census[a]

Species	Guereza area	Badius area
Colobus badius	2.90	3.80
Colobus guereza	3.30	0.84
Cercopithecus ascanius	2.90	2.77
Cercopithecus mitis	1.70	1.80
Cercopithecus lhoesti	0.10	0.32
Cercocebus albigena	0.30	0.39
Pan troglodytes	0	0.14

[a] Fifteen censuses were made by Oates in the guereza area along a census route 4,000 meters long. Forty-four censuses were made by Struhsaker in the badius area along a census route 4,020 meters long.

Inter-individual Spacing

In an attempt to describe the spacing patterns between members of a social group, a series of samples was taken in which the distance between a specific individual monkey and its neighbors was estimated. Results from 155 such samples on seven individual badius have been compared with the results from 170 samples on ten individual guereza. This comparison clearly shows how much more cohesive were guereza groups than badius groups. For example, the mean number of neighbors within 2.5 meters of all ten guereza sampled was 2.8. In contrast, there was an average of only 0.86 neighbors within 2.5 meters of the seven badius sampled. These differences are highly significant ($0.01 > p > 0.002$, Wilcoxon two-sample test for the unpaired case). This pattern persisted even when the comparison was broken into specific age-sex classes. For example, the adult male guereza had an average of 2.6 members of his group within 2.5 meters of him, and the three adult males of the badius group had averages of 0.32 to 0.68 members of the group within 2.5 meters. The differences between the females of the two species were not of the same order of magnitude, but still the guereza adult females averaged 1.6 to 3.7 more neighbors within 2.5 meters of them than did the badius adult females, whether they were in possession of a newborn infant or not. Thus, the members of the guereza group were more closely spaced than were the members of the badius group.

Grooming Relations

In both species the adult females performed the majority of grooming. Adult female guereza not only groomed more than would be expected on the basis of their proportional representation in the group membership, but also received more grooming than would be expected by chance. This differs from the badius females who, although they groomed more than would be expected, received about as much grooming as would be expected by chance. Adult male badius were groomed more than would be expected by chance. The single adult male guereza in the main study group received about as much grooming as would be expected by chance, but he groomed others very rarely. Juvenile badius appeared to perform more grooming than did juvenile guereza, although part of this difference may be attributable to possible differences in aging criteria in the two studies.

Social Relations of Newborn Infants

Newborn infant badius were neither handled, groomed, nor carried by other monkeys in the group. Juveniles and young juveniles often approached the mother and infant and tried to touch the neonate, but the mother always responded with threat gestures. Infant badius first made contact with monkeys other than their own mothers when they were about 1 to 3.5 months old. In sharp contrast to this, neonate guereza were often carried and handled by other females in the group. This attention was given primarily by adult females and to a lesser extent by immature monkeys. Newborn guereza were most attractive to other group members during the first month after birth.

Relations among Adult Males

The adult males in the main study group of badius formed a kind of subgroup. They characteristically maintained close spatial proximity and stable, long-term membership in the group, and they displayed cohesive and united effort in aggressive episodes with other badius groups. In addition, there was a stable and linear dominance hierarchy among them, which was defined as priority of access to space, food, and grooming position. Correlated with this dominance hierarchy were fixed roles in a stylized encounter called the present type II. Dominant males gave the present type II to their subordinate males. This display seemed to reinforce their dominance relations and may have also strengthened the ties of their subgroup and at times reduced agonistic tension. The manner in which membership in this subgroup was achieved is not clearly understood. However, it seemed to involve a slow familiarization between young males and adult males. Initially, the young males harassed the adult males, but as they grew older, the roles reversed, with adults harassing the old juvenile and subadult males. Eventually, it would seem, the maturing male either attains membership in the subgroup or leaves the group altogether.

No such subgroup was observed among the one-male social groups of guereza, and it appears that maturing males either leave or are forced out of the group before reaching full physical maturity. It also seems possible that a young and developing male may, on occasion, force the harem male out of the group.

Mating Systems

Neither badius nor guereza is panmictic. Assortative mating is, however, achieved in different manners and to different extents in the two species. The single adult male in the guereza study group performed all heterosexual mounts. In the main study group of badius, there was a good correlation between dominance and copulatory success. The dominant male performed at least 84 percent of all heterosexual mounts during the first part of the study. The other two adult males copulated very little. During the second half of the study, a young male attained sexual and physical maturity and eventually became the dominant male in the group. In this transitional period, the previously dominant male performed only 42.7 percent of the mounting, the remainder being divided unequally among the other three males (30.0, 19.1, and 5.5 percent; in 2.7 percent cases the male was not properly seen and therefore not identified). Clearly, badius males reproduce differentially, but apparently not to the extent of guereza males. However, one should not exclude the possibility that the degree of assortative mating in guereza is, in fact, very nearly the same as that in badius, because guereza live in smaller groups, so that each harem male may serve fewer females than do some badius males. In a hypothetical case, single guereza males in four different harems containing five, four, three, or two females respectively would be equivalent to one badius group containing fourteen females in which the mating was assorted among four males on a 35.7, 28.6, 21.4, and 14.3 percent basis.

Intergroup Relations

The intergroup relations of these two species are quite different. The main study group of badius had very extensive, if not complete, overlap in its home range with two other badius groups. The relations of these three groups were usually aggressive, involved only the subadult and adult males, and seemed to be based on an intergroup dominance hierarchy, which was independent of spatial parameters. No matter where any two of the three groups met in their home ranges, one usually supplanted the other. Direction of successful supplantation varied enough, however, to suggest that in some cases the outcome depended on which particular males of the two groups happened to meet. These three groups used an area of approximately 50 hectares* to the virtual exclusion of all other badius groups. The four or five other badius groups who entered this 50-hectare area did so very infrequently and were usually chased out immediately.

Guereza groups had preferred areas in their home ranges which they defended against, and from which they could readily chase, other guereza groups. However, these preferred areas were not used to the exclusion of other groups. The element of exclusive use is often considered to be the prime factor defining territoriality (Pitelka 1959)—on this basis, the guereza groups in the Kibale Forest do not have territories. One might think of their intergroup relations as being based on dominance relationships which are dependent on spatial parameters, i.e. the boundaries of the preferred areas of each group. However, Burt's (1943) definition of territoriality, which emphasizes the defense of a specific area, would apply to the guereza situation.

COMPARISON OF ECOLOGY

Home Range

The main study group of badius had a home range of about 35 hectares, approximately twice that of the guereza group (15 to 16 hectares). However, consideration of the differences in size and biomass between the two groups reveals that the biomass density was about the same, i.e. 118 kilograms/35.3 hectares for the badius group (3.34 kilograms/hectare) and 61 kilograms/15.5 hectares for the guereza group (3.94 kilograms/hectare). Nonetheless, the overall biomass density of badius was considerably greater than that of the guereza because of the much more extensive overlap in home ranges of the badius groups and the much larger size of the typical badius group. In the vicinity of the main study group of badius, the biomass density of badius was estimated at about 1,760 kilograms/square kilometer and that of guereza at only 64.7 kilograms/square kilometer. Even in areas of greater density, the estimated guereza biomass density did not exceed 570 kilograms/square kilometer.

Daily Ranging Patterns

The distance traveled by both study groups was extremely variable, but in general the badius group traveled further each day than did the guereza group. During fifty-four days in which the badius group was followed continuously from sunrise to sunset, it traveled an average of 648.9 meters/day, but this daily distance varied from 222.5 to 1,185 meters. In sixty days the

*It is unlikely that all 50 hectares were used by these three groups, because some were unsuitable habitat for badius, but it is also unlikely that any other badius groups used these areas.

guereza group moved an average of 535.1 meters/day, varying from 228 to 1,004 meters. A t-test comparing the day ranges of the two species showed that on the average the badius group moved significantly further each day than did the guereza group (t = 3.12, df = 112, 0.01 > p > 0.001).

Distance traveled per day is only one way of viewing ranging patterns and certainly tells little of how the animals utilize their home range or how they distribute their time in space. Our most detailed analysis of ranging patterns consisted of superimposing a grid system over the daily range maps and then plotting the amount of time the groups spent in specific 0.25-hectare quadrats of the grid system. This analysis gives us several measures of ranging patterns.

Our monthly samples were collected during five continuous days of observation. During these five-day periods the badius group used an average of 45.1 quadrats per five-day sample, varying from 27 to 65 (N = 7). The guereza group tended to use fewer quadrats during the five-day sample, averaging 29.8 quadrats per sample and varying from 14 to 44 (N = 12).

A more refined analysis of ranging patterns measured the dispersion of time in space. For this analysis, we used the Shannon-Wiener information measure (Wilson and Bossert 1971) to compute a diversity index that reflected the diversity in ranging patterns. In months when the group distributed its time evenly among quadrats, its index of ranging diversity was high. Months when the group spent the majority of its time in a few quadrats gave a low index of ranging diversity. Again, it is apparent that the badius group generally had a more diverse pattern of ranging than did the guereza group. The monthly index of ranging diversity averaged 3.240 over seventeen months (range 2.742 to 3.779) for the badius group and 2.719 over twelve months (range 2.098 to 3.174) for the guereza group.

On both a daily and five-day basis, the badius group had a more diversified pattern of ranging than did the guereza group, as indicated by daily travel distance, number of quadrats used per month, and monthly indices of quadrat utilization diversity. These differences are almost certainly related to the differences in the number of monkeys and biomass in the two groups. For example, although the guereza group used only 66.6 percent as many quadrats during the five-day sample as did the badius group, their group size was only 55 to 60 percent and their biomass 51 percent that of the badius group.

One final analysis concerns the annual ranging pattern. We have already shown that the badius group had a larger home range than the guereza group, but we have not yet considered distribution of time in these ranges. The Shannon-Wiener information measure was again used, but in this case the computation was made on all the ranging data for a twelve-month period. In contrast to the preceding analysis, the indices of quadrat utilization diversity for the two species were very similar: badius 5.025 and guereza 5.465. Thus, although the badius group had a larger home range and a more diverse monthly ranging pattern, their proportional distribution of time in space during a twelve-month period was much like that of the guereza group.

Time Budgets

There were very pronounced differences between the two species in the allotment of time to various activities. The badius spent considerably more of their time feeding (44.5 percent) and moving (9.2 percent) and less of their time resting (34.8 percent) than did the guereza (19.9, 5.4, and 57.3 percent respectively). These differences in time budgets could be related to dietary differences. For example, the more diversified diet of the badius may require that species to feed and move more than the guereza, with its more monotonous diet. The nature of these different foods may also involve different rates of digestion and may thereby necessitate different amounts of time devoted to resting, presumably when much of the digestion occurs. These data also suggest that the two species may have some very basic metabolic differences which are related to the energy of their respective diets, but in the absence of physiological data, further speculation seems unwarranted. *because more diverse diet*

Vertical Stratification

Although the heights preferred by the two species in the forest are not strikingly different, several thousand height estimates of the monkeys tabulated at half-hourly intervals suggest that the guereza prefer slightly lower heights than do the badius. In terms of ecological niche separation, feeding heights are probably most important. Of the feeding observations for badius, 25 percent occurred from 3 to 13.5 meters above the ground, 50 percent between 13.5 and 25.5 meters, and 25 percent from 25.5 to 37.5 meters or higher. In contrast, 50.4 percent of the feeding observations for guereza in the same forest compartment occurred below 13.5 meters, only 38.2 percent between 13.5 and 25.5 meters, and 11.4 percent above 25.5 meters. These differences may contribute to niche separation between the two species, but they are relatively insignificant when compared with dietary differences.

Diets

Although both species are herbivorous and justly qualify as leaf eaters, their diets differ in many ways.* The guereza diet was found to be much more monotonous than that of badius. Indices of food-species diversity were

*Data on monkey food items were collected by recording frequency, rather than duration, of feeding.

computed each month for the two species using the Shannon-Wiener information measure. In seventeen months this index for badius averaged 2.606 and varied from 1.973 to 3.051. In fourteen months the same index for guereza averaged only 1.720, ranging from 1.205 to 2.152. In only two months was the badius index as low as the highest guereza index. The monotony of the guereza diet was due primarily to its heavy concentration on one common tree species, *Celtis durandii.* This species provided the major source of food for guereza in all fourteen months, sometimes comprising as much as 68.5 percent of the monthly diet.

Both species fed heavily on young leaves, but the blades of mature leaves played a much more important role in the diet of guereza than in badius. In one month mature leaf blades were the most important dietary item for guereza, constituting 32 percent of the food eaten. The badius, in contrast, rarely ate the blades of mature leaves. During a twelve-month period only 2.3 percent of the badius diet contained mature leaf blades. The badius did, however, regularly eat the petioles of mature leaves and in the same twelve-month period this item constituted 18.5 percent of their diet. Guereza, by contrast, rarely ate leaf petioles. Fruits were more important in the guereza diet than in that of the badius, and they were among the five most important food items in thirteen of the fourteen months. In one month fruits comprised as much as 34.2 percent of the guereza diet. On the other hand, fruit accounted for only 4.8 percent of the badius annual diet.

A more precise way of comparing the diets of these two species is to measure the percentage overlap in the specific food items eaten each month. In this analysis we employed the same methods as those used by Holmes and Pitelka (1968). In any given month, the percentage overlap in diet is the sum of the shared percentages of specific food items common to the two species in that month. For example, if in January, 1972, the badius diet contained 50 percent leaf buds of species A, 25 percent mature leaf petioles of species B, 25 percent floral buds of species C, and none of species D, and the guereza diet contained 25 percent leaf buds of species A, none of species B, 25 percent mature leaves of species C, and 50 percent young leaves of species D, then the percentage overlap in the diets of guereza and badius for this month would be 25 percent (leaf buds of species A). Percentages of overlap in diets were computed for the last twelve months of the study (April, 1971, to March, 1972; Table 3). Dietary overlap between badius and guereza was generally very low. The mean monthly overlap was 7.1 percent with a range of 2.0 to 15.7 percent. This is much lower than the inter-monthly overlap in diet of badius during seventeen months (136 monthly pair combinations): the mean overlap in diet between months was 24.3 percent with a range of 9.3 to 50.0 percent. In only 13 of these 136 monthly combinations was the overlap in the badius diet lower than the maximum overlap in the badius-guereza comparison, i.e. 15.7 percent. None of the intermonthly comparisons for badius had overlaps in diet that were as low as the mean overlap in the badius-guereza diet.

Table 3
Percentage overlap in diet (specific food items) of *Colobus badius* and *Colobus guereza*

Month	Percent overlap
April, 1971	7.29
May	3.64
June	4.70
July	3.77
August	4.54
September	15.68
October	12.71
November	2.00
December	13.25
January, 1972	3.36
February	9.81
March	4.30

We conclude from this that overlap in food habits was very slight between these two species. Furthermore, the little overlap that did occur involved very common species and common food items. In the four months when diet overlap was greatest, most of this overlap was accounted for by *Celtis durandii* (7.2 percent in September, 8.9 percent in October, 10.0 percent in December, and 3.0 percent in February). In all of these cases, the young leaves, leaf buds, and/or fruits of this species accounted for most of the overlap. All of these items were extremely abundant. It will be recalled from Table 1 that *Celtis durandii* was one of the most common trees in the forest and as such probably does not constitute a food source that limits the population of either colobus species. Consequently, we conclude that the slight overlap in diet between these two species did not constitute competition for food. Additional support for this conclusion was obtained by observing the food habits of badius living within the home range of the main study group of guereza. Their food habits did not differ substantially from those of the main study group of badius.

DISCUSSION

How can we account for these interspecific differences from an evolutionary standpoint? There are at least three obvious possibilities: (1) the social structures may be specialized adaptations to the Kibale habitat; (2) the social structures originally may have been adaptations to other habitats which nevertheless have permitted a certain amount of success in Kibale; and (3) the third possibility concerns the phylogenetic proximity of the two species:

just how closely are they related? Obviously, these three possibilities are not mutually exclusive.

Adaptation to Kibale-like Habitat

In the first hypothesis we suggest that the small and cohesive social groups of guereza which defend fixed areas ("territories") against neighboring groups are adapted to the efficient exploitation of a monotonous diet whose chief component (*Celtis durandii*), has a high density, a high cover index (a function of crown size and tree density), and a uniform dispersion (Struhsaker i.p.). In addition, the phenology of this species provides an abundance of preferred guereza food at most times of the year. The concentration of guereza on a food species of this nature permits partitioning of the habitat into territories and, in combination with the small group size, also permits relatively restricted ranging patterns on a daily and monthly basis. An important feature of guereza social behavior which probably contributes to both stability in group membership and group defense of territories is the attention given to newborn infants by other members of the group. Such attention, we suggest, is important to the rapid and complete integration of neonates into the social group. The close spatial cohesion between members of guereza groups is probably related to their emphasis on a monotonous diet and a high-density food source (in terms of both tree density and item density on a given tree) and not to their territoriality, because territoriality in other primate species, e.g. vervets (*Cercopithecus aethiops* [see Struhsaker 1967]), is not necessarily accompanied by close inter-individual spacing within the group. Differential mating is achieved in guereza by the one-male social groups, from which other adult males are excluded. The ecological significance of this exclusion, if any, is not apparent, particularly because the extra-group males utilize the same food resources in the same areas as do the social groups. The significance of this exclusion of all but one adult male from social groups is more probably involved with outbreeding, gene flow, and assortative mating, rather than with feeding ecology *per se*.

The ecological adaptation of the badius social structure is less apparent. There is no obvious reason for having large, multi-male groups in order to exploit a large home range and a great diversity of foods. Although it is tempting to suggest that a greater diet diversity indicates a greater diversity in ranging pattern, the data for the main badius study group do not support this suggestion (Struhsaker 1974). Nor was any correlation found between the pattern of food dispersion and ranging diversity. In fact, the badius ranging-pattern diversity was more closely related to the frequency of intergroup conflicts and intergroup proximity (Struhsaker 1974). Greater interindividual distance between members of a badius group might, however, be related to their more diversified diet. By feeding on a greater variety of foods, the badius encounter a greater variety of food dispersion patterns, many of which are not densely spaced. The absence of group territoriality in badius may be related to their great dietary diversity, which in turn may dictate a home range larger than can be efficiently defended in a rain forest habitat. The extensive overlap in the home ranges of three groups may permit each group to use a larger area than would be possible if, given the same group density, they defended territories. Their intergroup dominance relations effectively determine which group has priority of access to any food source, but the distribution of food sources is such that the supplanted group never appears to have difficulty in immediately locating another ample food source. The fact that these three groups appear to defend and exclude other groups from an area of about 50 hectares may be viewed as a kind of territoriality and is probably related to food allocation.

Several of our results and conclusions have been substantiated by Clutton-Brock (1972) during his two-month study of badius and guereza in the Kibale Forest and his nine-month study of badius in the Gombe Park of Tanzania. He found, for example, that the guereza had a less diverse diet, moved less each day, and had a smaller monthly range than did the badius. He also speculated that guereza might be able to subsist primarily on mature leaves at certain times of the year, a speculation substantiated by the studies of Oates (1974). In an attempt to explain the adaptive significance of the badius social structure, Clutton-Brock (1972) suggested that a single large group having relatively little overlap in home range with other groups was advantageous:

1) because food is heavily clumped, so that each food source (usually a group of trees in shoot or flower) can usually support the greater part of the troop at one time, 2) because it permits an integrated use of the animals' food supply. If the whole area was used by several troops, troops would be likely to visit areas where the supply of shoots had been cropped immediately previously by another troop.

Although this hypothesis may be applicable at Gombe, which seems to have a much lower density of badius groups than Kibale, it clearly does not apply to the Kibale Forest, where there is extensive and nearly complete overlap in home ranges of badius groups. Our study found that groups supplanted one another, and on some occasions these supplantations were clearly over clumped food sources. In such cases it seemed that because of their dietary diversity and abundance of food, the supplanted group had no difficulty in readily locating another suitable food source. Furthermore, badius groups usually left a clumped source of food before they had depleted it, even without being supplanted by another group. Clutton-Brock (1972) goes on to suggest that the higher proportion of young growth in the diet of badius, combined with their larger group size, results in a day range larger than that of guereza. However, it is clear from our studies that not only does young growth usually constitute a high proportion of the guereza diet, but when one considers the differences in biomass between groups of the two species, the differences in day ranges are virtually eliminated (see above).

The lack of attention to newborn infants or the lack of opportunity to express this interest by other group members may result in a looser attachment of juvenile badius to their group. Although the data are sparse, there is some suggestion that the juvenile class is the most mobile element in the badius society and that juveniles frequently shift between groups. This could be the way in which most outbreeding is achieved by badius.

Adaptation to Habitat Different from That of Kibale

Surveys in many other parts of East Africa indicate that guereza is most successful in relatively young secondary forest and that it often succeeds in narrow strips of riparian forest and very small patches of relict forest. Guereza can successfully occupy old and mature rain forests, but they are much less numerous there than in younger forests. The social structure of guereza is more clearly adaptive in an ecological sense for a species inhabiting secondary and riparian forests than for one inhabiting mature rain forest. Many of the habitats in which guereza seem most successful are characterized by few tree species and pronounced dry seasons, when young, succulent plant growth is usually absent. Such an environment would clearly select for an herbivore that could survive on a monotonous diet and, at certain times of the year, on mature leaves alone. The small area of some of these habitats would also select for small group size and close interindividual spacing because of the small and clumped nature of the food resources. Territoriality would be one means of allocating food resources among groups in such a situation. The suggestion that small group size is related to small- or low-density food resources is indirectly supported by evidence from studies of other species. For example, two subspecies of red colobus, *C. badius temminckii* and *C. badius rufomitratus,* live in relatively dry woodlands where food is considered to be more seasonal and generally less abundant than in the rain forest. Both of these subspecies live in considerably smaller social groups than do the subspecies living in rain forests. Another example is the vervet monkey (*Cercopithecus aethiops*) of Amboseli, Kenya, whose average group size declined in apparent response to a reduction in food supplies (Struhsaker 1973). Clutton-Brock (1972) has also suggested that guereza may, in fact, be better adapted to a drier habitat than to the rain forest, again because of their more monotonous diet and ability to subsist exclusively on mature leaves at certain times of the year.

Considering the overall distribution of badius, we conclude that the species is adapted primarily to the rain forest. Some populations and subspecies, however, are able to succeed in relatively dry and seasonal habitats, including savanna woodland and riparian and relict patches of

which kind of habitat each species were adapted to.

forest. As noted above, these populations generally live in smaller social groups than do those living in rain forest. Presumably, this is a function of food availability. Clutton-Brock's (1972) suggestion that the dietary dependency of badius on young growth limits them to habitats that provide young growth throughout the year, thus restricting them to rain forest habitats, seems tenable, with the possible exceptions of *C. badius temminckii* and *C. badius rufomitratus.*

Phylogeny

As mentioned in the introduction, the taxomony of the African Colobinae has not been worked out to the satisfaction of all. As far as this article is concerned, the major question is whether or not badius and guereza are rightfully members of the same genus. Because species of different genera are phylogenetically more distant than species of the same genus, the former are assumed to have been separated over a relatively longer period of time and to have had evolutionary histories more different from each other than have species of the same genus. The extent of these differences in the duration and nature of their evolutionary histories may be of considerable importance in explaining the current differences in the behavior and ecology of badius and guereza. Our behavioral and ecological data demonstrate that these two species are very different. Marler's (1969, 1970) and our studies on guereza and badius vocalizations also show pronounced differences. Whether the magnitude of the differences described by these studies warrants placing guereza and badius in different genera cannot be stated with certainty at this time. However, the osteological evidence that has been used to place them in the same genus is no more compelling than our data suggesting that they should be placed in different genera. Clearly, more data of a different sort are needed before a satisfactory conclusion can be reached. Detailed and comparable data on karyology and serology could be extremely useful. Barnicot and Hewett-Emmett (1972) have examined red cell and serum proteins of badius, and Sarich (1970) presents information on the immunology of *Colobus polykomos,* but comparable data for badius and guereza are not available.

Comparison with Other Colobinae

There are few data on other Colobinae which permit the kind of comparison we have been able to make with guereza and badius. However, it appears that guereza is much more like the majority of Indian and Asiatic Colobinae than is badius. *Presbytis johnii* (Poirier 1970), *P. cristatus* (Furuya 1961-1962; Bernstein 1968), *P. senex* (Eisenberg, Muckenhirn, and Rudran 1972), and some populations of *P. entellus* (Yoshiba 1968; Vogel 1971) all live in small social groups having only one fully adult male. Furthermore, these species are all territorial and, as among guereza, other members of the group handle and carry newborn infants. These studies support the conclusion that attention to infants enhances closer cohesion among group members, which in turn facilitates their united effort in defending territories against neighboring groups. Exceptions to this are some populations of *P. entellus* (Jay 1965; Vogel 1971) in which the groups are large, have several adult males, and are not territorial but still practice "aunt" behavior toward neonates. In some ways these exceptional entellus groups resemble the badius more closely than they resemble the guereza.

Ecological comparisons with other Colobinae are equally difficult because of the paucity of data. Hladik and Hladik (1972) described the food habits of *P. senex* and *P. entellus* at Polonnaruwa, Ceylon, and their results suggest some parallels with the guereza-badius comparison. For example, like guereza, senex eats more mature leaves and fruit than does entellus, which is more like badius in its food items. Furthermore, senex has a less diverse diet than does entellus, also consistent with the parallel between guereza and badius.

We feel that one of the main lessons emerging from this comparison is that gross classifications of habitat and food habits are poor predictors of social behavior and social organization among primates. The forest-dwelling leaf eaters can have very different societies, as demonstrated by *Colobus badius* and *Colobus guereza*.

REFERENCES

Barnicot, N. A., D. Hewett-Emmett 1972 Red cell and serum proteins of *Cercocebus, Presbytis, Colobus* and certain other species. *Folia Primatologica* 17:442-457.

Bernstein, I. 1968 The lutong of Kuala Selangor. *Behaviour* 14:136-163.

Burt, W. H. 1943 Territoriality and home range concepts as applied to mammals. *Journal of Mammalogy* 24:346-352.

Clutton-Brock, T. H. 1972 "Feeding and ranging behaviour of the red colobus monkey." Unpublished doctoral dissertation, Cambridge University, Cambridge, England.

Dandelot, P. 1968 *Primates: Anthropoidea*. Smithsonian Institution Preliminary Identification Manual for African Mammals 24:1-80. Washington: Smithsonian Press.

Eisenberg, J. F., N. A. Muckenhirn, R. Rudran 1972 The relation between ecology and social structure in primates. *Science* 176:863-874.

Furuya, Y. 1961-62 The social life of silvered leaf monkeys (*Trachypithecus cristatus*). *Primates* 3:41-60.

Groves, C. P. 1970 "The forgotten leaf-eaters and the phylogeny of the Colobinae," in *Old World monkeys*. Edited by J. R. Napier and P. H. Napier, 555-588. New York: Academic Press.

Hladik, C. M., A. Hladik 1972 Disponibilités alimentaires et domaines vitaux des primates à Ceylan. *La Terre et la Vie* 26:149-215.

Holmes, R. T., F. A. Pitelka 1968 Food overlap among coexisting sandpipers on northern Alaskan tundra. *Systematic Zoology* 17:305-318.

Jay, P. 1965 "The common langur of North India," in *Primate behavior.* Edited by I. DeVore, 197-250. New York: Holt, Rinehart and Winston.

Kuhn, H.-J. 1967 "Zur systematik der Cercopithecidae," in *Neue Ergebnisse der Primatologie.* Edited by D. Starck, R. Schneider, and H.-J. Kuhn, 25-46. Stuttgart: Gustav Fischer.

1972 On the perineal organ of male *Procolobus badius. Journal of Human Evolution* 1:371-378.

Marler, P. 1969 *Colobus guereza:* territoriality and group composition. *Science* 163:93-95.

1970 Vocalizations of East African monkeys I. Red colobus. *Folia Primatologica* 13:81-91.

Napier, J. R., P. H. Napier 1967 *A handbook of living primates.* New York: Academic Press.

Oates, J. F. 1974 "The ecology and behaviour of the black-and-white colobus monkey (*Colobus guereza* Rüppell) in East Africa." Unpublished doctoral dissertation, University of London.

Pitelka, F. A. 1959 Numbers, breeding schedule, and territoriality in the pectoral sandpipers of northern Alaska. *Condor* 61:233-264.

Poirier, F. E. 1970 "The nilgiri langur (*Presbytis johnii*) of South India," in *Primate behavior: developments in field and laboratory research.* Edited by L. A. Rosenblum, 251-383. New York: Academic Press.

Rahm, U. 1970 "Ecology, zoogeography, and systematics of some African forest monkeys," in *Old World monkeys.* Edited by J. R. Napier and P. H. Napier, 589-626. New York: Academic Press.

Sarich, V. M. 1970 "Primate systematics with special reference to Old World monkeys: a protein perspective," in *Old World monkeys.* Edited by J. R. Napier and P. H. Napier, 175-226. New York: Academic Press.

Sokal, R. R., F. J. Rohlf 1969 *Biometry: the principles and practice of statistics in biological research.* San Francisco: W. H. Freeman.

Struhsaker, T. T. 1967 Social structure among vervet monkeys (*Cercopithecus aethiops*). *Behaviour* 29:83-121.

1973 A recensus of vervet monkeys in the Masai-Amboseli game reserve, Kenya. *Ecology* 54:930-932.

1974 Correlates of ranging behavior in a group of red colobus monkeys (*Colobus badius tephrosceles*). *American Zoologist* 14:177-184.

i.p. *Behavior and ecology of red colobus monkeys.* Chicago: University of Chicago Press.

Verheyen, W. N. 1962 *Contribution à la craniologie comparée des primates.* Annales de la Musée Royale de l'Afrique Centrale, série 8vo, Sciences Zoologiques 105. Tervuren, Belgium.

Vogel, C. 1971 Behavioral differences of *Presbytis entellus* in two different habitats. *Proceedings of the Third International Congress of Primatology, Zurich* 3:41-47. Basel: Karger.

Wilson, E. O., W. H. Bossert 1971 *A primer of population biology.* Stamford, Connecticut: Sinauer Associates.

Yoshiba, K. 1968 "Local and inter-troop variability in ecology and social behavior of common Indian langurs," in *Primates: studies in adaptation and variability.* Edited by P. C. Jay, 217-242. New York: Holt, Rinehart and Winston.

Terrestrial Old World Monkeys

10

Ecological Relations and Niche Separation between Sympatric Terrestrial Primates in Ethiopia

1974

R.I.M. DUNBAR AND E.P. DUNBAR

Abstract. *Theropithecus gelada, Papio anubis* and *Cercopithecus aethiops* are commonly sympatric in Ethiopia. It is suggested that niche separation would be more marked among terrestrial open country species than among forest primates. The ecological relationships between these three species in an Ethiopian valley where they coexist are analysed. Quantitative data are presented on density and biomass, size of home ranges and day ranges, activity patterns, use of habitat, diet and feeding patterns and on interspecific interactions. These are compared across the species to determine to what extent ecological competition could occur and in what ways it is reduced. The data are discussed with reference to studies of forest primate communities where *niche* overlap has commonly been reported.

Key Words Synecology Niches *Theropithecus gelada Papio anubis Cercopithecus aethiops*

In many habitats several primate taxa are known to coexist. It is generally thought that such coexistence can only occur where there is *niche* separation between the species concerned (Gause, 1934). Observations by Gautier and Gautier-Hion (1969) and Gartlan and Struhsaker (1972), however, suggest that in some forest habitats where food is superabundant, *niche* separation

Reprinted from *Folia Primatologica, 21*:36-60 (1974) with permission from S. Karger AG, Basel.

may be minimal, and different species can coexist without significant competition occurring. Such a situation is presumably less likely to occur in savanna habitats where the production of the environment is generally lower and marked seasonal variation in food resources may occur.

While the synecology of forest primate communities has been receiving much attention in recent and current field studies (e.g. Gautier and Gautier-Hion, 1969; Gartlan and Struhsaker, 1972; Rodman, 1973; F.P.G. Aldrich-Blake, personal commun.), open country populations have been relatively little studied. Some details of the synecology of populations of terrestrial, open country primates have been given by Hall (1965), Struhsaker (1967) and Crook and Aldrich-Blake (1968), but a detailed quantitative study of the way in which the different species utilise the available resources has yet to be undertaken. The only recent attempt to carry out such an analysis is that of Nagel (1973).

Three species of baboons are found in Ethiopia (*Papio anubis, Papio hamadryas* and *Theropithecus gelada*), as well as the semi-terrestrial monkey *Cercopithecus aethiops*. With the exception of *P. hamadryas*, these species are commonly sympatric (Fig. 1). Nagel's (1973) analysis of *P. hamadryas* and *P. anubis* ecology at the species border suggests that the major reduction in ecological overlap results from the former's social and morphological adaptation to a semi-desert habitat too arid for the latter species to colonise successfully. Such a marked habitat preference is less clear between the anubis, gelada and vervets, and reduction of potential competition must occur at more subtle levels. Although Crook and Aldrich-Blake (1968) suggest that anubis and gelada occupy different *niches*, the two species are commonly sympatric (Fig. 1). A more detailed quantitative study of their synecology is of interest in view of Tappen's (1960) suggestion that the gelada's present distribution represents a retreat habitat to which the species has become confined as a result of competition from the more successful *Papio* baboons. *C. aethiops* supersp. and *P. cynocephalus* supersp. are commonly sympatric throughout their ranges in Africa, and, although the baboons are able to make greater use of the savanna grasslands than the primarily forest-fringe vervets, competition between them may be high since their dietary preferences are at least superficially similar. Furthermore, baboons are known to prey on vervets in some areas at least (Altmann and Altmann, 1970), and coexistence may place the smaller vervets at considerable risk.

The present article reports an analysis of the ecological relations between vervets, anubis baboons and gelada baboons in an area where the three species coexist in Ethiopia. Our aims were to determine, firstly, how much competition could occur, and secondly, in those cases where overlap is extensive, how the potential competition is reduced.

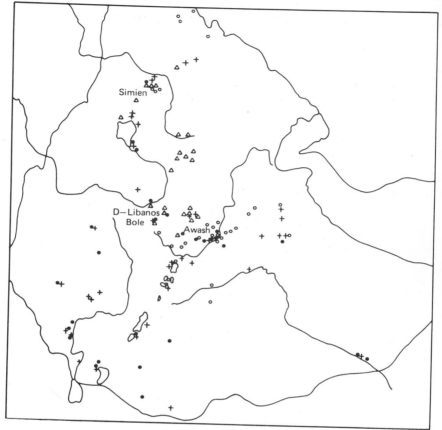

Fig. 1. Distribution of hamadryas (○), anubis (●) and gelada (△) baboons and Vervet (+) monkeys in Ethiopia. The locations shown are those where the various species are known to occur based on information in the literature, personal observation and personal communications from other scientists.

METHODS

The observations reported below were made during the course of three extended visits to the Bole Valley (9°25'N, 38°00'E) in May 1971, May-July 1972 and September-October 1972. In all, gelada were under observation for 82 h, anubis for 137 h and vervets for 58 h.

Observations were not made systematically by walking regular transects or by distributing the observation time equally among the three species. Rather, a search would be made for monkeys in a particular part of the habitat, and once a party was located, irrespective of species, it would be followed for as long as was possible or convenient. During periods of observation, activity counts of all visible animals (Crook and Aldrich-Blake, 1968) were taken at 10-min intervals recording the activity of each animal and, if

feeding, what it was feeding on. The locations of all parties and their movements were recorded on mimeographed maps of the study area.

In May 1972 a botanical transect of the study area was carried out,
recording the amount of cover at three levels in m^2 plots spaced 15 m apart
on a compass bearing. 83 plants specimens were collected in all and these
were identified by the botanists of the Haile Selassie I University, Addis
Ababa.

DESCRIPTION OF THE STUDY AREA

The Bole Valley (1,700 m altitude at the river bed) is a box canyon some
600 m deep, and forms part of the Blue Nile drainage system. The valley
may be divided into four vertical strata as illustrated in Fig. 2. Zone 1, the
river bed and gallery forest, consisted of a stony river bed some 100 m
across at its widest point, bordered by a narrow gallery forest with trees up
to 30 m in height. The predominant trees species included *Ficus salicifolia*,
F. sycamorus, *F. vallis-chaudae*, *F. plattyphylla*, *Albizia grandibracteata*, *Syzigium
guineense*, *Celtis africana* and an unidentified Sapotaceae species (*sp. indet.* No.
2). The understorey, which included *Combretum molle*, *Calpurnia subdecendera*,
Myrica saliafolia and *Carissa edulis*, was luxuriant and dense, and beneath it
grew a variety of shrubs, herbs and grasses (including *Chenopodium* sp.,
Vernonia sp., *Cyperus digitatus* and coarse reedy grasses such as *Setaria chevalieri*).
In one part, the undergrowth had been cleared and coffee planted beneath
the canopy. Imported *Grevillea* sp. trees had been planted in one section over
the coffee to replace the canopy trees which had been removed. Included
in zone 1 were strips of forest running up small gulleys into the zone above
(zone 2). In the heads of these gulleys, permanent springs supported luxuriant stands of forest. The gallery forest was separated from the next zone up
by a sheer sandstone cliff face some 60 m high.

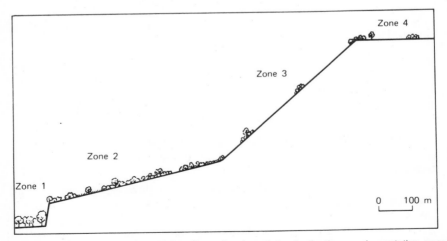

Fig. 2. Cross-section of the Bole Valley illustrating the relative inclinations and vegetation cover
in zones 1-4.

The flatter, lower slopes of the valley above the gallery forest (zone 2) supported a dense cover of *Olea africana, Acokanthera schimperi, Rhus* sp., *Combretum glaucescens* and *Euclea schimperi.* The thicket was interspersed by areas of more open grassland (predominantly tall grasses such as *Panicum maximum* and *Cymbopogon validus,* with shorter species such as *Sporobolus pyramidalis* and *Andropogon pratensis* in some areas). The steep upper slopes of the valley (zone 3) consisted almost exclusively of open grassland (*Pennisetum schimperi, Hyparrhenia tubercolosum, Andropogon pratensis*) with scattered *Olea, Rhus* and *Dodonea viscosa* bushes and small trees. Herbs and small plants, such as *Coreopsis* spp. and *Otostegia integrifolia,* were common. The plateau top, zone 4, consisted of open grassland with scattered thickets of *Rosa abyssinica, Carissa edulis, Rumex nervosus, Solanum* sp. and occasional *Acacia* spp. trees. This zone has been much affected by cultivation and was almost certainly more densely covered with thicket in the past.

The relative slopes and vegetational profiles of these zones are shown in Fig. 2, which is drawn along the line of the transect about half-way down the eastern side of the valley.

At the northern end of the valley floor, the gallery forest gave way to open grassland with occasional *Ficus* and *Acacia* spp. trees. Herbs such as *Trifolium ruepellianum* and other shrubs and small bushes were common. This zone has also been much affected by cultivation, and may have been more densely covered with vegetation in the past.

Table 1 shows the proportions of ground (up to 30 cm), bush (30 cm to 6 m) and tree (more than 6 m) level cover in each of zones 1-4 taken from the transect data.

The climate of the area can be divided into two main seasons: a wet season from July to early September, and the dry season from mid-September to June. A slight increase in rainfall around March can usually be detected giving rise to the "short rains" at this time of year. Annual rainfall averages some 2,000 mm, most of it falling between July and October, but

Table 1

Vegetational composition of zones 1–4 based on data from the botanical transect (see text for details)

Zone	Mean percent ground level cover[a]	Plots with bush level cover %	Plots with tree level cover %	Number of plots
1	44.0	90.0	70.0	10
2	42.3	89.5	13.2	38
3	39.4	28.2	—	39
4	82.0	20.0	—	15

[a]Average of percentage ground level cover in all plots.

some rain is to be expected in all months of the year. During the wet season, heavy mists are common on the upper slopes and may reduce visibility for several hours at a time to distances of 50-100 m. Daytime temperatures are generally in the region of 35°C.

Among the other animals to be found in the valley are *Colobus guereza* and *Galago senegalensis,* bushbuck, common duiker and klipspringer. Potential predators include leopard, at least one other small felid, jackal, hyena, native dogs and several large birds of prey (Verraux's eagle, Crowned Hawk eagle and Black-chested harrier eagle). The puff adder (*Bitis arietans*) was the only other potential dangerous animal in the valley. During the crop-growing season (September-November), the local villagers harassed the baboons and vervets, using sling-shots and occasionally traps to keep them off the crops. Some deaths may have occurred as a result of this, but none were recorded in the 1972 season.

RESULTS

Composition of the Population

The valley supported 7 troops of *P. anubis* totalling 140 animals, 243 *T. gelada* and at least four groups of *C. aethiops* totalling at least 60 animals (estimated total ca. 75). The seven anubis groups had an average size of 20 individuals (range 15-24). Only one vervet group could be accurately counted, and this consisted of 29 animals. Partial counts of two other groups gave figures of 13 + and 15 +. On the bases of these figures, the total vervet population in the valley was estimated at between 70 and 80 animals. All groups of both species were multimale in composition. The gelada were divided into some 17 one male reproductive units and a number of small all male groups. These were in turn organised into groups of 3-5 units which shared a common home range: these clusters of units were termed bands (Dunbar, 1973; Dunbar and Dunbar, in preparation). There were five such bands in the valley. In this area, the gelada did not form large herds when foraging: rather, individual units foraged independently of each other for most of the time, small herds of two or three units being formed only very occasionally (compare Crook, 1966).

anubis/vervet - multi ♂   gelada - uni ♂ + all ♂

Habitat Preferences

Fig. 3 shows the location of the ranges of the various groups of the three species insofar as these are known. The ranges shown for gelada are those of the bands. These ranges, of course, represent only the areas each group was known to use during only half of the year, and the annual range in each case might be greater. However, the period from which the data were obtained covered both the dry and the wet seasons, and it seems doubtful that the annual range would have been much greater in any given case.

It can be seen from Fig. 3 that the gelada show a marked preference for zones 3 and 4. The vervets, on the other hand, occur throughout zones 1, 2 and 5, while the anubis occur in all zones, although rather more groups are based in the lower zones 1, 2 and 5 than in the upper zones 3 and 4.

The gelada only rarely came down to zone 2, when they would spend the night on the sandstone cliffs behind the gallery forest. They were never observed to enter the gallery forest itself. On the whole they tended to remain confined to the steep slopes of zone 3, rarely moving more than 100 m from the cliff edge when they moved onto the flat plateau top (zone 4). Although the anubis used all zones in the valley to some extent, they made rather less use of the more open zones 4 and 5 than was expected. This was due to the fact that they were constantly chased off these zones when the crops were being grown. This resulted in a tendency for them to use zone 4 in particular much less than would otherwise probably have been the case.

Home range sizes for those groups whose ranges were known with sufficient accuracy can be estimated from Fig. 3. Three anubis groups' ranges can be measured, and these average 93.7 ha (range 74.5–112.0 ha). The ranges of the two gelada bands on the eastern face of the valley covered 90 ha and 78 ha, respectively (mean 84 ha), while the ranges of the two

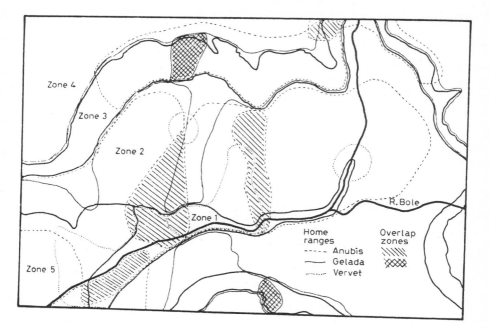

Fig. 3. Locations of home ranges of groups of each species.

vervet groups for which adequate data are available had a mean of 29.5 ha
(19.5 and 29.4 ha, respectively). The larger of these two ranges is that of
the large group of 29, and a sizeable portion of their range (7.5 ha) fell in
the relatively open zone 5. Comparatively little of their time was spent in
this zone, and, except when crossing it, they tended to remain close to the
gallery forest edge. The large size of their range may, therefore, reflect the
fact that this part of the habitat was somewhat unproductive as far as they
were concerned, but happened to lie between the forested part of their
range and some cultivated fields which they raided on a number of occasions.

   These figures give densities of $26/km^2$ for anubis, $73/km^2$ for vervets
and $82/km^2$ for gelada (taking into account any overlap in the ranges of
neighbouring groups). Despite their equivalence in size, gelada are able to
maintain more than three times the density of anubis baboons in their
preferred habitats. This implies that gelada habitat is more productive as far
as the gelada are concerned than anubis habitat is for anubis. The exact
sympatry of several groups of gelada and anubis on the upper slopes further
allows us to see that gelada habitat is absolutely more productive for gelada
than it is for anubis. The same area of gorge face on the eastern side of the
valley supported 125 gelada (two bands) and only 46 anubis (two troops).

   Table 2 gives these data and the biomass values in each of the two major
sections of the habitat. The biomass values were estimated from the average
weights of adult males and females as given by Napier and Napier (1967)
and Hill (1966): all subadults were counted as equal to females in weight,
and all juveniles and infants as equal, on average, to half the weight of an
adult female. In the case of the vervet population, the total has been taken
as 75 and the composition was estimated from that of the composition of
the three groups for which counts are available. Although the vervets had
a substantially higher numerical density than the anubis baboons on the
basis of home range size, their smaller physical size and the fact that they
did not make use of all parts of zones 2 and 5 results in a lower overall
biomass over the whole area. If these calculations are made on the basis of

Table 2
Density and biomass of each species in the Bole Valley

Species	Density/km² [a]	Number of animals		Biomass, kg [b]	
		zones 1, 2, 5	zones 3, 4	zones 1, 2, 5	zones 3, 4
Anubis	26	94	46	718.0	444.5
Gelada	82	—	243	—	1,361.5
Vervets	135	ca. 75	—	207.0	—

[a] Density is based on home range sizes.
[b] Biomass is based on total area in each section of the habitat.

Table 3

Means and ranges of day range length for each species

Species	Day range length, m		N
	mean	range	
Anubis	1,210	300–2,000	13
Gelada	630	500–1,000	8
Vervets	700	600–800	4

home range size only, the vervets biomass approximates more closely that of the anubis (201 kg/km² for vervets, 206 kg/km² for anubis).

These data show that the anubis were unable to maintain as high a biomass in zones 3 and 4 as they were in zones 1, 2 and 5. This correlates with the differences in vegetational composition between these two sections of the valley, the upper zones (3 and 4) having considerably less tree and bush cover than the lower ones (Table 1).

Use of Habitat

Activity Patterns and Day Ranges. Day ranges for each of the species can be calculated from a sample of 13 days in the case of the anubis, 4 days for vervets and 8 days for gelada units. The means and ranges of these are given in Table 3. Vervet day ranges are consistently smaller and of more constant length than those of the anubis, whose ranging patterns were much affected by the local availability of food. Gelada day ranges are comparable in length to those of the vervets (Mann-Whitney U test, p > 0.10), and both species' day ranges are significantly smaller than those of the anubis (Mann-Whitney U test, p < 0.01 and < 0.02, respectively).

The temporal structure of the day ranges of each species differs markedly. This can be seen by comparing the activity patterns of each species. These are given in Fig. 4-6, based on the activity count data. Each graph shows the percentage of the total records for each hour accounted for by each of the major categories of activity.

The anubis (Fig. 4) show a series of alternating peaks of activity early in the morning, at midday and in the late afternoon. Generally speaking, a peak in movement is closely followed by a peak in feeding, these peaks of activity being separated by period of inactivity during which social behaviour takes place. Peaks of sitting and social activity are then followed by movement to another place where feeding recommences.

The gelada (Fig. 5) are in marked contrast. The morning is spent almost exclusively engaged in social activity, but thereafter feeding predominates throughout the day, averaging some 50% of the records of these hours. A smaller peak in social activity is observed again in the evening. Movement

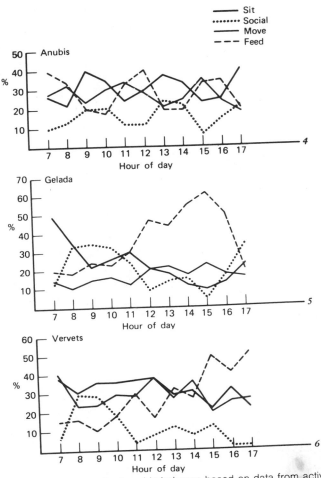

Fig. 4. Diurnal activity patterns of anubis baboons based on data from activity counts.

Fig. 5. Diurnal activity patterns of gelada baboons based on data from activity counts.

Fig. 6. Diurnal activity patterns of vervet monkeys based on data from activity counts.

shows a fairly steady level throughout the day during feeding periods reflecting a tendency for the animals to drift from one clump of grass to
another (p. 00).

The generally sendentary nature of *Cercopithecus* monkeys is reflected in
the relatively high level of sitting throughout the day in the data for vervets
(Fig. 6). Social activity is common early in the morning, but rare thereafter.
Feeding, on the other hand, shows a tendency to increase gradually as the
day goes on, most feeding being done later in the day. Movement shows a
tendency to peak first thing in the morning when the animals move out of
their sleeping positions to more exposed places to sit and groom in the sun.

Table 4
Frequencies with which animals of each species were recorded on the ground or
in trees and bushes during activity counts

Species	Number of animals recorded		Total
	in trees	on ground	
Anubis	690	1,784	2,474
Gelada	42	2,644	2,686
Vervets	874	671	1,545

A second peak occurs early in the afternoon when the animals move from
the vicinity of the night's sleeping trees to feed. This latter move generally
accounted for the greater part of each day's range length.

Use of Trees. Table 4 shows the number of animals of each species recorded
in activity counts as being on the ground or in trees and bushes. Height
above ground was not distinguished further since it seemed to be relatively
unimportant. The canopy was on the whole discontinuous, and no distinct
levels could readily be distinguished.

Of the three species, the vervets are clearly the most arboreal: 56.6%
of the animals sampled were in trees or bushes, compared with 27.9% for
anubis baboons and only 1.6% for gelada. The differences between the
species are all statistically significant (χ^2 tests, $p < 0.001$ in each case).

The gelada very seldom climbed into trees or bushes. When they did
so it was usually only to obtain a better view of some disturbance in the
distance or, less often, to obtain fruits that could not be reached from the
ground. Jolly (1972) notes that their limb proportions are better adapted to
terrestrial quadrupedal locomotion than to moving in trees, and they are
indeed noticeably poorer at climbing and moving in trees and bushes than
the anubis baboons. One other factor which might account for their terrestriality is the fact that their diet is almost exclusively to be found on the
ground (see below), and their preferred habitat has few or no trees in which
to climb. Observations in other areas, however, indicate that even when they
do enter well wooded areas (in this case, 6-10 m *Erica arborea* forest in the
Simien Mountains) they rarely climb into the trees.

On the other hand, the vervets invariably slept and rested in trees and
bushes, while the anubis made use of both trees and ledges on the rock
faces. While almost all movement by the anubis took place along the
ground, the vervets frequently moved through the arboreal vegetation,
descending to the ground primarily to cross areas devoid of vegetation.

Food and Feeding Habits

Diet. Table 5 gives the percentage distribution of feeding records among the
various plant species for each species of monkey. It is clear from these data
that the gelada differ markedly from the other species in their dietary prefer-

ence. They are almost exclusively graminivorous, grasses accounting for more than 95% of their diet. Both the anubis and the vervets are much more catholic and eat a wide variety of species.

Close examination of the data reveals that there are more marked differences between the anubis and vervets than appears to be the case at first sight. Table 6 gives the data in terms of the parts of plants eaten, irrespective of species. While both anubis and vervets fed predominantly on soft fruits and seeds, vervets tended to feed in addition mostly on flowers, whereas the anubis baboons preferred leaves.

Table 5

Percentage of each species' diet accounted for by various plants based on data collected during activity counts

Plant type	Species	Part eaten	Diet, %		
			anubis	gelada	vervets
Grass	*(Andropogon pratensis, Bothriochloa*	L	10.3	91.4	3.1
	pertusa, Cynodon dactylon, Eragros-	S	8.9	5.0	0.3
	tis cylindriflora, Heteropogon	R	0.4	0.5	—
	contortus, Hyparrhenia tuberculosum,				
	Rhynchelytrum repens, Sporobolus				
	pyramidalis)				
	Eragrostis teff (cultivated)	L	0.2	—	—
	Maize (cultivated)	L	2.7	—	—
		S	—	—	0.3
Herbs	*Trifolium ruepellianum*	L, Fl.	12.3	—	14.8
	Coleus sp.	E	—	—	
	spp. indet. (various)	L	—	0.3	18.7
		Fl.	—	0.3	—
Fungi	mushroom (sp. indet)	E	—	—	
Shrubs	*Lantana trifolia*	Fl.	0.7	—	—
	Chenopodium sp.	L	—	—	0.3
		R	0.4	—	—
	Coreopsis spp.	Fl.	—	—	0.3
		F	18.8	0.7	—
	Achyranthes aspera	L	—	—	0.6
	Clerodendron myricoides	E	—	—	
	Aloe sp.	Fl., stem	—	—	0.5
	spp. indet. (various)	L	0.4	—	1.1
		stem	0.7	—	—
		F	0.9	—	—
		bulb	0.7	—	—

Table 5 (continued)

Plant type	Species	Part eaten	anubis	gelada	vervets
Bushes	Acokanthera schimperi	F	0.8	—	0.6
	Rumex nervosus	F	0.4	—	—
	Pittosporum viridiflorum	L	0.2	—	—
	Combretum spp.	L	0.9	—	—
	Ricinus communis	L	—	—	0.3
	Rhus sp.	F	—	0.7	0.3
	Euclea schimperi	L	0.4	—	—
		F	0.2	0.1	2.3
	Carissa edulis	F	—	—	0.3
	Vernonia sp.	Fl.	—	—	E
	Dodonea viscosa	F	0.4	—	—
	coffee (cultivated)	F	—	—	0.3
	sp. indet	L	0.2	0.1	—
Creepers	spp. indet. (various)	L	—	—	0.6
		F	—	—	0.3
Trees	Albizzia grandibracteata	L	0.7	—	2.3
		Fl.	6.7	—	2.6
		F	—	—	0.3
	Ficus spp. (platyphylla, salicifolia,	L	—	—	0.3
	sycamorus, vallis-chaudae)	F	17.3	—	17.3
	Combretum molle	L	—	—	0.3
		F	—	—	0.6
	Syzigium guineense	F	1.6	—	5.7
	Celtis africana	F	0.7	—	—
	Mayternus undatus	F	—	—	18.7
	Olea africana	L	0.2	—	—
		F	0.2	—	—
	Sapotaceae (sp. indet. No. 2)	F	4.9	0.5	1.4
	spp. indet. (various)	L	1.1	—	—
		F	—	—	0.8
	Albizzia, Ficus spp., Combretum and Acacia spp.	bark	0.6	—	5.7

L = Leaves; Fl. = flowers; F = fruits; S = seeds; R = roots, rhizomes. E indicates species which were eaten occasionally, but were not recorded in the quantitative sample. Sample sizes were 447 records for anubis, 764 for gelada and 352 for vervets.

Table 7 shows the data analysed in terms of the levels at which the animals were feeding as defined by gross plant types (grasses and herbs; plants and small shrubs; bushes; trees). Most of the vervet's feeding was concentrated on the tree layer (63.9%), whereas the anubis concentrated more on the lower strata (39.3% on the ground and 21.9% in the plant/-shrub layer), making rather less use of the tree layer (35.2%). Gelada, as to

Table 6
Analysis of diet preferences by part of plant eaten

Part of plant	Anubis		Gelada		Vervets	
	number	%	number	%	number	%
Leaves	147	32.9	701	91.8	66	18.7
Flowers	33	7.4	6	0.8	62	17.6
Fruits, seeds	245	54.9	53	6.9	178	50.6
Roots, bulbs	7	1.6	4	0.5	—	—
Bark	3	0.7	—	—	20	5.7
Insects	12	2.7	—	—	26	7.4
Total	447		764		352	

be expected, fed almost exclusively on the ground layer vegetation where the bulk of their diet is to be found. These differences are all statistically significant.

These data are, of course, representative only of a small proportion of the year, and the situation could be markedly different at different times of year. The sample period did, however, cover both the rainy season and the dry season, and it would seem to be unlikely that the situation would be markedly different during the post-rains pre-dry-season period from November to April.

Feeding Methods. There are marked differences in the ways in which the three species of monkeys typically fed. Gelada, for instance, sit on their haunches harvesting grass blades with both hands, shuffling forwards a few metres at a time in the sitting position. When digging for rhizomes and other roots, they show a similar specialization. The hands are held at right angles to the forearm with the fingers extended, and both hands are used in rapid alternation to dig away the earth. Anubis, on the other hand, when feeding on grass blades, stand on three legs harvesting the blades with one hand. They are consequently able to collect grass blades at only half the rate at which gelada do, and indeed do so in a qualitatively more leisurely way.

Table 7
Analysis of diet preferences by height of plant eaten

Layer	Anubis		Gelada		Vervets	
	number	%	number	%	number	%
Ground (grasses and herbs)	171	39.3	744	97.6	92	28.2
Shrubs and plants	95	21.9	9	1.2	9	2.8
Bushes	16	3.7	7	1.0	17	5.2
Trees	153	35.2	4	0.5	208	63.9
Total	435		764		326	

Similarly, when digging for rhizomes, they tend to do so by scraping away the earth with one hand and then pulling the pieces out of the ground with hands or teeth. In contrast, the gelada's hand action rapidly loosens the earth around the roots allowing them to be easily removed.

Comparable differences in feeding methods distinguish the anubis and the vervets. When feeding on figs, for instance, vervets tend to pluck a single fruit, then sit and bite small pieces from it, whereas the anubis tend to place all or most of the fruit in the mouth at once. Consequently, the anubis tend to feed more rapidly than the vervets and consume a greater bulk of food per unit time. These differences were not, however, quantified directly.

Time Spent Feeding. The amount of time which the three species spent feeding does not differ to any extent. By averaging the amount of time spent feeding across all hours of the day in Fig. 4, 5 and 6, we find that the gelada spent 31.7% of the time feeding, anubis 27.7% of the time, and vervets 27.4%.

However, these data almost certainly conceal marked differences in the amounts of food consumed. Gelada, although spending about the same proportion of time feeding as do anubis, almost certainly consume more food by weight than the latter species since their rate of feeding is much higher. Likewise, anubis feed faster than vervets, and probably therefore consume a larger bulk weight of food each day than the latter.

The ranging patterns of the species in part reflect their dietary preferences. Although the group sizes were similar, anubis groups travelled consistently further each day than did vervet groups (Table 3). The diets of these two species are probably sufficiently similar for their ranging patterns to be subject to the same kinds of ecological factors. That the anubis travelled on average twice as far per day as did the vervets can perhaps be attributed to the fact that a given clump of fruiting trees can support a group of vervets for longer than it can a comparably sized group of anubis baboons, whose biomass is four times greater.

On the other hand, the fact that the anubis also ranged twice as far as the gelada did during the day can perhaps be attributed to the fact that, whereas "clumps" of grass are generally contiguous, clumps of fruiting trees and bushes tend to be rather more dispersed. Reference to Table 1 shows that in zone 3 ground cover was much denser than bush-level cover. Presumably, therefore, anubis groups in this zone would have to travel further each day to obtain its nutritional needs than a comparably sized gelada group with about the same biomass. This would account for the fact that the section of steep gorge side on the eastern half of the valley was able to support three times as many gelada as anubis (Table 2).

Reactions to Predators

No actual cases of predation or attempted predation were observed. Vervets were, however, observed to react with chutter vocalizations and rapid movement down from the canopy to Crowned hawk eagles and Verraux's eagles. All three monkey species, were, in addition, harrassed by the local farmers during the crop-growing season.

Struhsaker (1967) has pointed out that vervets, because of their smaller size, are vulnerable to four time as many predators as baboons, and that this makes it necessary for them to remain close to the safety of trees, whereas baboons are able to make more use of the open savanna and its resources. In the Bole Valley, at least 11 species can be listed as potential predators of the vervets, against only five for the gelada and anubis baboons.

Although both gelada and anubis rely primarily on flight to safety (the former down steep gorge sides, the latter into trees) as a means of defense against predators, both species are known to rely to some extent on the presence of large adult males who can actively drive off at least the smaller predators (Altmann and Altmann, 1970, in the case of *Papio* baboons; Crook, 1966, and personal observation in the case of gelada). In contrast, vervet males are too small to be able to take on any but the smallest preda- tors, and their main role in the group seems to lie in "look-out" behaviour and the detection of predators (Gartlan, 1968).

Polyspecific Associations

Table 8 shows the frequencies with which groups of each species were observed alone (main diagonal) and in mixed-species parties with each of the other species. The data are based on samples at 10-min intervals throughout periods of observation on individual groups. Groups of two species were considered to be in a mixed-species party if the distance sepa- rating the two nearest individuals of each species was no greater than the greatest distance separating any two adjacent individuals of the same spe- cies.

Thus, for instance, in a total of 736 samples, gelada groups were in association with anubis baboons in 86 (11.6%) and in none at all with

Table 8
Frequencies with which parties of each species were observed alone and in mixed-species parties, based on records made at 10-min intervals during periods of observation

| | Number of animals recorded | | | Total | alone, % |
	gelada	anubis	vervets		
Gelada	650	86	—	736	88.3
Anubis	86	715	25	826	86.6
Vervets	—	25	315	340	92.6

vervets. On the whole, the proportion of time each species spent in polys-pecific associations was relatively small (approximately 10% in each case). Since groups of each species spent most of their time in monospecific parties, direct competition between the species is likely to be relatively insignificant.

Vervets generally reacted nervously when in association with anubis baboons, and would often withdraw if the anubis moved into the trees they were in. Altmann and Altmann (1970) likewise note that vervets at Amboseli tended to be nervous of baboons and might vacate the trees they were in and move elsewhere. They also recorded that the baboons killed and ate vervets in this area. Mixed-species parties between these two species at Bole consequently tended to be of short duration, and no direct social interac-tions between them were observed in any of the nine mixed parties re-corded. The necessity to be on their guard and if necessary move out of the way of the baboons must affect the vervets' feeding efficiency. The fact that mixed parties occur relatively rarely, however, reduces the potential disrup-tion of the vervets' daily routine that would otherwise result. It is, of course, possible that the vervets actively reduced the frequency with which mixed-species parties occurred by moving away if the baboons showed any signs of moving in their direction. No evidence for this was observed, however; the vervets seldom bothered to move if the baboons passed them by without actually moving into the trees they were in at the time.

Mixed-species parties between gelada and anubis, on the other hand, tended to last longer, sometimes for the greater part of a morning. Eleven such parties were observed in all, and social interactions between members of the two species were fairly common: 23 friendly interactions and nine agonistic interactions were recorded. Gelada threatened and/or displaced anubis on five occasions, and anubis displaced gelada on four occasions. The outcomes of all these interactions, however, appeared to be size deter-mined: in each case, the larger animal was able to displace the smaller, irrespective of species. Play between juveniles was observed several times, but sexually oriented interactions accounted for most of the non-agonistic interactions. These mostly involved subadults and older juveniles, and ap-pear to have resulted in a limited degree of hybridization between the two species (Dunbar and Dunbar, in press).

On the whole, however, the more rapid movement and longer day range of the anubis tended to carry them beyond the gelada after an hour or so. On several occasions, the gelada were observed to follow the anubis when the latter began to move, but were unable to keep up with them and ceased following after moving a few hundred metres. This fact undoubtedly contributed greatly to the relatively small amount of time that these two species spent in polyspecific associations with each other.

Indirect Competition and Niche Separation

While in general there was little direct competition between the various species, considerable indirect competition might result from use of the same area in those cases where dietary preferences are similar. The dietary preferences of the gelada make it unlikely that they would suffer from indirect competition either from the anubis or from the vervets to any significant extent. The vervets and anubis, on the other hand, have rather more overlap in diet, and indirect competition may become a critical factor in their case.

The actual amount of competition between vervet and anubis groups, however, is likely to be limited by the fact that the latter had larger home ranges and day ranges than the former. This would make it unlikely that the anubis would spend more than a small proportion of the day in any one vervet group's range. It is possible to determine just how much overlap in use there was between the large vervet group and the two anubis groups whose ranges overlapped the former's range.

The presence or absence of these two groups from the vervets' range could be determined on 30 days each. During these periods (360 h of daylight), they spent 97.5 h (26.5%) and 50.8 h (14.1%), respectively, in the vervets' range. This area, however, included a large section of overlap between the ranges of these anubis groups. On 24 of the days the movements of both groups are known, and during this time one or both groups were present for 125.8 h (43.6% of the time). Thus, for at least half of the time, the vervets had unrestricted access to all parts of their range.

Some seasonal variation in the amount of time that the baboons spent in the vervet's range was observed. Table 9 gives the number of hours they were present during May-June 1972 (the end of the dry season) and in July and September (during the rains). It can be seen that the baboons spent significantly more time there during the dry season than during the wet season (χ^2 corrected for continuity = 16.050, p < 0.001, d.f. = 1). The main fruiting season for the *Ficus* and *Syzigium* spp. in 1972, however, was in May and June. By July, only about half the crop remainded, and by our return to the study area at the end of August, only a few *Ficus* trees remained in fruit. Although this change was not quantified directly, it is indicated by the fact that violet-backed starlings (*Cinnyricinlus leucogaster*) were present in the area only during May to early July.* Thus, the period when the baboons made most use of the forest (where the bulk of the vervets' home range lay) coincided with the period of maximum soft-fruit production. During and immediately after the rains (July and September), more food was available outside the gallery forest, primarily in the heavily bushed zone 2, and the baboons spent correspondingly less time in the forest.

---

*These birds are found only in the vicinity of fruiting *Ficus* trees, apparently moving in large flocks from locality to locality following the local fruiting cycles (Mackworth-Praed and Grant, 1960).

Table 9
Number of hours during which at least one anubis group was present in the home range of the 29 vervet group during the dry season (May—June) and the rains (July and September)

	Number of hours		Total	Number of days
	present	absent		
Dry season	86.5	69.5	156	13
	(68.2)	(87.8)		
Rains	39.2	92.8	132	11
	(57.5)	(74.5)		
Total	125.7	162.3	288	24

Expected values based on marginal totals are shown in parentheses in the body of the table. x corrected for continuity = 16.050 (p < 0.001; d.f. = 1).

The generality of this conclusion is indicated by the fact that in May 1971 when we first visited the valley, the fruiting season had barely started and only a few trees had fruit on them. During this period, the baboons were rarely observed in the gallery forest: on a sample of 9 days, they spent only 12 h (11.1%) in the vervets' range.

Thus, the actual amount of indirect competition between the two species is lower than at first sight might seem to be the case. This is particularly true at that time of year when the vervets' preferred foods are least abundant. At this time of year, it may be supposed, both direct and indirect competition for access to the few trees remaining in fruit would be likely to reach critical proportions. Being smaller, the vervets might suffer more from the competition as a result of continual displacement by anubis groups. That this does not occur seems attributable to two factors: firstly, the quantity and dispersion of soft fruits in the forest is such as to force the baboons to move further and cover a larger area to obtain enough nutrients each day, and, secondly, the fact that the more open country beyond the forest provides a rich source of nutritious foods, not merely in terms of the seeds and hard fruits available after the rains, but also grass rhizomes and other roots during the height of the dry season. Struhsaker (1967) has pointed out that rhizomes are among the most nutritious sources of food in the savanna in the dry season, and that they are a source of food which is denied to the vervets because of their smaller size and physical strength.

DISCUSSION

Although these species are sympatric throughout much of their geographical ranges, close examination of their ecological adaptations reveals that there is relatively little overlap between them even in areas where they coexist.

The gelada are clearly differentiated from the other two species in habitat and dietary preferences. Indeed, Jolly's (1972) reconstruction of theropithecine palaeo-ecology suggests that it is unlikely that they have ever been in competition with the *Papio* baboons to any extent, contrary to Tappen's (1960) supposition. The gelada's major adaptation as a terrestrial graminivore clearly puts them at an advantage over the *Papio* baboons in their preferred habitat where grasses provide the bulk of the primary production. The situation is analogous to that between anubis and hamadryas at a more overt level (Nagel, 1973).

Although anubis and vervets would seem to have a great deal in common, detailed analysis reveals that there are a number of ways in which the amount of overlap is reduced which, between them, make it possible for the two species to coexist. Firstly, there are differences in the parts of plants most commonly eaten. Secondly, there are differences in the height at which they prefer to feed, with vervets being the more arboreal. Thirdly, the amount of actual overlap in use of habitat is relatively small, especially at that time of year when the vervets' preferred food is relatively scarce. Finally, direct competition between the two species, which is potentially severe for vervets when it does occur, is uncommon.

These differences have been noted in other areas. Both Struhsaker (1967) and Hall (1965) noted that the baboons at Amboseli and Chobi tended to use the open grassland beyond the forest more so than the vervets. Likewise, at Awash in Ethiopia, the baboons spent much of the day in the *Acacia* scrub beyond the gallery forest and approximately 60% of their feeding took place there (Aldrich-Blake *et al.*, 1971). Parties of vervets were observed on 24 occasions at Awash, and all sightings occurred in the gallery forest or bordering strip of grassland (identified as zones 1 and 2 in that article). Out of a total of 18 days (243 h of daylight), the main study group of baboons spent only 54.5 h (22.5%) in the gallery forest and bordering grassland. Likewise, in Senegal, the *P. papio* study troop spent only 3.6 and 9.9% respectively, out of a total of 12 days in the ranges of two groups of green monkeys (Dunbar, in preparation). In both these latter areas, the baboons fed predominantly on the ground (91% of 3,092 feeding records from activity counts at Awash; 65.8% of 363 feeding records from Senegal) whereas the vervets were more arboreal.

The ecological separation between the baboons and vervets at Bole thus seems to be fairly typical of the situation elsewhere in Africa. A tendency for baboons to feed on low-level vegetation and an ability to utilise the resources of the more open parts of the habitat result in relatively little use of those parts which vervets occupy. As a result of this, potential competition between them is reduced, and vervets are able to coexist with the baboons.

Actual competition between anubis and gelada baboons seems likewise to be negligible, although a number of localities are known where the two species coexist (Fig. 1). At Debra Libanos, for instance, Crook and Aldrich-Blake (1968) found that the two species never came into direct contact since the gelada rarely entered the *Juniperus-Olea* forest inhabited by the anubis baboons. The general absence of the gelada from the forest both at Debra Libanos and Bole can probably be attributed to the relative (or perhaps total) lack of suitable grasses on the forest floor: those species that were present at Bole were predominantly of the coarse type never eaten by the gelada. Tall, coarse grasses such as *Cymbopogon validus* which predominated in zone 2 at Bole were likewise not eaten, and this probably accounts for the fact that they only rarely entered this zone.

Both Gautier and Gautier-Hion (1969) and Gartlan and Struhsaker (1972) suggest that polyspecific associations may improve predator detection. They also suggest that increased efficiency in finding food sources, especially in the case of those species with larger home ranges, may lead to the formation of such associations. Neither of these accounts would seem to be relevant to the Bole Valley. Gartlan and Struhsaker (1972), however, do remark that the breakdown of ecological boundaries between some species due to human interference may have resulted in more frequent formation of polyspecific parties in the Cameroon than would normally be the case, and this may be relevant in the case of the gelada and anubis in the Bole Valley. Here the anubis on the upper levels tended to make more use of the plateau top than the gelada, but the fact that much of this zone was under cultivation resulted in the groups being forced down onto the cliff face. This has probably resulted in much more use of this area than would normally be the case.

These observations suggest that there tends to be rather more marked ecological separation between terrestrial, open country species than between the forest-living species studied by Gautier and Gautier-Hion (1969) and Gartlan and Struhsaker (1972). Although food distribution in forest may be seasonally patchy (Aldrich-Blake, 1970; Clutton-Brock, in press), it is probably less so than in savanna and forest-edge habitats, and less marked ecological separation would be expected among forest-dwelling species than among open country ones.

The situation at the hamadryas-anubis border at Awash described by Nagel (1973) runs somewhat counter to this argument at first sight in that there was relatively little ecological separation between the two species where they occupied comparable habitats in the Awash Canyon. Nagel (1973), however, suggests that the hamadryas are more successful than the anubis in colonising the more arid habitats, since their social organization is better adapted to the ecological requirements of the semi-desert than that of the anubis. They would likewise appear to be morphologically better adapted to existence under marginal conditions. These facts seem to have prevented a more extensive invasion of hamadryas habitat by the anubis (Nagel, 1973). Thus, although these two species have similar dietary preferences, a clear geographical separation occurs between them, with the result that they rarely come into contact. Thus, while the anubis-gelada-vervet situation can be described in terms of a division of the habitat among them at a micro-level, that between these species and the hamadryas represents a division of the macro-habitat.

SUMMARY

Habitat and dietary preferences clearly demarcate gelada baboons from both anubis baboons and vervet monkeys. Gelada are exclusively graminivorous and prefer the steep sides of gorges. Both anubis and vervets prefer the forest-edge, and although they share a number of similarities in their diets, detailed analysis reveals that the anubis make rather more use of the ground-level vegetation than the vervets. The former also make more extensive use of the non-forested parts of the habitat, and the actual amount of overlap in range use is minimal, especially at those times of the year when competition is likely to be highest. Thus, although vervets are at a potential disadvantage in areas where they coexist with *Papio* baboons, actual competition is limited. In comparison with populations of forest monkeys, ecological separation between the species of a savanna-based primate population is more marked, varying from total geographical separation between closely related species to division of the habitat at the micro-level in other cases.

ACKNOWLEDGEMENTS

We are grateful to the committee and members of the Bole Valley Society (in particular Messrs. R. Sandford and L. Melville) for permitting us to work in and use the facilities of the valley. The Haile Selassie I University, Addis Ababa, kindly sponsored us as Visiting Research Scientists during our stay in Ethiopia. We would like to thank Dr. J. H. Crook for his continued support and encouragement, and Mr. M. Gilbert, Mr. S. Gilbert and Dr. Tewolde B. G. Egziabher for identifying our plant specimens. This research was undertaken while the senior author was in receipt of a Science Research Council NATO Scholarship, and the field study was financed by grants from the Science Research Council and the Wenner-Gren Foundation for Anthropological Research.

REFERENCES

Aldrich-Blake, F. P. G.: Problems of social structure in forest monkeys; in Crook Social behaviour in birds and mammals, pp. 79-101 (Academic Press, London 1970).

Aldrich-Blake, F. P. G.; Bunn, T. K.; Dunbar, R. I. M., and Headley, P. M.: Observations on baboons, *Papio anubis,* in an arid region in Ethiopia. Folia primat. *15:*1-35 (1971).

Altmann, S. A. and Altmann, J.: Baboon ecology: African field research. (Karger, Basel 1970).

Clutton-Brock, T. M.: Red colobus: feeding and ranging behaviour. (In press.)

Crook, J. H.: Gelada baboon herd structure and movement: a comparative report. Symp. zool. Soc. Lond. *18:*237-258 (1966).

Crook, J. H. and Aldrich-Blake, F. P. G.: Ecological and behavioural contrasts between sympatric ground dwelling primates in Ethiopia. Folia primat. *8:* 192-227 (1968).

Dunbar, R. I. M.: Social dynamics of the gelada baboon, *Theropithecus gelada;* Ph. D. thesis Bristol (1973).

Dunbar, R. I. M.: Observations on the ecology and social organization of *Cercopithecus (aethiops) sabaeus* in Senegal. (In preparation.)

Dunbar, R. I. M. and Dunbar, E. P.: On hybridization between *Theropithecus gelada* and *Papio anubis* in the wild. J. Human Evol. (in press).

Gartlan, J. S.: Structure and function in primate society. Folia primat. *8:* 89-120 (1968).

Gartlan, J. S. and Struhsaker, T. T.: Polyspecific associations and niche separation of rain-forest anthropoids in Cameroon, West Africa. J. Zool. Lond. *168:* 221-266 (1972).

Gause, G. F.: The struggle for existence (Williams & Williams, Baltimore 1934).

Gautier, J. P. et Gautier-Hion, A.: Les associations polyspécifiques chez les cercopithécidae du Gabon. Terre Vie *23:* 164-201 (1969).

Hall, K. R. L.: Ecology and behaviour of baboons, patas and vervet monkeys in Uganda; in Vagtborg The baboon in medical research, vol. 1, pp. 43-61 (Univ. of Texas Press, Austin 1965).

Hill, W. C. O.: The primates, vol. 6, (Univ. of Edinburgh Press, Edinburgh 1966).

Jolly, C. J.: The classification and natural history of *Theropithecus (Simopithecus)* (Andrews, 1916), baboons of the African Plio-Pleistocene. Bull. Brit. Mus. (Nat. Hist.), Geol. *22:* 1-123 (1972).

Mackworth-Praed, C. W. and Grant, L. H. B.: Birds of eastern and northeastern Africa, vol. 2 (Longmans, Green, London 1960).

Nagel, U.: A comparison of anubis baboons, hamadryas baboons and their hybrids at a species border in Ethiopia. Folia primat. *19:* 104-165 (1973).

Napier, J. R. and Napier, P. H.: A handbook of living primates (Academic Press, London 1967).

Rodman, P. S.: Synecology of Bornean primates. I. A test for interspecific interactions in spatial distribution of five species. Amer. J. phys. Anthrop. *38:* 655-660 (1973).

Struhsaker, T. T.: Ecology of vervet monkeys (*Cercopithecus aethiops*) in the Masai-Amboseli Game Researve, Kenya. Ecology *48:* 891-904 (1967).

Tappen, N. C.: Problems of distribution and adaptation of the African monkeys. Curr. Anthrop. *1:* 91-120 (1960).

11

Forest Living Baboons in Uganda

1966

T. E. ROWELL

The climate and vegetation of the study area is described: in many ways it was not typical of baboon habitats in Uganda, but had the advantage and interest of offering a clear choice of a variety of habitats. The structure and composition of the baboon population is described— this was a rapidly expanding population with no seasonality in breeding. Their use of the habitat and some aspects of their organization are discussed. The ways in which their behaviour differs from that of previously described baboon populations are stressed, and these are related to differences in habitat. In general these baboons tended towards behaviour regarded as typical of arboreal primate species as compared to terrestrial ones. It is suggested that direct environmental effects may be responsible for many differences at present regarded as specific, so that caution and intraspecific comparative studies are needed before behaviour is described as species typical.

INTRODUCTION

Baboons in many areas live in more or less open country, which makes them an excellent subject for behavioural observations, in contrast to the monkeys of tall dense forest. Thus it is not surprising that field studies to date (the most recent are summarized by DeVore & Hall (1965)) have been made in grassland habitats with good visibility. This accumulation of data has given the impression that the baboon can be described as a savannah animal, and many of the features peculiar to the genus *Papio* have been accepted as

Reprinted from *Journal of Zoology, London 149*: 344-364 (1966) by permission of The Zoological Society of London.

adaptations to this habitat, these ideas in turn being incorporated into theories of primate evolution (e.g. Washburn & DeVore (1961)). It is not the purpose of this article to question such theories directly, but rather to show that the picture has been oversimplified as far as the baboon is concerned: over a large part of their range baboons live in forested country, and this description of a population in such a habitat brings new evidence of the variability of behaviour of these animals.

In the first place, the general ecology itself has been much oversimplified in this context. There is, of course, in Africa no simple choice for the primate between high rain forest and open, short grass plains: there are large areas of every intermediate type of "bush" giving different amounts of cover and richness of food supply. Nor is it known how far the present picture is due to human influence, especially through clearing and burning, and therefore extremely recent in evolutionary terms. In present day Uganda, for example (where matches are universally available) there are very few areas which are not burned at least once a year, and in this the number of natural fires is negligible. The lake forest and other damp forests do not burn, but nearly all of these have undergone extensive felling of timber trees in the last 60 years. Clearly the human population is one of the largest single influences on the vegetation in this country, and it is against a background of this degree of disturbance that an assessment of the age and prehuman extent of short grass plains and other habitats must be made. (In the study area, and in other parts of the Queen Elizabeth National Park, regular burning has been a feature of park management. Comparison of present conditions with aerial photographs taken 10 years ago shows that the number of trees outside areas of continuous forest has been much reduced in that time.)

In Uganda there are no extensive short-grass plains. Where there is open grassland the dominant grass species grow to about five feet, and it is impossible to study animals two or three feet tall. This study was primarily of social behaviour and reasonable observation conditions were essential, so an extensive preliminary search for a good area was made, and that eventually chosen was in many ways not typical of the habitats chosen by Ugandan baboons in general. The majority of populations considered lived in high forest or extremely dense bush, and were pointed out because they were in the habit of sitting on a roadside or an exposed rock for a short time each day. In some ways, however, the area chosen for the study was of particular interest, because it offered a series of very clearly and sharply separated vegetation types, including both forest and short grass. Thus the baboons had a choice of habitat, and their behaviour in different vegetation could be compared.

In the following sections, the habitat in which the observations were made is first described in some detail, because it is becoming apparent that insufficient attention has been paid to the direct effects of habitat on primate organizations, and there is need for much more intraspecific comparison with this factor in mind (Rowell, in press (*a*)). The density and structure of the baboon population is described, their economy, and finally some aspects of their social interactions which might be directly related to properties of their habitat.

(A preliminary account of some of the findings was read to the East African Academy in 1964 (Rowell, 1966).)

THE STUDY AREA

1. Geography and vegetation

The study was carried out in the area around the Ishasha river camp, in the southern tip of the Queen Elizabeth National Park, Uganda (0° 37' S. 29° 40' E). The Ishasha river is permanent, fast flowing, cold, and runs north to Lake Edward, forming the boundary between Uganda and the Congo (Parc National Albert) at this point. It flows in a slight trough in the surrounding grassland, in which there is a strip of gallery forest, on average about a quarter of a mile wide. In some places the forest fills the trough, in others there is a strip of coarse grass or patchy low bush between the edge of the forest and the edge of the trough (Fig. 1). Within the forest the river meanders and frequently changes course, undermining old trees and leaving new areas to be recolonized, so that the vegetation is very varied, ranging from newly abandoned riverbed and oxbows to tall forest. Because of the trough the forest looks unimpressive from the surrounding grassland, but the canopy trees reach 80 to 100 feet in the older established parts.

The river is bridged every few yards by fallen trees, undermined as the bank is cut away. The banks from which the river is retreating are colonized by a lush grass *Paspalum conjugatum,* and by tangled thickets of a rambling bush, *Alchornea cordifolia.* There is an abrupt boundary between the forest and the grassland. In the wetter areas the forest is adjoined by areas dominated by a tall coarse grass, *Imperata cylindrica,* and in drier places by sparse short grasses heavily grazed by hippo, with frequent bare sandy areas and

Fig. 1. A diagrammatic cross section of the trough in which the Ishasha river flows, to show the distribution of vegetation. From left to right: grassland, forest to trough edge, river bridged by newly undercut tree, forest, forset edge, coarse wet grass, short grass with sandy areas, low bush, termite mound, solitary tree, grassland.

some big termite mounds. Here also there are patches of low bush, up to about five feet high, including berry-bearing species like *Securinega virosa*, among which young forest trees may become established. Along lines of seepage draining the surrounding grassland (sometimes streams in wet weather) there are patches of high bush—shrubby trees up to 15 feet high, and especially at the northern end of the area as the river nears the lake there are large areas of this type of vegetation. In the open grassland, especially not far from the forest, a bramble-like fruit-bearing shrub, *Capparis tomentosa* frequently colonizes old termite mounds; again in the grassland, especially outside the river trough a low shrub *Maerua edulis* is very common, and forms an important part of the baboon diet. Isolated trees, mainly fig and acacia are fairly common in the northern and southern edges of the area's grassland, and some acacias grow near the edge of the forest but separated from it by grass; a series of dead trees in similar positions suggests that such trees were more common in the past.

The distribution and relative area of vegetation types is shown in Fig. 2.

2. Climate and seasons

The area is within half a degree of the equator, and there is very little predictable seasonal variation. Midday shade temperatures are usually between 30 and 32°C, at night the temperature falls to about 19°C. Variation depends on cloud cover, which is in turn related to rainfall. Fig. 3 shows the rainfall recorded at Ishasha camp during part of the study period. There are two annual "wet seasons" at the equinoxes, but rain falls throughout the year and it is exceptional to have more than two weeks without rain. The difference between the same month in two successive years can be as large as the difference between "wet" and "dry" seasons. The soil drains very rapidly after rainfall which usually occurs in short heavy bursts. The seasonality of the plants is complex, as might be expected in such a climate, and is discussed further in section 2(c). Animal species in the area also varied in this respect—of the two most numerous antelopes, living in partly mixed herds, the topi has a sharply defined breeding season, while the kob breeds throughout the year.

Burning was carried out over most of the grassland once or twice in the year during dry spells. It affected the movements of animals, including baboons.

3. Other large mammals

Colobus abyssinicus, *Cercopithecus ascanius* and chimpanzees live in the gallery forest, and *Cercopithecus aethiops* in the grassland. Elephant, buffalo and hippo

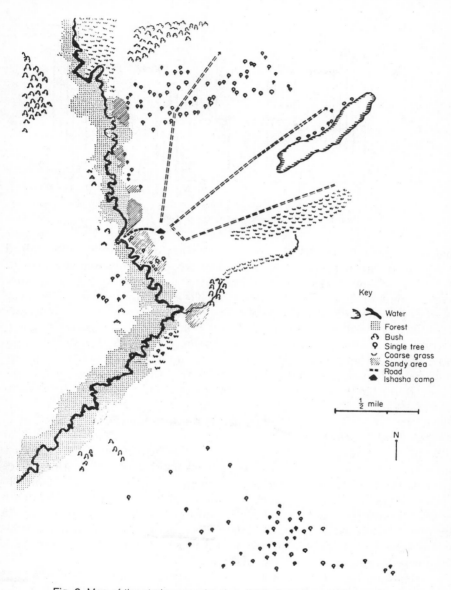

Fig. 2. Map of the study area, showing distribution of vegetation types.

are all common, and their grazing, browsing and trackmaking have considerable effect on the baboons habitat, especially in maintaining open tracks through the forest which are used by all other species. Elephants and baboons also share many food plants, and so are potential competitors.

Potential predators include lions, leopards, and possibly serval cats.

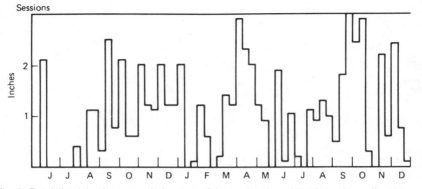

Fig. 3. Rainfall at Ishasha camp during part of the study, plotted in 10-day intervals. Above, thick lines show the timing of observation sessions during this period.

Baboons alerted and made alarm noises when non-hunting lions walked near the troop in the early morning on two occasions. Alarm noises of baboons, "singing" of colobus and sawing of leopard were heard together in the night on two or three occasions. No other interactions with these predators were seen or deduced. Antelope include topi, kob, waterbuck and bushbuck but interaction between the baboons and these, and warthog and forest hog seemed to be minimal. Ungulates and baboons, though they took note of each others' alarm calls, rarely acted on them; there was none of the interspecific co-operation reported by DeVore (1963). African hares (*Lepus capensis*) lived in the open grass and were preyed on by baboons (see section 2(c)).

ANIMALS AND METHOD — *time sampling*
The baboons were of the "olive" type—large, heavily-built animals, dark grey in colour, some individuals having black hands and/or feet; their noses protruding beyond the mouth. *Papio anubis doguera* would be a reasonable name to choose from the present taxonomic confusion. Troops living in the study area were strung along the river: most of the observations were made on two adjacent troops, S and V, about 90 animals altogether, but observations were also made on the troops upstream (F) and downstream (O), comprised of 50 or so animals each.

No attempt was made to watch the animals continuously; instead a time sampling method was used. The study area was visited for roughly 2 weeks in every 2 months, so that nearly all times of the year were covered (no observations were made in May or October). Table 1 gives the days worked (a 2- or 3-day preliminary survey in April 1963 is not included here) and the relation of these sessions to seasonal changes in rainfall is shown in Fig. 3. It was not possible, unfortunately, to follow the animals continuously during daylight hours, because there was too much thick cover. Table 2 shows the number of hours at different times of day in which observations on the three

Table 1
Observation sessions in the study area

	Dates	No. of days
1.	25. 6.63- 5. 7.63	9
2.	9. 8.63-21. 8.63	13
3.	27.11.63-12.12.63	16
4.	29. 1.64-11. 2.64	14
5.	27. 3.64- 8. 4.64	13
6.	25. 6.64-15. 7.64	21
7.	13. 9.64-23. 9.64	11
8.	8.11.64-15.11.64	8
9.	26. 3.65- 4. 4.65	10
	Total	115 days
	Total available daylight	1380 h

best-known troops were made in sessions 2 to 9 inclusive. The S troop's position was known for about a third of the available time, the V troop's for about a quarter. For slightly more than half of this time the baboons were in such a position that useful observations on social behaviour could be made. On the other hand, there was of course a lot of negative information not included here—even when a troop's exact position was not known, its possible whereabouts could be determined, by a process of elimination.

Most observations were made on foot, accompanied by a Park Ranger, and were recorded in writing.

RESULTS

1. Population composition, structure and growth

Recognition of individuals was difficult, and though eventually many of the adults were known, juveniles were rarely identified. On the other hand,

Table 2
Hours during which the troops' positions were known (excluding 12 trips to southern fig trees)

Hour beginning:	07	08	09	10	11	12	13	14	15	16	17	18	Totals
Troop S	26	36	44	50	54	52	29	19	19	21	20	16	386
V	54	46	29	22	16	12	11	14	19	34	38	44	339
F	0	5	16	12	5	2	0	1	2	0	1	0	44
Totals	80	87	89	84	75	66	40	34	40	55	59	60	769

Table 3

Composition of the best known troops at the final session (March 1965)

	S Troop	V Troop
Adult ♂	5	14
Subadult ♂	0	3
Adult females:		
with swellings	1	5
pregnant	1	3
lactation interval	2	4
others	1	4
Total adult ♀♀	5	16
Large juvenile ♂	8	2
Large juvenile ♀	2	2
Large juvenile ?		1
Small juvenile ♂	3	
Small juvenile ♀	2	
Small juvenile ?	1	7
6 month–1 year (grey babies)	3	9
4–6 month (intermediate babies)	1	1
4 months black babies	2	3
Totals	32	58

immediate classification of baboons by age, sex, and reproductive state is relatively easy. Babies up to about six months could be aged to within a week or so (using comparison with babies born in a caged group) by colour changes in the fur and naked skin areas. After that the young were classed on an estimate of size, four sizes of females and five of males corresponding roughly to year classes (these classes are condensed in the following analysis). Colour changes in the naked rump of the female indicate pregnancy, lactation, or cycling: this skin is normally black, but begins to redden in early pregnancy, is bright red in the second half of pregnancy, fades to a pale pink soon after birth and then gradually regains the black pigment during lactation. In addition, there is, of course, the perineal "sexual swelling" which accompanies oestrus in the cycling female, and menstruation could also be observed. Males were considered adult when they showed full canines and a mane: they varied in build and mane development, but this was not obviously correlated with the social role.

(a) Troop Composition. Table 3 shows the composition of the S and V troops as far as it could be determined at the last session. A notable feature is the equality of the adult sex relation. In this, the population differed from those reported by other workers, who have all found a preponderance of females. In V troop there was in fact a slight excess of male over female adults at the

start of the study, but then several young females matured and the balance was restored.

Throughout the study period, S troop had an almost exclusively male group of large juveniles. One female in this class began to cycle at the end of the last session, thus joining the adult females. The sex ratio of the babies born in this troop during the study was about equal.

The female to juvenile ratio in this population (1: 2·5) was higher than the average found by DeVore & Hall (1965) (1:2), which suggests a slightly higher survival rate; but the figures for the two troops are very different, and the actual figures recorded seem to be something of a historical accident. Thus that for V troop, (1: 1·8) was greatly increased when a group of young females matured but were still too young to have contributed to the juvenile group; at the beginning of the study the ratio would have been nearer 1:3. On the other hand the ratio was unusually low for S troop (1:4·2) because the large juveniles were nearly all males, and so were still classed as juveniles, at an age when females would have already contributed an infant to the population.

It should be stressed that a composition table as given is an abstraction from a continuum which changes and develops with time. In a small population random fluctuations in a birth sex-ratio which may be unity in the long run can produce quite large variation in the population structure—as in the unusual structure of S troop as this time. We have no idea of the effect of such variation on social organization as yet, or whether it persists or is altered towards a mean population structure by migration of individuals, for example. However, it would seem an important first step that the existence of the statistical possibility of such variation should be recognized.

(b) Mortality. During the study the body of one animal was found, too late to discover the cause of death—a known subadult male from S troop. One female of V troop probably lost an infant at birth. No other traces of dead baboons were found, and no other known animals disappeared. Thus there is no information about population control factors, and indeed the population increased steadily over the two years.

(c) Births. Table 4 gives the months in which births were recorded. Some of them were observed directly, most are estimated from the age of the babies when first seen and are judged accurate to within two or three weeks. No birth season was detectable. Pregnant and cycling females, as well as young babies, were always present in the population. Of course in a small troop, for example S troop with only five adult females, there could not be all stages of the reproductive cycle always present, and there were periods in this troop when no female was sexually active for several months. These data do not preclude a seasonal fluctuation in births, perhaps with biennial peaks, as opposed to a sharply demarcated breeding season. Less than 30 females are represented here, and to detect or disprove such a fluctuation would require enormous numbers of animals or a very lengthy observation period. Females in this population gave birth at much shorter intervals than

Table 4

Changes in the baboon population. Brackets indicate unchecked counts

		Births			Troop counts			
		V	S	F *(incomplete)*	V	S	S+F	F
1963	Jan.	1	2					
	Feb.							
	Mar.		1					
	Apr.							
	May			1				
	June	1	1		29			
	July	1		1			(50)	
	Aug.	1	1		45	30		
	Sept.	1						
	Oct.	1		1				
	Nov.				48			
	Dec.	1						
1964	Jan.	1		1				
	Feb.			1				
	Mar.							
	April				29		(75)	
	May	1	1	1				
	June							
	July		1		49	32		
	Aug.	1	1					
	Sept.	1			54			
	Oct.		1					
	Nov.				55	31		
	Dec.	1	1	2				
1965	Jan.			2				
	Feb.	3		2				
	Mar.		1	1	58	31		(62)

estimated by DeVore & Hall (1965). Pregnancy lasted six months, and was followed by a lactation interval of about five months. There were then one to three menstrual cycles of about five weeks before the female became pregnant again. In the S troop the birth interval was between 12 and 16 months. Babies were suckled until replaced by the new infant. DeVore & Hall (1965) estimated a lactation interval of 12 to 15 months, and hence a minimum birth interval of 18 months, extended to two years in populations with an annual breeding season. The shorter interval found here was also found in a caged group kept in Uganda, which included both "olive" Uganda and "yellow" Kenya baboons.

(d) Numbers. Opportunities for really reliable troop counts were rare, be-

cause of the dense vegetation. Table 4 shows the numbers obtained. The V troop increased steadily, and the increase could be related to the number of births fairly well, except for the addition of two adult males between July and September 1964. Excluding these males, the troop increased at nearly 15% per year over the two years. This is clearly not a trend which can continue indefinitely, and it is hoped to follow this population for a year or two to see what happens. S troop also increased but much more slowly. The picture is complicated by movements between troops, which were particularly difficult to follow since the two best known troops did not exchange members with each other, but with the outlying troops in which few individuals were known. S troop especially, associated fairly often with its neighbour, F. The number of adult males in S troop fluctuated between four and five and once reached six; two individuals were always present, and a third for most of the time, but the remaining one or two had often changed between sessions. This troop also did not increase in accordance with its own birth rate, and some older juveniles may have left it. At the same time F troop, or counts of S and F together, increased so rapidly that it was taken as more probable that the disappearing members of S were moving to F rather than dying without trace. These movements were not accompanied by social disturbances, and their frequency was probably underestimated until the last session, when assistance was available for making daily counts and attention was concentrated on this aspect. At this time S and F troops were in contact quite frequently, and it was apparent that some individuals were moving to and fro from day to day. Only by knowing every individual in the population could such movements be detected in ordinary circumstances, and of course, it is the animals which do not move which are easier to learn to recognize. As an example of the tolerance of troops towards new members, the case of the adult male vervet may be mentioned. This animal lived with the V troop for at least two years (its origin is unknown) and then moved to the next troop but one, F, where it was equally accepted in grooming and sexual interactions.

(c) Troop identity. With this amount of movement between troops—the occasional joining together of two troops in foraging expeditions (S and F), and the fact that as the V troop increased in size it occasionally divided into smaller parties which followed different routes during part of the day—the validity of the troop, as a population unit, might be questioned. This is a theoretical query, there was no subjective doubt in the field as to the entry of the troops, probably because each had a stable nucleus of known individuals; in particular there was no indication that any female, and certainly no adult female, ever exchanged troops. The difficulty in this and other field studies is the short time available relative to the life-span of the animals. We assume that in an expanding population, troops subdivide, but this may well be a slow process lasting several years, and even after it is complete the adults of the two new troops will retain some memory of previous social relationships. A short study effectively takes a cross section through this process, and may find it at any stage. In this population, for example, it is possible that S and F troops had divided relatively recently, and that V troop was in the very early stages of division while still retaining some familiarity with its downstream neighbour. In this type of dynamic situation it is clearly a waste of time to aim for any more precise definition of the troop than is useful in the field. In the short term, it is probably the conservatism of the females that is the basis of troop identity.

2. Use of Habitat

(a) Range. Fig. 4 shows the estimated area used by the two main troops (the range in the Congo on the left bank of the river is estimated from the time the animals spent there, and assuming their speed of movement remained about the same, when out of sight). The areas are roughly 1·5 square miles for V troop and two square miles for S troop. There was about 4500 square yards of overlap in the two ranges, so that the overall population density provided by these two troops is about 28 baboons per square mile (the true figure is rather higher, since there was also overlap with the troops on the other sides, especially between S and F). These estimates are very far from the average home range of 15 square miles and a population density of ten per square mile given by DeVore (1965). In fact they fall within the range of figures given by him for typically arboreal species, with much more restricted range and higher density.

The total area used, however, does not give a very useful picture of the animals relation to their environment since they did not use all parts of the range equally. The area was divided into grid squares of approximately 200 yards square, and the number of hours in which each troop was observed in each square was recorded. If they moved during the hour, the square in which they spent the most time was recorded. This grid-hour system gave a reasonable picture of the animals' movements in most places, but it could not be usefully applied in the south-east part of the range of S and F troops, because their behaviour there was quite different. The troops would walk briskly to the scattered fig and acacia trees in the south-east, moving through perhaps ten squares in less than an hour. They then fed heavily, mainly on ripe figs, and walked rather more slowly back to the forest. On these excursions the time spent in any one square was too small to be reasonably compared with movements in the rest of the range. This behaviour by S troop also explains their larger home range than that of V troop, although their numbers were smaller. Nearly half of the area which they were known to use, was effectively a corridor for reaching this single occasional food source.

Each troop concentrated much of its activity in a very small part of the available area. These could perhaps be regarded as "core areas" of the ranges, but they were not the exclusive "property" of any one troop. There

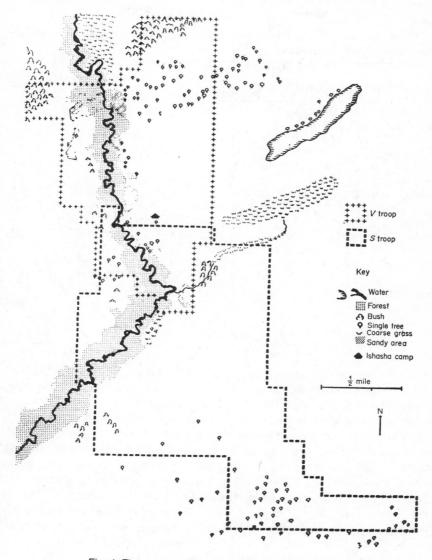

Fig. 4. The ranges of the S and V troops, on a 200-yd grid.

was a suggestion, though no good evidence, that S and V troops avoided each other in their area of overlap in so far as each troop usually used it whenthe other was at the far end of its range. F and S troops, as has already been noted, sometimes moved together in their area of overlap. No behaviourwas seen which could be interpreted as territorial defence.

(b) Range in relation to vegetation types. Table 5 analyses the grid-hours

Table 5
Time spent observing S and V troops in different habitats (excluding 7 trips by S troop to southern fig trees)

Hour beginning:	07	08	09	10	11	12	13	14	15	16	17	18	Totals
Forest	78	53	36	26	20	14	7	6	6	7	18	50	321
River-edge bush	0	10	10	10	3	3	4	2	1	2	2	0	47
Forest/grass edge	0	8	6	13	12	9	3	2	2	2	8	3	68
High bush	0	0	0	1	5	3	4	4	6	4	2	0	29
Low bush	0	0	4	1	5	4	2	4	2	5	6	0	33
Around solitary tree	0	0	1	0	3	6	2	4	3	5	5	0	29
In coarse grass	0	3	3	3	4	6	3	1	2	2	3	0	30
Short grass	1	7	8	12	16	16	12	7	10	16	7	3	115
Sandy areas	1	1	5	5	0	3	3	3	6	12	7	4	50
	80	82	73	71	68	64	40	33	38	55	58	60	722

available for the S and V troops according to the type of vegetation in which they were observed. The proportion of time in different vegetation types was not related to their availability in the area. Most strikingly, the time spent in forest was very high: in the home ranges of S and V troops, 18% of the grid squares contained forest, river edge bush, or forest edge, but the animals spent 60% of the time in them. The home ranges themselves already represent a selection of forest—the proportion of forest in the study area as a whole is about 12% (Fig. 2). Part of this preference is related to the fact that they always returned to the forest to sleep, choosing particularly tall trees as resting places. On gloomy days they had settled to sleep as early as five o'clock, very occasionally they did not reach the chosen trees until after dark (7.00–7.30 p.m.). The relation, shown in Table 5, between vegetation type occupied and time of day, gives an overall picture of daily movements. Baboons are not early risers, and they rarely left the sleeping trees before 8.00 am (dawn 6.00–6.30 a.m.). They usually fed in the sleeping tree and neighbouring trees until mid morning, feeding gradually giving way to play and grooming and finally dozing. After this, a move would be made to another part of the forest, or out to feed in the open grass or one of the areas of bush or isolated trees—hence observations in the middle of the day were in a wide variety of vegetation types, with no one type especially frequent. Towards evening they drift back to the forest and commonly sit about playing and grooming in the open sandy areas at this time before moving in to the sleeping trees. The timing of these moves could of course vary a great deal, this is only a general picture. The distance travelled in any one day was not very great; DeVore & Hall (1965) give three miles as an average day's journey, but here four miles was the longest, a mile or a mile and a half typical, and on some days they moved only a few hundred yards. There were relatively few routes used, but a troop never took exactly the same path on two successive days.

This discussion of range use is based on a time sample and not a complete record, so possible sources of bias must be considered.

(i)   Observations were not evenly distributed throughout the day (Table 5).

(ii) The observer's own learning processes meant that the animals were looked for most in places where they had been observed previously.

(iii) Visibility was not equal in all areas. It was impossible to follow movements more than a few days from the river on the Congo side, and in some areas of bush they were also invisible. They could be spotted much further away in open grass but behaviour differences in cover and open ground tended to cancel this advantage (see later).

(iv) It is possible that in spite of efforts to the contrary, the observer was "herding" the baboons by her own movements. This certainly occurred occasionally, but on the whole the animals' behaviour did not suggest it was happening.

All these factors would perhaps tend to overemphasize the time spent in or near the forest. Against them can be set the subjective impression that the picture obtained was not unduly disturbed, and the fact that after the first session (not included in the analysis) the troops rarely turned up in an entirely unpredicted place. Though too much attention should not be paid to the details of the figures given, the qualitative picture they illustrate is probably valid, and is sufficiently clear cut to survive, after allowing for the biasing factors.

(c) Food. The food supply was striking in its abundance and variety. Table 6 gives a list of important food plants together with the type of vegetation in which they occur. A complete list of plants eaten would probably be approximate to the botanical species list for the area, for instance only one fruit seen was not eaten, and that an extremely bitter and poisonous Solanum.

(i) Seasonality. Some food plants had a co-ordinated and relatively brief annual season, but even these formed part of the diet for several months (e.g. Cynometra seeds from November to March). Others, like Pseudospondias, or the fig Ficus guaphalocarpa followed clear cycles as individual trees, of leaf growth, flowering and fruiting, but the trees in the area provided edible fruit at different times. It was in relation to such trees in particular, that the baboons' local knowledge seemed important, and each tree was exploited

Table 6

Important food plants of Ishasha baboons

Plant	Form	Part eaten	Habitat
Cynometra alexandri (C.H. Wright)	Tree	Seed from pod	Forest
Pseudospondias Microcarpa ((A. Rich.) Engl.)	Tree	Fruit pulp (seed undigested when swallowed)	Forest
Parkia filicoidea (Wilw. ex Oliv.)	Tree	Flower and seeds	Forest
Treculea africana (Decne)	Tree	Flower and fruit	Forest
Sorindeia sp.	Tree	Fruit	Forest
Ficus barteri (Sprague)	Tree	Fruit	Forest
Ficus sp.	Tree	Fruit	Forest
Ficus guaphalocarpa (Mik.) (A. Rich.)	Tree	Fruit	Forest edge or solitary tree
Acacia sieberiana (D.C.)	Tree	(Flower and) seeds	Solitary tree
Euclea latidens (Stapf)	Small tree	Fruit	Forest edge or high bush
Maerua duchesnii ((De Wild) F. White)	Small tree or bush	Fruit, tubers	Forest
Tavenna graveolus ((S. Moore) Brem)	Large bush	Fruit	High bush
Alchornea cordifolia ((Schum.) Müll. Arg.)	Rambling bush	Fruit	Riverside, forest
Erythrococcus bongensis (Pax.)	Low bush	Fruit	Forest edge and low bush
Securinega virosa (Roxb. ex Willd.) Baill	Low bush	Fruit	Forest edge and low bush
Capparis tomentosa (Lam.)	Low bush	Fruit	Low bush and grassland
Maerua edulis (Gilg. and Bened.) De Wolf	Shrub	Buds, fruit, root	Grassland
Paspalum conjugatum (Berg.)	Grass	Seeds	Forest
Imperata cylindrica (L.) Beauv.	Coarse grass	New shoots	Wet grassland

Table 6 (continued)

Plant	Form	Part eaten	Habitat
Other grasses		New shoots (after burn) and storage leaf bases	Grassland
Rhamphicarpa montana (N.E.Br.)	Herb	Flowers	Grassland
Cucumis aculeatus (Cogn.)	Creeping herb	Leaves, seedlings	Grassland (seedlings from elephant dung)
Sansevieria sp.	Aloe	Root	Under bushes
Mushroom spp.	Mushroom	Fruit bodies	Grassland

as it became edible. Occasionally a baboon, usually an adult male, was seen to leave the main party to inspect a tree, taste the fruit, and if it was not ready, to spit it out and return to the troop. Other plants seemed to have biennial seasons, and yet others were eaten at nearly every stage—e.g. *Maerua edulis,* whose new shoots, roots buds and fruit at all stages of ripeness provided a year round part of the diet. With these differences in the seasonality of the plants, and their varied rates of development, there was no season at which a wide variety of food plants was not available.

(ii) Proteins. Many of the plant foods were protein-storing seeds, like *Parkia* and Acacia Fruits were, for the most part, eaten long before they would be considered ripe by human standards, green and often extremely bitter (though the sweet, insipid, ripe figs were relished as well as the green). At this stage the fruit flesh presumably contains a much higher proportion of protein than later in ripening. Of grasses, the parts eaten were new shoots, green seedheads, and the storage leaf bases of some species, which are again high in protein content. It is suggested that by careful selection of diet from a wide range of species and stages, the baboons obtained a much higher protein intake than might be expected—and much higher than is normally offered to them as a diet in captivity.

The mainly herbivorous diet was supplemented by animal protein. Insects (mainly grasshoppers and butterflies) were grabbed in the open grass; land snails (*Limicolaria*) were suspected but never confirmed as food (a baboon found and bit open a dead shell and expressed disgust when he found it full of mud); birds' nests were investigated and a passing bird grabbed at, but again eating was not seen. Occasional lizards were caught and eaten. On six occasions, hares (*Lepus capensis*) were coursed by baboons after being flushed in open grass, and on four of these they were successfully caught and eaten. This represents a catch rate of one every 30 hours that baboons were observed in the appropriate vegetation, which, over the year, would represent both considerable protein input into the population, and predation pressure on the hares. (One captor was a female, one a juvenile male, and two adult males.)

(iii) *Exploitation.* Different classes of baboons did not eat the same diets. For example, small juveniles climbed slender saplings and ate leaves inaccessible to larger animals; large juvenile males tended to climb in the highest and flimsiest branches of the forest trees, eating bark, insects and leaves that were not exploited by other classes; females with young babies typically sat together on the ground, under low bushes, and ate a higher proportion of grassheads than other classes. These are unsystematic observations, and there were many exceptions to the rule, but they do suggest an interesting possibility: that in situations where food is not quite so superabundant, the members of a baboon troop may compete less for the available supply than might be expected from their density and gregarious foraging.

3. Behavioural Aspects

At any time, the overall activity of a troop could be classed as resting (sleeping, grooming, playing), walking (a brisk movement, usually short, often more or less in a single file, from one resting or eating place to the next), eating (usually all on the same food), or foraging. A foraging troop walks well spread out through the grass, individuals frequently stopping to pick up food; they may do a round trip from a base, where some may sit and wait for the rest to return (mainly females with young). The animals give the impression of looking for something—insects and hares were always caught during this activity. Vegetation types were associated with different activities: open sandy areas, and open places on the river bank, were favoured as resting places; foraging was seen mainly in short or course grass. Isolated trees were typically eating places, abandoned as soon as the troop was satiated. In forest, and high bush, the pattern was eating and resting, followed by a very short walk to the next site. This pattern, also reflected in the short daily journey, is different from that of open country baboon troops, which seem to keep moving fairly steadily through the day, foraging as they go.

(a) *Flight Distance.* The response to the approach of people, on foot or in cars, was dramatically different in different parts of the range. In the forest, it was possible to watch the animals across the river, at a distance of 15 yards. In grass within the river trough, 30 to 40 yards was acceptable. Outside the trough the distance suddenly lengthened to nearer a 100 yards, and the S and F troops, when at the southern fig trees, would flee from a car at over 200 yards, although there was a park track there and cars were not infrequent. Flight was always towards the forest, and, if followed, they allowed us to approach closer and closer and would settle to feed and groom at the forest edge within 30 yards of the car. Within this pattern there were some modifications: any attempt to move between the baboons and the forest at any distance, provoked panic flight back to the forest; and they would approach much closer in a few places where they were used to seeing the observer, than at similar but less-used observation points—there seemed to be little general acceptance of the observer, mainly a limited acceptance of a few specific situations.

In respect of flight distance, then, the baboons behaved like animals that "felt at home" in the forest, but made potentially dangerous sorties into the open, rather than like grassland animals.

(b) Baboons in the forest. Since the baboon has frequently been cited as adapted to savannah life, his performance in the forest is of some interest. In this strip of gallery forest lived chimpanzees, black-and-white colobus (*Colobus abyssinicus*) and red tail monkeys (*Cercopithecus ascanius*) whose loco-motion and other behaviour could be compared directly with that of the baboon. The species are of course very different in weight, so that the same tree presents different problems to each. Both chimpanzees and baboons would move at any level of the forest in their own time, but came to the ground and used the hippo tracks when alarmed, while the lighter colobus and red tails fled into the trees. Chimpanzees were cautious climbers, and never seemed to jump; the distances which the other three jumped were about the same (i.e. relatively much wider for the smaller red tails), and they hesitated, and carried large babies, across the same gaps. Red tails were very light and nimble, colobus seemed clumsy in handling the trees, hurling themselves at whole clumps of branches, compared with the baboons which would manipulate, using a thin branch to pull a thicked one towards them from the next tree, for example. A colobus was seen to fall, but never a baboon. Chimpanzees, of course, brachiated, as did young baboons in play; older baboons occasionally hung by their arms for single moves, colobus were only seen to move on the top of branches. Vertical stratification be-tween the species was not marked when the animals were not frightened; for instance, colobus would travel through a tree below a baboon group, or sit amongst them.

In summary, whatever the baboons' adaptation to life in grassland, they do not seem any less competent as a forest animal than other primates, particularly if the mechanical problems related to size differences are taken into account.

(c) Troop organization. Some points of general troop organization which differ from published accounts seem worthy of mention since they may be related to habitat.

(i) Troop Movements. Previous workers (e.g. Washburn & DeVore, 1961) have been impressed by the organized deployment of baboons by age/sex class

as they move from place to place. By protecting mothers of young babies in the centre of the group and subjecting expendable young males to periferal dangers, this deployment has been taken to have strong survival value. Such organization was not observable in this population. Only two frequent placement patterns were detected: an adult male often sat, glancing back down the trail, until the whole troop was ahead of him before moving on; and the outlying animals, following the most divergent paths, were most usually pregnant females. Young males moved with the main group. There was great concern for stragglers. The whole group would wait for half an hour for a female to finish grooming her baby and catch up, and there was no question of any animal being left behind, even those with injuries that made walking difficult.

Possibly the lack of organized deployment was related to the short distances usually travelled between long stationary periods. DeVore (1963) also described fleeing troops forming a bodyguard of adult males between the source of danger and the rest of the troop. This deployment was seen on occasion, but only when the cause of alarm was so slight that the more confident adult males did not respond to something that set the juveniles running; a stronger stimulus produced precipitate flight, with the big males well to the front and the last animals usually the females carrying heavier babies.

(ii) Noises. DeVore (1963) cited frequent contact noises as characteristic of forest species, but it seems likely that this again is a direct effect of the environment—a response to thick cover. Baboons moving through head high grass kept up a continuous chorus of quiet grunts, audible only for a short distance. Of particular interest was the use of the loud two-syllabled bark of the adult male, usually described as an alarm bark. The alarm bark common to all classes of baboons is a shrill single bark similar to that of the rhesus (Rowell & Hinde, 1962). The male double bark was given *after* a source of danger was recognized, and used to "comment on" its movements or the movements of the troop relative to it, usually by a single male which had moved to a good vantage point. Thus if the observer surprised a troop, the first noise would be a shrill bark, then a male might come and make a few double barks. Troop and observer would then sit quietly together for an hour or so, though movement by the observer would elicit a double bark. When the troop finally moved, a male might come to a vantage point and double-bark before following the last member. The bark was also used when the troop moved through forest: some baboons would move away, then a male with them would bark, answered by a male with the remaining part of the troop. These two would continue to bark alternately while the rest gradually drifted from one to the other, the barking male staying till last. This noise seemed to convey information about position of the barker relative to some focus of interest; although often heard in the general

context of flight or withdrawal, its function was more than the mere registration of alarm, for which other noises were available.

(iii) The role of the big males. The above examples have already shown some of the big males' activities in the group. The general impression was, above all, one of mutual co-operation between them as they policed the environment. (There is a striking contrast between these animals, living in a largely male-oriented interaction pattern, and the bored solitary male caged with females with little social activity available except the sexual, which normally occupies only a small fraction of such an animal's time). The males' remaining social interactions chiefly involved mutual grooming with females and acting as focus of juvenile play. In this pattern the "dominance ordered society" considered typical of baboons since the descriptions of Zuckerman (1932) could not be observed. Inter-adult interactions which might have displayed it (threat, fighting, supplanting) were extremely rare—in contrast to a caged group of the same species (Rowell, in preparation). The typical male co-operation immediately complicated any dispute as supporters were recruited and changed sides, so that a conclusive outcome never ensued. Differences in the environments may be related to this difference between behaviour observed and that previously reported by other workers. In particular the abundance of food and its scattered distribution meant that there was rarely competition for it; disputes were seen over some isolated prized items—seedlings growing in elephant dung, and mushrooms. In an environment where the food competition situation occurs more frequently (feeding by tourists is an extreme case of this) it may be that tensions can be sustained until a hierarchical social structure develops. Another factor to be considered is the density of cover. In thick bush disputes are often ended when visual contact is lost—perhaps in an open environment they continue longer and have a more lasting effect.

DISCUSSION

Baboons are usually regarded as animals of open country, yet this population spent 60% of its time in forest, though open grassland was available to them. Their behaviour also was in many respects outside the range regarded as typical of baboons. Some of the differences were clearly directly dependent on their environment: forest is a more productive plant community than savannah, so there could be a higher population density and less time spent searching for food. It is relatively easy to accept that home range size is not a psychologically-based specific character, but an environmental variable. Other differences have less obvious environmental causes, but it seems economical to attempt explanations of them on this basis as a first approach. The structure of the population, with its equal adult sex ratio and its high proportion of juveniles, may be the result of unusually low predation pressure—which may in turn reflect a greater efficiency in avoiding predators in thick cover or forest. The rapid breeding rate of the females may be correlated with the rich food supply, and the relative lack of seasonal changes. The relatively huge growth of the population during the study is presumably only a temporary feature; it may be that it is still responding to political changes—the National Park was only extended to include the study area in 1959; or there may be long term cyclical changes in numbers, perhaps regulated by disease, as appears to be the case for the monkeys of South America which are periodically decimated by yellow fever (Balfour, 1914). This is purely speculative, but some generalizations must already be discarded because of this population—for example the correlation between a marked size difference between males and females and an unequal sex ratio.

Yet more speculative, but of greatest theoretical importance, is an attempt to explain differences in social behaviour as the result of differences in environment. At the moment adequate data is not available for the necessary quantitative comparisons, and apparent differences may turn out to be differences in observer approach. However, the general impression given by this population was not that of rigid organization and hierarchical structure stressed by other workers, and comparison is easier with the descriptions available of some arboreal species which also seem to form more "relaxed and friendly" groupings. Perhaps availability of cover is an environmental factor which may be relevant to these differences.

Baboons have probably been studied in a wider variety of habitats than any other primate, except man, but even so it is clearly too early to generalize about specific behaviour, except at the level of simple communicative units. Comparisons between larger taxa should be approached with even greater caution, especially where they are studied in different environments. It may be that species differ more in the range of environments they will tolerate than in their behaviour in equivalent environments; baboons seem to have an extremely wide tolerance and will probably be especially useful in investigating this point. Another approach is the comparison of species in similar artificial environments in captivity, which tends to limit the range of behaviour observable but has some merits of precision. Bernstein (pers. commun.) has made some interesting comparisons in this way.

Until we know more about the general effects of environmental differences on primate behaviour, we may also be misled about evolutionary mechanisms in the group, especially since knowledge of past environmental changes is also inadequate. As an exercise in this field, the baboon might be considered not as a primate adapted to life in grassland, but as an unusually versatile forest dweller which managed to survive where forest cover was (recently) lost, but which still uses patches of trees where they are available, and which, given the opportunity for choice, is more at home in forest. This picture is also inadequate, but has the advantage of resting on a different set of assumptions from those currently accepted, and so may be of value in provoking discussion.

The author was a D.S.I.R. research assistant during this study. Thanks are also due to the staff of the Queen Elizabeth National Park and of the Nuffield Unit of Tropical animal ecology, especially M. Beadle who provided the rainfall figures, and M. Lock who advised on botanical questions.

REFERENCES

Balfour, A. (1914). The wild monkey as a reservoir for the virus of yellow fever. *Lancet* **1914:1**: 1176-1178.

DeVore, I. (1963). A comparison of the ecology and behaviour of monkeys and apes. In *Classification and human evolution* (Ed. Washburn, S. L.). Viking Fund Publications in Anthropology No. 37: 301-319.

DeVore, I. & Hall, K. R. I. (1965). Baboon ecology. In *Primate behavior: field studies of monkeys and apes.* (Ed. DeVore, I.) New York: Holt.

Hall, K. R. L. (1963). Variations in the ecology of the Chacma baboon. *Symp. zool. Soc. Lond.* No. 10: 1-28.

Rowell, T. E. (In press (*a*)). Variability in the social organisation of primates. In *Primate ethology.* (Ed. Morris, D.). London: Weidenfeld & Nicholson.

Rowell, T. E. (1966). The habit of baboons in Uganda. *E. Afr. Acad. Symp.* No. 2 (1964) 121-127.

Rowell, T. E. & Hinde, R. A. (1962). Vocal communication in the rhesus monkey. *Proc. zool. Soc. Lond.* 138: 279-294.

Washburn, S. L. & DeVore, I. (1961). Social behavior of baboons and early man. In *Social life of early man* (Ed. Washburn, S. L.). Viking Fund Publications in Anthropology No. 31: 91-104.

Zuckerman, S. (1932). *Social life of monkeys and apes.* London: Kegan Paul.

12

Patterns of Movement in Open Country Baboons

1977

ROBERT S. O. HARDING

Abstract. In the early accounts, baboons are described as invariably assuming a particular troop formation when moving across open country. This movement pattern is an integral part of a model of social organization in which behavior is a response to predator pressure. The model, in turn, has been applied to describe populations of early hominids moving onto the savannah from the forest. The special formation was not observed during a year's study of free-ranging baboons in Kenya, nor has it been reported by other field workers since its original description. Thus at least one aspect of an early model of baboon social organization lacks empirical support and this model should not be used uncritically in reconstructing the behavior of early human groups.

Key Words Papio anubis, baboons, social organization, hominid evolution.

Several early studies of baboons reported that these animals invariably maintained a particular spatial arrangement when travelling from one feeding area to another during the day, or to a sleeping site at dusk. This pattern was said to be led by low-ranking adult males and older immature males, followed by pregnant females, estrous females and their consorts, and younger juveniles. In the center of the troop were the most dominant adult males, females with infants, and the youngest juveniles. The rear portion of

previous belief

235

the troop was reported to be a mirror image of the front, with low-ranking adult males and older immature males bringing up the rear (DeVore and Washburn '63: p. 343). Use of this pattern of movement has not been reported for other baboon groups in Africa, nor was it used by baboons at Gilgil, Kenya, whose ranging behavior was the subject of this author's research. This article represents an analysis of movement patterns among Gilgil baboons.

MATERIALS AND METHODS

Between November 1970 and October 1971, 1,032 hours were spent observing a troop of olive baboons (*Papio anubis*) on a 182-km^2 cattle ranch near Gilgil, Kenya, on the floor of the Rift Valley. The central part of the ranch, about 45 km^2, was open grassland interspersed with patches of dense scrub, divided into parallel valleys by lines of cliffs which provided sleeping sites for the seven baboon troops inhabiting the area. Rainfall at Gilgil, 595.1 mm during the study year, approximated that at the "arid" site of Aldrich-Blake et al. ('71) in Ethiopia, although the area was somewhat cooler (mean maximum daily shade temperature was 25.5°C).

No open watercourses existed in the study troop's range, but troughs of water provided for cattle were also used by baboons. Predator populations were reduced, with lions all but eliminated by shooting. Leopards had been live-trapped for removal to National Parks but had not been exterminated, for several cows were killed by leopards in the baboons' home range during the study year. No encounters between baboons and predators (other than feral dogs) were seen. A more detailed description of the study site is available elsewhere (Harding, '76).

To monitor fluctuations in the study troop's composition, the troop was censused as often as possible, using methods adapted from Altmann and Altmann ('70). As the troop entered large open areas, where each animal could be seen, the identity of each baboon was called off into a tape recorder as soon as it passed a preselected point. Only one such count was taken during any troop movement, and thus each count represented an independent sample. The shortest interval between two censuses was five hours; most were taken days apart.

Data from 76 troop movements, summed and averaged, resulted in a mean troop size of 49.3 animals, divided as follows:

3.8 adult males (AM)	6 + years
18.6 adult females (AF)	4 + years
1.3 subadult males (SAM)	4-6 years
15.4 juveniles (JUV)	1-3 years
10.2 infants (INF)	0-1 years

Using these figures, I calculated the number of times a baboon of each

Table 1
Observed vs. expected position of baboon age-sex classes during troop movement

	AM	AF	SAM	JUV	INF
Observed first	38	33	3	2	0
Expected first	5.9	28.7	2.0	23.7	15.7

AM, adult males.
AF, adult females.
SAM, subadult males.
JUV, juveniles.
INF, infants.

age-sex class could be expected by chance to be first animal in a troop movement. Thus, for instance, in 76 movements of a troop averaging 49.3 animals, 3.8 of them adult males, this class of baboon could be expected to be first animal

$$76 \times \frac{3.8}{49.3},$$

or 5.9 times. Table 1 compares the expected figures with those actually observed.

Table 2
Location of baboon age-sex classes by section during troop movement

		Section of troop					
		1st	2nd	3rd	4th	5th	Total
AM[a]	Obs	137	52	35	28	37	
	Exp	58.6	58.6	58.6	58.6	54.8	289
AF	Obs	355	279	262	242	276	
	Exp	286.5	286.5	286.5	286.5	268	1,414
SAM	Obs	38	32	21	7	4	
	Exp	20.7	20.7	20.7	20.7	19.3	102
JUV	Obs	124	211	280	311	247	
	Exp	237.7	237.7	237.7	237.7	222.3	1,173
INF	Obs	106	186	162	172	147	
	Exp	156.6	156.6	156.6	156.6	146.5	773
Total		760	760	760	760	711	3,751

[a] Abbreviation as in Table 1.

To determine the location of each census class during troop movements, each individual census was then divided into five sections, as follows: the first ten animals passing the counting point were assigned to the first section, the next ten to the second section, the next ten to the third (or center) section, and the next ten to the fourth section. Since there were anywhere from 47 to 52 animals in the troop during the study year, the fifth, or last, section did not always equal each of the other four, but might contain as few as seven or as many as 13 animals. Data from all 76 censuses were then combined by section and compared with the figures to be expected if each age-sex class were evenly distributed throughout every section of the troop (Table 2).

As Table 1 indicates, an adult male baboon led the troop in 50% of all troop movements, more often than would be expected by chance ($X^2 = 189.34$, 1 d.f., p < 0.001).

Table 2 makes it clear that adult males, adult females and subadult males were in the leading section of the troop, juveniles and infants were in the middle and toward the rear, and adult females were in the last section of the troop, more often than expected. The probability of this particular troop configuration occurring by chance is less than 0.001 ($\chi^2 = 283.70$, 8 d.f.).

Furthermore, these figures may be combined for analytic purposes so that the data from the first section of the troop may be tested again the data from all other sections combined. Thus, for instance, adults of both sexes plus subadult males tended to be in the first section when compared with juveniles and infants, who tended to be in the other four sections ($\chi^2 = 178.38$, 1 d.f., p < 0.001). Similarly, adult and subadult males were more likely to be in the first section than adult females ($\chi^2 = 57.04$, 1 d.f., p < 0.001).

Finally, while no stable dominance hierarchy or central coalition was observed during this study, adult male Carl was clearly the animal with the highest status in the troop by nearly every measure of dominance. Instead of moving at the center of the troop, however, Carl was in the first section in 59% of all censuses, and indeed led the troop in 16 of the 38 instances in which an adult male was first animal. Under the assumption that at any time there were four males in the troop, each equally likely to take the lead, Carl's behavior is a significant departure from chance expectations ($\chi^2 = 5.93$, 1 d.f., p < 0.05).

probability — *the lower the # the more significant the find* — *p is less than 500 in 1000 times*

degree of freedom

DISCUSSION

A baboon troop that moves with most of its males in front and its females at both front and rear obviously differs from those described in some of the first reports on baboon social behavior (DeVore and Washburn, '63; DeVore and Hall, '65; Hall and DeVore, '65). Unfortunately, no direct

when p > .05 not significant

comparison can be made between the data presented here and these early reports, since the observations of baboon movement which they contain are quantitative and no data have been published to support them.

Gilgil is not the only study site where observers have failed to see the male-centered troop arrangement of the previous accounts. Stoltz and Saayman ('70: p. 131) did not see "clearly-defined formations of this nature" in chacma baboons (*Papio ursinus*) in the Transvaal, nor did Ransom ('71), who worked with forest-living olive baboons in the Gombe National Park, Tanzania.

In 20 counts each of two different olive baboon troops in Ishasha, Uganda, Rowell ('69) also saw a different type of baboon progression. Breaking down each troop into blocks of five positions, she calculated the proportion of animals of each age-sex class in each block. As was the case at Gilgil, adult males were significantly more often in the first block of positions (p < 0.001), adults of both sexes were seen more often toward the front and rear, and small juveniles toward the middle. Infants carried by their mothers were found more often at front and rear as well.

Rhine and Owens ('72) found that adult male and black infant baboons were associated more often than would be expected by chance at the Gombe National Park, but their results are not comparable with those presented here, since: "the data apply to animals entering a clearing rather than to the full progressions discussed by Washburn, DeVore and others (p. 279)." However, Rhine ('75) also found that adult male yellow baboons (*papio cynocephalus*) at the Masai-Amboseli Game Reserve in Kenya tended to be in the front or rear sections of a troop, and that an adult male was more often first (64.6% of all troop movements) or last animal (55.7%) than expected by chance (p < 0.001). More dominant males tended to be first and subordinate males tended to bring up the rear (p < 0.001). Black infants were spread equally throughout the troop, as was the case at Gilgil.

Finally, in their first Amboseli study, Altmann and Altmann ('70) could not confirm the male-centered model:

"The progressions that we observed, even those of a single group, did not reveal any invariable order of progression, nor was it true that the front and back of the progression invariably consisted of adult males and older juvenile males (p. 188)."

In an exhaustive statistical analysis of progressions observed between 1963 and 1972 at Amboseli, S. A. Altmann ('77) has been unable to detect any indication of a fixed progression order, either by age-sex class or by individual.

In short, when compared with observations made elsewhere in Africa, the Gilgil data are not at all anomalous.

CONCLUSIONS

To put the present findings in perspective, it is necessary to consider the importance of baboon progression formations to the theory of baboon social organization proposed by DeVore and Washburn ('63). The male-centered, "invariable" formation was seen as only one of a series of interlocking adaptations evolved to cope with the dangers of a terrestrial life on the open savannah. Other adaptations included large size, great strength, sharp canine teeth, and the aggressive temperament of the adult males, as well as a strict dominance hierarchy thought to keep intratroop aggression in check. The threat of predation on the open savannah was such a central factor in this view of baboon social life that Hall and DeVore ('65: p. 49) stated:

> "The relations between baboons and animals that prey upon them is (sic) the most important single fact in the interpretation of baboon ecology and social behavior."

The Gilgil baboons have few large cats to contend with, and thus may not need to arrange themselves in a defensive formation while moving. However, the "invariable" defensive formation has not been reported in any other study, whether the risk of attack by big cats is assumed to be low (Gilgil, Gombe, Transvaal) or high (Amboseli, Ishasha). Furthermore, as Rowell ('72) has emphasized, humans are and probably have been for a long time the major predators upon all primates, including baboons.

Given the variability in movement patterns described here, what causes baboons to arrange themselves in a particular spatial configuration while moving? With so many different observers using different methods of observation and analysis, it may be that no one theory can satisfactorily account for every reported variation. Nonetheless, baboon movement patterns can still be seen as a response to two factors: the experience of a troop with its habitat and the temperament of different classes of individuals.

Many observers have noted that baboon troops behave differently in different parts of their range. Altmann and Altmann ('70: p. 111) reported that their yellow baboons aggregated before going through "a critical pass in the foliage" or when passing through undergrowth. Maxim and Buettner-Janusch ('63) describe olive baboons clustering into a triangular formation when approaching a dangerous area, with "dominants" at the apex of the triangle closest to the danger zone. At Gilgil the baboons drew together in heavy scrub, in places where humans were often encountered, and in areas frequented by other troops (Harding, '76). Behavior which changes with troop location was also reported by Rowell ('66). In short, baboons change their dispersion patterns in response to past experiences in different parts of their range.

Temperament also plays a part, in that adult males, twice the size of females and equipped with long, sharp canine teeth, are more self-assured and bolder than their smaller conspecifics, hence less reluctant to enter potentially dangerous areas.

With the advantage of hindsight it is easy to say that no group of mammals is likely to move as systematically as baboons were once thought to move, and indeed, according to Washburn (personal communication), followup visits to Nairobi Park showed that even these baboons alter their behavior when entering different areas. Nevertheless, the variability of this behavior is stressed here because the male-centered formation was described as "invariable" in the early reports, and as such was a key link in an adaptive pattern that made survival possible for baboons in the open grassland. This adaptive pattern in turn was then used as the basis for theories which described early hominids venturing onto the African savannah.

However, other aspects of the proposed baboon adaptive pattern have also been undercut by recent long-term field studies. For instance, it now appears that male dominance hierarchies are transitory and stable only over short periods of time (Hausfater, '75), that virtually every male spends part of his life in a troop other than the one in which he was born (Packer, '75), and that female kinship groups, rather than male dominance hierarchies, provide the stable core of the troop (Rowell, '66; Strum, '75, and personal communication; Hausfater, '75). These findings and the ones reported here run counter to earlier descriptions of baboon social organization (e.g., Hall and DeVore, '65), and suggest that if baboon troops are to be used as models for early hominid groups, such theorizing should be continually revised as new information becomes available.

ACKNOWLEDGMENTS

The study on which this article is based was supported by NIMH Research Fellowship MH 47679-01 CUAN. I am grateful to the Government of Kenya and Mr. and Mrs. Arthur G. Cole of Kekopey Ranch for permission to carry out the research, to R. J. Rhine for a preliminary analysis of part of these data, to W. Ewens and J. deCani for statistical help, and to P. C. Dolhinow and S. L. Washburn for their support during the study.

LITERATURE CITED.

Aldrich-Black, F. P. G., T. K. Bunn, R. I. M. Dunbar and P. M. Headley 1971 Observations on baboons, *Papio anubis,* in an arid region in Ethiopia. Folia Primat., *15*:1-35.

Altmann, S. A. (1977, in manuscript) Baboon progressions: order or chaos? A study of one-dimensional group geometry.

Altmann, S. A. and J. Altmann 1970 Baboon Ecology. Univ. of Chicago Press, Chicago.

DeVore, I., and K. R. L. Hall 1965 Baboon ecology: In: Primate Behavior:

Field Studies of Monkeys and Apes. I. DeVore, ed. Holt, Rinehart and Winston, New York, pp. 20-52.

DeVore, I., and S. L. Washburn 1963 Baboon ecology and human evolution. In: African Ecology and Human Evolution. F. C. Howell and F. Bourliere, eds. Aldine, Chicago, pp. 335-367.

Hall, K. R. L., and DeVore 1965 Baboon social behavior. In: Primate Behavior: Field Studies of Monkeys and Apes. I. DeVore, ed. Holt, Rinehart and Winston, New York, pp. 53-110.

Harding, R. S. O. 1976 Ranging patterns of a troop of baboons (*Papio anubis*) in Kenya. Folia Primat., *25*: 143-185.

Hausfater, G. 1975 Dominance and Reproduction in Baboons (*Papio cynocephalus*): a Quantitative Analysis. S. Karger, Basel.

Maxim, P. E., and J. Buettner-Janusch 1963 A field study of the Kenya baboon. Am. J. Phys. Anthrop., *21*: 165-180.

Packer, C 1975 Male transfer in olive baboons. Nature (London), *255*: 219-220.

Ransom, T. W. 1971 Ecology and Social Behavior of Baboons (*Papio anubis*) in the Gombe National Park. Unpublished Ph.D. dissertation, University of California, Berkeley.

Rhine, R. J. 1975 The order of movement of yellow baboons (*Papio cynocephalus*). Folia Primat., *23*: 72-104.

Rhine R. J., and N. W. Owens 1972 The order of movement of adult male and black infant baboons (*Papio anubis*) entering and leaving a potentially dangerous clearing. Folia Primat., *18*: 276-283.

Rowell, T. E. 1966 Forest living baboons in Uganda. J. Zool. (London), *149*:344-364.

1969 Long-term changes in a population of Uganda baboons. Folia Primat., *11*: 241-254.

1972 The Social Behaviour of Monkeys. Penguin Books, Baltimore.

Stoltz, L. P., and G. S. Saayman 1970 Ecology and social organization of chacma baboon troops in the northern Transvaal. Ann. Transvaal Mus., *26*: 99-143.

Strum, S. C. 1975 Life with the "Pumphouse Gang." Nat. Geogr., *147* (5): 672-691.

13

Baboons, Space, Time, and Energy

1972

STUART A. ALTMANN

Synopsis. How are social organization and ecology related to each other? Yellow baboons, hamadryas baboons, and gelada monkeys are all large, terrestrial African primates, but they have three different patterns of social organization, and they live in three, markedly different habitats: savannah, steppe-desert, and alpine heather-meadowland, respectively. An attempt is made to provide testable hypotheses and heuristic principles that can relate these two classes of phenomena.

" . . . There is an adaptation, an established and universal relation between the instincts, organization, and instruments of animals on the one hand, and the element in which they are to live, the position which they hold, and their means of obtaining food on the other. . . . "
—Sir Charles Bell,
Bridgewater Treatises,
1833

Reprinted from *American Zoologist, 14*:221-248 (1974) by permission of the American Society of Zoologists.

Based on the Philip J. Clark Memorial Lecture, delivered at Michigan State University, May 18, 1972. Field research on primates was supported primarily by grant GB27170 from the National Science Foundation and MH19,617 from the National Institute of Mental Health. I am grateful to J. Altmann, M. Slatkin, and S. Wagner for a critical reading of the manuscript. I would also like to thank P. Olindo, Director of the Kenya National Parks, and D. Sindiyo, former Game Warden of the Ambosali Reserve, Kenya, for enumerable courtesies.

INTRODUCTION

There are two complementary, but quite different approaches to "explaining" social organization. One is reductionistic: an examination of immediate behavioral or motivational causes in the individuals that make up the social group. In baboons, this approach has been pursued most vigorously by Kummer, in a series of illuminating papers on the nature of social bonds and repulsions in hamadryas and their relatives (Kummer, 1967a,b, 1968b, 1971 b). The alternative approach, and the one that we will pursue here, is to study the adaptive significance or ecological function of group processes. The distinction between what these two approaches seek is essentially that made by Baker (1938) between proximate and ultimate causes.

Animal species differ in their habitat, and they differ in their social organization. The question is, are these differences related? To what extent do group processes represent adaptations to the exigencies of the environment? To what extent are differences in the success of a species in two habitats—or of two species in a single habitat—attributable to differences in group size, numbers of adult males and females per group, group responses to predators, spatial deployment of group members when progressing or foraging, differential use of parts of home range, simultaneous occupancy—by different groups—of zones of home range overlap, and defense of territories? If a population of animals were transplanted from the habitat in which they live to the kind of habitat occupied by a related species with a different social organization, would they survive? Would their social organization converge on that of the related species?

In the last decade these questions have begun to attract the attention of primate field workers and anthropologists (e.g., Chalmers, 1968a,b; Crook, 1967, 1970a,b; Crook and Gartlan, 1966; Denham, 1971; DeVore, 1963; Dyson-Hudson, 1969; Eisenberg et al., 1972; Fisler, 1969; Forde, 1971; Gartlan, 1968; Hall, 1965a,b, 1966; Kummer, 1971a; Rowell, 1967; Struhsaker, 1969; Vayda, 1969). The major tack taken in most previous attempts to relate primate ecology and social organization has been simple correlation, combined with *post hoc* explanation. First, a correlation is noted between a habitat and certain characteristics of primates living in it. For example, forest frugivores live in small multi-male groups, whereas savannah vegetarian omnivores live in medium to large multi-male groups (Crook and Gartlan, 1966). Then an attempt is made to "explain" the correlation in a pre-eminently reasonable fashion, e.g., greater visibility in open savannah country leads to larger groups. The most ambitious attempt so far to provide such a scheme of classification is that of Crook and Gartlan (1966). They classify primates into a series of five grades representing levels of adaptation in forest (nocturnal or diurnal), tree savannah, forest fringe, and arid environments, respectively. This scheme of classification has been revised by Crook (1970a), by Jolly (1972), and by Eisenberg et al. (1972). The latter make its basis explicit:

> "When a group of allopatric species shares the same relatively narrow range of adaptation, then this group begins to exhibit a predictable 'adaptive syndrome' with respect to feeding, anti-predator behavior, spacing mechanisms, and social structure."

While such classifications may help organize large bodies of data, they must fail as explanatory devices. The reason is this: there are many different ways to exploit any habitat. Thus, food in a rain forest may be widely dispersed for one species and markedly clumped for another, as a result of dietary differences. Social communication and group coordination over distance may be easy or difficult, depending on sensory modality used, height of the animals in the forest, the particular plant complex, and so forth. Consequently, as we learn more about primates in the wild, we continue to find exceptions to the classifications. For example, contrary to the correlation noted in the last paragraph, several forest *Cercopithecus* monkeys are forest frugivores that live in one-male groups (Struhsaker, 1969), whereas patas monkeys live in savannahs in large one-male groups (Hall, 1965a). At this point an attempt is made to change the classification (e.g., grassland species that inhabit partially wooded areas live in multi-male groups, those that inhabit more open grassland live in multi-male groups). The final result of such revisions will be either a classification that continues to lump together species that share some characteristic while ignoring species-specific characteristics, or one that copes with species diversity by describing social organization and ecology in a manner that is unique for each species, so that there is just one species per lot. Even then, there is no assurance that the particular chacteristics of the habitat that have been used to make the classification are in fact the ones that are important to the animals.

STRATEGIES FOR COMPARISON

Comparative and correlational studies can be more informative if we take into account the taxonomic relations between the groups that are being studied. When we attempt to relate differences in group behavior to ecological factors, the populations that are compared may be either closely related or distant ("unrelated"), and their habitats or niches may be similar or dissimilar. This results in four basic types of comparison.

Type I

Comparison of unrelated organisms in dissimilar habitats is probably the least informative type, with one notable exception: convergent traits may illuminate constraints imposed by common factors in the two environments.

For example, in a wide variety of habitats, large social groups, extending beyond the immediate family, may represent an adaptation to predation. Another class of common traits (e.g., suckling in new-world and old-world primates) consists of patterns that were no doubt present in the most recent common ancestor of both populations. Such ancient patterns (*plesiomorphous characters*, in the terminology of Hennig, 1966) delimit a broad adaptive zone that is characteristic of a major taxonomic group, but provide little information about ecological aspects of behavioral differences in contemporary species. Distinguishing between these two classes of characters—convergent and ancient traits—must be made on the basis of other types of comparison (see Hennig, 1966).

Type II

Similarities in unrelated populations that occupy similar niches (e.g., behavioral and anatomical similarities in colobus monkeys and howler monkeys in old-world and new-world tropical rain forests) offer essentially the same information as in Type I comparisons, except that the greater number of common niche parameters can be expected to increase the number of convergent traits. Such comparisons are the mainstay of the "grades" of Crook and Gartlan (1966) and the "adaptive syndrome" of Eisenberg et al. (1972). Furthermore, where the species being studied occupy the same biome, one has the opportunity to study interspecific competition, niche separation, character displacement, and related phenomena. Such comparisons sometimes turn up striking cases of convergent or parallel evolution. An example from birds has been given by Cody (1968). In short grass, birds can probe the ground (e.g., meadowlarks, *Sturnella* spp.), launch into the air for flying insects (e.g., larks, *Alaudidae*), or forage from the foliage (most sparrows); and each shortgrass area of the world contains probers, launchers, and foliage-gleaners, usually one of each.

Types III and IV

Studies of closely related populations provide some of the most illuminating comparisons. Particularly valuable are comparative studies of closely related species or populations in a group that has recently undergone a major adaptive radiation (Tinbergen, 1960). In such groups, species differences primarily reflect adaptations to the differing characteristics of the habitats,* and most species similarities will be homologs. For this reason, the hamadryas/cynocephalus baboon contrast that will be used in this article is particularly revealing.

---

*One alternative is that they are different solutions to a problem that is faced by every species in the group.

Konner
adaptibility
vs
adaptation
Sociobio.

   Within a single generation of one population, and thus with virtually
no genetic change, one can sometimes observe changes in behavior and
social organization that can be related to ecological factors. Such studies on
a single population have the advantage of minimizing the effects of genetic
differences and thus suggesting the extent to which intrapopulation variability reflects the plasticity of the individual animals. The changes in behavior
and ecology may occur naturally and repeatedly, as a result of daily, lunar,
annual, or seasonal cycles. In some cases, longer range secular changes in
the environment occur, such as those that have been observed in our main
study area for yellow baboons, the Amboseli Reserve of southern Kenya
(Western and Van Praet, 1973). Alternatively, changes in the habitat may be
artificially induced for experimental purposes. Planned interventions are of
particular value in studies of limiting factors, of the sort that will be described below. If, for example, the spatial distribution of water is believed
to limit the area exploited by a group, artificial water sources could be
established. Watson and Moss (1971) have altered territorial behavior in
grouse by increasing productivity of their heath, through locally applied
fertilizer.

   In addition to such temporal changes, geographic variability and habitat variations over space may sometimes be correlated with intraspecific
variability in group processes. On a large scale, such comparisons involve
macro-geographic differences. For example anubis baboons live in rain
forests on the slopes of Mt. Meru, in verdant areas of the Kenya highlands,
and in arid short-grass savannahs in the northern frontier district of Kenya.
An illuminating study could be based on a sample of comparable ecological
and sociological data in each of these habitats. On a much smaller scale
there often are differences in the ecological conditions in the home ranges
of groups in a local population, and these may be related to differences in
group behavior and composition. Here again, one has the advantage of
populations that have relatively small genetic differences.

   Whichever type of comparison is made, the initial effort will usually be
simply an observed correlation of some ecological variable with a biological
or behavioral characteristic. Such correlations are more convincing if populations or species that live in an intermediate habitat exhibit an intermediate
form of the character. Particularly revealing comparisons can be made by
making use of the following principle: A trait that is convergent for two
species (or species groups) and that represents, for at least one, divergence
from the corresponding trait of its closest relatives is probably an adaptation
to distinctive features that occur in the environment of the two convergent
species. For example, howlers of the new-world tropics and colobines of the
old-world have specialized stomachs that are adapted to digesting high-
cellulose leafy diets, and both of these groups of monkeys differ in this
respect from their closest relatives. Douroucoulis, the only nocturnal primates of the New World, have many traits that diverge markedly from those

of other new-world monkeys but are convergent with traits found in several nocturnal primates of the Old World.

Correlation by itself is not adequate. For example, group size in baboons is correlated (inversely) with aridity. But many other characteristics of the environment co-vary with this environmental factor. The test of any putative ecological determinant rests on a demonstration of its mode of operation. As Williams (1966) has put it, adaptation requires a mechanism.

ECOLOGY VERSUS PHYLOGENY

What about those traits of a species that are also found in closely related forms, despite differences in habitats? A recurrent question in recent discussions of the social ecology of primates is the extent to which primate social structure represents the phylogenetic heritage of the species rather than specific adaptations to the local environment. Struhsaker (1969) has emphasized that:

> "In considering the relation between ecology and society . . . each species brings a different phylogenetic heritage into a particular scene. Consequently, one must consider not only ecology, but also phylogeny in attempting to understand the evolution of primate social organization. The interrelations of these . . . variables . . . determines . . . social structure. In some cases, the immediate ecological variables may limit the expression or development of social structure, and, with other species and circumstances, variables of phylogeny may be limiting parameters."

As an example of the latter, Struhsaker points out that heterosexual groups with one adult male seem to be typical of most *Cercopithecus* species (excepting vervets) and of the closely related patas monkeys, despite the fact that there are considerable differences in the habitats and niches of these species.

Any "ecology vs. phylogeny" controversy over determinants of social behavior and group structures may turn out to be as sterile as the "hereditary vs. environment" controversy that has plagued the behavioral sciences, and for exactly the same reason. Indeed, these two controversies seem to be two aspects of the same problem. It can be avoided by asking every question in a way that suggests a verifiable answer, for example: To what extent are observed differences in the social behavior of two populations attributable to genetic differences between them? A number of research strategies can be used to answer such questions, including regression of variance in offspring against parents (Roberts, 1967), cross-fostering experiments (Kummer, 1971*b*), and studies of hybrid zones (Nagel, 1971; Müller, unpublished; Kummer, 1971*a*).

THE STUDY OF ADAPTATIONS

An alternative to this purely correlational approach is to analyze the adaptive aspects of group processes in each species, to propose testable (and thus falsifiable) hypotheses about the relations between specific aspects of social organizations and ecology in primates, and to test these on the basis of data from observational and experimental research. Such an approach seems to me to be essential if we are to achieve something more than a "species-in-dot" comprehension of the ecology of primate societies.

In what follows, I shall present a number of organizing principles and hypotheses about relations between group organization and ecology. They are consistent with what is now known about the social ecology of baboons and geladas. I leave it to others to decide whether they have broader application than that.

The basic concept that will be developed here is that of an adaptive distribution of baboon activities:

> "For any set of tolerable ecological conditions, the adaptive activities of baboons tend in the long run toward some optimal distribution away from which mortality rate is higher, or reproductive rate is lower, or both" (Altmann and Altmann, 1970, p. 201).

The so-called "principles" that will be presented below are intended to be heuristic guides in the search for such optima. They will be based largely on methods for analyzing intra-specific variations in adaptive behavioral processes.

A more deterministic viewpoint is expressed by Denham (1971, p. 78), who assumes that in energy acquisition, "the most efficient strategy compatible with the structure of the organism is used by members of a primate population in a natural habitat." Similar claims about the perfection of nature have been made by others. Denham's statement is either a tautology (because any other course of action would not be compatible with the structure of the organism) and thus analytically true, or else it is intended to be verifiable, in which case I believe that it is false. It seems more likely that in the efficiency of foraging strategies, as in many other traits, natural primate populations include a wide range of variability, and that there is a genetic component to this variability. Such variability is the raw material on which natural selection works and is a rich source of material for the study of adaptive aspects of behavior.

BABOON SOCIAL ECOLOGY

Baboons are among the best primates for analyzing ecological aspects of social organization. They live in habitats that range from evergreen tropical rain forest, through various types of woodland and savannahs, to semi-desert steppe country. In parts of Africa they are abundant and readily observable. Their groups include one-male harems and multi-male groups. They have been studied at several locales in Africa by a number of investigators.* For no other genus of non-human primates do we have a comparable body of information on behavior, social relations, population dynamics, and ecological phenomena.

Our discussion will center on two markedly different baboon populations, the hamadryas baboons (*Papio hamadryas*) of Ethiopia and the yellow baboons (*P. cynocephalus*) of east Africa. Further contrasts will be provided by gelada "baboons," a large, terrestrial primate of the Ethiopian Highlands, whose relationship to the other cercopithecine primates is uncertain. These three primate species—yellow baboons, hamadryas, and geladas—center on three quite different habitats: savannah, sub-desert steppe, and alpine heather-meadowland, respectively, and there are equally marked differences in their social organizations.

Hamadryas in eastern Ethiopia have been studied primarily by Kummer and his students (Kummer and Kurt, 1963; Kummer, 1968*a*, *b*, 1971; Nagel, 1971, unpublished observations by Kummer, Abegglen, Goetz, Müller, and Angst) and were observed by M. Slatkin and me during September, 1971. Yellow baboons have been studied most extensively in southern Kenya by us and our associates (Altmann and Altmann, 1970; Cohen, 1971, 1972; unpublished studies by the Altmanns, Slatkin, G. Hausfater, S. Hausfater, and Fuller). Both populations are still being studied.

Major publications on ecology and naturalistic behavior in other baboons include those by DeVore (1962), DeVore and Hall (1965), Hall (1965 *b*), Rowell (1966, 1969), and Aldrich-Blake et al. (1971) on anubis baboons, and by Hall (1962*a,b*) and Stoltz and Saayman (1970) on chacma baboons. A number of other publications describing the results of recent studies are currently in preparation. Our knowledge of geladas in their natural habitat comes primarily from observations by Crook (1966), Crook and Aldrich-Blake (1968), and several recent unpublished field projects by Altmann, Dunbar, Nathan, and Slatkin.

The basic contrast in social organization and ecology between hamadryas and yellow ("cynocephalus") baboons* have been described by Kummer (1968*b*, 1971*a*) and will only be summarized here. Yellow baboons live in social groups that usually contain more than one adult male, several adult females, and associated immature off-spring. In the Amboseli Reserve,

*In 1972 alone, no less than 16 people were involved in field studies of baboons: J.-J. & H. Abegglen, S. & J. Altmann, R. & P. Dunbar, M. Fuller, R. Harding, C. & S. Hausfater, S. Malmi, W. Müller, N. Owens, O. Oyen, R. & D. Seyfarth.

*Under *P. cynocephalus*, Kummer includes not only yellow baboons (*P. cynocephalus, sensu stricto*), but the closely related chacma, anubis, and guinea baboons. The difference is immaterial for our present purposes.

Kenya, yellow baboon groups ranged in size from 18 to 198 during 1963-64 with a mean of about 51. Mean group composition was 14.8 adult males, 16.5 adult females, 12.7 juveniles, and 10.5 infants. Within each group of yellow baboons a female often mates with any of several males during each menstrual cycle. These groups are virtually permanent: they do not routinely break up, either daily or seasonally. With very few exceptions, only males move from one group to another, and then, only as adults. Females generally spend their entire lives in the social group in which they were born.

Although yellow baboons live in a wide variety of habitats in central and eastern Africa, perhaps the largest area occupied by them consists of moderately arid savannah country, areas with variable tree cover but in which the dominant ground cover is grass. In areas to the north and to the west, yellow baboons are replaced by anubis baboons, which often inhabit areas of higher rainfall and with a larger proportion of non-grass flowering plants as ground cover.

Hamadryas live in a much more arid region. They are sometimes called desert baboons, which I thought to be a misnomer until I observed them in Ethiopia, during a 1971 field trip. Their habitat represents an extreme environment for non-human primates; it is a region of high temperature, low seasonal rainfall, little soil, rapid erosion, and sparse vegetation.

Yellow baboons (*Papio cynocephalus*) in acacia woodland (Amboseli Reserve, Kenya).

Hamadryas baboons (*Papio hamadryas*).

    A hamadryas population has three levels of group organization. Several bands—which in many respects resemble the groups in yellow baboons—join together in using the same sleeping rock, thus forming a large herd or troop. The membership of the band appears to be consistent, whereas that of the herd is not. On the other hand, the band may split into one-male units that forage independently during the day (Kummer, 1968*a*).

    Geladas, too, have harems that amalgamate into enormous herds, but for those that we observed in Ethiopia, the daily cycle of fission and fusion was exactly the opposite of hamadryas: the harems slept separately or in small clusters, each on a cliff ledge, and then amalgamated into herds on the upland feeding ground during the morning. They foraged *en masse* until late afternoon, then again broke up into harems. The heart of the geladas' range is the alpine meadowland of the Ethiopian Highland. They generally occur well above 7000 feet; my observations and those of Slatkin, Dunbar, and Nathan were made in the vicinity of Sankaber, at about 11,000 feet.

SOME PRINCIPLES OF PRIMATE SOCIAL ECOLOGY

Many of the ideas presented here have grown out of our early field work on baboons, and some were published in *Baboon Ecology* (Altmann and Altmann, 1970). In some cases, similar ideas have been developed by others, including Denham (1971), Kummer (1971*a*), and Schoener (1971). The section on time budgets was stimulated primarily by a recent study of time budgets in yellow baboons and gelada monkeys by Slatkin (unpublished).

Gelada baboons (*Theropithecus gelada*) on an alpine meadow.

1) Resource distribution and group size

A group of vertebrate animals can grow as a result of just three processes: births, immigrations, or amalgamations with other groups. It can decrease in size as a result of deaths, emigrations, or group fission. Thus, any attempt to account for the immediate causes of the group size distribution in an area must be made in terms of these processes. This approach has been very successfully pursued with data on primate populations by Cohen (1969, 1972). However, such an account tells us nothing about the adaptive significance of group size. What would happen, for example, if there were twice as many groups of baboons in Amboseli, but each was half as large? Why is it that so many primates live in groups that are considerably larger than an immediate family?

Primate group sizes appear to be adaptations to two major classes of selective forces: the distribution and density of essential resources, such as food, water and sleeping sites, and patterns of predator attack and antipredator behavior. These factors will be discussed in this and the next section.

A resource will be referred to as *sparse* if it occurs at a low density both locally and throughout the home range. Thus, if a food resource is both sparse and has a patchy or clumped distribution, no clump or patch will contain an abundance of food. Each will be small relative to the daily food consumption of the animals. We can now state a relationship between resource distribution and group size:

Principle One. *If a slowly renewing resource is both sparse and patchy, it can be exploited more effectively by small groups. Conversely, large groups that are simultaneously using a single resource will occur only if the supply is adequate, either because of local abundance, or because of rapid renewal of the resource. Large groups will be more effective if a resource has a high density but a very patchy distribution and the patches themselves occur with low density, so that the resources tend to be concentrated in a few places ("supermarkets").*

In several cases, when the abundance of food that is available to a primate population has been artificially increased, the average group size has increased. This has happened, for example, with the rhesus monkeys on Cayo Santiago, Puerto Rico (Koford, 1966), and the Japanese macaques on Mt. Takasaki, Japan (Mizuhara, 1946; Itani, 1967). Conversely, in the Amboseli Reserve, Kenya, recent marked changes in the habitat, including the death of many of the fever trees (*Acacia xanthophloea*) and the transformation of the plant community from a hydrophytic to a xeromorphic form (Western and Van Praet, 1973), have been accompanied by a decrease in the mean group size of vervet monkeys (Struhsaker, 1973) and of baboons (personal observations). A decrease in food abundance probably was a major contributor to these changes in group size. Unfortunately, in none of these cases can one separate out the relative importance of food density and food dispersion. We predict that group size will increase or decrease according to the dispersion of the food, and independent of its abundance. This does not rule out the possibility that food density *per se* is a major factor controlling mean group size.

If some resources are abundant but only locally and others have a sparse and patchy distribution these can be effectively exploited by a population that aggregates at the concentrated resources and breaks up into small units to exploit the sparse resources. Hamadryas baboons provide an excellent example of such a social system. The small feeding units probably represent an adaptation to sparse, patchy resources. On the other hand, safe sleeping rocks are few and far between in the Gota Region of Ethiopia, but those that do occur can hold hundreds of baboons. Thus, the night-time sleeping aggregations on these rocks enable large numbers of individuals simultaneously to utilize an essential resource that is abundant locally.

Geladas, too, alternate between large herds and much smaller clusters of animals, but contrary to the hamadryas system, geladas sleep in small groups, consisting of one or a few harems. In the Simien Mountains, sleeping ledges are abundant, but each is fairly small, and they tend to be scattered along the cliff faces; of course, new ledges are formed very slowly.

Thus, they are a sparse, slowly renewing resource that is best exploited by small, dispersed groups. The immense herds that we observed in the Simien at the end of the rainy season were feeding on luxuriant alpine meadowland that was continuous over large areas of the geladas' range.* I believe that these herds are an adaptation to occasional predation (Principle Two), rather than to the resource distribution.

We will argue below that predation selects for baboon groups with at least one male. Sparse resources may select groups with *at most* one male, or very few males, because male baboons, with their greater food requirements and their ability to displace other members of the group from food, compete with adult females and other members of the group for already scarce resources. Thus, both males and females can reduce competition by associating in groups that include a single fully adult male.

2) Predation and group size

Predation on baboons affects the size and composition of their groups: for baboons, there is safety in numbers and safety in proximity to adult males. There are several reasons for this. Any member of a baboon group that sees a predator gives an alarm call to which all other members of the group respond. This means that each individual takes advantage of the predator-detecting ability of all other members of his group. As a group of baboons forages across the open grassland, each individual glances around occasionally. But if the group is large the total rate of such visual scans is high. "To live gregariously is to become . . . the possessor . . . of eyes that see in all directions . . . " (Galton, 1871, quoted in Hamilton, 1971). Thus, in open terrain it is almost impossible for a predator to approach a group of baboons undetected. The importance of this predator-detection system probably overrides the disadvantage of the predator's increased ability to locate baboons in large groups.

Another advantage of affiliation with a large group is protective hiding (Williams, 1966). If a predator appears at a random position in the area and strikes at the nearest individual, those individuals that are near other animals are less likely to be preyed upon (Hamilton, 1971), regardless of any predator detection, evasive action, or anti-predator behavior. In fact, the benefits to sociality accrue to animals that are subject to any pattern of predator attack in which the predator neither selects prey animals with equal probability, nor favors those that are near others, i.e., have a higher-than-average density of neighbors. Baboons are preyed on almost entirely by terrestrial predators, seldom by raptors. Thus, a solitary baboon, one in a

*Our observations on geladas were made possible by the hospitality of Patricia and Robin Dunbar and Elizabeth and Michael Nathan, to whom we are most grateful.

small group, or one on the periphery of a group probably is not as safe as one in the midst of a large group.

For baboons, another predator-selected advantage of living in large groups is that the baboons of a group sometimes react *en masse* to predators, similar to the mobbing reactions that are given by many passerine birds to owls and other raptors. The effectiveness of such mob responses probably depends on the size of the group. In addition, they are probably more effective because of the presence in the group of adult males: Baboons are highly dimorphic, and many of the special characteristics of the adult males make their anti-predator behavior particularly effective. In these highly dimorphic primates it is very rare to find females without a male, though the converse is not the case. Doubtless it is no coincidence that even the small minimal foraging groups of hamadryas baboons virtually always contain an adult male.

These factors, in combination, lead to the following principle for baboons:

Principle Two. *Predation selects for large groups and for groups containing at least one adult male.*

No claim is made here that aggregations are the only predator-defense responses that are available to primates, or that the above factors are the only predator-induced sources of aggregation. [For example, patas monkeys live in relatively small groups and rely for defense on crouching in the grass while the one adult male of the group puts on a "distraction" display (Hall, 1965c)]. For animals that hide in this manner, large groups may actually be disadvantageous.) What is maintained here is that in animals such as baboons that retain conspicuous aggregations in the presence of a predator, the size of a group and the responses of its members to a predator contribute to the safety of the animals in it.

3) Localized resources and home range size

The maximum distance that an animal can go from an essential resource without replenishment will be referred to as the animal's "cruising range" for that resource. At about half the cruising range, the animal reaches a "point of no return" beyond which he must locate a different source if he is to survive. An upper limit on the size and location of home ranges and the maximum length of day-journeys is established by the distribution of resources and the animal's cruising range for each. The relationship may be stated as follows:

Principle Three. *Home ranges are limited to areas that lie within cruising range of some source of every type of essential resource.*

However, for many baboon resources, such as grass, resource points are sufficiently close together that natural variations over space in nearest-neighbor distances never approach the distance beyond which baboons can walk. Such resources are therefore irrelevant for our present purpose, which is to establish an upper limit on *habitable* areas—though the density of grass or other dispersed resources vis-à-vis the animals' foraging strategy may be significant in determining the *inhabited* areas. In contrast, other resources have a more restricted distribution, and thus the foregoing principle may be sharpened as follows:

Home ranges are limited to areas lying within cruising range of the essential resources with the most restricted distribution. Such resources act as limiting factors determining home range size and site.

For many baboon populations in arid regions of Africa, water is the essential resource with the most spatially restricted distribution. Baboons are probably obligate drinkers: No baboon group that we have observed or that has been described in the literature live in an area without some source of permanent water. (It is possible that in some rain forest areas, baboons can obtain enough moisture from succulent foods.) This dependence of baboons on permanent water sources was strikingly demonstrated to us in September 1972 near the end of the dry season, when J. Altmann and I traveled over many miles of the arid Northern Frontier District of Kenya, without seeing a single baboon. The few places where baboons (anubis) were observed during that trip all have some permanent source of water, sometimes nothing more than a pit in a sandy wadi.

For hamadryas baboons, too, water sources are essential and scarce. During September 1971, Slatkin and I observed hamadryas in the arid region near Gota, Ethiopia.* The rainy season had just ended, yet the porous alluvial soil retained virtually no moisture. There was surface water only in the larger rivers. Elsewhere, hamadryas obtained drinking water from holes in sandy river beds. In 1960-61, Kummer studied hamadryas baboons in this same area. He writes:

"During the dry season, each troop had 2 to 4 permanent watering places within its range, mostly at pools under small chutes in the otherwise dry river beds. The hamadryas frequently dug individual drinking holes in the sand of the river bed . . . " (Kummer, 1968a, p. 164).

In Amboseli, the yellow baboons occur only in the vicinity of the permanent water sources—a series of waterholes and swamps that are fed by underground water from Mt. Kilimanjaro. One might ask, why don't the baboons in Amboseli conform to the seasonal pattern of many large mammals in Amboseli and elsewhere in East Africa (Lamprey, 1964), and move completely out of the woodland/waterhole region during the rains, when drinking water is widely available in rain pools and food is abundant elsewhere, then move back in as peripheral areas dry out? Such a migratory

*These observations were made at the study site of J.-J. and H. Abegglen. We are indebted to the Abegglen's for their hospitality during our visit.

pattern would minimize the risk of over-utilizing the woodland area, which must sustain a large population through the dry season on the standing crop available at the end of the rains. Perhaps the explanation is this. During the rains, when the home range size and location is no longer restricted by water sources, sleeping trees become the essential resource with the most restricted distribution. In Amboseli, baboons sleep in fever trees (*Acacia xanthophloea*), which occur only where there is adequate year-round ground moisture, i.e., in the vicinity of permanent water sources, but for reasons that are not known to us they do not sleep in umbrella trees (*Acacia tortilis*), which occur in more arid parts of the Reserve, and this despite the fact that at certain times of the year the baboons may spend many hours of the day in umbrella trees, feeding on the green pods. (A large umbrella tree that is contiguous with one baboon group's favorite sleeping grove of fever trees is used by that group for feeding but not for sleeping. In the morning the group has been observed to move from the sleeping grove to the adjacent feeding tree without first descending to the ground.) These two species of trees are the only ones in the Reserve that are large enough for baboons to use as sleeping trees.

4) Sparse resources and home range size

Unless the size and location of a home range is such that it includes an adequate supply of every essential resource, it will not sustain the animals in it. We can therefore state the following principle:

Principle Four. *The essential, slowly renewing resources whose distributions are sparsest relative to the needs of the animals set a lower limit on home range size and site.*

Principle Four can be made clearer by citing some resources to which it does *not* apply. Oxygen is an essential resource for baboons and is sparse —the amount available in, say, a cubic meter of air could not sustain a baboon for long—but it renews rapidly through diffusion and convection. In Amboseli, water is locally abundant, and if baboons only had to satisfy a water requirement, they could spend the entire day in the vicinity of the waterholes. Thus, neither of these resources sets a lower limit on home range size.

Food in the semi-desert area inhabited by hamadryas baboon populations is the most striking case of a sparse, slowly renewing baboon resource. In the Gota region, the abundance of their food is severely limited by a combination of factors: (i) an unconsolidated alluvial upper horizon from which soil or other fine particles readily erode, leaving a barren, rocky terrain; (ii) ground of high porosity and low water-retaining ability; (iii) strong slope combined with a short, but torrential rainy season, thus making erosion even more rapid; and (iv) hot, dry climate the rest of the year. The area appears to be a recent sediment basin which is now an area of rapid erosion, as water from the high plateau to the south rushes northward toward the Danakil Depression.

As a result, hamadryas baboons must cover an enormous amount of terrain in order to obtain a subsistence amount of food. Hamadryas day-journeys are far longer on the average than those of any other baboon and, so far as we know, longer than those of any other species of non-human primate. They average about 13.2 km per day (Kummer, 1968a). By comparison, chacma baboons averaged 4.7 km in the Cape Reserve of South Africa (Hall, 1962a) and 8.1 km in the Northern Transvaal (Stoltz and Saayman, 1970); anubis baboons in Nairobi Park, Kenya, averaged about 3 miles (4.8 km) per day (DeVore and Washburn, 1963); yellow baboons in Amboseli, Kenya, averaged about 5.5 km per day (Altmann and Altmann, 1970). The sizes of hamadryas home ranges are not known but we predict that they, too, will turn out to be exceptionally large.

At present, little information is available on other sparse resources for baboons. It is not often that the sparseness of resources is as conspicuous as it is for hamadryas in the Gota area, though it is often possible, even fairly early in a field study, to make a reasonable guess about the sparsest resources. In some cases, plant mapping, combined with biochemical analysis of nutritional components, may be required to identify sparse resources: There may be situations in which the size and location of home ranges and the length of day journeys are critically dependent on certain sparsely distributed vitamins, amino acids, or other essential nutrients.

According to Principles Three and Four, sparse resources and those with a restricted distribution act as limiting factors controlling home range size and site, and the length of day-journeys. For example, in an area in which sleeping trees, water, and other resources are abundant, but food plants are sparsely distributed, we would expect the lower limit on home-range size to be determined primarily by available food. In such a situation, the hypothesis can be checked by observing changes in the home range that result from naturally or artificially induced changes in food productivity (cf. Watson and Moss, 1971).

There are many areas in the Amboseli Reserve, Kenya, which are not occupied by baboons, but in which grass and other baboon foods are more abundant at the end of the dry season than they are in those parts of the Reserve that the baboons occupy; however, these areas are without permanent water and are beyond cruising range of permanent water sources. A shortage of suitable sleeping sites may also restrict utilization of these areas. Another example of the limiting influence of spatial distribution of water in arid regions is the following. In October, 1963, which was the last month of the dry season, the home range of one group of yellow baboons in the Amboseli Reserve was a relatively small circumscribed area. But at the beginning of the next month, the group moved far beyond the limits of their October range (Fig. 1). We offered the following explanation for this expansion:

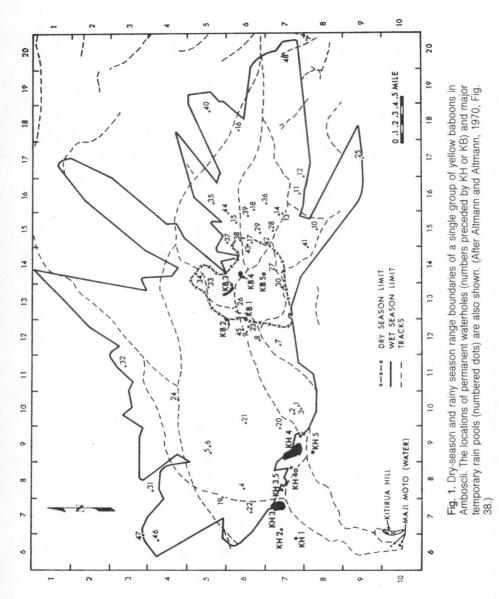

Fig. 1. Dry-season and rainy season range boundaries of a single group of yellow baboons in Amboscli. The locations of permanent waterholes (numbers preceded by KH or KB) and major temporary rain pools (numbered dots) are also shown. (After Altmann and Altmann, 1970, Fig. 38.)

"The onset of the rains on November 1 of that year brought about a major ecological change: thereafter, the baboons could get drinking water almost anywhere and thus were no longer tied to the vicinity of permanent waterholes. That this change in range size cannot be attributed to alterations in vegetation is evident from the fact that the change was apparent as soon as the rainy season began, during the first week of

November, before the new grass had a chance to grow" (Altmann and Altmann, 1970, p. 116).

5 and 6) Resource distribution and home range overlap

The spatial distribution of resources also affects the extent to which home ranges of adjacent groups overlap. The natural tendency of animals to occupy all available parts of the habitat while minimizing competition with conspecifics, combined with the advantages of an established, familiar area, tends to produce a mosaic distribution of home ranges, with contiguous or minimally overlapping boundaries. Such a situation cannot prevail, however, if one or more essential resources are not well distributed throughout the habitat. When this is the case, we propose the following two general hypotheses to explain home range overlap.

Principle Five. *Home range overlap depends primarily on those essential resources with the most restricted spatial distribution: it will be low in relatively uniform habitats and will be extensive if several essential resources have very restricted distributions.*

Under the latter conditions, overlap zones will include a local concentration of essential resources that are not readily available elsewhere.

Principle Six. *The amount of time that groups with overlapping ranges will simultaneously be in the shared portion of their ranges depends primarily on those essential resources in the overlap zone that can only be utilized slowly and whose availability is most restricted in time.*

As a result, simultaneous occupancy of overlap zones will be long wherever slowly utilized resources have a restricted period of availability, and conversely, will be brief if those resources that require longest to utilize are continuously available.

> "Thus an essential natural resource is a restrictive factor in home range separation, in time or space, to the extent that increasing its dispersion in time or space will reduce home range overlap" (Altmann and Altmann, 1970, p. 202).

In the Gota region of Ethiopia, overlap of home ranges of the one-male hamadryas units was imposed by the small number of available sleeping cliffs in the region and resulted in large aggregations or herds (Kummer, 1968*a*, 1971). During our 1963-64 study of the yellow baboons in Amboseli, Kenya, numerous groups used each waterhole, though usually at different times of the day. In the evening, several of these large groups converged on areas of the woodland in which sleeping groves were particularly abundant; each group moved into a separate grove (Altmann and Altmann, 1970). During 1969, when there was a drought in Amboseli, the Masai tribesmen had large herds of livestock in and around the waterholes during most of the day. In the late afternoons, after the livestock left the area, several baboon groups simultaneously converged on a waterhole. At such times, intergroup conflicts were frequent.

Such aggregations at waterholes, sleeping cliffs and tree groves illustrate the effects on home range overlap of essential resources that are spatially and temporally restricted. In contrast, during DeVore's 1959 study of anubis baboons in Nairobi Park, Kenya, home range overlap was much less extensive (DeVore and Hall, 1965). In that area, water sources and sleeping sites were much more widely dispersed than in either Amboseli or the Gota region, and no other resource in Nairobi Park is known to have a comparably restricted distribution. Some of the effects of resource distribution and abundance on group size and home range overlap in hamadryas, anubis, and yellow baboons (Principles One, Five and Six) are summarized in Fig. 2.

7) Time-limited resources

As constituents of the body, many essential resources can be characterized by (i) a critical minimum below which the animal is incapacitated or dead, and (ii) a limited "reservoir" or storage capacity. Such a resource can be referred to as "time-limited." Energy is one of the most important time-limited resources, and we can immediately state an energy limitation: No animal's energy expenditure can exceed its energy input for very long. Because of ongoing metabolic processes, energy resources must be renewed before degradation either kills the animal or renders it incapable of meeting

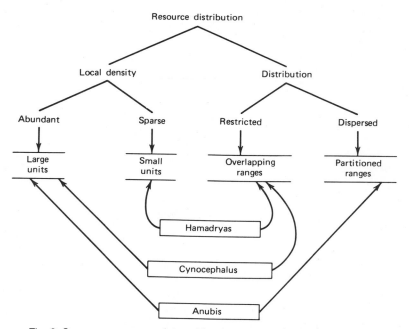

Fig. 2. Some resource correlates of home range overlap and group size.

the metabolic requirements of foraging and other vital activities. (Animals that become dormant, e.g., by hibernating, postpone the problem, but they cannot avoid it.) This suggests the following ecological principle:

Principle Seven. *Energy will place constraints on the time budget whenever it is the essential resource with the shortest exhaustion time.*

Clearly a similar restriction applies to water or to any other vital body component that has a critical minimum and a finite reservoir. For each such time-limited resource, the basic restriction on "foraging" for that resource is that access to the resource must occur before exhaustion of reserves. Critical time allocation decisions involve resources for which reserves are small compared to the rate of use, so that the time to exhaustion is small. The requirements of some time-limited resources, such as oxygen, are readily met by baboons. Others, such as water, present much greater challenges.

Time budgets are further shaped by "scheduled" activities, that is, activities that must occur at a particular time, place, or other contextually defined situation (Hockett, 1964). Examples from Amboseli baboons include the necessity to get into sleeping trees before darkness, and the special alerting reactions given when the baboons go through a critical pass in dense foliage (Altmann and Altmann, 1970).

8) Resource distribution and time budgets

Consider a record of an animal's activities that (i) is obtained continuously over one or more time periods and that includes (ii) the time (real or lapsed) between transitions from one activity state to the next and (iii) the state that the animal is in between each transition and the next. Each such sample of the time course of activity states is a focal-animal sample, in the sense of J. Altmann (1974).

Such a record explicitly or implicitly includes at least the following five types of information: (i) rate and relative frequency at which each state is entered, (ii) distribution of the durations of "bouts" in each state, (iii) percent of time spent in each state, (iv) transition probabilities: for all states i and j, if the animal is in state i, the probability that the next state it enters will be state j; and (v) the time correlation functions: the probability that if the animal is engaged in activity i at time t, it will be engaged in activity j at time $t+\tau$, for all i, j, t, and τ. We will attempt to provide ecological interpretations of some of these parameters and provide a single analytic framework for them. The third item, proportion of time spent in each activity, is often referred to in the literature as a "time budget," though one or more of the other types of information may also be included under that rubric. Some of this information can be obtained from other types of samples. For example, repeated instantaneous samples can be used to estimate proportion of time spent in each state (J. Altmann, 1974). Slatkin (unpublished) used both of these sampling methods in his comparative study of time budgets in gelada monkeys and yellow baboons.

Many food items (berries, grass plants, etc.) and some other essential resources occur in small discrete packets whose positions in space can be regarded as points. Because the "processing" or utilization of food items and the movements of an individual from one resource point to another each take a certain amount of time, the spatial distribution of resource points that are utilized is reflected in the time budgets of the animals. For example, at various places in the Amboseli grassland, there are patches of a special soil type on which few plants except *Leucas stricta* grow, and those are not eaten by the baboons. When the baboons encounter such a patch while foraging, they stop feeding, walk across the patch, and resume feeding on the other side. Thus, the spatial array: *edible grass→ Leucas stricta→ edible grass* is reflected in the time sequence: *feed→ walk→ feed.*

Some feeding activities (chewing, reaching, etc.) can be carried out while the animal is moving to the next food item, but if the next food item is too far away, a period of progression (for baboons, walking) will occur between feeding bouts. We define *food patches* as the most inclusive sets of food items that are situated in such a manner that is possible for an animal to go from one to another without interrupting his feeding activities. Clearly, the maximum possible distance from any food item to the nearest other food item in the same patch will depend on the animal's processing time and the rate at which the animal travels from one food item to the next. If the latter is relatively constant, or at least, if it has a non-zero minimum, then the length of foraging bouts will be limited by the number of food items per patch. Bout length will be further reduced by the presence of other individuals that feed concurrently on the same patch, by the failure of the animals to utilize all of the items in a patch, and by activities that interrupt feeding.

The proportion of time spent feeding probably reflects relationships between several factors, including the richness of the food sources, their spatial proximity, the processing time required to utilize them, and the metabolic requirements of the animals.

Suppose that we plot a conditional probability function for feeding activities: for all moments t at which an animal was observed feeding, we plot the probability that the animal will be feeding at each subsequent moment $t+\tau$, plotted as a function of the lapsed time τ, regardless of whether feeding is continuous between t and $t+\tau$ (Fig. 3). This time correlation function must begin at 1 (where $\tau = 0$) and converges on a limit which is the proportion of time spent feeding. Slatkin (unpublished) has pointed out that the "rate of decay" of this function for feeding may reflect aspects of food distribution and has suggested some possible ecological correlates. A somewhat different interpretation is as follows: The function should decay slowly under any of the following conditions: (i) very large food patches, so that the animals can feed continuously; (ii) medium or very small patches that are sufficiently close together that they form a "super-patch," in which the time required to move from patch to patch is small; (iii) medium sized

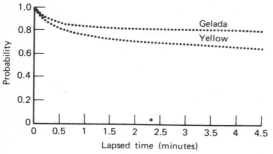

Fig. 3. Time correlation functions for feeding in adult male gelada monkeys and yellow baboons. For explanation, see text.

patches that are far apart, so that the animal is better off staying where he is and utilizing some of the less accessible food items in the patch. The function will decay more rapidly if each patch is very small, so that the animal quickly exhausts the local supply, and if the patches are over-dispersed, so that when the animal does move to another patch, he must move a considerable distance.

The first systematic analyses of primate time budgets that has included time correlation functions and length of feeding bouts, as well as the conventional proportion of time per activity, was carried out by Slatkin (unpublished). In a comparative study of adult male time budgets in yellow baboons (Amboseli Reserve, Kenya) and geladas (Simien Mountains, Ethiopia), he discovered that gelada males spent more time feeding than did yellow baboon males, that their feeding bouts were longer, on the average, and that the correlation function for feeding decayed much more slowly than did that of yellow baboons. The fact that geladas spent more time feeding per day, "reflected the fact that during the study periods, the geladas were eating low nutrient food items (grass, leaves, flowers) while the yellow baboons were eating higher quality foods (seeds, rhizomes, berries). The average duration of a feeding bout was greater for the geladas because the food items tended to be closer to each other in the geladas' habitat, thereby allowing an individual to feed continuously for a longer period of time without moving to another feeding site." The amount of time for the function to drop by 90% "was much greater for the yellow baboons than for the geladas (1 min vs. 3 min) even though the average feeding bout for the yellow baboons was shorter," probably because "the food resources that are actually utilized are much more patchily distributed for the yellow baboons than for the geladas" (Slatkin, personal communication).

If food items can only be processed one at a time, the amount of food obtained per minute of foraging time will be limited by processing time. One of the advantages of cheek pouches is that they avoid this limiting effect of processing time (especially for mastication) on harvest rate. It is surely no coincidence that cheek pouches occur in all cercopithecine primates (baboons, macaques, mangabeys, vervets, etc.) but not in the colobines (langurs, colobus, etc.). Cercopithecines tend to feed on relatively small quantities of concentrated foods that have a more patchy distribution than the food of the leaf-eating colobine monkeys. It is also noteworthy that among cercopithecines, cheek pouches are smallest, relative to body size, in geladas (Murray, 1973): Over much of their range these animals live in habitats in which food is more uniformly distributed than is the food of baboons in *their* habitats. Where the two occur in the same region, as at Debra Libanos, Ethiopia, the geladas spend much more time feeding (Crook and Aldrich-Blake, 1968).

Now let us try to include the various temporal characteristics of an animal's state time course in a single, testable model. Consider the following two characteristics of these time courses. (i) The transitions from one state to another are not rigidly fixed sequences. However, for each two-state sequence i,j we can estimate the conditional probability p_{ij} that, given the animal is in state i, the next state he enters will be state j. As a first approximation, it seems reasonable to suppose that within any one habitat type, the distribution of these probabilities is stationary. By a habitat type, we refer here to a (plant) community in which each species that affects the animal's foraging can be represented by a single spatial distribution. (ii) The duration of stays ("bouts") in each state is a random variable, not a constant. It seems reasonable to suppose, again as a first approximation, that within any one habitat type, the conditional probability that an animal will go from state i to state j by time t given that he was in state i at time 0 depends only on the pair of states i,j and the amount of time t that has been spent in i. For each pair of states i,j the distribution of times between transitions can be described by the function $F_{ij}(t)$ specifying the probability that the animal will go from state i to state j at or before time t, given that he entered state i at time 0 and that state j is the next one he will occupy.

A system that can be described by the properties assumed in (i) and (ii) above is called a semi-Markov process.* Such a process contains two sets of parameters, the set p_{ij} of conditional transition probabilities and the parameters of the distribution F_{ij}.[†] Once these two parameters are known, many other characteristics of the time process can be derived.

Focal-animal state samples, which we described at the beginning of this section, contain the data necessary to estimate these parameters. The transition probabilities p_{ij} are estimated from the number of i-to-j moves, in the usual manner: $p_{ij} = n_{ij} / \sum_k \eta_{ik}$. The observed times between i-to-j moves are used to estimate the distributions $F_{ij}(t)$.[‡]

---

*For a lucid exposition of many fundamental theorems of semi-Markov processes, as well as a guide to the literature, see Ginsberg (1971, 1972).

[†]When the F_{ij} are all exponential and independent of j, the semi-Markov process reduces to a Markov process. According to Slatkin (personal communication), the distributions are not exponential for either cynocephalus baboons or geladas, but are close to it.

[‡]In practice, beginning and terminating sample periods produces end effects. the methods for treating these truncated intervals is given by Moore and Pyke (1968).

In summary, we give:

Principle Eight. *The time course of an animal's activities within any habitat that can be represented by a single spatial distribution can be approximated by semi-Markov processes, the parameters of which are related to characteristics of the spatial distribution of resources.*

9) Foraging strategies

We assume that natural selection will favor those individuals that utilize their forage time in such a manner that they obtain the largest yield of energy (or other nutritional component) per unit of time.*

For a variety of reasons, a baboon's movement from one food item to the next will usually take place at a rate that can only vary within fairly narrow bounds. Energy expenditure probably is a step function of gait, with running consuming far more energy per unit distance or unit time, thus reducing feeding efficiency as well as making it more difficult for the animal to maintain a stable body temperature. Consequently, an animal that ran from plant to plant would lose more than he gained: feeding efficiency is not synonymous with feeding rate. As a result of this progression-rate restriction a baboon cannot make any appreciable improvement in his yield rate by going more rapidly from one food item to the next. On the other hand, a slower pace would reduce the rate of food intake and increase the amount of time necessary to obtain a sustaining amount of food. For baboons in hot and arid regions, it would keep them on the feeding ground and away from shade and water for longer periods of time.

These relations can be summarized as follows:

Principle Nine. *For animals that forage on slowly renewing, stationary food items and that move from one food item to the next at an essentially fixed rate, the foraging pathway with the shortest average distance per usable food calorie will be optimal in the sense of yielding the most energy per unit time.*

Consider a species that feeds on a slowly renewing, stationary resource that occurs as a set of point sources. Suppose we know the mean gross caloric value of a food type, the maximum rate of movement from item to item, the caloric cost of moving a unit distance, the spatial array of the food items (assumed to be stationary point sources and to be slowly renewing), and the place in the habitat where foraging begins. What is the optimal foraging pathway for the animal to follow?

Consider first a "one-step evaluator," that is, an animal that can see and estimate the distance to each of the various food items within its perceptual range, but cannot take into account their spatial relationships to each other.

---

*Schoener (1971) distinguishes between those animals ("time minimizers") that accomplish this by minimizing time required to obtain a subsistence amount of food and those that do it by maximizing food input for a set amount of foraging time ("energy maximizers").

Assume for the moment that the animal cannot estimate the caloric yield of
the food items, but can only treat them as of equal value. Such an animal
can do no better than go to the nearest food item at each move. If he is lucky,
he will never have to go so far that he depletes his energy reserves. If, in
addition, the animal can, at a distance, estimate the gross caloric content of
each food item (i.e., can act on the basis of a perceived correlate of caloric
content) as well as the caloric "cost" (energy required in moving that far,
then harvesting and processing food), his optimal strategy is to deduct from
the gross caloric value of each food item the cost (in calories) of using that
food item, then go to the food item with the highest net yield—of course
eating any other food that he passes en route.

Now suppose we have a somewhat more skillful animal, one that can,
like the last one, estimate the distance from its present position to any food
item in its perceptual range, but one that can also estimate the distance from
that point to the food item nearest to that point. Such an animal is able to
detect a minimal "clump," consisting of but two food items, and will be
called a "two-step evaluator." Suppose also that the animal cannot at a
distance estimate the value of a food item (or that the items are essentially
identical). For such an animal, there exists an optimal strategy: (i) choose
among the pairs of points (food items) on the basis of which pair has the
shortest total distance from animal to nearer point and from there to the
farther point; (ii) begin with the closer member of the pair, then go to the
other one. If on the other hand the animal can evaluate the caloric content
of each food item and the caloric cost of each point-to-point pathway, he
should find that ordered pair of points for which the total net caloric yield
is highest.

A two-step evaluator may be able to increase his yield (and cannot
decrease it) if he re-evaluates his strategy in the same manner at each point
(i.e., at the first point of each selected pair).

This type of analysis can be continued through a series: three-step
evaluators, fourstep evaluators, etc. For all such animals the goal is the
same: minimize the total distance traveled* or, for an animal that can evaluate calories, maximize the total net caloric yield. But the expected value is
not: the advantages of an n-step evaluator over an $(n-1)$-step evaluator are
the ability to detect larger clumps, thereby reducing costly clump-to-clump
movements, and the ability to reduce the risk of following a chain of points
out to the tip of a "peninsula" of points that would require excessive
walking in order to get back to a foraging area.

We do not know, for baboons or any other animal, whether the fine-

*For an animal that cannot re-evaluate as it goes along and that cannot evaluate caloric
content of the food items, the problem is closely related to a well-known problem in graph
theory, often worded in terms of a traveling salesman who leaves the office, visits each of a
specified set of factories, then returns to the office (or in an alternate version, then goes home).
His problem is to minimize the amount of traveling that he does. In our case, however, the
terminal point is not fixed and the set of points to be visited is not specified in advance.

grain geometry of their nutrient distribution is a major determinant of their movements nor whether individual differences in foraging movements are related to the chance of survival. It seems extremely likely that both effects exist and are of sufficient magnitude that they are measurable.

Denham (1971) writes: "I assume that a specific primate population will occur only where food of a kind usable by that species is present. Thus we can control for, or disregard, the 'food supply composition' variable." But the nutritional value of food items can be disregarded only if the nutrient compositions of all food items are identical. It is precisely the varied distribution of nutrients that presents a challenge to the selective feeding ability of animals. If animals can select nutritious foods and can thereby "balance their diets," any adequate model of foraging strategies must treat food items as something other than indistinguishable points in space. Just how good the foraging strategies of baboons are remains to be seen. Certainly the available evidence suggests a very considerable ability:

"In reviewing our records of baboon's food plants . . . we have been impressed by the apparent capacity of these animals to feed selectively on the most nutritious parts of the plants available in their habitat at each time of the year" (Altmann and Altmann, 1970, p. 169).

10) Resource distribution and foraging formation

For animals that live in groups, sparse resources present a special spacing problem: How can the members of a group remain together and still avoid excessive competition? One part of the answer is to be found in the deployment of the members relative to each other while foraging for such a resource. Consider a group of animals on a plane surface, e.g., a group of baboons foraging on flat, open grassland. Under those conditions a parallel, in-line formation of individuals, moving in a rank, is the unique foraging pattern that enables each animal to maintain a continuous, exclusive forage-swath, thereby maximizing the harvest per unit distance traveled, while simultaneously minimizing each animal's mean distance to his neighbors. Thus, we have:

Principle Ten. *A rank foraging formation will be favored whenever there is an advantage to remaining in a group and the group is foraging on slowly renewing resources that are of low overall density in the home range and that are not locally abundant.*

When yellow baboons move away from their sleeping grove toward a distant foraging area, they move in a long file; if they then forage in an area of dry, open grassland, the file formation is transformed into a long rank (Altmann and Altmann, 1970). In a sparse habitat this tendency to forage in a rank formation may be partially counteracted by several local factors: (i) a long resource processing time, so that an individual must remain in one place for an appreciable time (e.g., a baboon digging a grass rhizome out of the ground); combined with (ii) a dispersed food distribution, so that neighboring individuals may have to move ahead of stationary ones before encountering another food item. In addition, the forage-swaths may not be completely private: displacement of one yellow baboon by another at food sources sometimes occurs (Altmann and Altmann, 1970).

But wouldn't the animals be better off foraging separately? Aren't they competing for food? The answer that would usually be given is, yes, except that the group is a predator-protection mechanism (see Principle Seven): a solitary baboon is a dead baboon (DeVore, 1962). Another source of sociality under these conditions is this: the animals may not, in fact, be better off foraging independently. In a sparse habitat, foraging efficiency is probably limited by the distance between plants. To the extent that animals encounter used forage-swaths of other individuals, their foraging efficiency will be reduced. By feeding together in foraging parties that are as large as is compatible with the local food supply, the animals may be better off, because an individual will not encounter the foraging path of a member of his group until the whole group returns to the same spot. This advantage would be of particular significance in sparse habitats. " . . . When food becomes scarce, . . . participation in . . . social groups apparently results in a maximally efficient apportionment of available resources with a minimal level of hostile interactions" (Morse, 1971).

This characteristic of a rank—that it enables an individual to minimize the mean distance to neighbors while maintaining an exclusive and continuous forage-swath—is not provided by any other configuration, though under some conditions, alternative configurations may have other, overriding advantages. In particular, a cluster or herd formation offers central individuals greater protection from predators that attack the periphery of the group. Where food is sufficiently abundant, foraging clusters may prevail. When feeding in the acacia woodland of Amboseli, yellow baboons tended to be closer to their nearest neighbors and to be more clustered than when feeding in the open grassland. Such areas were, in general, areas in which predator attacks were more probable (Altmann and Altmann, 1970). They also were areas in which food tended to be locally concentrated and in which visibility was reduced by foliage. All of these factors would contribute to the increased cohesiveness of the group.

At this point, one might ask, why can't a very large group forage in a "thin" habitat? Why couldn't the several hundred hamadryas of one sleeping rock forage as a unit, moving in one enormous rank? There are at least three factors that militate against this: (i) Those at the far ends of such an enormous phalanx would have to walk an inordinate distance whenever the group turned; (ii) hamadryas food has a patchy distribution and no patch is large enough for several hundred baboons to feed simultaneously; (iii) the irregular hamadryas habitat would repeatedly sever visual contact, thereby effectively subdividing the herd.

Geladas forage on the rich alpine meadows of the Simien Mountains,

Ethiopia, in herds of several hundred individuals. If food were sparse and not locally abundant, those at the back of such a herd would find little to eat. Crook and Aldrich-Blake (1968) note that the geladas in the vicinity of Debra Libanos, Ethiopia, did not occur in large herds of the sort that had been observed in the Simien, "namely because areas of unbroken grassland were few. In fact herds of any size only congregated at the one large area of grazing near Chagal." Unfortunately, little information is available on gelada foraging formations under these conditions.

11) Macro-strategy: area occupation density

Suppose one records where an animal goes as it moves about in its home range. For animals like baboons that forage in a group, one can keep track of the mean position or center of mass of the group (Altmann and Altmann, 1970). If such records are accumulated over many days, it becomes apparent that the animals do not move about at random. The areas in which members of a baboon group spend much of their time tend to be areas in which several resources co-occur (Table 1). Conversely, areas that the animals seldom enter and in which they remain only briefly tend to be areas in which resources are sparse and hazards are high.

None of this is surprising. If the animals are to survive in an area, they must go where the resources are, and they must avoid hazardous areas as best they can:

Principle Eleven. *The survival of an animal depends on its ability so to allocate the distribution of its activities among the parts of its home range that it gains access to the essential resources therein, while avoiding excessive exposure to hazards.*

This statement is not meant to be taken as a truism, asserting only that animals move about in such a way that they manage to survive. Rather, it is meant to be testable. An animal that spent a large part of its time in areas in which predators were abundant and resources were sparse probably would not fare as well as one that spent more time utilizing abundant, safe resources. But how large are the actual individual differences in strategies for exploiting localized resources and avoiding localized hazards? And how much difference do they make, in terms of survival and reproduction? Unfortunately, we do not yet know, but I believe that these questions can be answered and in what follows I shall discuss one approach to this problem.

Suppose that we evaluated the resources and hazards that occur in the various parts of an animal's home range. We could then see how accurately we could predict the animal's occupancy density distribution and localized activities from such data about the habitat, and whether differences in rates of survival or reproduction could, in turn, be related to different patterns of land use and to differences in the net value of home ranges.

No attempt will be made here to describe, even in principle, what a "real estate assessment" for a baboon should look like. Indeed, I suspect that it would be different, depending on the question at hand—for example,

Table 1

Resources, other than food, in the 14 most-used quadrats of one group of yellow baboons in Amboseli. These 14 quadrats, each 0.4 × 0.4 mile, included about one-quarter of the home range area of the group, but accounted for about three-quarters of its time. (After Altmann and Altmann, 1970, Table XX.)

Rank of quadrat	Per cent of time	Number of sleeping groves	Per cent of nights	Number of water-holes	Per cent of drinks	Number of rain-pools	Per cent of drinks	
1	20.67	3	36.0	—	0	1	1.7	to 5 sleeping groves
2	9.39	—	0	2	16.7	4	6.7	to 5 sleeping groves and 2 waterholes
3	6.33	2	13.6	1	5.8	1	2.5	to 3 sleeping groves
4	5.97	—	0	3	10.8	1	0.8	to all waterholes in eastern woods
5	5.96	2	26.4	—	0	1	0	to 4 sleeping groves
6	5.61	—	0	—	0	9	10.0	to eastern woods
7	3.57	2	12.8	2	4.2	0	0	to 2 sleeping groves, 2 waterholes
8	3.52	—	0	1	8.3	2	4.2	to 3 sleeping groves, 1 waterhole
9	2.65	2	4.0	—	0	1	0	to 3 sleeping groves, 1 waterhole
10	2.45	1	2.4	2	0	1	1.7	to 1 sleeping grove, 2 waterholes
11	2.33	—	0	—	0	1	0.8	between eastern and western woods
12	2.23	—	0	—	0	1	1.7	to 2 waterholes, 2 sleeping groves
13	2.13	—	0	—	0	0	0	to 1 waterhole, 2 sleeping groves
14	1.84	2	4.8	2	0	0	0	to 1 sleeping grove
15-87	25.35	—	0	1	1.7	25	22.5	

on whether one was attempting to predict the occupancy density of a particular area, was trying to find out how the requirements for a particular nutrient were being met, or was trying to relate resource utilization to survival and reproduction. But some common components of such evaluations can be described.

The choice of a time scale for the evaluation is important. In the short run, the value of an area to an animal is not constant. It depends on variations in the environment and on the past behavior of the animal.

But in the long run, each area will have a net utility to the animals, namely, the difference between what they gain from the area, in terms of access to those resources that will enhance growth, survival, and reproduction, and what they lose, as a result of hazards to life therein. For example, the value of water sources to a group of baboons is very different, depending on whether it has been a cool or a hot day, and on whether the baboons have recently drunk. The value of a forage area depends on how much it has been used since the last crop matured. Over an extended period of time, however, the baboons of an area have a total water loss that must be restored if they are to survive, and this in turn establishes an overall value to water resources of the area. The value of any particular waterhole will depend on whether it lies between, say, mid-day foraging areas and evening sleeping groves. The predation risk of any area on any particular day depends on whether a predator is nearby at the time. In the long run, however, the predation risk of an area will depend on the frequency with which predators attack animals in that area.

The evaluations can be carried out on the basis of density of resources and hazards in areas (e.g., quadrats) or of proximity to sample points. The former procedure has the disadvantage that the results depend in part on the initial choice of quadrat size (cf. Greig-Smith, 1964) Another difficulty with quadrats for this purpose is that they cannot be evaluated independently: the value of a quadrat would depend on resources (e.g., the nearest sleeping grove or waterhole) that might be in another quadrat. Beyond that, it may be that proximities to resources and hazards are more important than their densities, and thus that an evaluation on the basis of proximities to sample points comes closer to evaluating environmental factors that are significant to the animals. One is asking, if the animal found himself at this point in his home range, how far would he have to go to the nearest edible plant? To the nearest source of water or shade? How close is he to places where leopards hide? Of course, this evaluation cannot be carried out without a knowledge of the animal's capabilities for exploiting various resources. A food that the animal cannot reach, or cannot digest, is of no value, whatever a biochemical analysis might show to be present. In hilly or mountainous areas, allowance would have to be made for the fact that walking up and down slopes represents a different cost than moving horizontally along a slope or across an alpine meadow.

The risk at a point can be estimated from frequency and proximity of predator sightings, predator attacks, alarm calls, from measurements of proportion of shade cover, local mosquito or tick density, and so forth. For baboons, there are several localized hazards, including: (i) intense insolation, lack of shade, and remoteness from water with attendant dangers of dehydration and thermal imbalance; (ii) intergroup competition; (iii) mosquitoes, schistosomiasis, and coxsackie B_2 (endemic to Amboseli baboons), all of which are probably most abundant or most readily transmitted in or near permanent waterholes; and (iv) carnivores, which for Amboseli baboons include leopard, lion, and silverback jackal, and probably also cheetah, spotted hyena, striped hyena, hunting dog, marshal eagle, and hawk eagle. Of these the leopard is probably the baboon's major predator.

Predators are the hazards that are the most spectacular and have received the most attention, although they may not represent the greatest risk in a baboon's life. Alternatively, others have argued that attacks by predators on baboons have seldom been seen, despite many man-hours devoted to watching them in the wild, and thus that predation cannot be a very important selective factor acting on baboon societies. But the birth of infants is also seldom seen. In neither case does the frequency with which these events are observed—or actually occur—provide an adequate measure of their biological importance. The relationship between these two classes of rare events is this: In a species in which the reproductive rate is low, the survival of each individual is of particular significance in maintaining the population, and the selective impact of a predator killing a single individual is much greater than it is in a species in which the number of off-spring per individual is high.

The space-specific risk of predation for baboons results from the fact that the baboon's predators tend to concentrate their hunting to particular habitats within the baboon's home range. For example, leopards stalk from cover and are seldom seen in areas of low, open grassland.

The baboon's problem, then, is to avoid areas of high risk, and yet still get at areas with concentrated resources. However, the predators face exactly the same problem, and to the extent that baboons are a resource for the predators, the latter will go where the baboons go. Fortunately for baboons, they are not a major food source of any predator (cf. Schaller, 1972, Tables 33, 36, 37, 63, and Kruuk, 1972, Tables 11, 12, 22).

Because the value of an area within the home range depends on both resources and hazards, a particular net value of a point or area can result from various combinations of risks and resources. For example, an area might be avoided either because it contained few resources, or because the risk of entering the area was not sufficient to compensate for the resources that were present. Another possibility, and one of importance in trying to evaluate differences in adaptation is that the utility value of an area within the home range is high, but the animals are not taking full advantage of it. In any case, we should not completely ignore areas that the animals do not enter frequently, nor remain in long, but should try to find out why that is so.

> "Thus, in analyzing the utilization of home range, we must consider not only those parts of the home range that the animals enter frequently or remain in for long periods of time, but also those that are seldom and briefly entered. More precisely, we must consider the spatial distribution of home range utilization and its relationships to the distribution of both hazards and natural resources among the parts of the home range" (Altmann and Altmann, 1970, p. 200).

At the other extreme, if all the resources were in one area of the home range, and all the hazards in another, the best strategy in terms of area occupancy would be to stay in the former and avoid the latter.

Hazards and resources will affect differential occupancy of areas within the home range only if neither is uniformly distributed in time and space. Because recources and hazards often co-occur in time and space, and be- *p+.* cause the animals must sometimes go through a hazardous area to get from *1∅* one resource area to another, the problem of finding a viable distribution of activities is not trivial.

SUMMARY

For the convenience of the reader, we list here the eleven principles that we have developed in this article.

1. *Resource distribution and group size.* If a slowly renewing resource is both sparse and patchy, it can be exploited more effectively by small groups. Conversely, large groups that are simultaneously using a single resource will occur only if the supply is adequate, either because of local abundance, or because of rapid renewal of the resource. Large groups will be more effective if a resource has a high density but a very patchy distribution and the patches themselves occur with low density, so that the resources tend to be concentrated in a few places.

2. *Predation and group size.* Predation selects for large groups and for groups containing at least one adult male.

3. *Localized resources and home range size.* Home ranges are limited to areas that lie within cruising range of some source of every type of essential resource. More importantly, home ranges are limited to areas lying within cruising range of the essential resources with the most restricted distribution.

4. *Sparse resources and home range size.* The essential, slowly renewing resources whose distributions are sparsest relative to the needs of the animals set a lower limit on home range size and site.

5. and

6. *Resource distribution and home range overlap.*

    5. Home range overlap depends primarily on those essential resources with the most restricted spatial distribution: it will be low in relatively uniform habitats and will be extensive if several essential resources have very restricted distributions.

    6. The amount of time that groups with overlapping ranges will simultaneously be in the shared portion of their ranges depends primarily on those essential resources in the overlap zone that can only be utilized slowly and whose availability is most restricted in time.

7. *Time-limited resources.* Energy will place constraints on the time budget whenever it is the essential resource with the shortest exhaustion time.

8. *Resource distributions and time budgets.* The time course of an animal's activities within any habitat that can be represented by a single spatial distribution can be approximated by semi-Markov processes, the parameters of which are related to characteristics of the spatial distribution of resources.

9. *Foraging strategies.* For animals that forage on slowly renewing, stationary food items and that move from one food item to the next at an essentially

fixed rate, the foraging pathway with the shortest average distance per usable food calorie will be optimal in the sense of yielding the most energy per unit time.

10. *Resource distribution and foraging formation.* A rank foraging formation will be favored whenever there is an advantage to remaining in a group and the group is foraging on slowly renewing resources that are of low overall density in the home range and that are not locally abundant.

11. *Macro-strategy: Area occupation density.* The survival of an animal depends on its ability so to allocate the distribution of its activities among the parts of its home range that it gains access to the essential resources therein, while avoiding excessive exposure to hazards.

POSTSCRIPT

A few years ago, when I first began to think about the ways in which various activities of baboons depend on the spatial and temporal distribution of resources and hazards in their environment, I came across a posthumous paper by Milne (1947) on East African soils. Milne had a remarkable ability to predict plant communities on the basis of edaphic, climatological and topological features. The corresponding problem for animals is no doubt more complicated, but if the composition and distribution of plants in an area depend to a considerable extent on the local geology and meteorology, and if many activities of animals are adaptations to the structure of that plant community, it should ultimately be possible to combine these relationships into a single ecological theory of baboon social systems. Such a theory could be tested, in part, by making observations on the physical ecology of two or more areas and (in order to avoid a confounding problem in zoogeography) on the plants and animals of the surrounding region. The adequacy of the theory would then be judged on how well one could predict, on the basis of these observations, the behavior, social relations, group structure and population dynamics of the animals in these areas. We are a long way from being able to produce such a theory, but I hope that this article will contribute to that end—to an ecological theory of social systems, literally "from the ground, up."

REFERENCES

Aldrich-Blake, F. P. G., T. K. Bunn, R. I. M. Dunbar, and P. M. Headley. 1971. Observations on baboons, *Papio anubis,* in an arid region in Ethiopia. Folia Primatol. 15:1-35.

Altmann, J. 1974. Observational study of behavior: sampling methods. Behaviour 48:1-41.

Altmann, S., and J. Altmann. 1970. Baboon ecology: African field research. Univ. Chicago Press, Chicago, and S. Karger, Basel (Bibliotheca Primatologica No. 12).

Baker, J. R. 1938. The evolution of breeding seasons, p. 161-171. *In* G. R. deBeer [ed.], Evolution. Oxford Univ. Press, Cambridge.

Chalmers, N. R. 1968a. Group composition, ecology and daily activities of free living mangabeys in Uganda. Folia Primatol. 8:247-262.

Chalmers, N. R. 1968b. The social behaviour of free living mangabeys in Uganda. Folia Primatol. 8:263-281.

Cody, M. L. 1968. On the methods of resource division in grassland bird communities. Amer. Natur. 102:107-147.

Cohen, J. E. 1969. Natural primate troops and a stochastic population model. Amer. Natur. 103:455-477.

Cohen, J.E. 1971. Social grouping and troop size in yellow baboons, p. 58-64. In H. Kummer [ed.], Proc. 3rd Int. Congr. Primatol., Zurich, 1970. Vol. 3. S. Karger, Basel.

Cohen, J. E. 1972. Markov population processes as models of primate social and population dynamics. Theort. Populat. Biol. 3:119-134.

Crook, J. H. 1966. Gelada baboon herd structure and movement. A comparative report. Symp. Zool. Soc. London 18:237-258.

Crook, J. H. 1967. Evolutionary change in primate societies. Sci. J. (London) 3:66-70.

Crook, J. H. 1970a. The socio-ecology of primates, p. 102-166. In J. H. Cook [ed.], Social behaviour in birds and mammals. Academic Press, New York.

Crook, J. H. 1970b. Social organization and the environment: aspects of contemporary social ethology. Anim. Behav. 18:197-209.

Crook, J. H., and P. Aldrich-Blake. 1968. Ecological and behavioural contrasts between sympatric ground dwelling primates in Ethiopia. Folia Primatol. 8:192-227.

Crook, J. H., and J. S. Gartlan. 1966. Evolution of primate societies. Nature 210:1200-1203.

Denham, W. W. 1971. Energy relations and some basic properties of primate social organization. Amer. Anthropol. 73:77-95.

DeVore, I. 1962. The social behavior and organization of baboon troops. Doctoral Diss., University of Chicago.

DeVore, I. 1963. A comparison of the ecology and behavior of monkeys and apes, p. 301-319. In S. L. Washburn [ed.], Classification and human evolution. Aldine, Chicago.

DeVore, I., and K. R. L. Hall. 1965. Baboon ecology, p. 20-52. In I. DeVore [ed.], Primate behavior: field studies of monkeys and apes. Holt, Rinehart & Winston, New York.

Dyson-Hudson, R., and N. Dyson-Hudson. 1969. Subsistence herding in Uganda. Sci. Amer. 2202:76-89.

Eisenberg, J. F., N. A. Muckenhirn, and R. Rudran. 1972. The relation between ecology and social structure in primates. Science 176:863-874.

Fisler, G. F. 1969. Mammalian organizational systems. Contrib. Sci. L. A. Co. Mus. 167.

Forde, D. 1971. Ecology and social structure: the Huxley Memorial Lecture 1970. Proc. Roy. Anthropol. Inst. 1970:15-29.

Gartlan, J. S. 1968. Structure and function in primate society. Folia Primatol. 8:89-120.

Ginsberg, R. B. 1971. Semi-Markov processes and mobility. J. Math. Sociol. 1:233-262.

FIELD STUDIES

Ginsberg, R. B. 1972. Critique of probabilistic models: application of the semi-Markov model to migration. J. Math. Sociol. 2:63-82.

Greig-Smith, P. 1964. Quantitative plant ecology. 2nd Ed. Buttersworth, London.

Hall, K. R. L. 1962*a*. Numerical data, maintenance activities, and locomotion in the wild Chacma baboon, *Papio ursinus.* Proc. Zool. Soc. London 139:181-220.

Hall, K. R. L. 1962*b*. The sexual, agonistic and derived social behaviour patterns of the wild Chacma baboon, *Papio ursinus.* Proc. Zool. Soc. London 139:283-327.

Hall, K. R. L. 1965*a*. Social organization of the old world monkeys and apes. Symp. Zool. Soc. London 14:265-289.

Hall, K. R. L. 1965*b*. Ecology and behavior of baboons, patas, and vervet monkeys in Uganda, p. 43-61. *In* H. Vagtborg [ed.], The baboon in medical research. Univ. Texas Press, Austin.

Hall, K. R. L. 1965*c*. Behaviour and ecology of the wild patas monkey, *Erythrocebus patas,* in Uganda. J. Zool. 148:15-87.

Hall, K. R. L. 1966. Distribution and adaptations of baboons. Symp. Zool. Soc. London 17:49-73.

Hamilton, W. D. 1971. Geometry for the selfish herd. J. Theoret. Biol. 31:295-311.

Hennig, W. 1966. Phylogenetic systematics. Univ. Illinois Press, Urbana.

Hockett, C. F. 1964. Scheduling, p. 125-144. *In* F. S. C. Northrop and H. H. Livingston [ed.], Cross-cultural understanding: epistemology in anthropology. Harper & Row, New York.

Itani, J. 1967. Postcript by the editor. Primates. 8:295:296.

Jolly, A. 1972. The evolution of primate behavior. Macmillan, New York.

Koford, C. 1966. Population changes in rhesus monkeys, 1960-1965. Tulane Stud. Zool. 13:1-7.

Kruuk, H. 1972. The spotted hyena: a study of predation and social behavior. Univ. Chicago Press, Chicago.

Kummer, H. 1967*a*. Tripartite relations in hamadryas baboons, p. 63-71. *In* S. A. Altmann [ed.], Social communication among primates. Univ. Chicago Press, Chicago.

Kummer, H. 1967*b*. Dimensions of a comparative biology of primate groups. Amer. J. Phys. Anthropol. 27:357-366.

Kummer, H. 1968*a*. Social organization of hamadryas baboons. Univ. Chicago Press, Chicago; S. Karger, Basel (Bibliotheca Primatologica No. 6).

Kummer, H. 1968*b*. Two variations in the social organization of baboons, p. 293-312. *In* P. C. Jay [ed.], Primates, studies in adaptation and variability. Holt, Rinehart and Winston, New York.

Kummer, H. 1971*a*. Primate Societies. Aldine-Atherton, Chicago.

Kummer, H. 1971*b*. Immediate causes of primate social structures, p. 1-11.

In H. Kummer [ed.], Proc. 3rd Int. Congr. Primatol., Zurich, 1970. Vol. 3. S. Karger, Basel.

Kummer, H. and F. Kurt. 1963. Social units of a free-living population of hamadryas baboons. Folia Primatol. 1:4-19.

Lamprey, H. F. 1964. Estimation of the large mammal densities, biomass and energy exchange in the Tarangire Game Reserve and the Masai Steppe in Tanganyika. E. Afr. Wildlife J. 2:1-46.

Milne, G. 1947. A soil reconnaissance journey through parts of Tanganyika Territory December 1935 to February 1936. J. Ecol. 35:192-265.

Mizuhara, H. 1946. Social changes of Japanese monkey troops in the Takasakiyama. Primates 5:27-52.

Moore, E. H., and R. Pyke. 1968. Estimation of the transition distributions of a Markov renewal process. Ann. Inst. Math. Stat. 20:411-428.

Morse, D. N. 1971. The insectivorous bird as an adaptive strategy. Ann. Rev. Ecol. Syst. 2:177-200.

Murray, P. 1973. The anatomy and adaptive significance of cheek pouches (*bursae buccales*) in the Cercopithecinae, Cercopithecoidea. Doctoral Diss., University of Chicago.

Nagel, U. 1971. Social organization in a baboon hybrid zone, p. 48-57. *In* H. Kummer [ed.], Proc. 3rd Int. Congr. Primatol., Zurich, 1970. Vol. 3. S. Karger, Basel.

Roberts, R. C. 1967. Some concepts and methods in quantitative genetics, p. 214-257. *In* J. Hirsch [ed.], Behavior-genetic analysis. McGraw-Hill, New York.

Rowell, T. E. 1966. Forest-living baboons in Uganda. J. Zool. London 149:344-364.

Rowell, T. E. 1967. Variability in the social organization of primates, 219-235. *In* D. Morris [ed.], Primate ethology. Weidenfeld & Nicolson, London.

Rowell, T. 1969. Long-term changes in a population of Ugandan baboons. Folia Primatol. 11:241-254.

Schaller, G. B. 1972. The Serengeti lion: a study of predator-prey relations. Univ. Chicago Press, Chicago.

Schoener, T. W. 1971. Theory of feeding strategies. Ann. Rev. Ecol. Syst. 2:369-404.

Stoltz, L. P., and G. S. Saayman. 1970. Ecology and behaviour of baboons in the northern Transvaal. Ann. Transvall Mus. 26:99-143.

Struhsaker, T. T. 1969. Correlates of ecology and social organization among African cercopithecines. Folia Primatol. 11:80-118.

Struhsaker, T. T. 1973. A recensus of vervet monkeys in the Masai-Amboseli Game Reserve, Kenya. Ecology 54:930-932.

Tinbergen, N. 1960. Comparative studies of the behaviour of gulls (Laridae): a progress report. Behaviour 15:1-70.

Vayda, A. P. [Ed.], 1969. Environmental and cultural behavior. Ecological studies in cultural anthropology. Natural History Press, Garden City.

Watson, A., and R. Moss. 1971. Spacing as affected by territorial behavior, habitat and nutrition in red grouse (*Lagopus l. scoticus*). *In* A. H. Esser [ed.], Behavior and environment: the use of space by animals and men. Plenum, New York.

Western, D., and C. Van Praet. 1973. Cyclical changes in the habitat and climate of an East African eco-system. Nature 241:104-106.

Williams, G. C. 1966. Adaptation and natural selection. Princeton Univ. Press, Princeton, N.J.

APES

introduction
field studies on apes

There are two families of apes: Hylobatidae and
Pongidae. The Hylobatidae ("lesser apes") are
found in the far east, ranging from Assam to
Borneo. The family is comprised of two genera,
Hylobates (the "gibbons") and *Symphalangus* (the
"siamangs"). The Pongidae ("great apes") are
comprised of the African genera *Gorilla* (highland and
lowland gorillas) and *Pan* (Chimpanzees), and the
South-East Asian genus *Pongo* (Orang-utans) of
Sumatra and Borneo.

The apes first appear in the fossil record in
the Oligocene (about 30 million years ago).
Gibbonlike fossils have been discovered in the
Oligocene and, thus, the Hylobatidae formed a
separate lineage quite early in the evolution of the
Hominoidea. Modern genera of both greater and
lesser apes, however, do not appear until the
Pleistocene or more recently. The hylobatids live in
family groups with the adult males and females
forming pair-bonds. They are highly arboreal, and
are perhaps the most agile and acrobatic of all the

primates. Their typical mode of locomotion is brachiation—arm swinging beneath the branches. Gibbons are highly frugivorous, whereas siamangs are mainly folivorous (Selection 14).

The great apes differ from each other (and from the lesser apes) in general ecology, diet, and social structure. For example, orang-utans are mainly frugivores and have a social organization that in many ways resembles that of the nocturnal prosimians; it could be called "solitary but social" (Selection 15). Gorillas are strict herbivores and live in groups with many adult females and more than one adult male ("multimale" groups), whereas chimpanzees are frugivorous or omnivorous and have a "fission-fusion" type of social organization (Selection 16).

Historically, the African apes were among the first primates to be studied during the 1920s and 1930s by a group of Yale psychologists (see Southwick 1963; Reynolds 1967). A resurgence in interest in these animals recurred in the late 1950s and, during the past 15 to 20 years, there have been many behavioral and ecological studies of the African apes. However, these studies have concentrated mainly on eastern Africa; very little is known about the populations in central and western Africa. Furthermore, the lowland gorilla and the pygmy chimpanzee remain unstudied (Baldwin and Teleki, 1973). Selection 16, on gorillas and chimpanzees, was originally written in 1965, and many studies of these animals have been completed since it was written. However, there are no other selections available that so succinctly compare the differences in behavior and ecology of the African apes (for more recent references see Baldwin and Teleki 1973, and relevant articles in Clutton-Brock 1977).

The gibbon was first studied by Carpenter in the 1930s but, until recently, the Asian apes have been relatively ignored as subjects of primate field research. This was especially so of the orang-utans and siamangs. However, within the last 10 years, there have been numerous long-term projects completed or which are still in progress on all three

genera of Asian apes (see Baldwin and Teleki
1974). Because of this rapid increase in interest in
these forms, their general natural history is now
relatively well known. Selection 14 by Chivers is not
only an excellent study of the synecological
relationships of the gibbon and siamang, but it also
gives a detailed overview of the general behavioral
and ecological differences between the two genera
of lesser apes. The article by Horr (Selection 15)
briefly summarizes our present knowledge of the
natural history of orang-utans.

REFERENCES CITED

Baldwin, L.A. and G. Teleki, 1973. Field research on chimpanzees and
    gorillas: An historical, geographical, and bibliographical listing. Pri-
    mates, 14: 315-330.
Baldwin, L.A. and G. Teleki, 1974. Field research on gibbons, siamangs
    and orang-utans: An historical, geographical, and bibliographical list-
    ing. Primates, 15: 365-376.
Clutton-Brock, T.H. (ed.). 1977. Primate Ecology: Studies of Feeding and
    Ranging Behavior of Lemurs, Monkeys and Apes. Academic Press,
    London.
Reynolds, V. 1967. The Apes. Dutton & Co., New York.
Southwick, C.H. 1963. Primate Social Behavior. Van Nostrand Reinhold,
    New York.

14

The Siamang and the Gibbon in the Malay Peninsula

1972

DAVID J. CHIVERS

I. INTRODUCTION

The aim of this article is to compare the behaviour of the siamang and the white-handed gibbon in the Malay Peninsula (West Malaysia) in order to question the recent tendency to assume that the siamang will resemble the gibbon in each characteristic under study. In the present investigation of behaviour no such assumptions were made, and some surprising differences were found between the siamang and the white-handed gibbon.

While the siamang is anatomically similar to the gibbons, it nevertheless differs in characteristics such as body size, hair density and form, karyotype (Schultz, 1933; Hamerton *et al.*, 1963). Certain less well-known gibbon taxa, however, do seem to some extent to bridge the gap between the siamang and the lar gibbon. Although both gibbons and siamang live in monogamous family groups in relatively small territories and have loud calls, there are important differences in other aspects of behaviour and ecology. Siamang groups occupy smaller territories which they use and maintain in a different way. Social organisation within the group is different particularly with regard to grooming, sleeping, and adult/sub-adult and male/infant relations. Calls are not used in the same way, and although qualitatively similar the diets are very different quantitatively.

Reprinted from *Gibbon and Siamang* by D.M. Rumbaugh, Editor, 1972, by permission of S. Karger AG, Basel.

285

The siamang of Malaya.

The white-handed gibbon of Malaya.

Groves, elsewhere, reviews in detail all available information on each gibbon species and subspecies. Such information is primarily morphological at present; the lack of other data hinders a real understanding of the systematics and phylogenetic relationships of the gibbons. In this article, therefore, taxonomic problems are avoided. Until gibbon species are accurately identified it is difficult to assess relationships at the subgeneric or generic level. Groves's inclusion of the siamang in the genus *Hylobates* will be more acceptable when it can be demonstrated that the Malayan white-handed (lar) and black-handed (agile) gibbons really are subspecies, i.e. that sympatric populations and hybridisation occur. As Groves indicates, ecological and behavioural data on gibbons are surprisingly limited; only the white-handed gibbon and the siamang have been the subjects of systematic study, although there are numerous brief reports on most taxa.

In Thailand, Carpenter (1940) made the first detailed study of the free-ranging white-handed gibbon, the invaluable basis for all subsequent field work. Ellefson [1967] confirmed Carpenter's findings in his recent study of the white-handed gibbon in West Malaysia, and so our attention here will be mainly confined to the Malay Peninsula. By drawing on Ellefson's unpublished dissertation, and his one published paper (1968), the author plans to compare the ecology and behaviour of the white-handed gibbon with that of the siamang, as revealed by his own findings and the brief but pertinent observations of McClure (1964).

Ellefson made preliminary population surveys at Jerangau, Trengganu (4.52 N, 103.12 E), and in Ulu Gombak, Selangor (3.20 N, 101.47 E). His detailed study of 4 groups of white-handed gibbon for 2,091 h during 1964-65 was at Tanjong Triang, near Mersing, Johore (2.36 N, 103.46 E). In the second half of 1968, Chivers visited 99 localities in all states in search of suitable study areas, and attempted total population estimates of gibbons and siamangs. He spent at least 3 days in Ulu Gombak each month for 2 years making close discontinuous observations on siamang family group dynamics, and investigating seasonal variations in the calling of gibbons and siamangs.

For this study, from January 1969 for 17 months, at least 10 days in each month were spent in the 2 main study areas. Four groups of siamang were observed for 585 h on the eastern slopes of the main range of mountains in Ulu Sempam, near Raub, Pahang (3.46 N, 101.45 E). The detailed daily behaviour of 1 group of siamangs was studied for 1,807 h in the lowlands around the Pahang River at the foot of Gunong Benom, at Kuala Lompat Post (eastern end of the Krau Game Reserve, 3.43 N, 102.17 E) near Temerloh, Pahang. Six siamang groups were observed in Ulu Gombak for 69 h. This article will compare the results of these studies, with reference to other gibbons and apes where relevant.

II. HABITAT

A. Vegetation

The vegetation of the Malay Peninsula is Indo-Malaysian, and it can be distinguished from the northern deciduous monsoon forests of India, Burma, Indo-China, and Thailand, and from the eastern evergreen rain-forest of the Philippines, Celebes, New Guinea, and northeast Australia (Corner, 1952). This central evergreen rain-forest of the Sunda Shelf (Malaya, Sumatra, Borneo and Java) can be divided into 5 climatic climax formations (Wyatt-Smith, 1952).

Siamang and gibbons are found in lowland, hill, and upper dipterocarp forest. Siamang, in particular, do not seem to be affected by the floral changes that take place at about 2,500 ft, but both gibbon and siamang seem to be restricted generally by the change to montane oak forest at about 4,000 ft. Siamang do not usually occur in lowland dipterocarp forest, i.e. below about 1,000 ft above sea level. The main exceptions to this are the central lowlands of Pahang, southwards from Gunong Benom towards Tasek Bera, and low isolated hills near Sungei Bernam to the *west* of the main road and railway. The significance of these exceptions, and of occurrences at high altitudes, is not yet clear, but some light may be shed on this problem by a careful comparison of the ecology and behaviour of siamang in Ulu Sempam (2,000 ft) and Kuala Lompat (150 ft)(Chivers, in prep.).

Tanjong Triang is a promontory of about 1,000 acres extending into the South China Sea; about two-thirds is lowland forest rising up to 200 ft above sea level. The area has been logged intermittently over the last 60 years so that the white-handed gibbons there are isolated. The study area in Ulu Sempam has been selectively logged at least twice in the last 15 years. The territories of the study groups were logged in 1964 and 1970, and observations were made which illustrate the strong affinity siamang develop for their territory. In both cases relatively little timber was removed and the population was neither isolated nor greatly disturbed, if breeding is a useful indicator. It is a feature of siamang and gibbon ecology and behaviour that they can adjust well to selective logging. This secondary forest is more dense, and the canopy is lower and more broken, but important food trees such as figs, *Ficus* spp., are more numerous.

Logging and associated activities are more disturbing than the human activities in the lowland area at Kuala Lompat, where primates are effectively protected from hunting in the Game Reserve. Thus there is little fear of man, which greatly assists habituation for behavioural observations. Disturbance to the forest has been in the form of ancient cultivation and present limited harvesting, such as the collection of rattan and fruits and, until recently, the collection of sap from the jelutong, *Dyera costulata,* for the manufacture of chewing gum. The forest is characterised by emergent giants (tualang, *Koompassia excelsa*, and kempas, *K. malaccensis*) and a canopy in which *Dipterocarpus grandiflorus*, *Intsia palembanica*, *Parkia speciosa*, *Dialium platysepalum*, *Sindora coriacea*, *Cynometra malaccensis*, and species of *Ficus*, *Dillenia*, *Shorea*, *Parinari*, and *Alstonia* predominate. In the lower layer *Sloetia elongata* is common, and there are good numbers of *Endospermum diadenum*, *Randia scortechinii*, *Xylopia caudata*, and *Knema laurinum* (in swampy areas).

The Ulu Gombak study area commenced at the University of Malaya Field Studies Centre, 15 miles from Kuala Lumpur at 850 ft above sea level. The forest at the top of the valley is Virgin Jungle Reserve; in between is Forest Reserve that has been quite heavily logged this century. The area is

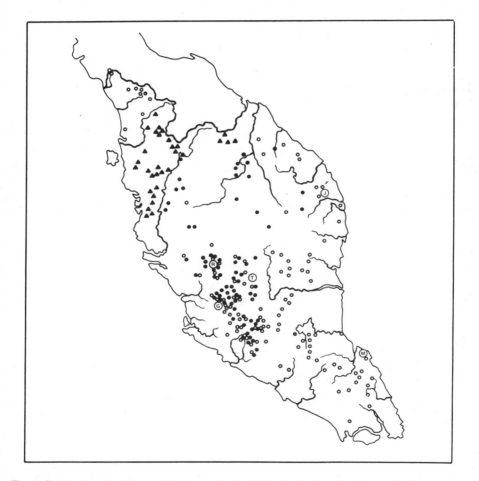

Fig. 1. Distribution of gibbons and siamang in the Malay Peninsula. J = Jerangau, Trengganu; M = Tonjong, Triang, Mersing, Johore; G = Ulu Gombak, Selangor; R = Ulu Sempain, Raub, Pahang, and T = Kuala Lompat, Temerloh, Pahang. ● = siamang, ○ = white-handed gibbon, and ▲ = black-handed gibbon.

much disturbed by logging and other activities because of its proximity to the capital. Such activities help to explain the pattern of group dispersion (see below and Fig. 2). Among the large number of trees of the Dipterocarpaceae, the distinctive crowns of the large *Shorea curtisii* are prominent along the ridge tops.

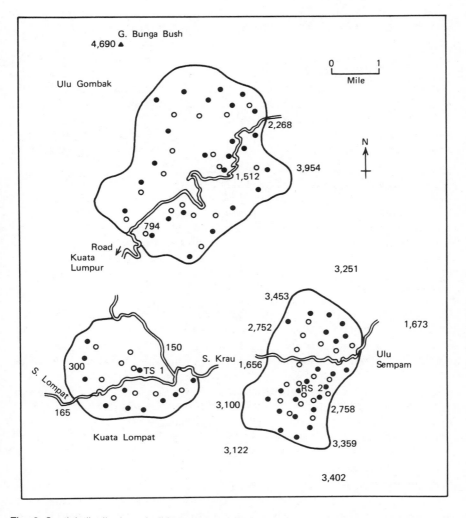

Fig. 2. Spatial distribution of gibbon and siamang groups. Roads, rivers, altitude, and study groups (RS2, TS1) marked. ● = siamang group, ○ = white-handed gibbon group.

B. Distribution

Gibbons are found almost everywhere in the Malay Peninsula where there is still forest—about three-fifths of the area of West Malaysia. In contrast, the siamang occurs in about one-fifth of the forested area, mainly in the mountain range which runs the length of the west side of the country, and has extended east in the mountains of the Pahang/Kelantan border into the National Park. But the siamang does not appear to have become well established in the eastern mountains of Trengganu; it is absent from Johore in the south, and excluded from east Pahang by the Pahang river (Fig. 1).

Sungei Perak, on the west of the main range, and Sungei Kelantan, on the east, are important barriers to gibbons and siamang, at least in their lower reaches. The siamang is restricted to the south of the Perak river, which forms the boundary between white- and black-handed gibbons. On the east the siamang extends northwards across the Thai border, but the white-handed gibbon again gives way to the black-handed gibbon. The significance of this distribution is very interesting; insurgent activity has so far prevented an investigation of gibbon populations in the upper reaches of these rivers, where there can be no effective barrier between these 2 gibbon taxa. This problem is fundamental to the understanding of gibbon taxonomy and phylogeny: is there a hybrid zone, or is there sympatry or allopatry between 2 biological species? In Kedah, north of the Sungei Mudah, one finds the white-handed gibbon again, but its relationship to the southern form is not yet clear. Henceforth, in this article, "gibbon" is used to denote the white-handed gibbon of the southern part of the Malay Peninsula as studied by Ellefson, and observed by the author.

C. Climate

Such seasonality as there is in the Malay Peninsula is most marked on the east coast because of the rain-laden northeast monsoon at the turn of the year, which is followed by a dry season. Even on the west this pattern is distinct. The weather is less predictable from June until November, with the south-west monsoon on the west coast sometime after April and a dry period around October. Central Pahang, in the rain-shadow of the mountains, is the driest part of the country, with the most marked dry season. East Johore is one of the wettest areas, and was particularly so in 1964-1965; 1969 was relatively dry. The heaviest fruiting years of the decade were 1963 and 1968.

Comparative data on temperature, sunshine, and rainfall for the 4 study areas throughout the year are given in Table 1. Ulu Gombak is on the west side of the main range, Ulu Sempam on its more sheltered eastern slopes, and Kuala Lompat at the eastern foot of Gunong Benom. Tanjong Triang, on the east coast lowlands, is much affected by the daily sea breezes, which reach a peak around mid-day. Here, also, rainfall is more seasonal than in the other study areas, and there is less sunshine on average. At Kuala Lompat and Ulu Sempam sunshine, cloud and rain were sampled for every 15 min between 0600 and 1800 h; rainfall was measured every 24 hours at 0700 h. In both areas a large part of the rainfall occurred between 1600 and 0600 h, i.e., the hours of dusk and darkness, so that the effect of rain on siamang behaviour was minimal. Moderate and heavy rain cause siamang to rest.

III. POPULATION STRUCTURE AND DENSITY

Carpenter (1940) reported that the mean grouping tendency of *Hylobates lar* was a family—male, female, and their young. Furthermore, he postulated that the entire genus *Hylobates* was characterised by the monogamous family group. Certainly this is true for the gibbons and siamang of the Malay Peninsula. These groups live in small territories, which are comparable in size with those of other arboreal primates. Adjacent territories have little overlap, and conspecifics are excluded by ritualised calling and chasing behaviour. Ellefson claimed that the daily behaviour of gibbons revolves around this territorial defence. This is not really true for the siamang, as will be described below.

Table 1

Comparative climatic data for 4 study areas

	1964–1965 Tanjong Triang Johore, ca. 50 ft	1969 Ulu Gombak Selangor, 900 ft	1969 Ulu Sempam Pahang, 1,500 ft	1969 Kuala Lompat Pahang, 150 ft
Temperature, °F				
Absolute maximum	98	92	92	97
Absolute minimum	71	66	61	68
Average maximum			84.8	90.8
Average minimum			68.2	72.7
Sunshine, h Jan.	5.9		5.9	8.5
Feb.	6.3		6.5	8.0
March	5.8		8.3	7.0
April	6.7		8.7	5.7
May	6.3		8.5	6.5
June	6.9		6.6	5.2
July	6.6		6.2	6.4
Aug.	7.0		5.5	5.1
Sept.	5.7		6.4	7.3
Oct.	5.1		7.7	7.4
Nov.	3.6		2.1	3.2
Dec.	2.9		3.0	4.8
Mean	5.5		6.4	6.3

Table 1 (continued)

		1964–1965 Tanjong Triang Johore, ca. 50 ft	1969 Ulu Gombak Selangor, 900 ft	1969 Ulu Sempam Pahang, 1,500 ft	1969 Kuala Lompat Pahang, 150 ft
Rainfall, in	Jan.	0.7	4.41	1.92	1.84
	Feb.	3.4	5.30	5.02	0.53
	March	2.0	6.10	8.09	1.28
	April	4.8	4.23	8.40	3.42
	May	6.8	11.77	9.60	10.42
	June	4.1	11.52	6.75	3.81
	July	8.6	8.60	1.51	4.69
	Aug.	4.9	11.51	8.87	12.12
	Sept.	3.9	9.14	4.88	3.13
	Oct.	7.4	17.65	12.52	6.19
	Nov.	23.8	12.02	10.65	9.24
	Dec.	57.0	5.29	13.00	6.07
Total		127.0	107.54	91.21	62.74
per cent rain days		47	67	62	54
in/rain day		0.93	0.44	0.42	0.40
Jan.–March		6.1[a]	47[b] 15.81	47 15.03	18 3.65
April–June		15.7	57 27.52	53 24.75	61 17.65
July–Sept.		17.4	79 29.25	59 15.26	67 19.94
Oct.–Dec.		88.2	86 34.96	90 36.17	69 21.50

[a] Number of inches.
[b] Percent rain days in 3-month period.

The gibbon or siamang territory is here defined as the area used exclusively by a family group throughout the year. These areas, which are defended against conspecifics, have borders clearly defined by behaviour and topography. Data collected by Carpenter, Ellefson, and Chivers for gibbon and siamang in different areas are presented in Table 2. Comparisons are made between the area available for each group and actual territory size. Where both species are present in an area it is assumed that there is no exclusion of one species by the other. As Ellefson suggests, siamang can and will displace gibbons at a given locus, usually a preferred food source such as a fig tree. But Chivers found no evidence of siamang displacing gibbons permanently from any part of their territory; rather, overlap generally appeared to be complete between groups of the 2 species.

Carpenter's average territory size of 25 ha for the Thailand white-handed gibbon is based on a relatively short period of observation in

Table 2
Population density

Author	Location	Number of groups			Area, ha	Number of ha/group			Estimated territory size	sample size
		S	G	t		S	G	t		
Carpenter	Doi Dao		18		777		43		25	3
			3		194		65			
	Doi Intonan		9		1,036		113			
Ellefson	Jerangau		67		6,788		104			
	Mersing		4		243		61		39	4
	Ulu Gombak			43	3,303			77		
	Gibbon	23					165			
	Siamang 20					144				
Chivers	Ulu Gombak			42	2,745			65		
	Gibbon	17					161			
	Siamang 25					110			24	6
	Ulu Sempam			41	1,531			34		
	Gibbon	18					83			
	Siamang 23					67			18	4
	Kuala Lompat			20	1,098			55		
	Gibbon	9					122			
	Siamang 11					100			34	1

deciduous forest. Ellefson gives an average territory size of 101 ha, but the only figures he gives of gibbon territory size at Tanjong Triang average out at 39 ha. The most likely explanation for this disparity appears to be that he failed to distinguish between space available and space used. It would seem either that the gibbon and siamang populations in all study areas are below saturation, or that parts of these areas are unsuitable for these species. The latter may be a factor affecting the apparent clustering of groups in each area (Fig. 2). At Kuala Lompat, gibbons and siamang were not found within about 400 m of the rivers; this part of the forest has a denser shrub layer and fewer tall trees, so that the canopy is very open. Similarly, there may be ecological differences in the valley bottoms at Ulu Sempam; certainly parts of the forest have been made unsuitable for gibbons and siamang by human activities (described above). But the proximity of conspecific groups may not exert simply a repellent effect on each other. Ellefson's description of the gibbon's territorial behaviour leads one to think that groups are to some extent attracted to each other, so that a community of groups is formed. The apparent clustering observed might then reflect an equilibrium between these 2 forces.

Ellefson concludes that the size of gibbon territories varies between 40 and 300 ac (16-122 ha), and that Carpenter's estimate is too low. But the available data suggest that gibbon territories in these parts of the Malay Peninsula are about 40 ha. Estimates of territory size for 12 groups of siamang in the 3 study areas average at 23 ha. Certainly gibbons have a territory that is larger than the siamang territory, perhaps twice as large. But the siamang group represents twice the biomass of the gibbon group, so it is remarkable that the siamang territory is at most half the size of a gibbon territory relative to biomass. This difference points to some fundamental differences between the 2 species, which will be investigated below. Because of existing behavioural differences, it seems unlikely that the larger territory of the gibbon can be to any real extent the result of displacement by siamang.

IV. TERRITORY

A. Use

Because of the spatial stability of siamang and gibbon territories, they differ from the overlapping home ranges of primate populations such as howler monkeys *Alouatta* (Chivers, 1969), where only the space about the group at a given time—"mobile territory"—is defended against conspecifics. The size and nature of these areas have already been described. At Tanjong Triang, 4 gibbon groups had territories of 20, 44, 46, and 44 hectares; group size was 2, 3, 3, and 5 respectively. The siamang study group in Ulu Sempam, which had 2 groups of 4 adjacent, had a territory of 15 hectares and was composed of male, female, sub-adult male, and infant male (RS2). The Kuala Lompat siamang study group (TS1)—male, female, sub-adult-male, juvenile male, and infant female—had a territory of 35 ha with no conspecific groups adjacent. A gibbon group of 5 animals (TG2) ranged throughout this siamang territory and out to the north where their territory abutted on to those of 2 other gibbon groups. TG2 had a male (golden above, brown below), a female who carried an infant throughout the 17 months, a juvenile, and a sub-adult male who was not often seen with the rest of the group.

At Tanjong Triang the daily routine of the gibbons begins soon after dawn with the group giving "morning calls" as they leave their night position, which is situated centrally in the territory, and forage towards a boundary. Within the next hour or so on every other day on average, the group will have a "battle" with one of its neighbours. This involves calling (male and female) and chasing to and fro across the boundary (males only). Thereafter the group forages and rests about its territory, several times coming together to feed in the same tree ("preferred food source"). Between 1500-1600 h the group settles for the night, scattered in different trees, having travelled an average of 1,600 m.

The siamang travels about its smaller territory more slowly in single file along well-known arboreal pathways at about 25-30 m, from food tree to food tree where all animals feed together, if only briefly. The female usually leads, followed by older young, and then male and dependent young with any sub-adult at the rear. Ellefson makes no comment on gibbon group travel order, but Carpenter remarks that females without infants are often seen at the head of the group. Brief sightings at Kuala Lompat usually revealed the male in the lead. Occasionally one siamang is more than 30 metres or more from the rest of the group; occasionally the group is scattered in 2 or more trees when feeding (perhaps containing branches of the same vine). In contrast, Ellefson reports that gibbon sub-adults are often up to 300 m from the rest of the group. Thus the siamang group shows greater cohesion in its daily ranging.

The siamang daily activity commences at dawn. If a preferred food is close at hand, feeding may start at once. Alternatively, after stirring to change positions and urinate and defaecate in concert, the group will continue resting (usually without grooming) for at least 30 min. For the rest of the day long feeding periods (with or without travel between food trees) alternate with resting and grooming either in pairs or a huddle of most group members. Feeding decreases as the day passes, grooming increases; the group settles for the night together in a well-known sleeping tree between 1600 and 1800 h. In each tree each animal has one or more usual sleeping positions—infant with female, juvenile with male, and sub-adult male alone. At the end of the dry season the average day range was about 1,600 m, and the group would circle its territory in a day, whereas during the rest of the year it would take more than 3-5 days with day ranges of about 750 m. If the siamang territories are divided into hectare squares it transpires that TS1 utilises about 50 %, and RS2 about 70 % of the squares per 10-day period. Ellefson does not give much data on gibbon ranging patterns, but it appears they are similar to the longer day ranges of the siamang.

These long day ranges reflect a relative abundance of widely dispersed preferred fruits, whereas the shorter day ranges are associated with a more general abundance of lower-energy foods, i.e. leaf shoots, and so forth. The relative significance of these 2 patterns has still to be evaluated. It became apparent that most trees through which the siamang group travels will provide food at some time, and that by continually quartering their territory and recognising visual and vocal signals, such as bird concentrations around specific trees, siamang efficiently utilise the available resources. Despite the richness of the tropical forest growth, the great diversity of plant species results in wide dispersion of primate foods. The more specialised the diet of a species, the farther it must range; this appears to be one of the fundamental differences between gibbon and siamang.

Some plant species produce food as often as twice a year, more usually once a year, but not infrequently only once every 2-4 years. It will be seen that these variable floral cycles exert a major influence on siamang behaviour (Chivers, in prep.). Both species are adapted to the small branch niche, which involves a large proportion of the time being spent moving or suspended beneath small branches. The gibbon, however, appears to be more specialised morphologically, which helps to explain some of the differences in ecology and behaviour. Gibbons flit rapidly through the trees, and perch on small branches in a hunched position; siamang are slower and more deliberate in their movements (although capable of the spectacular locomotor feats frequently performed by gibbons), and rest draped across large branches or in forks where they can prop themselves.

B. Maintenance

Ellefson (1967, 1968) repeatedly stresses the importance of inter-group territorial conflicts in gibbon behaviour and phylogeny. His report helps to explain the relative lack of such behaviour in the siamang today. But first it is necessary to recognise the 2 main components of inter-group relations. The occupation of an area may be maintained by the group emitting signals that may be recognised at a distance. Such signals may then be reinforced by contact at the boundary of the area. Gibbons use distance signals to locate the position of their neighbours so that they may then interact at the territorial boundary in the elaborate way that Ellefson describes. Siamang, on the other hand, do not appear to have to reinforce such distance signals to such an extent, and this is a further indicator of fundamental differences between the species.

Chivers noted a tendency in most areas at certain times of the year for gibbon males to call at dawn as the group began its daily activities. It would be a further hour or more before female "great calls" would be heard. Mackinnon (personal communication) observed this pattern in Bornean gibbons. Groups tend to chorus together; this calling precedes conflicts along territorial boundaries. Siamang group calls start about an hour later than those of the gibbons and pass sequentially around the forest, one group ceasing calls as the next starts. It seems that groups only approach each other if their territory is threatened. Contacts at the boundary are reduced because siamang show a greater awareness of the territory of their neighbours, and perhaps because there is less competition between groups for some reason.

At Tanjong Triang gibbons called for an average of 15 min on 85 % of mornings, whereas siamang groups called for the same duration on 30 % of mornings. At Kuala Lompat and Ulu Sempam, gibbon groups called for 18 and 29 min on average, respectively (but this included conflict hooing, which presumably Ellefson did not). 60 % of gibbon calls at Kuala Lompat and Ulu Sempam occurred between 0700 and 1000 h with a peak around 0800, whereas 50 % of siamang calls occurred between 0900 and 1200 h,

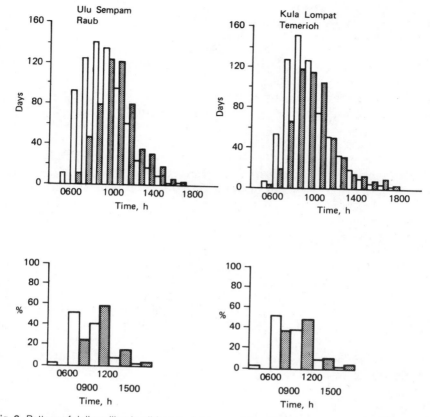

Fig. 3. Pattern of daily calling in gibbon and siamang. ■ = siamang, □ = white-handed gibbon.

with a peak around 0900 in Ulu Sempam and 1000 h at Kuala Lompat (Fig. 3).

Calling by siamang is a response to some form of disturbance, usually the calling of a nearby group (which need not be visible or in the adjacent territory). The chorus starts with loud barks (or hoots), chatters, and booms. The female leads the chorus by giving a double series of barks about every 3 min at the climax of which the group swings about with the male giving loud screams and the female chattering. The form of the call varies if the group is alarmed by the disturbance, e.g. an actual or potential predator. The female gibbon also leads group calls with her "great calls" during which the male is silent. Gibbon calls also vary according to their functions in spacing, alarm, and conflict.

Gibbon inter-group conflicts occurred for each group at Tanjong Triang on 50 % of days, and lasted for about 50 min on average. The males chase each other to-and-fro after a prolonged period of pre-conflict "hooting," during which the female calls. At intervals the males return to their females for grooming. In contrast, in Ulu Sempam in early 1970, conflicts between siamang groups (RS2 and RS1 or RS4) occurred on less than 20 % of 30 days, in silence with young and female hiding, and males (including sub-adult) chasing vigorously to-and-fro. Such conflicts may be preceded by calling between the 2 groups across the boundaries with the female bark series of each group alternating. But groups will often move apart, back into their territories, after calling with no chasing at close quarters. Groups may also approach each other closely without one or even either group calling. Thus siamang groups appear to spend less time in maintaining their territories than gibbons; such territorial behaviour that they do exhibit is mainly in the form of distance signals in contrast to the inter-group conflicts across boundaries. Siamang day ranging pattern, however, does involve a large amount of circling which could be interpreted as a passive boundary patrol.

In the lowland study area early in 1970, siamang calling increased, particularly in TS1. Sexual behaviour was observed for the first time, and the sub-adult male was being eased out of the group more rapidly. The group was more sensitive to disturbance and the sub-adult male appeared to be advertising his availability as a mate by calling before and after the rest of the group had stopped. In May a sub-adult female was seen in the southeast corner of the territory at the beginning of May apparently having responded to this calling. By October these 2 siamang were establishing a territory in this area (Hunt, personal communication). Ellefson suggests that the young male leaves his parental group in search of a mate; this is facilitated by the hostility of the adult male. But it appears that in the siamang the female plays a greater part in group formation.

V. FEEDING

At Tanjong Triang gibbons fed for an average of 30 min in 5 trees per day. This feeding composed 25 % of the group daily activity. The siamang group (TS1) spent 55 % of the day feeding, but it does not feed scattered while travelling which is called "foraging" by Ellefson. This accounts for a further 40 % of the gibbon day. Because of his role in inter-group conflicts the adult male gibbon loses feeding time which he makes up at the end of the day when the rest of the group are in their night positions. Even on days with conflicts the adult male siamang is usually the first to reach his night position. Furthermore, the adult female siamang eats faster and for about 30 min longer each day than the male, who usually initiates resting and grooming bouts.

It appears that the main qualitative difference between the diets of the 2 species is that the siamang eats fewer insects (both in variety and quantity), but there are major differences in the quantities of each food type ingested. Food is comprised of fruit (of varying stages of ripeness), leaf shoots, new leaves (newly opened leaves of immature size and colouration), leaf stems

Table 3

Comparative feeding behaviour

	Gibbon, trees %	Siamang, trees %	TS 1–148 days per cent time 51,378 min	per cent trees 1,912	Av. number trees/day	Av. number min/day	Av. min/visit
Fruit	74	26	29	30	3.9	101	27
Leaves, new stems shoots	34	67	58	60	7.8	202	25 33 35
Flowers	12	8	9	8	1.0	30	30
Buds							29
Insects	some	5	2	3	0.4	5	14
Total	120[a]	106[a]	98[b]	102[b]	12.9	347	55% daily activity
Fig trees		13	25	20	2.6	86	33

[a] total more than 100 because some trees provide more than 1 type of food; either because leaves and fruits are eaten at different times, or because there is more than 1 species of plant within the tree.

[b] total not exactly 100 because of summation of sub-totals and correction to whole number.

(and midribs), flowers, buds and insects. Siamang were only seen eating termites and caterpillars, whereas gibbons ate stick insects also. Table 3 gives comparative data on diet and feeding behaviour.

At Tanjong Triang, 82 trees were marked and 140 specimens were collected; 142 trees were marked at Ulu Sempam, and 539 at Kuala Lompat. It is not clear what sample size Ellefson used in his calculations of dietary proportions. It is perhaps as true for gibbons as it is for siamangs that the preferred food trees are the easiest to record, and these are usually trees in which fruit are eaten. For example, in the first 12 months at Ulu Sempam when the siamang were very elusive and could not be observed for more than a few hours each day, they ate fruit in 56 % of the 66 trees in which they were seen feeding. In the last 4 months of dawn-to-dusk observations fruit were eaten in only 29 % of food trees. This variation is primarily a reflection of observer bias because of an inadequate sample, although there are seasonal differences in diet. But the pattern of ranging, territory size, and inter-group relations provides further evidence for a real difference in diet between the 2 species.

Three-quarters of the gibbon diet are fruit, whereas two-thirds of the siamang diet are new leaf derivatives. To check the relative proportions of different food types in the siamang diet, as revealed by counting the food trees which produced each type of food, the amount of time spent in, and the number of visits to, food trees of each type were calculated. Table 3 shows the results, which are in close agreement with the initial figures except with regard to fig trees, *Ficus spp.* 13 % of food trees are figs, but 20 % of visits were to fig trees, and 25 % of feeding time was in these trees. In Ulu Sempam 19 % of RS2 food trees were figs.

Gibbons have been described as unusual among primates in their frugivorous habits (80 % Carpenter, 74 % Ellefson), and this has been related to their marked territorial behaviour. Ellefson suggests that a dispersion of groups into small non-overlapping territories is the most efficient way for this species (living in the terminal branch niche where agility has been achieved at the expense of mobility) to harvest a relatively widely dispersed and specialised diet. With fruit forming less than 30 % of the siamang's diet it should be regarded as folivorous. That the siamang shares morphological characters with the gibbon, and shows a milder pattern of territorial behaviour, suggests that this ability to subsist on a wide variety of plants is not a relatively recent adaptation but a fundamental ability of ancestral gibbons (and other hominoids).

VI. GROUP SOCIAL BEHAVIOUR

A. General

The daily activity pattern, widespread among most tropical animals including primates (Chivers, 1969), of early and late feeding with midday resting is not apparent in gibbons (Ellefson, 1967) and siamang (Fig. 4). Instead, they slowly decrease time spent feeding, and increase time spent resting as the day passes. Neither are there early and/or late travel peaks. The average amount of time spent each day in each activity is given for the 2 species in Table 4. As has already been indicated, the different pattern of ranging

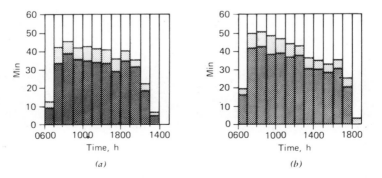

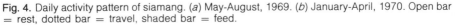

Fig. 4. Daily activity pattern of siamang. (*a*) May-August, 1969. (*b*) January-April, 1970. Open bar = rest, dotted bar = travel, shaded bar = feed.

Table 4

Comparative daily activity

Activity	Gibbon days	Siamang TS 1, 148 days	RS 2, 18 days
Rest	0.5	2.8	3.2
Feed	2.8	6.0	4.9
Forage		—	—
Travel	4.5	1.0	1.2
Social	1.7[a]	0.7[b]	0.4[b]
			0.5[a]
Total, hours	9.5	10.5	10.2

[a] Gibbon social interactions are mainly inter-group, i.e. territorial.
[b] Siamang social interactions are mainly intra-group during rest periods.

makes a detailed comparison of the two species difficult. Because the gibbon group forages, i.e. individuals disperse more and feed briefly in numerous trees as they travel between major food sources, Ellefson was not able to obtain accurate measurements of the time spent feeding or the amount of food ingested. It is not even clear that the siamang spends more time feeding than the gibbon and consumes more food, although it seems likely.

Perhaps because the gibbon group is more widely dispersed in its daily ranging, individuals have a larger repertoire of calls and gestures. In the siamang the relative lack of vocal and visual signals within the group appears to be compensated for by animals constantly watching each other and the forest about them. In this way the group members act together so that there is a high degree of cohesion in the group. Throughout the day the average distance between individuals is less than 10 m (Chivers, in press), the spatial pattern varying a little according to activity. How this cohesion is achieved may be better understood by studying the maturation of the young, and the way in which group members come to interact with each other.

Intra-group calls in the siamang are limited to the distress bleat and squeal of the infants, the submissive squeal of (usually) the sub-adult males in scraps, and the grunts of immature animals during intense play. The only observed facial expression was an open-mouth threat given by any class of animal before and during a scrap. It is likely that there are other less distinct expressions involved in communication, which could not be detected in this black animal in the rain forest habitat at a distance. Further, even if one can detect that 2 animals are interacting in this way, it is almost impossible to observe their 2 faces simultaneously. The gibbons have more submissive and aggressive gestures, a submissive grimace prior to embracing and grooming, a rapid open-and-close mouth threat, and calls used in the group indicating mild upset, alarm, submission, and threat.

Play is an important social interaction in both species. It takes a different form from other primates because there is never more than one individual of each age class in a family. Thus young animals play most with objects in their environment, e.g. by modifying their locomotion and by elaborations of usual postures. In gibbons, play is characteristic of the juvenile stage, whereas in siamang it most frequently involves the older infant (infant-2, see below) and sub-adult, rarely involving adults. Play bouts are longer and more frequent when food is abundant.

B. Maturation

Ellefson defines the infant stage as being from 0 to 2-2½ years, which is until it stops sleeping with the female. This is usually 2 months before the birth of the next infant. At the age of 2 months the infant is showing better coordination and is manipulating leaves and twigs; by 4 months it is ingesting solid food, and beginning to move away from the female. In the second year it increasingly travels independently, and sometimes engages in play initiated by the juvenile. At all times it is completely tolerated by the adults.

In the siamang it is necessary to recognise 2 stages of infancy. The infant-1 is dependent on the female for the first 12-16 months of life, but the infant-2 is carried by the male until becoming completely independent during the third year of life for group activities. Thus the male plays an important part in instructing the infant in these activities. At the beginning of the second stage the female keeps well away from the male and infant during the day, but by the time the infant requires the male's help only for the more difficult crossing, she allows the infant to follow her more closely. Hence the pattern emerges of independent, enthusiastic young following the female closely about the day range. The infant continues to sleep with the female at night, presumably until near the time of the next birth. Although the female is thus prematurely freed from the care of the infant, she does not appear to come into oestrus again for a further year. This may be a recent response to increased population pressures, and it may be that this behavioural mechanism has had an important survival value during the siamang's evolution. Possibly this behaviour is partly or wholly a response to the increased physical burden the infant presents as it grows to the smaller female. Also such prolonged care of the infant may interfere with the female's role in group leadership, should this be important to the group.

The juvenile gibbon (2-2½ to 4-4½ years) is completely independent of the female, much interested in play, and increasingly involved in agonism with the adults over food. Little agonism occurs between the juvenile siamang and the adults over food. Although the juvenile is spatially close to the adults during the day range, it is socially distant, i.e. it interacts least with other animals. During rest periods it may get some attention from the female for grooming, but its repeated presentations are usually ignored, since as grooming during the day is an activity of adults and sub-adults.

Because the sub-adult is most active with the male in grooming during the day, it is with the sub-adult that the infant-2 plays when the female comes to the male for grooming. The juvenile is the last member of the group to get to know the infant well, and this makes sense because such a relationship has the least value to the growing infant. At night the juvenile sleeps with the male, which contrasts with the more widely dispersed pattern of the gibbon group.

In the gibbon, adolescence (4-6 years = sub-adult) is a period of peripheralisation. Intolerance of its presence in food trees by the adults results increasingly in the adolescent foraging alone, often 300 m away from the rest of the group for long periods. The adults respond less to its contact calls as time passes. Ellefson describes the animal as sub-adult at 6 years when sexually mature, and as involved in group formation essentially as a lone animal. Chivers treats these 2 classes as 1 in the siamang; the subadult sleeps alone at night from the age of about 5 years. It soon reaches full size and appears to be sexually mature; it should leave the group within 3-4 years.

Although the 213 brief scraps observed in the 148 days at Kuala Lompat usually involved the sub-adult male siamang, it is much better tolerated by the rest of the group. Scraps usually occurred when the sub-adult approached too close to, or was closely approached by, an adult (usually male) in a food tree. The effect was immediate, resulting in the increase of distance between agonists within the food tree. The other predictable occasion for a scrap was when the juvenile came to the male to settle for the night if the sub-adult was grooming with the male. The juvenile appeared to initiate these scraps.

Towards the end of the sub-adult male's life with the group the male adopted a new display during the breeding period. This occurred in part in calling and inter-group interactions and involved a short circular vigorous swinging, terminating with the male facing away from the sub-adult, as the sub-adult entered the same food or night tree as the male. Thereafter the sub-adult male increasingly fed after the others had left each tree, and slept in a different tree. He had been calling alone for a month or so before this final stage; within 5 months he was establishing a new territory with a sub-adult female. The observations of Fox (in press) on a captive siamang group suggest that the sub-adult female has similar relationships with the adults, even with regard to peripheralisation; thus the process cannot be explained in terms of intolerance between sexually mature animals of the same sex.

Even allowing for differences in terminology, it seems clear that there are differences between the gibbon and siamang in the timing, speed, and mode of departure of the sub-adult from the group. In the gibbon it is more clearly an expulsion starting at a relatively early age, but in the siamang the cohesion of the social unit delays the departure, possibly until it is largely voluntary.

C. Grooming

Sitting close to, and grooming with, another individual is the most important social activity of siamang in terms of time and frequency. In both species self-grooming is frequent, but in the siamang self-grooming (rare) is distinguished from self-scratching (frequent, particularly prior to a change in activity). Adult gibbons groom an average of 15 min/day in a nearly reciprocal manner, which reflects dominance in that the dominant animal receives more grooming than he gives. In contrast to siamang, Ellefson indicates that gibbon sub-adults do not participate much in this activity.

Two or more siamang (mainly adults and sub-adult) spend on average 87 (TS1) or 71 (RS2) min sitting together, of which they groom for an average of 36 (TS1) or 23 (RS2) min/day. For TS1 1041 grooming sessions were recorded in 148 days (about 7/day) of which 83 were self-grooming. The details are given in Table 5 by means of the proportion of the total number of minutes of actual grooming in which each class member grooms the others. In contrast to the adult female leading the group about its day

Table 5
Grooming in the siamang

(a) April–December, 1969						99 days 3,956 min
	GROOMED, %					
	male	female	sub-adult	juvenile	infant	Total
Male	0.7	9.3	22.9	9.8	1.1	43.7
Female	6.5	1.5	5.8	5.2	2.3	21.2
Sub-adult	19.7	8.6	0.8	2.5	0.5	32.6
Juvenile	1.4	0.4	0.4	0.2	0.1	3.5
Infant	0	0	0	0	0	0
Total, %	28.2	20.8	29.8	17.2	4.0	

(b) January–May, 1970						49 days 1,431 min
	GROOMED, %					
	male	female	sub-adult	juvenile	infant	Total
Male	0	23.0	12.9	9.1	2.1	47.1
Female	14.0	0.2	4.6	4.5	6.8	31.3
Sub-adult	8.9	5.5	0.1	1.9	2.0	18.4
Juvenile	1.5	0.7	0.3	0	0.1	2.6
Infant	0	0	0.2	0	0	0.2
Total, %	25.7	29.4	18.2	15.4	11.2	

The label "GROOMS" appears vertically along the left side of each table section.

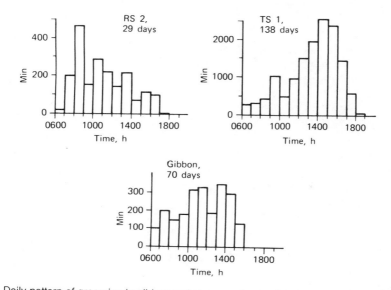

Fig. 5. Daily pattern of grooming in gibbon and siamang. Accumulated time (min) over specified number of days for 2 groups of siamang and the gibbon (from Ellefson, 1967). For the siamang, figures represent the time 2 or more animals are together for grooming, rather than the number of minutes of actual grooming.

range, the adult male is the focus for resting and grooming. Thus he participates most in grooming, especially with the sub-adult male. The adult male grooms both female and sub-adult rather more than each groom him. In the breeding season the grooming pattern is reversed with the male following the female more closely and repelling the sub-adult. In Ulu Gombak (S16B) a sub-adult female was seen associating with the adult male often, and this included grooming.

The daily pattern of grooming in the gibbon shows a peak around mid-day, but the patterns for the siamang in Ulu Sempam and at Kuala Lompat differ from the Tanjong Triang pattern and from each other. The larger sample from Kuala Lompat shows a small peak in the morning after the first main feeding session, with the main peak in the afternoon as other activities decrease. In Ulu Sempam, however, the mid-morning peak is the largest, with a considerable amount of grooming around mid-day (Fig. 5). Comparison with Table 4 shows that resting bouts are distributed surprisingly evenly throughout the day at Kuala Lompat, at least at most times in the year. This is also true at Tanjong Triang, and probably true in Ulu Sempam. It seems, therefore, the timing of grooming might be explained by individual (and group) preference, certainly within each species, possibly even between species once major differences have been allowed for.

Juveniles and infants are generally neglected during this daytime grooming despite their efforts to solicit grooming, but the siamang male grooms the juvenile, and the female grooms the infant as they settle for the night. The infant does not reciprocate, and the juvenile is remarkably slow to do so.

VII. INTER-SPECIFIC RELATIONS

Ellefson describes gibbons as showing only mild consternation at the close proximity of siamang, whereas monkeys were chased away. Flying lemurs, fruit bats, binturong and hornbills were observed feeding on the same foods as gibbons. At Kuala Lompat the siamang would not tolerate gibbons in the same tree, and the male would chase them out. But the siamang reacted if the gibbons did not move *right* away. On 2 occasions the persistance of the gibbons and their calling caused the siamang to climb up out of the disputed food tree and call. Strangely, the force of this display was minimised by the siamangs facing away from the gibbons as they called.

Siamang showed much more tolerance to monkeys which came into their feeding or resting trees. Sometimes groups of sjamang and a species of leaf monkey would intermingle as they travelled along the same arboreal pathway. Occasionally infant siamang played with young leaf monkeys; this was also observed by McClure (1964). Once an infant male siamang (RS2) swung vigorously on the tail of a female banded leaf monkey, with no visible response from the monkey, even when he bit her tail. It was nearly 5 minutes before the monkey moved on. Sometimes the siamang slept near 1, or even 2, groups of leafmonkey; perhaps each species finds the close proximity of the others advantageous, presumably against predators. Certainly leaf monkeys are sensitive to disturbance, particularly in the dark. Macaques were not usually encountered by siamang as they seemed to remain nearer the river banks in the denser lower foliage which characterises this habitat, and also they seemed to utilise different layers of the forest when they did range into the siamang territory. Hornbills, giant and tree squirrels, pigeons, and many other small birds, and binturong were often seen feeding in siamang food trees.

As had been observed in captivity, siamang generally showed more interest in the animals they encountered than gibbons, which tended to avoid or omit such interactions. The response of the 2 species to man goes some way towards illustrating this. In the wild, Ellefson reports that the response of gibbons to intrusion by man is always fright and flight, and that they alternately hide and flee while pursued. Curiosity appears to play a more important part in the reaction of siamang, unless they have been harmed by man. Certainly the siamang vanish from view, but not in fearful flight as repeated attempts at pursuit reveal. If one waits instead of pursuing the siamang reveal themselves, having moved into a position where they can see but not be seen. Once the pathways of the siamang group are known, they may be found carrying on their usual activities just outside the range of vision (about 50 m in tropical rain forest).

The long day ranges of TS1 in the early stages of the study were initially thought to be a form of flight from the persistance of the observer, particularly as day ranges decreased in length as the months passed. But when they reverted to a similar pattern of ranging 12 months later it seemed that their behaviour had not been a response to the observer, but to varying ecological conditions. There had certainly been no other indications of fear. On the one occasion I appeared in the next tree to the group at 30 m there was no rapid flight, but after staring for a moment the group moved round the tree and on about their day range. Unlike gibbons, siamang do not usually call, even softly, when meeting man in the forest. If nothing else, this description reveals the difficulties of interpreting behavioural observations in the tropical rain forest. But siamang are thought to differ from gibbons and resemble the great apes in being more sociable.

VIII. CONCLUDING DISCUSSION

The main differences between gibbon and siamang ecology and behaviour are summarised in Table 6. They are concerned with territory size and pattern of use, feeding behaviour, and social relations within and between family groups. Usually these differences are ones of degree rather than of kind, but such differences occur between families, and not only between species. Differences in group cohesion, communication, diet, group formation, and infant care appear to be particularly relevant to elucidating the relationship between the 2 species. That they are closely related becomes increasingly clear, but the relationship between the siamang and the great apes has still to be investigated in detail.

Three alternative patterns of relationship reflecting the phylogeny of the Hominoidea are theoretically possible. First, the siamang may be more closely related to the gibbons than to the great apes (usual assumption). Thus similarities that the siamang shares with the great apes are explained by either convergence because of some factor such as large body size, or by the retention of primitive hominoid characteristics. Second, the siamang may be basically a great ape, which has converged with the gibbon into the small branch niche. Third, the siamang may be a distinct group from both gibbons and great apes, but for the present, following the principle of Occam, this alternative will be rejected.

The usual assumption is that the siamang and the gibbon had a common origin with the great apes (and man) in about the early Miocene from an ancestral hominoid. In the light of recent studies of fossil and living apes this model needs further examination, in terms both of finding more fossils and of studying living taxa in greater depth. It may become necessary to think in terms of a much later emergence of living ape species.

Table 6
Summary of ecological and behavioural differences between the gibbon and the siamang

	Gibbon	Siamang
Social grouping	monogamous family	monogamous family
Group formation	SA-male initiative	SA female initiative
Territory size	39 ha	23 ha
Territory use	forage	feed/travel
leader	male?	female
day range	1,600 m	970 m TS 1 150 days
		800 m RS 2 30 days
locomotion	fast, light — jump	slow, heavy — bridge
Feeding		
time		longer
fruit	74%	TS1 26% RS2 47%
leaves	34%	67% 60%
other	12+%	13% 7%
Social	loose	compact
	mother-infant-2	father-infant-2
		male-SA male
		rest more, travel less
Night positions	scattered	1 tree — young with adult
Grooming	15 min/day	30 min/day
	male-female	male-female-sub-adult
		adults-young — dusk
Play	juvenile	subadult-infant
Intra-group calls	much	little
Calling	early	about an hour later
	85% days/group	30% days/group
duration	13 min	15½ min
leader	female	female
	male quiet during female great call	male screams towards end of female bark series
Conflicts	frequent	infrequent
	male assisted by female	male assisted by SA male others hide
Inter-specific relations	not sociable	sociable

The solution of the present problem essentially lies in enumerating the similarities and differences between the gibbon and the siamang, and by comparing them with the great apes. This has been the approach used by Groves, and it reveals that while the siamang exhibits striking differences from the lar gibbon, so does the concolor gibbon. Thus, at this stage, one must recognise 3 taxa of equal separation. Perhaps when ecological and behavioural data are obtained for the latter, the pattern will revert to the traditional division into 2 genera—*Hylobates* and *Symphalangus*. But as Groves indicates, this appears unlikely, and the recognition of 3 taxonomic groups of gibbons (siamang, concolor, and lar) decreases the likelihood of the siamang having a closer relationship with the great apes than with the gibbons. Instead it has probably retained the ancestral features to a greater extent, and avoided the more extreme specialisations of the gibbons. The siamang is unlikely to be phyletically closer to the great apes because it possesses such typical gibbon specialisations, e.g. thumb. Phenetic similarities to the great apes have still to be fully explained.

Comparisons with the great apes are prompted by the large body size of the siamang, different karyotype, hair density and appearance, and behaviour patterns. But behavioural comparisons prove to be rather unsatisfactory. Although there are close similarities between the siamang and great apes in some features of ecology and behaviour, there are equally striking differences in others.

The orang-utan, the only other southeast Asian ape, is mostly a solitary nomadic animal with apparently neither the family group of the gibbon nor the large community group of the chimpanzee. The orang-utan is mainly frugivorous, spending much time in the trees; it gives loud spacing calls (Mackinnon, in press).

The African great apes—the chimpanzee and gorilla—spend much time on the ground, although the chimpanzee takes to the trees when alarmed (Reynolds, personal communication), as do all Malaysian primates except for the pig-tailed macaque. Gorillas live in groups averaging between 6 and 17 animals (according to locality) in a home range of about 6 mi^2 (Schaller, 1963). They are herbivorous, whereas chimpanzees are frugivorous. Groups of the latter are much larger and more loosely organised into unstable noisy bands within a home range of about 7 mi^2 (Reynolds, 1965). All 3 species of great ape build nests, which distinguishes them from gibbons and siamang which exhibit territorial behaviour. The gibbon is frugivorous, the siamang is folivorous; both species live in family groups in small territories. These differences can be explained by divergence into different niches within (and without) the tropical rain forest, thereby permitting a greater variety of phenetically similar, but ecologically and behaviourally different, phyletically related apes.

In the Malay Peninsula, the main question is whether the differences described between the gibbon and the siamang denote an early separation, or are only recent adaptations allowing the 2 species to co-exist in the same habitat. The problem is to distinguish environmental effects, such as those producing differences between lowland and highland populations of siamang, from the genetically controlled traits which identify different species. The problems are increased when dealing with patterns of behaviour which are composed of elements affected by the genes and the environment in varying degrees.

The siamang's diet is mainly composed of the young parts of vines (including figs), whereas the gibbon consumes a wide variety of fruit. This suggests a difference in gut morphology and physiology, and Kohlbrugge (1891) has demonstrated that the siamang has a longer large intestine relative to small intestine and head and body lengths. If such a difference in gut proportions, presumably in association with other factors, is involved in the dietary differences, it has a far wider indirect effect. For it results in a different pattern of ranging, and of utilising the territory, in order to obtain necessary nourishment. It is also necessary to maintain these resources, and to protect them if they are in short supply—hence territorial behaviors. It is only feasible to defend against conspecifics. In this way it is possible to explain most of the differences we have observed between gibbon and siamang behaviour by relating them to 1 source. In so doing, the case for an earlier phyletic separation may be weakened, and one becomes more interested in the recent timing and nature of the events which produced gibbons and siamangs.

Schultz (1933) suggested that the siamang was further from the ancestral gibbon than the gibbon, because the differences in the siamang represented increased specialisations. The small size of their territories is in accord with this, their diet and size (possibly) are not. Schultz also noted the similarities between the siamang and the great apes. It now seems more reasonable to suggest that the siamang represents a more conservative stage than the gibbon in morphology, dentition [Frisch, 1965], and behaviour. The less specialised diet, the lower level of territorial behaviour, and milder social behaviour within the group can be explained in this way. Although Carpenter suggests that the adult male gibbon plays an important part in guarding the young, this role is developed in a unique way in the siamang. It is also difficult to explain the greater cohesion of the siamang family group, which is characterised by lack of calls, much grooming, and little spatial separation. But taking several factors into account it is probably understandable that the main feature of gibbon inter-group relations is contact, and of intra-group relations is calls, whereas the converse is true for the siamang.

It is clear, however, that at this level of analysis the siamang has more in common with the gibbon than it has with the great apes in terms of ecology and behaviour. Thus they provide an excellent example of the diversification of species into differing niches within the same habitat. The next step is to pursue this study in greater detail, and to examine the relationship of the black handed gibbon to the other 2 ape species in the Malay peninsula. Further light can be shed on the whole problem by investigating similar phenomena within and between other ape genera. Much work still remains to be done, both in the field and the laboratory, to elucidate the primates which are gibbons and siamang, in relation to each other, and to the great apes and man.

SUMMARY

Data are presented on the ecology and behaviour of the siamang in the Malay Peninsula, and compared with the recent study of the white-handed gibbon by Ellefson. Both species live in tropical rain forest, where seasons are not very marked. Climatic data are given for the 4 study areas described.

The siamang is mainly found in the west-central mountain range, whereas gibbons occur wherever forest still stands. The white-handed gibbon of Thailand is separated from that of Malaya by the black-handed gibbon between Sungei Perak and Mudah on the west, and northwards from Sungei Kelantan on the east. Potential areas of overlap have still to be investigated to ascertain the status of these 2 taxa. Siamang and gibbon populations are partly clustered in small territories of family groups. There is virtually no overlap between territories of the same species, which contrasts with the tendency for almost complete overlap between groups of the 2 species. Gibbon territories are about 40 ha, which is nearly twice the size of siamang territories (representing twice the biomass). Not all the available space is apparently used.

Siamang groups use and maintain their territories in different ways from gibbon groups; they have shorter day ranges and fewer interactions with neighbouring groups. Gibbon groups call more often and earlier on average than siamang groups. Siamangs spend 25% feeding time in fig trees; only 29% feeding time is spent consuming fruit. This contrasts with gibbons, which feed on fruit in 74% food trees. Gibbons have a different major mode of obtaining food—foraging.

Siamangs have a smaller repertoire of intra-group calls and gestures; yet they have a closer spatial organisation. Play is important in both species. The siamang is unique among primates in that the adult male carries the infant during the day from when it is weaned from the female at the beginning of its second year until it becomes independent. The adolescent or sub-adult siamang is better tolerated by the adults; there are differences from the gibbon in the timing and form of the peripheralisation process. The sub-adult female siamang appears to take greater initiative in the formation of new groups than in the gibbon. Grooming is a more important part of siamang daily activity, and is most frequent between male and sub-adult (male); the female siamang leads the group about its day range. Gibbons and siamang have differing relations with other animals, including man; the siamang appears to be more sociable.

The relationship of the siamang and gibbon to each other is discussed, and an attempt is made to assess the position of each in relation to the great apes. On balance it seems that the siamang is closer phyletically to the gibbon than to the great apes. Such similarities as occur between siamang and great apes may be explained in part by the siamang representing a more conservative stage in gibbon evolution, thereby retaining ancestral (primitive) characteristics. More detailed work is required on fossil and living primates to elucidate the precise phylogeny of the apes and the relationships between living taxa.

ACKNOWLEDGEMENTS

The field study of the siamang was carried out under a Malaysian Commonwealth Scholarship, a Goldsmiths' Company Travelling Studentship, and a Science Research Council Overseas Studentship. Grants for equipment and field assistance were received from the Boise Fund (Oxford), the Emslie Horniman Fund (London), the New York Zoological Society, and the Merchant Taylors' Company (London). I am indebted to all these organisations for their encouragement and financial support.

I am grateful to Enche Abdul Jalil Bin Ahmad, State Game Warden of Pahang, for his assistance in my study in the Krau Game Reserve, to Dr. E. Soepadmo, University of Malaya, and Dr. T.C. Whitmore, Forest Research Institute, for their help in plant identification.

I acknowledge the basis for present and future studies of gibbon ecology and behaviour laid by Prof. C.R. Carpenter, and am grateful for the opportunity to draw on Dr. J.O. Ellefson's dissertation, which was produced by University Microfilms.

The manuscript has been read by Lord Medway, Dr. C.P. Groves, and Mr. R. Tenaza, and I am grateful to them for discussion and helpful critical comments. Dr. D. R. Pilbeam helped to initiate this study, and Lord Medway's advice and encouragement during the last 3 years have been invaluable. I am indebted to my wife for her help in every aspect of this study.

REFERENCES

Carpenter, C. R.: A field study in Siam of the behavior and social relations of the gibbon. Comp. Psychol. Monogr. *16* (5): 1-212 (1940).

Chasen, F. N.: A handlist of Malaysian mammals. Bull. Raffles Mus. *15:* 1-209 (1940).

Chivers, D.J.: On the daily behaviour and spacing of howling monkey groups. Folia primat. *10:* 48-102 (1969).

Chivers, D.J.: Spatial relations within the siamang family group. Proc. 3rd. Int. Congr. Primat., Zurich (in press).

Chivers, D.J.: The social organisation of the siamang. University of Cambridge; Ph.D. dissertation (in prep.).

Corner, E.J.H.: Wayside trees of Malaya, Singapore (1952).

Ellefson, J. O.: A natural history of Gibbons in the Malay Peninsula. University of California, Berkeley; Ph.D. dissertation (1967).

Fox, G.: Some comparisons between siamang and gibbon behaviour, Folia primat. (in press).

Frisch, J.E.: Trends in the evolution of the hominoid dentition. Bibl. primat. vol. 3 (Karger, Basel 1965).

Hamerton, J.L.; Klinger, H.P.; Mutton, D.E., and Lang, E. M.: The somatic chromosomes of the Hominoidea. Cytogenetics *2:* 240-263 (1963).

Kohlbrugge, J.H.F.: Versuch einer Anatomie des Genus *Hylobates;* in Weber Zool. Ergebn. einer Reise in Nied. Oest-Indien, vol. 2, pp. 139-254 (Leiden 1890-91).

McClure, H. E.: Some observations of primates in climax dipterocarp forest near Kuala Lumpur, Malaya. Primates *5* (3-4): 39-58 (1964).

Mackinnon, J.: Orang-utan today. Fauna Preservation Society, London (in press).

Napier, J. R. and Napier, P. H.: A handbook of living primates. (Academic Press, London 1967).

Reynolds, V.: Budongo. A forest and its chimpanzees (Methuen, London 1965).

Schaller, G. B.: The mountain gorilla (University of Chicago Press, Chicago 1963).

Schultz, A. H.: Observations on the growth, classification, and evolutionary specialisations of gibbons and siamangs. Hum. Biol. 5: 212-255 and 385-428 (1933).

Wyatt-Smith, J.: Malayan forest types. Malay. Nat. J. 7: 45-55 (1952).

15

The Borneo
Orang-Utan   1972

DAVID AGEE HORR

The orang-utan has been the most enigmatic of the apes. Despite our long knowledge of this animal's existence and its large size and spectacular appearance, very little has been known of its behavior and social organization in the wild.

Between September, 1967, and November, 1969, my wife and I undertook the first long-term study of wild orang-utans in the Segaliud-Lokan Forest Reserve in Sabah. Our overall study site was some 8 square miles of primary jungle on the Lokan River near Pintasin. This study was funded by the National Institutes of Mental Health and was done in conjunction with the Game Branch of the Sabah Forest Department. Over 1,200 hours of observation were made on some 27 orang-utans, but due to the dispersed nature of these animals, most work was done on a few animals in a 1 ½ square mile area.

Following this initial study, the project was relocated in the Kutai Reserve in Eastern Kalimantan, Indonesian Borneo, by Mr. and Mrs. Peter S. Rodman, where they conducted a study of the synecology of all higher primate species in lowland Borneo rain forest. I spent three months in the summer of 1971 there, and observations on orangs in the Kutai confirm the findings of the Sabah study.

Reprinted from the *Borneo Research Bulletin* 4(2):46-50 (1972) by permission of the Borneo Research Council.

GENERAL BEHAVIOR PATTERN

Orang-utans are primarily creatures of the lower jungle canopy. Although they do climb high in the trees, most of their time is spent between 20 and 60 feet off the ground where there are a lot of continuous tree crowns. This is not only the part of the forest canopy most abundant in food, but travel from tree to tree is easiest here where trees are small and close together.

We also found that orang-utans spend a surprising amount of time on the ground. Since their feet are so much like hands, it had been thought that orangs never came to the ground. In fact, they often come to the ground for water and to eat a wide variety of foods. I have seen them come to the ground and walk for some distance when there were breaks in the forest canopy, and when orangs really want to get away from humans, they come to the ground and run away into the undergrowth or into a swamp. This latter happened to us several times — though at first I did not believe these reports myself.

Each night, wild orangs make new nests to sleep in. Although juveniles make their own nests, they like to sleep in their mother's nest, and we have seen huge fights between mothers and juveniles over the issue of whether the juvenile would be permitted to sleep with its mother. This nest construction is very important, partly because it allows these large animals to live continuously in the trees, but also because it gives young infants a place to leave their mothers and move around without constantly clinging to her body.

The typical orang-utan day is outwardly relaxed. Orangs may get up early, but often it is 8 or 9 A.M. before the older ones leave the nest. Breakfast is first on the agenda, then a period of rest around mid-day. In the afternoon orangs either return to eating what they were feeding on in the morning, or they begin a slow amble through the canopy, snacking as they go. A leaf here, a flower there, perhaps a few termites or some bamboo shoots, and by evening they normally have settled into some larger food source, either a tree in fruit or perhaps some bark which they gnaw off the tree limbs like corn-on-the-cob. As dusk falls, a nest is built, though juvenile orangs may continue to feed after sundown. Usually orangs sleep throughout the night, though in some instances they have been observed to move some distance through the trees before making a new nest and bedding down. Although adult males and adult females may form consort pairs for several days during breeding, adult animals apparently never occupy the same nest.

SOCIAL BEHAVIOR AND SOCIAL ORGANIZATION

Orang-utans do not live in large social troops as do most other higher primates. Their semi-solitary existence has often been described (Carpenter, 1938; Schaller, 1964; Harrisson, 1962; Davenport 1967; and others), but the true nature of their social organization has never been fully understood largely because orang-utans are seldom found and are difficult to follow in the jungle.

From the Sabah study it has been possible to derive a picture of the nature of normal orang-utan social organization and behavior in the wild as well as a possible explanation for their rather unique social system.

POPULATION UNITS

Orang-utans are found in three kinds of basic units which usually forage independently in the jungle. (1) The only long-term social unit is the *adult female and her dependent children.* As many as two offspring may forage with their mother, and slightly older offspring may remain near her. These female-offspring units live in more or less permanent areas of about ¼ square mile in size. (2) *Adult males* forage as solitary individuals over a much larger area, perhaps as much as two miles or more. (3) *Juveniles* of both sexes forage with increasing independence of their mothers, probably starting in their third year. Although this is merely a transition stage to adult patterns, nonetheless they do form independent population units.

Not only do orangs move about in these small, isolated units, but contacts between these groups are infrequent. When orangs do meet, very often they seem to ignore each other, and contacts between orang-utan units usually last from a few minutes to only a day or so.

LIFE CYCLE

For the first year of life, orang infants cling to their mother's bodies throughout the day, leaving her only when in sleeping nests or when she is resting in a large tree crotch. By the end of the second year, young orangs are taking solid food and moving away from the female for increasing periods of time. Infants of this age are beginning to copy their mother's behavior patterns, and for example, may wave tiny twigs at an observer to threaten him. In their third year, young orangs are spending a lot of time away from their mothers and can make their own sleeping nests though they may still prefer sleeping with their mothers.

Juvenile females stay in the vicinity of their mothers for several years. They probably first breed about age 7 years and at that time they set up their own mother-offspring unit in a conservative range, perhaps overlapping that of their mother.

Juvenile males apparently range further away from their mothers at an early age, since we find solitary juvenile males in the jungle but seldom any near the adult mothers except for brief encounters.

Adult males and females assume the ranging pattern described earlier, though old adult males abandon the wide ranging pattern and live in much smaller areas. These also spend increasing times on the ground as they lose the agility required to keep their large bulk in the trees.

BASIS OF ORANG-UTAN SOCIAL ORGANIZATION

What might produce this unusual isolated mode of existence in orang-utans? It is probably largely due to the character of their jungle habitat and to their breeding pattern. Orang-utans are largely vegetarian. The nutritionally important parts of their diet are fruits, but orangs eat a great amount of leaves, inner bark, and bamboo shoots, as well as orchids, termites and other insects, and even dirt from termite mounds. No direct evidence of egg or meat eating was seen in the wild. Thus some species of plant is in fruit in nearly any month, but usually there are no great quantities of fruit available at any given time. The other diet items are everywhere available throughout the year. Since orang-utans are large animals, they can soon consume most of the fruit in a particular place, and bark and leaves probably do not have all of the nutrition required for survival.

Another important aspect is the absence of any serious natural predators for orang-utans. Although clouded leopards might prey on isolated juveniles, no predators (except man) are a major threat to adult orangs — even females with babies.

In view of the above, orang social organization might easily be explained as follows: In order not to overload the food supply, orangs disperse themselves in the jungles. Females carrying infants or tending young juveniles can best survive if they don't have to move far. Young orangs could also best learn the jungle in a restricted, familiar area. Apparently ¼ square mile can support a female with one or two dependents for an indefinite period of time. Adult males are unencumbered by young and can more easily move over wider areas. This means that they compete with females for food only for short periods of time, and thus they do not overload her food supply and force her to move over wider areas. Since there is no predator threat, males do not serve any function for females other than reproduction.

If orangs formed large groups, they would have to move over large areas to get enough food. In fact, MacKinnon found just such a situation in the Segama (1971), where orangs may have been crowded together due to logging activities.

The other factor which contributes to orang dispersal is their breeding pattern. Orang-utan females breed only once every 2 ½ to 3 years. If a male is to maximize his breeding potential, he is best advised to travel over as wide an area as possible so that he will have the greatest chance of being with a female when she is sexually receptive. By moving over large distances, the male breeds more frequently than if he stayed with one female, and also he does not overload the female's food supply. Since the general location of a female is pretty predictable, males range through the jungle and to announce their presence they give a loud bellowing vocalization using their throat sacs as resonating chambers. If females are receptive they will move towards these sounds. If they are not interested, I have observed them to move away from the male. If males persist, I have seen females threaten them away. Since receptive females are such a scarce resource, males compete for them, and this has probably resulted in the large size, heavy beards, and big cheek flanges on the males' faces. I have observed males who were with females threaten away other males by bellowing at them and making large aggressive displays.

BIBLIOGRAPHY

Carpenter, C. R., 1938, A Survey of Wildlife Conditions in Atjeh, North Sumatra, *Netherlands Comm. Internat. Pro.* 12: 1-33.

Davenport, R. K., 1967, The Orang-utan in Sabah, *Folia Primat.* 5: 247-263.

Harrisson, B., 1962, *Orang-utan*, London, Collins.

MacKinnon, J., 1971, The Orang-utan in Sabah Today, *Oryx* 12: 140-191.

Schaller, G. C., 1961, The Orang-utan in Sarawak, *Zoologica*, 46: 73-82.

16

Some Behavioral Comparisons between the Chimpanzee and the Mountain Gorilla in the Wild

1965

V. REYNOLDS

INTRODUCTION

Two types of behavioral information may be distinguished in the literature on chimpanzees and gorillas: 1) that obtained in the wild, which may be subdivided into the reports of unarmed naturalists such as Garner (1896) or the Akeleys (1923), and the reports of collectors or hunters such as Aschemeier (1921); and 2) that obtained in captivity, in zoos (e.g. Budd *et al.* 1943), laboratories (e.g. Yerkes 1943), and by people who kept these apes as pets (e.g. Hayes 1951, Lang 1961).

Following in the tradition of the unarmed naturalists, a series of field-workers (Schaller 1963, Goodall 1962, 1963, Kortlandt 1962, Reynolds 1963, 1964, 1965 and in press) has recently contributed greatly to our knowledge of the natural behavior of these anthropoids.

The aim of this article is to make some behavioral comparisons between the population of chimpanzees (*Pan troglodytes schweinfurthii*) studied by the author in the Budongo Forest, W. Uganda, and the population of mountain gorillas (*Gorilla gorilla beringei*) studied by Schaller (1963), chiefly at Kabara in E. Congo and also in S. W. Uganda. Unless otherwise stated, data referring to the mountain gorilla are taken from Schaller (1963) and those for the chimpanzee from the author's observations (Reynolds 1963, 1964, 1965, and in press).

Reprinted by permission of the American Anthropological Association from the *American Anthropologist, 67,* 1965.

DISTRIBUTION

The mountain gorilla occupies a smaller range than the chimpanzee (Fig. 1). The total area inside which *Gorilla g. beringei* occurs is about 35,000 sq. miles, from the equator to latitude 4°20′S and from longitude 26°30′E to 29°45′E. The total area within which *Pan troglodytes schweinfurthii* occurs is around 400,000 sq. miles, from latitude 5°N to 8°S and longitude 17°E to 32°E. While it is known that the gorilla occurs in isolated populations within its total range (Emlen and Schaller 1960), no evidence is available concerning the existence of isolated populations in the chimpanzees of the E. Congo forest, although it is probable that in some areas they do exist (Schaller, personal communication). In Uganda, at the eastern edge of the range, chimpanzee distribution coincides with the isolated patches of rainforest; likewise, isolated population units occur in the Sudan and in Tanganyika. The altitude of the gorilla's range is from around 1,500 feet to 13,500 feet, while that of the chimpanzee is from around 1,000 feet in the Congo basin to over 9,000 feet in the Ruwenzori Mountains (Wollaston 1908 and personal observation). In areas where the chimpanzee and gorilla

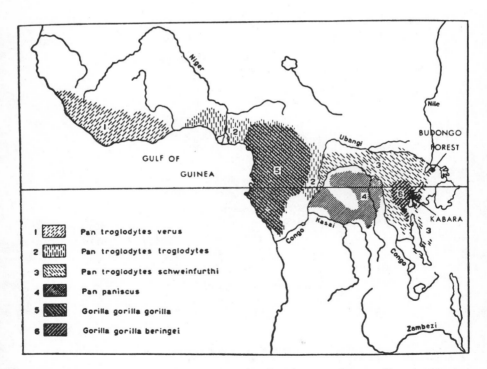

Fig. 1. Distribution of chimpanzees and gorillas, showing relevant study areas. (Based on Vanderbroek, 1958.)

occur together, there is some evidence that they do not mix freely. In the Kayonza Forest, Uganda, Pitman (1935 and personal communication) has observed that as gorillas moved into an area containing chimpanzees, the latter moved away.

HABITAT

Gorillas are confined to humid forests. Around three-quarters of all mountain gorillas occur in lowland rain forest, the remaining quarter occupying mountain rain forest. The latter quarter additionally occupy bamboo forest, and sporadically the Senecio zone occurring above 11,500 feet. Like mountain gorillas, chimpanzees are typically found in humid forests, but a smaller proportion of them occupies mountain rain forest. A small proportion of all chimpanzees, however, lives permanently in forest-savannah mosaic (Keay 1959) around the periphery of the rain forest zone. These areas are humid for part of the year, but have a definite dry season. Nissen (1931) and Goodall (1962) studied chimpanzees in this type of habitat, which typically consists of grassland, woodland, and gallery forest. Thus, while both chimpanzees and gorillas live in lowland and mountain rain forest, chimpanzees additionally live outside such forests and therefore show greater habitat diversity than gorillas. The temporary exploitation of high altitude zones by mountain gorillas is a specialization not found in chimpanzees.

FOOD HABITS

Both species are vegetarians, but while the gorilla feeds mainly on stems, leaves, and shoots, the chimpanzee is primarily a fruit-eater. Schaller estimated that around Kabara, in the Virunga Volcanoes, Albert Park, 1 vine and 3 herbs furnished at least 80% of the daily food supply, while in the Budongo Forest, fruits of four tree species dominated the chimpanzees' diet during the periods when those species were ripe. Both gorillas and chimpanzees show a marked adaptability to different foods in different parts of their range, and this is true whether the same foods are available in different areas or not; thus, Schaller found no overlap of actual foods consumed between the Utu region and the Virunga Volcanoes, although 55 food species were available in both areas, and in chimpanzees, none of the four major food species of the Budongo Forest occurs in the nearby Kibale Forest, which harbors a large chimpanzee population.

LOCOMOTION

Both gorillas and chimpanzees are quadrupedal. Their food requirements take them long distances on the ground and into the trees. But whereas the gorilla is essentially terrestrial, climbing trees with caution, chimpanzees are best characterised as arbo-terrestrial, being equally at home and skillful in the trees as on the ground; and, in fact, in the Budongo Forest they were observed to spend about 75% of their time in the trees. Brachiation is found in chimpanzees but not in gorillas. Swinging and leaping from branch to branch is a feature of chimpanzee behavior but not of gorilla. Hanging from thin branches by one or more limbs, with the body dangling down, is a typical feeding position of the chimpanzee, which is forced to exploit the outer periphery of trees where the fruit often grows; these postures were not seen in adult gorillas. On the ground both species support the front of the body upon the knuckles and the rear of the body on the flat soles of the feet; in trees the hallux is abducted to grip the branch or trunk in both species, on stouter branches the knuckles are used for support, while on thin ones and when climbing vertically, the long hand is used to grasp.

POPULATION DYNAMICS

The total population of mountain gorillas is between 5,000 and 15,000; for the chimpanzee, no census has been made and it is possible to guess what the population may be, but from the available evidence it seems that the population is very much greater. In the Budongo Forest, there are between 1,000 and 2,000 chimpanzees.

Population density of the mountain gorilla varies from around 1 per sq. mile in the Congo basin to 6.6 per sq. mile in the Kabara area. In an area of intensive study in the Budongo Forest there were 10 chimpanzees per sq. mile. The Budongo Forest seems, however, to have a denser population than most of the other Uganda forests and certain areas of the E. Congo (personal observation and Schaller, personal communication).

Mountain gorilla females give birth about every three and a half to four and a half years if the infant survives. In chimpanzees the commonest birth interval is slightly less, being about three years most often, commonly four years, while anything from one year onwards occurred.

Both gorillas and chimpanzees seem to be relatively free from predators. Leopards do not appear to constitute a serious threat to either species. The worst enemy of both is man.

Diseases — especially viruses, bacteria, and various blood and intestinal parasites — are probably the major cause of death in gorillas, and the same is probably true of chimpanzees. In the wild, both species appeared robust and healthy for the most part. Blindness in one eye was observed in both studies (one adult animal in each case), but whether this occurred through disease or accident is not known. Symptoms resembling the common cold were observed in some gorillas, and prolonged bronchial coughing was observed in an adult chimpanzee.

In gorillas, injuries of a superficial kind are fairly common and internal injuries such as bone fractures and breakages are occasional (see discussion in Schaller 1963). Bites account for some of the wounds. None of the Budongo chimpanzees showed any superficial injuries or bite scars, but one broken wrist was observed, and on two occasions chimpanzees fell from trees when branches broke. Schultz (1940) has drawn attention to the high frequency of fractured and repaired bones in old wild chimpanzees. Fighting is rare among gorillas, although it seems possible that on the occasions when fighting does occur, serious wounds may be inflicted. Possibly the teeth are used more than in the chimpanzee, which uses its limbs extensively in quarrels. However, perhaps a major factor limiting physical combat between the chimpanzees is the fact that they spend most of their time in the trees, where fighting cannot be fierce without the risk of a fall.

The oldest gorilla of known age in captivity was 34½ years old when he died. In captive chimpanzees, a male aged 35 years was in fine condition when he died accidentally (Mason, personal communication) and a 42-year old female survives at the Yerkes Laboratories (Riopelle and Daumy 1962). In both species senility was rarely seen in the wild, and it was impossible to judge the ages of the oldest animals seen; it seems possible that in both life expectancy is about 30 years in the wild.

In adult gorillas the ratio of females to males was 1.5:1.0, while in chimpanzees this ratio was probably about 1:1. In gorillas the proportion of animals in the population aged about 6 years or below was 45%, in chimpanzees the proportion observed was a little under 25%. However, the fact that in chimpanzees the mothers with infants and juveniles are the shyest of all groups and consequently the hardest to observe, may have contributed to the small percentage of infants and juveniles seen.

SOCIAL ORGANIZATION

A major difference exists between the social organization of gorillas and chimpanzees. Whereas in the gorilla, groups are fairly permanent in that most of the members of any given group keep together most of the time, giving the group temporal and spatial distinctness, no permanent groups which conform to the above conditions are found in chimpanzees. The gorilla social group has a scatter of 200 feet or less and consists of both sexes and all ages, the average constitution of a gorilla group at Kabara being as follows:

Adult male	3.2 animals
Adult female	6.2 animals
Juvenile	2.9 animals
Infant	4.6 animals

No comparable data can be obtained for the chimpanzee. In this species, most bands are so fluid and volatile that their constitution changes daily or hourly, as new animals join the band and others leave; or the band may split into two or more units which go their own ways and do not rejoin for a period of days, weeks, or months, if indeed they ever rejoin. Within the framework of this loose social organization, four types of bands may be distinguished: Mother-groups, Adult male groups, Adult groups (both sexes), and Mixed groups (both sexes and all age groups). Of these, the former two have the highest degree of permanence; nevertheless, all are temporary. The diameter of chimpanzee groups at any time depends entirely on the degree of dispersion of the forage.

Changes in group composition, the daily norm in chimpanzees, are rare in some gorilla groups, occasional in others. Adult males show the greatest mobility between gorilla groups. In chimpanzees, Male-groups (bands of about four adult males) travel more widely than other groups. Thus it seems that in both gorillas and chimpanzees, certain adult males cover more ground than females and young. In both speceis, however, there are less mobile adult males; in gorillas these are the ones which stay with the group, including the leader males, while in the chimpanzee these are the more timid and older males, which are most often found with Adult groups and Mixed groups.

Lone males, that is males leading a solitary life for a period of a month or more, are common in gorillas, but lone females were not observed. In chimpanzees it is probable that no animal spent as long as a month without joining up with other chimpanzees, although animals of both sexes and all age groups down to Juvenile 2 (average age = 4 yrs.) were seen alone from time to time in the Budongo Forest.

Intergroup interactions in gorillas are usually peaceful. Two groups will approach to within 100 or 150 feet of each other, then part, or will rest side by side, or occasionally mix briefly, or sleep together at one site, and on one occasion only was antagonism observed between the dominant males of two groups. Small bands of chimpanzees mingle freely; however, there is often excitement exhibited by leaping and noisy vocalization when bands meet, before they settle down feeding together.

GROUP RANGES AND MOVEMENTS
Neither gorillas nor chimpanzees have defended territories. Gorilla groups had an estimated average home range of 10-15 sq. miles, and there was great overlap between the ranges of neighboring groups. Owing to the lack of a cohesive social unit in chimpanzees, it is difficult to apply the term "home range." However, regions measuring 6-8 sq. miles each were found in the study area, within which there was a higher percentage of chimpanzee movements and interaction than there was between them. These regions may perhaps be considered as the home ranges of the 70-odd chimpanzees which apparently spent most of their time within them.

Seasonal movements, controlled by changes in the distribution of the food supply, were not important in gorillas, in which the only example of this was the more extensive feeding in bamboo zones during the rains, when tender young bamboo shoots provided an attractive source of forage. For most of the year, however, gorilla groups travel around the range without a definite pattern, forage being equally abundant almost everywhere. In this respect, the contrast with chimpanzees is very great, where the seasonal ripening of fruits in different forest types at different times of the year is the major factor determining the whereabouts of the animals. In the Budongo Forest, during the fig season, chimpanzees were found at sunny spots in Mixed Forest where fig trees grow; during the periods of food shortage, they were widely scattered over the whole forest; during the *Pseudospondias* season they were concentrated in Swamp Forest where this species grows, and during the *Maesopsis* season they were concentrated in *Maesopsis* Forest. There is, however, no clear evidence that trees in equatorial forest ripen at a given time each year, so that there is probably no annual cycle of chimpanzee movements.

The daily pattern of feeding activity of gorillas is as follows. Within an hour of dawn they move away from the sleeping site, travelling slowly and feeding intensively as they go. There is a mid-day rest period, followed by renewed feeding activity in the afternoon, less intense and with faster movement than in the morning. The normal daily total distance covered by gorillas is 300 feet to 6,000 feet (average about 1,700 feet), although they occasionally travel 15,000 feet on one day. Chimpanzees have a less clear-cut daily cycle in the Budongo Forest. After rising at daybreak, there is an hour or two of high-intensity feeding, after which feeding is slower. At this point some animals move on to a new feeding area, perhaps a mile or two away, while others, especially mothers and their offspring, stay put and settle down to very slow feeding and grooming or play activities. The mobile animals feed again on arrival at the new area, and can often be found feeding around midday. This general pattern continues all day. In the late afternoon feeding becomes more intensive, and at dusk it reaches a peak comparable to the early morning peak. The total daily distance covered by the mobile Male and Adult groups may be 5,000 feet to 25,000 feet, while Mother groups probably travel less than 1,500 feet a day on many days. Daily travel distance is, however, greatly affected by the scarcity or abundance of food, being greater if food is scarce.

SOCIAL BEHAVIOR

In both gorillas and chimpanzees, posturing has communicative significance in the behavior of dominant males. Dominant male gorilla leaders move forcefully, and "stand and face" when they want the group to move off. Most chimpanzee groups have no leaders, but some Male groups have an especially dominant male which leads the way across roads and leaves feeding trees first; dominant males were seen in other types of group as well, being distinguished by forceful movements and fine physique. The role of facial expressions was difficult to determine in both species; each, however, exhibited teeth and gums in strongly disturbing situations.

Vocalizations present an area of extreme contrast between gorillas and chimpanzees. While gorillas are normally rather quiet animals, chimpanzees are among the noisiest animals in the Bugondo Forest. Most gorilla sounds are abrupt and of a low intensity when they are undisturbed; their most intense vocalization, the roar, is given to man, or when otherwise greatly excited. Twenty-one more or less distinct vocalizations were distinguished in free-living gorillas, of which eight were fairly common. In the chimpanzee, 11 more or less distinct vocalizations were common. Of these, one type — calls which are prolonged and high pitched and may be called hoots — are emitted by several animals together, a phenomenon which does not occur in gorillas. The resulting chorus is extraordinarily loud and has a carrying power of up to two miles. In response to man, chimpanzees individually emit a short loud bark. In both species, vocalizations exhibit great variability in pitch, pattern of delivery, and intensity, and in both species infants are the least vocal group, their vocalizations being confined to distress screeches.

A complex display is found in adult gorillas, comprising a total of 9 acts often in a definite sequence. In adult chimpanzees, a less stereotyped but similar display is found, and for comparison the two are outlined below.

Sequence of Events in Displays of Gorilla and Chimpanzee

Gorilla		Chimpanzee	
1.	2–40 clear hoots	1.	Panting hoots by one animal, slowly, low pitch
2.	"symbolic feeding"		
3.	Throws vegetation	2.	Panting hoots by nearby animals, with increasing pitch, tempo and volume
4.	Bipedal stance		
5.	2–20 chest beats		
6.	Leg kicking	3.	(on ground only) Drumming on tree buttresses
7.	Running sideways		
8.	Branch shaking and breaking	4.	Shaking of saplings or branches, running, and leaping.
9.	Thumps ground with one or both palms		

In chimpanzees, the display is primarily a communal activity; thus, if there is no response to stage 1 by other animals, the hoots of the initiating chimpanzee may trail off; although occasionally individual animals go from panting hoots to drumming, this is more commonly the case when many animals are vocalizing. This contrasts with the gorilla, where a single animal normally gives the display, irrespective of the response of other animals. A point of similarity is that chimpanzees frequently drum without accompanying vocalization, and chest-beating (perhaps the gorilla equivalent of drumming) may occur on its own unaccompanied by the display sequence. In both species the display results from excitement, caused by a wide variety of things.

Overt social interactions were few in both studies, as the following figures for certain kinds of interactions show:

	Gorilla		Chimpanzee	
	Total No. observations	Frequency obs/hr.	Total No. observations	Frequency obs/hr.
Dominance	110	0.23	25	0.08
Grooming	134	0.28	57	0.19
Social play	96	0.11	47	0.16

These figures indicate a somewhat greater proportion of dominance interactions in the gorilla than in the chimpanzee. In both cases it most often consisted of moving away by the subordinate at the approach of the dominant or pushing away of the subordinate by the dominant. Each gorilla group observed had a dominant silverbacked male who was dominant over every other member of the group. In chimpanzees, dominant males when they occurred were fully grown and black haired with greying rumps; other, apparently older, males with grey backs were not seen to be dominant. Between males, a linear hierarchy was present in the gorilla, not in the chimpanzee. In the gorilla, while silverbacked males were dominant over adult females, variable dominance relations were observed between blackbacked males and adult females; in the chimpanzees, all adult males were found to be dominant over adult females. In both species, females had no stable dominance hierarchy.

Three-quarters of all grooming interactions observed in the gorilla were between females and their offspring, and the remaining quarter were mainly grooming by juveniles of females, other juveniles, and infants. It was extremely rare for an adult male to groom, and no adult female was seen to groom an adult male. In chimpanzees, on the other hand, only a quarter of all grooming interactions observed were of a female grooming her offspring, while adult male:adult female grooming accounted for a little over 30% of the total. In two thirds of the cases of male:female grooming, the female was in oestrus.

In both gorillas and chimpanzees, juveniles and infants play. In gorillas, 43% of all play was solo, while in chimpanzees the proportion was 36%. Of play groups, 81% consisted of two youngsters in the gorilla, 83% in the chimpanzee.

Mother-infant ties are close in both gorillas and chimpanzees during the first three years of the infant's life. Thereafter, they persist in the chimpanzee for a further year or two, but not in the gorilla, which is usually independent by the age of three years, although associations with the female may continue to 4-4½ years. During the first three months of life, the gorilla baby clings to its mother's belly, and she supports it usually with one arm as she walks along; at around three months the infant begins to ride on its mother's back, at 6-7 months it occasionally walks and even climbs on its own, after the first year independent locomotion becomes more common until at 2½-3 years riding on the mother stops except in emergencies. In chimpanzees, no supporting of infants by the mothers was observed. The infant clings to its mother's belly ceaselessly for about six months, especially while she is moving around in the treetops. From 6 months to two years it makes forays of increasing length outwards from the mother, riding under her belly when she moves any great distance, say to a new tree. At 2-3 years juveniles are carried on the mother's back when she moves along the ground, but in trees, if they are carried, it is under the belly. This continues for a further 2-3 years, during which time juveniles are still carried during normal rapid group progression through the forest, although they trot behind the mother if progress is slow. On several occasions mothers were seen carrying an infant under the belly and a juvenile on the back. Thus in chimpanzees the pattern of mother:infant responses is more prolonged than in the gorilla. Possibly the stresses of arboreal life and the need for movement over long distances in order to exploit the fruit supply necessitate this prolongation of maternal care.

Sexual behavior was infrequently seen in both studies. Only two gorilla copulations were observed, while four chimpanzee copulations were seen. The only other form of sexual behavior seen was invitation to copulate, which occurred rarely in both species. All the wild copulations were dorsoventral, but the exact position adopted was rather variable; they included copulation of the series and single types.

The estrous cycle in captive gorillas is 30-31 days long, and the female exhibits no sexual swelling. There is a period of sexual receptivity during each cycle, lasting 3-4 days, when she may initiate sexual behavior. The estrous cycle in captive chimpanzees is $37(\pm0.14)$ days long (Young and Yerkes 1943), the female exhibits a very large and prominent pink or grey swelling, and, during the 10-day period when it is maximal, she initiates sexual behavior (Yerkes and Elder 1936). Thus both the sexual cycle and the period of sexual receptivity are longer in the chimpanzee than in the gorilla.

Observations on both species indicated the absence of any clear-cut breeding season or birth periodicity.

NESTING BEHAVIOR

Both species normally make a nest to sleep in every night. In chimpanzees, such nests were very rarely on the ground (2 out of 259), being found most often between 30 and 40 feet up (30% of all nests), while 15% were above 90 feet. In gorillas, nest heights vary according to locality, but ground nests are common: the lowest proportion of ground nests was found in lowland rain forest (21.8%), while the highest was 97.1% in Hagenia woodland, Kabara. In most areas, nests were lower than 20 feet, except in lowland rain forest, where 38% were above this height. The basic construction of nests by both species is similar, but the chimpanzee often finishes its nest by adding a lining of leaves and branchlets while the gorilla does not. The same fact has been reported by Reichenow (1920) concerning the chimpanzees and gorillas of Upper Njong, Cameroons.

Both gorillas and chimpanzees are wholly diurnal, resting throughout the hours of darkness. In gorillas the dominant male appears to select the nesting site and is the first to nest, but this sequence was not observed in chimpanzees. Gorillas often spend an hour or more in their nests after first light, but chimpanzees normally get up at dawn and after urinating and/or defecating, commence feeding. This fact has also been reported of the gorillas and chimpanzees of Gaboon (Aschemeier 1922). Gorillas nearly always soil their nests, whereas chimpanzees nearly always do not; infants normally sleep in the same nest as mother in both species; alarmed chimpanzees may hide in nests (Schaller, personal communication, and personal observation) while this has not been reported for the gorilla.

RELATIONS WITH OTHER SPECIES

In relations with man, gorillas are known to be more dangerous than chimpanzees; their bluff charges may, on rare occasions, lead to attack. Chimpanzees rarely charge a human. However, adult males, black- or grey-backed, often stay behind when the group they were with has fled as a result of the appearance of a human. They sit in a low branch, staring at the human, barking sometimes, and looking around. If the observer stays put, it may be many minutes before the male leaves, and during this time he may lie down along the branch with his back to the observer.

Gorillas were not observed to respond to golden monkeys (*Cercopithecus mitis kandti*) within 10 feet of them. In the Budongo Forest, chimpanzees came into contact regularly with four monkey species: redtail monkeys (*Cercopithecus ascanius*), blue monkeys (*C. mitis stuhlmanni*), black and white colobus monkeys (*Colobus abyssinicus*), and baboons (*Papio anubis*). Relations were normally peaceful, but on one occasion when a blue monkey came within three yards of an adult male chimpanzee, the latter jerked towards it and the monkey ran off. Chimpanzees did not react to the presence of baboons in the vicinity; however, on the only observed occasion when a party of baboons climbed into the crown of a tree in which chimpanzees were feeding, the latter moved away into neighboring trees. Our findings suggest that while chimpanzees are dominant over redtail, blue, and colobus monkeys, they are not dominant over baboons if the latter are in large numbers.

Relations with non-primate species, such as buffalo and elephant, appear to be neutral. No evidence of predation on either chimpanzees or gorillas was obtained in either of the present studies, but it may occasionally occur. With smaller mammals, such as duikers, no clear-cut interactions were observed in either study; however, Goodall (1963) has observed the eating of such animals by chimpanzees in a reserve in Tanganyika. There is no evidence that gorillas ever resort to meat-eating.

These apes live at peace with various bird species, they are not known to eat reptiles, and while chimpanzees regularly eat insects this is not known for the gorilla.

DISCUSSION

The chief morphological difference between gorilla and chimpanzee is one of size; their appearance, apart from the sagittal crest of the male gorilla, the longer coat and smaller ears in gorillas, and the pink skin colour of chimpanzees when it occurs, is rather similar. Their body proportions are very similar.

The manner of locomotion of these apes is largely similar, both on the ground and in trees; however, the chimpanzee brachiates while the gorilla does not, and shows greater skill and ease in moving about in trees than the latter, spending far more time in them.

They are closely related species, possibly being descended from a Miocene or Pliocene ancestor. Schultz (1927) has suggested, on the basis of foot structure, that gorillas have evolved from an ancestor with more arboreal specializations. One item of behavior supports this suggestion. Schaller noted that the gorillas "exhibit and retain vestigial nest-building behavior in an environment where nests seem superfluous — an anachronism which suggests arboreal ancestry" (1963:198).

The present day adaptation of the gorilla to terrestrial life, and its primarily herbivorous diet, compared with the more arboreal, frugivorous chimpanzee, may help us understand the present-day distribution of these species. The two extant gorilla types are widely separated from each other, occurring at the eastern and western ends of the equatorial forest but not between. Chimpanzees, on the other hand, are known to be distributed throughout this forest zone. Schaller speculates that in the past the gorilla population may have been continuous along a belt north of the present limit of rain forest, and that, in a dry period this population died out. In view of the dietary needs of the gorilla, depending as it does on the lush vegetation of humid zones, it is probable that the species cannot survive in woodland areas with a pronounced dry season and general desiccation at that time. Chimpanzees, however, are known to be able to exploit such areas (Nissen 1931 and Goodall 1962); Nissen made his study in French Guinea during the dry season, and throughout this period there was an abundance of fruits available to the chimpanzees. Thus one advantage of a fruit-eater over a shoot-eater is that its food supply is less seriously affected by the transition from forest into woodland caused by general climatic desiccation, and by the occurrence of a dry season in the annual cycle.

Secondly, assuming that the more open conditions of woodland increase the danger of predation by ground-living carnivores such as lion, the more arboreal chimpanzee would have a further advantage over the gorilla.

The extreme contrast between the social organization of the chimpanzee and the gorilla calls for explanation. Perhaps the most important factor underlying this difference is the distribution of food. For gorillas, the problem of finding sufficient food is negligible in areas such as Kabara. They are constantly surrounded by a variety of edible matter, and feeding decisions revolve, perhaps, more around the question of which of the available foods to consume than where to find food. This may be less true in Congo lowland rain forest, where large areas of primary forest provide little food for gorillas, which prefer secondary forest. In both regions, which in some ways represent the extremes of gorilla habitat, food is located in the same areas throughout the year, i.e. everywhere at Kabara, and in secondary forest in the Congo basin.

In the case of chimpanzees, the distribution of food is radically different. In the Budongo Forest, the successive ripening of food-fruits in different parts of the forest makes the finding of food a prime factor in chimpanzee survival. There are no small areas in the forest where food is available all year round, the minimum size of such an area being several square miles. At certain times of the year, food is highly concentrated in particular zones, while at others it is widely scattered over the forest.

In gorillas, permanent social groups may have an advantage over a looser form of social organization. For in this essentially ground-living and ground-feeding form, they may provide better defence against predators. The specialized role of the dominant male as protector of the group is seen in his spectacular display against enemies.

It is also possible to explain the difference in social organization by examining the advantage to the chimpanzee of having a very loose form. In the ecological setting of the Budongo Forest, this type of social organization enables a greater number of animals to survive because they are better able to exploit the food supply than they would be if they moved about in permanent groups. During the period of the year when food is scarcest, it is widely scattered over the forest, and clearly the most efficient way of exploiting such a food supply is to scatter widely, foraging in pairs and threes. Conversely, when food is extremely abundant but concentrated in, say, one hundredth of the total forest area, it can be exploited most efficiently only if the animals are all within this small area.

The problem of why the gorilla is a fairly quiet animal while the chimpanzee is exceedingly and noisily vocal may be related to the difference in social organization. Observations in the Budongo Forest showed that (a) during the period when food was scarce and widely distributed, the chimpanzees were not very vocal, hooting occasionally and chorusing rarely or not at all, while during each of the periods of concentrated food, they made the maximum noise, sometimes chorusing for stretches of minutes without interruption, at intervals of less than an hour throughout the day; and (b) that the response of chimpanzees in one area to strong chorusing from another part of the forest was to move towards the calling and join them. It thus seemed that the loud hooting choruses served to alert nearby chimpanzees to areas of plentiful food.

In the gorilla this entire signalling system is absent. Absent too are both the widespread, highly volatile population which could concentrate on sources of loud vocalizations, and areas which for a brief period have a superabundance of food. The conditions in rain forest, where the position of the best foods cannot be located visually, may have laid emphasis on the development of vocal cues in the system of food-finding of chimpanzees, while in the gorilla it seems reasonable to suppose that visual cues suffice in enabling animals to locate all the food they require.

If this interpretation is correct, the marked difference in development of vocalizations in the two species indicates that the difference in social organization has been present for a long time. Additionally, the larger size of the ears in chimpanzees than in gorillas is correlated with a more extensive reliance on vocal signals in the former.

A distinctive morphological difference between these two species is the occurrence in the chimpanzee of a large sexual swelling of the perineal area, which reaches it maximum development for a ten-day period during every oestrous cycle. Concurrent with her swelling, the female is sexually receptive and initiates mating with males by sexual presentation. During this period she ovulates, and thus matings may be fertile. In the gorilla such swelling of the female's sexual skin is totally absent. Here again, an adaptive advantage of the swelling in chimpanzees can be most clearly seen in the light of our knowledge concerning social organization. In the gorilla, the existence of permanent social groups means that females are near males at all times. There is thus no lack of opportunity for copulation. In chimpanzees there may be a greater need for some form of distance signal by females in oestrus, in order to increase the chances of matings occurring at the optimal time, and in order to attract and hold the attention of males which might otherwise move out of reach. Several males will follow a female with a prominent swelling, and may copulate with her one after the other. The swelling thus operates as a distance signal, visible by the male as soon as the female herself is visible, and acting as a source of attraction, drawing the male to her and causing him to follow her about. Gorillas neither have nor need such a signal.

In their displays, the gorilla's locomotor behavior is more stereotyped than that of the chimpanzee, and an explanation for this may be that gorillas display on the ground only, while all the parts of the chimpanzee's display may be exhibited in the trees except drumming. While a ground display gives the animal freedom to move its body in any direction it wishes, so that there is no inherent danger in such acts as standing bipedally, leg kicking, and running sideways, such acts could be dangerous if performed in a tree. Thus, the lack of interference of considerations of posture and balance have made it possible for the gorilla to evolve a stereotyped sequence of locomotor activities into its display sequence, while in the chimpanzee, such behavior has had to take constant consideration of the problem of support and balance, and has thus not developed much locomotor stereotypy.

The habitual soiling of its nest by the gorilla is in sharp contrast with the chimpanzee's habit of keeping its nest clean and defecating over the edge. Gorilla feces are firm and well knit, composed mainly of indigestible fibrous matter: "the usual consistency of the dung is such that it retains its shape when falling to the ground and that the animal's fur is not soiled when it sits on it" (Schaller 1963:88). Bingham (1932, Plate 14) has shown clearly that gorilla feces do not disintegrate when lain on. In contrast, the consistency of chimpanzee dung is normally soft, the feces being composed of the waste of a diet consisting in large part of soft fruits. A chimpanzee lying in its dung would soil its coat, and the resultant adhering feces might well attract insects and in other ways act as a source of infection. Thus it may be that, in the evolution of defecating behavior in these species, natural selection has favored defecation outside the nest in chimpanzees, while in gorillas this has not been the case.

Finally, we may briefly consider the contrast in temperaments between these two anthropoid species. Comparative behavior studies in the past often stressed this difference. Tevis (1921), for instance, wrote "In mental characteristics there is the widest difference between the two apes that we are considering. The chimpanzee is lively, and at least when young, teachable and tameable. The gorilla, on the other hand, is gloomy and ferocious, and quite untameable" (1921:122). It is possible to suggest an explanation for this contrast between the morose, sullen, placid gorilla, and the lively, excitable chimpanzee. The difference seems to be most clearly related to the difference in social organization and foraging behavior. The herbivorous gorilla is surrounded by food; the more intensively it feeds, the slower it travels; its survival needs are easily met, and it is protected from predators by the presence of powerful males. Here there is no advantage to any form of hyper-activity, except in threat displays and the charge of the big male, which is a hyper-aggressive behavior form. Chimpanzee survival, on the other hand, depends heavily on the fluidity of social groups and the ability to communicate the whereabouts of food by intense forms of activity (wild vocalizing and strong drumming). Moving rapidly about the forest, meeting up with new chimpanzees every day, vocalizing and drumming, and locating other chimpanzees by following their calls, are the basic facts of chimpanzee existence. Here an advantage may be seen in having a responsive, expressive and adaptable temperament. Hyper-activity is the chimpanzee norm in the wild, and with this goes a volatile temperament.

SUMMARY

Some behavioral comparisons are made between the population of chimpanzees (*Pan troglodytes schweinfurthii*) studied by the author in W. Uganda, and the population of mountain gorillas (*Gorilla gorilla beringei*) studied by Schaller (1963) in E. Congo and S. W. Uganda. The gorilla is less arboreal and less frugivorous than the chimpanzee; it lives in permanent groups while the chimpanzee does not, does not exhibit the vocal chorusing typical of chimpanzees, has a more stereotyped display than the chimpanzee. Chimpanzees are better able to exploit dry-season zones than gorillas. Their food supply is located in different places at different times of the year while the gorilla's is not, so that food-finding is a greater problem for the chimpanzee than it is for the gorilla. The looser social organization and development of group chorusing in chimpanzees may be a response to the difference in the pattern of food distribution. The sexual swelling of chimpanzees, absent in gorillas, may be a distance signal. The relative absence of display *stereotypy* in chimpanzees may be associated with their greater *arboreality*. The gorilla's habit of defecating in the nest, not found in the chimpanzee, may be related to the constitution of its *defecate*, in contrast with that of the chimpanzee. The more volatile temperament of chimpanzees than of gorillas is examined in the context of their ways of life in the wild.

NOTE

This article was written while the author was a Fellow of the Center for Advanced Study in the Behavioral Sciences, Stanford, California, and grateful acknowledgment is made for the research facilities provided by the Center. Dr. G. Schaller was also at the Center, and I wish to thank him for many helpful suggestions. Finally, I wish to thank Dr. C. Jolly for many useful criticisms.

REFERENCES CITED

Akeley, C. 1923 In brightest Africa. Garden City.

Aschemeier, C. R. 1921 On the gorilla and the chimpanzee. Journal of Mammalogy 2:90-92.

1922 Beds of the gorilla and chimpanzee. Journal of Mammalogy 3:176-78.

Bingham, H. C. 1932 Gorillas in a native habitat. Carnegie Institute, Washington, Pub., 426.

Budd, A., L. G. Smith, and F. W. Shelley 1943 On the birth and upbringing of the female chimpanzee "Jacqueline." Proceedings of the Zoological Society of London 113:1-20.

Emlen, J. T. and G. B. Schaller 1960 Distribution and status of the mountain gorilla (*Gorilla gorilla beringei*). Zoologica 45:41-52.

Garner, R. L. 1896 Gorillas and chimpanzees. London, Osgood, McIlvaine & Co.

Goodall, J. 1962 Nest building behaviour in the free-ranging chimpanzee. Annals of the N. Y. Academy of Science, 102, 455.

1963 Feeding behaviour of wild chimpanzees, a preliminary report. Symposium of the Zoological Society of London 10:39-47.

Hayes, C. 1951 The Ape in our house. New York, Harper.

Keay, R. W. 1959 Vegetation map of Africa south of the Tropic of Cancer. Oxford.

Kortlandt, A. 1962 Chimpanzees in the wild. Scientific American 206: 128-138.

Lang, E. M. 1961 Goma—das Korillakind. Zurich.

Nissen, H. W. 1931 A field study of the chimpanzee. Comparative Psychology Monographs, 8 (1):1-122.

Pitman, C. R. S. 1935 The Gorillas of the Kayonsa Region, Western Kigezi, S. W. Uganda. Proceedings of the Zoological Society of London 105, 2:477-499.

Reichenow, E. 1920 Biologische Beobachtungen an Gorilla und Schimpanse. Sitzber. Ges. naturf. Fr. Berlin, 1-40.

Reynolds, V. 1963 An outline of the behaviour and social organisation of forest-living chimpanzees. Folia Primatologica 1:95-102.
1964 The "Man of the Woods." Natural History 73, no. 1:44-51.
1965 Budongo: an African forest and its chimpanzees. Natural History Press, New York.

Reynolds, V. and F. Reynolds (in press) Chimpanzees in the Budongo Forest. *In* Primate behavior: field studies of monkeys and apes, Ed. I. DeVore, N. Y., Holt, Rinehart, and Winston.

Riopelle, A. J. and O. J. Daumy 1962 Care of chimpanzees for radiation studies. Proceedings of the International Symposium on Bone Marrow Therapy, 205-227.

Schaller, G. B. 1963 The mountain gorilla, ecology and behavior. Chicago, University of Chicago Press.

Schultz, A. H. 1927 Studies on the growth of gorilla and of other higher primates . . . Memoir of Carnegie Museum 11, 1:1-87.
1940 Growth and development of the chimpanzee. Contributions to Embryology, Carnegie Institute, 29:1-63.

Tevis, M. 1921 Gorillas, chimpanzees, and orang utans. Scientific American Monthly 4:121-125.

Vanderbroek, G. 1958 Notes ecologiques sur les Anthropoides. Society Royal Zoological Society, Belgium, 89:203-211.

Wollaston, A. F. R. 1908 From Ruwenzori to the Congo. London, John Murray.

Yerkes, R. M. 1943 Chimpanzees, a laboratory colony. New Haven, Yale University Press.

Yerkes, R. M. and J. H. Elder 1936 The sexual and reproductive cycles of chimpanzee. Proceedings of the National Academy of Science, Washington 22:276-283.

Young, W. C. and R. M. Yerkes 1943 Factors influencing the reproductive cycle in the chimpanzee: the period of adolescent sterility and related problems. Endocrinology 33: 121-154.

PART 2

THEORETICAL
PAPERS

introduction
relationships between ecology and social structure

Each of the articles in this section deals with the
question of relationships between ecology and
social structure. As we have seen in the preceding
articles, there are many different types of social
structure within the order Primates. These include
the following general types: "solitary but social"
(most nocturnal prosimians and orang-utans);
family groups (e.g., Callitrichidae, Aotinae, and
Hylobatidae); one-male groups (e.g., most species
of *Cercopithecus*, patas monkeys, black and white
colobus, and gelada and hamadryas baboons);
multimale groups (found in some species of almost
all higher taxa); and "fission-fusion" groups (e.g.,
Ateles, *Cercopithecus mitis*, chimpanzees and, at some
levels of organization, gelada and hamadryas
baboons). Of course, within each of these
superficial categories of social group we are not
really dealing with identical structures. For
example, multimale groups of lemurs may be quite
different from those of baboons in structure,
function, and selective advantage. This is even

more obvious when comparing species with a so-called "fission-fusion" group structure.

There is no simple relationship between social structure and phylogeny or ecology. However, as we read the following articles, we see that many of the relationships proposed earlier were entirely too simple. As more data were collected, it was found that the earlier categories were too gross, and the subtlety of these relationships became more and more apparent. For example, as is stressed in Selections 20 and 21, the earlier articles comparing the ecology and social structure of arboreal and terrestrial primates were written when practically nothing was known about arboreal primates. The recent increase in the number of quantitative, comparative field studies of primates has allowed more subtle categories of resource utilization to be revealed. This, in turn, has led to more sophisticated comparisons between animals that seem to have parallel or convergent roles in similarly structured ecological communities. These animals often share similar foraging and dietary adaptations, in their respective communities, and have certain features of social behavior and structure in common (see especially Selections 25, 26, and 27). It must be kept in mind, however, that direct and simple correlations between ecological variables and most aspects of primate behavior may never be found. We can assume that, at some level, there is a phylogenetic influence on social behavior (see especially Selection 20) and, because of this, different taxa may adapt to the same environment in different ways. Many features of social behavior may be sufficient but not necessary for any given environment. Furthermore, certain features may be adaptations to prior environmental conditions or may even be random (some of these factors are discussed in Selection 27). However, as more data of the kind described in the preceding studies are collected, relationships that do exist should become more apparent. Only then can more sophisticated hypotheses about and mathematical models of primate societies be formulated and tested.

17

A Comparison of the Ecology and Behavior of Monkeys and Apes*

IRVEN DEVORE

Comparisons of nonhuman primates have traditionally contrasted the behavior patterns of New World and Old World monkeys. The Platyrrhines of the New World are said to live in loosely organized social groups in which individuals are rarely aggressive and dominance behavior almost absent. The social group of the Old World Catarrhines is described as more rigidly organized by social hierarchies, based on dominance-oriented behavior and frequent fighting among adult males. To the extent that this distinction is valid, the behavioral differences being compared are not those between New and Old World monkeys, but between arboreal and terrestrial species. The only systematic field studies of New World monkeys have been on the howler monkey, *Alouatta palliata* (Carpenter, 1934; Collias and Southwick, 1952; Altmann, 1959), and Carpenter's (1935) brief observations on spider monkeys, *Ateles geoffroyi*. Both of these species are highly specialized, morphologically and behaviorally, for living in the tall trees of the South American jungle. They are seldom seen in the lower branches of trees and almost

Reprinted by permission from Sherwood L. Washburn (editor), *Classification and Human Evolution* (Chicago: Aldine Publishing Company); copyright © 1963 by Wenner-Gren Foundation for Anthropological Research, Inc.

*The preparation of this paper was supported by a National Science Foundation grant for the analysis of primate behavior, while the author was a fellow of the Miller Institute for Basic Research in Science, University of California, Berkeley. In addition to the conference participants, George Schaller and Richard Lee read and commented on this paper. The suggestions of all these persons are gratefully acknowledged, though sole responsibility for its final content rests with the author.

never come to the ground. On the other hand, behavioral studies of Old World monkeys, principally those of Zuckerman (1932) and Carpenter (1942), had until recently included only species in the baboon-macaque group. In both morphology and behavior the baboons-macaques are more terrestrially adapted than any other monkey or ape.

Comparisons between monkey and ape behavior have also been difficult, since Carpenter's gibbon study (1940) is the only long term study of an ape that has previously been available. This Asiatic brachiator is as highly specialized for arboreal life as the baboon-macaques are for life on the ground. If the monkeys and apes are arranged along a continuum with those that are terrestrially adapted at one pole and those with specialized arboreal adaptations at the other (Fig. 1), it is clear that long-term naturalistic observations have been confined almost entirely to species lying at the two extremes and that little has been known of the majority of species falling somewhere in between. The species shown in Fig. 1 are those for which some field data are available, and the intention is to suggest only the broad outlines of adaptation to life on the ground or in the trees.

Napier (1962) has discussed the habitats of monkeys in detail. In brief, spider monkeys, howler monkeys, gibbons, orangutans, and most species of colobus and mangabeys live in the higher levels of mature, tropical forests. The olive colobus lives by preference in the lower forest level, seldom higher than twenty feet from the ground, yet not descending to the ground (Booth, 1957); South American capuchin and African red-tail monkeys,

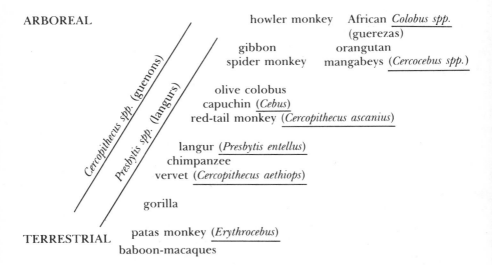

Fig. 1. Schematic representation of selected species of monkeys and apes along a continuum of relative adaptation to terrestrial or arboreal life.

though primarily tree dwellers, may also come to the ground to feed (Carpenter, 1958a, Haddow, 1952). Langurs, vervets, *Cercopithecus l'hoesti,* and chimpanzees seem equally at home on the ground or in trees; and gorillas, patas monkeys, baboons, and macaques are clearly adapted to terrestrial life. Field studies of nonhuman primates are expanding at a rapid rate, with more than fifty workers currently engaged in field research (DeVore and Lee, in press), and in a few years it will be possible to refine greatly the provisional conclusions discussed here. In the following pages the available evidence on home range, intergroup relations, population density, and social behavior is reviewed with respect to adaptation and taxonomy.

HOME RANGE

The size of the area which an organized group of animals customarily occupies, its "home range," varies widely among the primates. Current studies have revealed a high correlation between the size of home range and the degree to which a species is adapted to life on the ground—the more terrestrial the species, the larger its home range (Table 1). Arboreal gibbons occupy a range of only about one tenth of a square mile, and a group of howler monkeys range over about one half of a square mile or less of forest. Since the average gibbon group numbers only four and howler groups average about seventeen, the amount of range needed to support an equivalent amount of body weight in each species is comparable. These home ranges are small, however, by comparison to the home ranges of langur groups, *Presbytis entellus,* studied by Jay (unpub. ms.). Langurs, which frequently feed on the forest floor and raid open fields, range over an average area of three square miles in a year—even though their average group size (twenty-five) is much the same as howler monkeys.

Terrestrial species like the mountain gorilla and the baboon occupy much larger home ranges. Schaller and Emlen (in press) found that gorillas in the Virunga Volcanoes region of Albert National Park, Congo, have an average group size of seventeen, and customarily travel over an area of from ten to fifteen square miles during a year. An average troop of forty baboons, living on the East African savanna in the Royal Nairobi National Park, covers an area of about fifteen square miles during a year. In general, it is clear that a terrestrial adaptation implies a larger annual range; a baboon troop's range is 150 times as large as that of a gibbon group. Even if these figures are corrected to allow for the greater combined body weight of the individuals in a baboon troop, the baboon home range is still at least five times as large as that of gibbons. It is frequently stated that the habitat of arboreal monkeys is actually three-dimensional, and that a vertical dimension must be added to horizontal distances traveled before the true size of a group's range can be accurately determined. Many arboreal species, however, sel-

Table 1
Summary of Available Data on Group Size, Population Density, and Home Range for Monkeys, Apes, Wolves, and Human Hunting Groups Living in Relatively Arid, Open Country

Species	Groups		Sample		Population Density	Group Range	Source
	Mean	Extreme	Population	No. of Groups	(Indiv. per sq. mile)	(sq. miles)	
Orangutan	3	2–5	28	10			Schaller 1961
Gibbon	4	2–6	93	21	11	.1	Carpenter 1940
Black & white colobus	13			1		.06	Ullrich 1961
Olive colobus	12	6–20					Booth 1957
Spider	12	3–17	181	19			Carpenter 1935
Cebidae	15?	5–30					Carpenter 1958
Howler							
1933 census	17	4–35	489	28	82	.5	Carpenter 1934
1951 census	8	2–17	239	30	40		Collias & Southwick 1952
1959 census	18.5	3–45	814	44	136		Carpenter 1962
Langur	25	5–120	665	29	12	3	Jay Unpub. Ms.
Mountain gorilla (Virunga Volcanoes)	17	5–30	169	10	3	10–15	Schaller 1963
Macaca mulatta							
temples	42	16–78	629	15			Southwick 1961a
forest	50+	32–68	334	7			Southwick 1961b
M. radiata	32		65	2			Nolte 1955
M. assamensis	19		38	2			Carpenter 1958

Table 1 (continued)

Species	Groups		Sample		Population Density (Indiv. per sq. mile)	Group Range (sq. miles)	Source
	Mean	Extreme	Population	No. of Groups			
Baboon							
Nairobi Park	42	12–87	374	9	10	15	DeVore & Washburn (in press)
Amboseli	80	13–185	1203	15			
Human							Shapera 1930; D.
bushman	20				.03	440-1250	Clark pers. comm.
aborigine	35				.08	100-750	Steward 1936
Wolf	12	9–25+				260-1900	Stebler 1944; Murie 1944

dom come to the ground, or even to the lower levels of the forest. Howler monkeys are largely confined to the upper levels of primary forest, avoiding secondary growth and scrub forest; gibbons ordinarily stay just beneath the uppermost forest canopy; and most species of African colobus monkeys appear to stay above the shrub layer of primary forest. Limited to arboreal pathways through the forest, such species are far less free to exploit an area completely than are primates that do not hesitate to descend to the ground. In effect these arboreal species occupy a two-dimensional area above the forest floor. Some species which exploit both the arboreal and terrestrial areas of their range with facility, such as South American capuchin monkeys, Indian langurs, some African *Ceropithecus spp.*, and probably chimpanzees, may be accurately described as living in a "three-dimensional range." Even terrestrial species such as baboons and macaques spend much of their lives in trees, returning to them to sleep each evening and feeding frequently in them at certain seasons.

The ability to exploit both forested and open savanna habitats has had distinct advantages for the baboon-macaque group, enabling it to spread throughout the Old World tropics with very little speciation (DeVore and Washburn, in press). Similarly, the most ground-living Indian langur, *Presbytis entellus,* and the most ground-living *Cercopithecus* monkey, *C. aethiops,* have wider distributions than the other species in their respective genera. In this sense the home range of an organized group of monkeys can be said to be proportional to the geographic distribution of the species. However, this is not always true. The gelada baboon, *Theropithecus gelada,* and the gorilla are geographically restricted, while the gibbon is found throughout Southern Asia. Adaptation of the species to local conditions, the interference of human activities, and competition from other species may explain these contrary examples. Even so, the terrestrial gorilla represents a single, widely separated species while gibbons are divided into a number of well-defined species.

The ability to cross natural barriers and occupy a diversity of environments is reflected in both the size of a group's home range and the amount of speciation that has occurred within its geographic range. This is particularly true when man is considered with the other primates. Speech, tools, and an emphasis on hunting have so altered man's adaptation that direct comparisons are difficult, but if hunter-gatherers living in savanna country with a stone-age technology are selected for comparison, an immense increase in the size of the home range is apparent. Although detailed studies are lacking, the home ranges of two African bushman groups were estimated as 440 and 1250 square miles (Shapera 1930; Clark, personal communication). An average bushman band would number only about 20, or half the size of a baboon troop with a home range of 15 square miles. In Australia, depending on the food resources in a particular area, a band of aborigines has a home range of between 100 and 750 square miles; an average band numbers about 35 (Radcliffe-Brown, 1930; Steward, 1936). Many hunter-gatherer ranges average 100 to 150 square miles, and even the smallest are much larger than the biggest home ranges described for non-human primates. The land area it is necessary for human hunters to control is more comparable to that of group-living predators than it is to that of monkeys and apes. The home range of wolf packs, for example, is between 260 and 1900 square miles (Stebler, 1944; Murie, 1944). Consistent hunting rapidly depletes the available game in a small area, and in the course of human evolution an immense increase in the size of home range would have had to develop concomitantly with the development of hunting.

CORE AREA AND TERRITORIALITY

Within the home range of some animal groups is a locus of intensive occupation, or several such loci separated from each other by areas that are infrequently traversed. These loci, which have been called "core areas" (Kaufmann, 1962), are connected by traditional pathways. A core area for a baboon or langur troop, for example, includes sleeping trees, water, refuge sites, and food sources. Troops of babooons or langurs concentrate their activities in one core area for several days or weeks, then shift to another. Most of a baboon troop's home range is seldom entered except during troop movement from one of these core areas to another; a troop in Nairobi Park with a home range of just over fifteen square miles primarily occupied only three core areas whose combined size was only three square miles. Except that baboon range and core areas are larger, this description would apply equally well to langurs (Jay, unpub. ms.) or, for that matter, to the coati groups, *Nasua narica,* studied by Kaufmann (1962).

The home ranges of neighboring baboon troops overlap extensively, but the *core areas* of a troop very rarely overlap with those of another troop (DeVore and Washburn, in press). The distinction between a group's home range and the core areas within that range becomes important when the question of territorial defense is considered. Carpenter's review of territoriality in vertebrates concludes that " . . . on the basis of available data territoriality is as characteristic of primates' behavior as it is of other vertebrates" (1958a, p. 242). His description of "territory" in another review (1958b), however, conforms to what has here been described as "home range," following the distinction between "home range" and "territory" (as a defended part of the home range) made by Burt and others (e.g., Burt, 1943, 1949; Bourlière, 1956). Groups of baboons, langurs, and gorillas are frequently in intimate contact with neighboring groups of the same species; yet intergroup aggression was not observed in langurs (Jay, unpub. ms.), tension between baboon troops is rare (DeVore and Washburn, in press), and gorilla males only occasionally show aggression toward a strange group (Schaller and Emlen, in press). Clearly, no definite areal boundaries are defended against groups of conspecifics in these three species. On the other hand, organized groups of these species occupy distinct home ranges, and the fact that a strange baboon troop, for example, is rarely in the core area of another troop indicates that there are spacing mechanisms which ordinarily separate organized groups from each other. Different groups are kept apart, however, not so much by overt aggression and fighting at territorial boundaries, as by the daily routine of a monkey group in its own range, by the rigid social boundaries of organized groups in many monkey species, and, in some species, by loud vocalizations.

Intimate knowledge of the area encompassed in a group's home range is demonstrably advantageous to the group's survival. Knowledge of escape routes from predators, safe refuge sites and sleeping trees, and potential food sources combine to channel a group's activities into daily routines and seasonally patterned movements within a circumscribed area. Beyond the limits of a group's usual range lie unknown dangers and undiscovered food sources, and a baboon troop at the edge of its home range is nervous and ill at ease.

ORGANIZATION OF PRIMATE GROUPS

In addition to the spacing effect of relatively discrete home ranges, most monkeys and some apes live in organized groups which do not easily admit strangers. Gibbons, howler monkeys, langurs, baboons, and macaques all live in such "closed societies." Although Carpenter (1942) described numerous instances of individuals moving from one group to another in his study of the rhesus colony transplanted to Cayo Santiago, Altmann's later study (1962) revealed that almost no individuals changed troops during a two-year period. In retrospect it seems likely that when Carpenter first studied the rhesus colony the six heterosexual groups he described were composed of comparative strangers, and that organized troops had not yet had a chance to stabilize. During Altmann's study approximately the same number of animals were divided into two highly stable troops with almost no individuals' changing troop membership — despite the crowded conditions on the island, where natural spacing mechanisms probably could not operate normally.

Studies by the Japan Monkey Center on Japanese macaques, *Macaca fuscata*, confirm the fact that macaque societies are closed groups (Imanishi, 1960). During more than 1400 hours of observations on more than twenty-five baboon troops, only two individuals changed to new troops (DeVore and Washburn, in press). Jay found that langur troops were similarly conservative, but among langurs individual males or small groups of males may live apart from organized troops. Not all monkey and ape groups have such impermeable social boundaries. Some gorilla groups, for example, can be described as somewhat open or "fluid."

Although different gorilla groups vary considerably in the extent to which they accept individuals into the group, some of them have a high turn-over of adult males (Schaller and Emlen, in press). During a twelve-month study period, some dominant adult males remained with a group of females and young throughout, but other adult males frequently left one group, and either led a solitary existence or joined another group. Males who join a group have access to sexually receptive females, and group-living males usually make little or no attempt to repel them. This "relaxed attitude" toward nongroup members is reflected in intergroup relations. Schaller and Emlen report that "seven different groups were seen in one small section of forest during the period of study." In contrast, it is doubtful if any adult male baboon or macaque can join a new troop without some fighting with the dominant males of the troop.

At present there are not enough detailed studies of the social behavior of monkeys and apes to determine whether organized groups in most species are relatively closed, as in baboons, or relatively open, as in gorillas. A recent study of hamadryas baboons in Ethiopia by Kummer and Kurt (in press), for example, revealed an altogether different kind of group structure. The hamadryas were organized into small groups of from one to four, rarely as many as nine, females and their offspring accompanying only one adult male. These "one-male groups" aggregated into sleeping parties each night which numbered as many as 750 individuals, but during the day the small, one-male units foraged independently, and membership in them remained constant during the observation period, while the number of individuals gathering at sleeping places fluctuated constantly. Kortlandt's (1962) observations of chimpanzees indicate still a different group structure, with anywhere from one to thirty individuals gathered together at one time.

INTERGROUP VOCALIZATIONS

Loud vocalizations seem to aid in spacing groups of some species. Carpenter has suggested that "vocal battles" often substitute for physical aggression in howler monkeys and gibbons. Jay found that langurs also give resounding "whoops" which carry long distances and tend to keep langur troops apart; Ullrich (1961) found the same is true of black and white colobus. Since these loud cries invariably begin at dawn, when no other group is in sight, and often continue as the group leaves its sleeping place and moves to a feeding area, it seems likely that such vocalizations serve more often to identify the *location* of the troop than to issue a challenge to neighboring troops. Altmann's following observations of howling monkeys support this view:

> The morning roar, despite its spectacular auditory aspects, was not accompanied by any comparable burst of physical activity. While giving the vocalization, the male stood on four feet or sat. Occasionally, he fed between roars. . . . The vocalizations that are given by the males at sunrise are essentially the same as those that are given during territorial disputes, suggesting that these morning howls serve as a "proclamation" of an occupied area (1959, p. 323).

By advertising its position the troop reduces the likelihood that it will meet a neighboring troop. Such location cries can thereby function as spacing mechanisms, but usually without the directly combative connotation which "vocal battle" suggests. When a gibbon or howler group approaches another group, loud vocalizations increase. Two groups of gibbons, however, will mingle without agression after the period of vocalizing has passed (Carpenter, 1940, p. 156), and Jay saw langur troops frequently come together without aggression.

Loud vocalizations of the sort found in gibbons, howlers, colobus and langurs are conspicuously absent in baboons and macaques. The situation on the African savanna, where visibility frequently extends for hundreds or even thousands of yards, is very different from the very limited visibility of dense forests. Baboon ranges are large in Nairobi Park, and troops seldom come near each other. Adjustments in the direction of troop movement can easily be made by sight alone. Even when baboon troops do come together, as they frequently did in that part of the Amboseli Reserve where food and water sources were restricted, no intergroup aggression was observed. In view of the relatively high degree of dominant and aggressive behavior typical of baboons, the fact that intertroop relations are pacific may seem contradictory. If defense of territory is not thought of as fundamental in the behavior pattern of primates, however, the peaceful coexistence of neighboring troops is not inconsistent with a high level of agonistic and dominance behavior between individuals within the group.

Loud vocalizations, as a means of keeping groups apart when they are not in visual contact, are also absent in the gorilla. When two groups come together, or in the presence of man, adult males may give dramatic intimidation gestures (abrupt charges, chest beating, etc.). While some of these gestures may be accompanied by loud vocalizations, their impact is primarily visual. As in the baboons and macaques, visual communication tends to substitute for loud vocalizations, and it is suggested that methods of intergroup communication correlate with the degree to which a species is arboreal or terrestrial. Vocal expression as a spacing mechanism is very important in gibbons and howler monkeys, less important among langurs, and virtually absent in the gorilla, the baboons, and the macaques. It is also reasonable to suppose that vocalizations are more important in intragroup behavior in arboreal species, where foliage interferes with coordination of group activity, than they ordinarily are in terrestrial species. Baboon studies to date have concentrated on troops living in open areas, where troop members are usually in constant visual contact. If baboons become separated from the troop, however, contact is reestablished by vocalizations, and when the troop is feeding in dense vegetation soft vocalizations (e.g., low grunting) are apparently more frequent, and may serve to maintain contact between troop members. Many baboons live in forests and a study of these forest-living baboons could definitely determine whether vocal behavior is more frequent in this habitat than it is on the open plains.

POPULATION DENSITY

Pitelka (1959) has pointed out that the fundamental importance of territoriality lies not in the behavior, such as overt defense, by which an area becomes identified with an individual or group, but in the degree to which the area is used exclusively; that is, functionally, a territory is primarily an ecological, not a behavioral, phenomenon. Recent field studies indicate that if territoriality can be ascribed to most monkeys and apes it is in this more general, functional meaning of the term only. As a means of distributing the population in an available habitat, the core area of a nonhuman primate group's home range may serve the same function as literal, territorial defense does in some vertebrate species. At the human, hunter-gatherer level this is much less true. Although a hunting band may have core areas, around water holes for example, the band tends to consider most or all of its home range an exclusive possession. Human hunting activities require the use of large areas, and the hunting range is usually protected from unwarranted use by strangers. One of the important differences between home ranges which overlap and the exclusive possession of a range is reflected in the population density of the species. Although a howler monkey group ranges over about .5 of a square mile, the 28 troops counted in 1933 could not have exclusively possessed more than an average of .2 of a square mile per group; in 1959 the average per group could have been no higher than .136 of a square mile. In Kenya, about 50 per cent of a baboon troop's home range overlaps with that of adjacent troops, and a troop would seldom possess more than one square mile of its range exclusively. One result of this extensive overlapping is that the population density of the nonhuman primates is far higher than would be expected from a description of home range sizes alone (Table 1).

The population density of howler monkeys on Barro Colorado Island in 1959 was 136 individuals per square mile. Terrestrial species like baboons and gorillas have much lower population densities (10 and 3 per square mile), but they are nevertheless far more densely populated than many hunter-gatherers, who average only .03 to .08 per square mile in savanna country. That hunting man, like other large, group-living carnivores, will inevitably have a lower population density than the other primates can be illustrated by the numbers of lions in Nairobi Park. This area of approximately 40 square miles supports a baboon population of nearly 400, but it supports an average of only 14 lions (extremes:4-30—i.e., 4 to 30 lions) (Wright, 1960).

In summary, all monkeys studied to date live in organized groups whose membership is conservative and from which strangers are repelled. These groups occupy home ranges which may overlap extensively with those of neighboring groups, but which contain core areas where neighboring groups seldom penetrate. Rather than territorial defense of definite boundaries, monkey groups are spaced by daily routine, tradition, membership in a discrete social group, and the location of adjacent groups. Among the apes these same generalizations would seem to hold for gibbons, but they are much less true of gorillas, and, probably, orangutans and chimpanzees. The trend toward increasing size of the home range, however, from the very small range of arboreal species to moderately large ranges in terrestrial species, would appear to be true of all monkeys and apes. Means of identifying the position of adjacent groups shifts from loud vocalizations in arboreal species to visual signals in terrestrial ones. Man exemplifies his terrestrial adaptation in his enormously increased home range, his lower population density, and in his reliance on vision in the identification of neighboring groups.

GROUP SIZE

Mention has already been made of the average size of the social group in several species of primates; available data on group size are summarized in Table 1. When the size of the social group is compared to the degree of arboreal or terrestrial adaptation of the species, a trend toward larger groups in the terrestrial species is apparent, in both Old and New World monkeys and among the apes. Although there are no adequate data for the orangutan, the largest group ever reported is five, and the average size of a gibbon troop is four. Schaller found that the average gorilla group was seventeen, temporarily as large as thirty, and Kortlandt (1962) saw as many as thirty chimpanzees together in his study area. The average size of monkey troops varies from 20 or less in the olive colobus, spider, howler, and *Cebidae* group to an average of 25 (but as large as 120) in langurs, and an average of from 40 to 80 in the baboons and macaques (with some troops as large as 200).* Present observations suggest three central grouping tendencies in monkeys and apes: five or less for the gibbon and orangutan; from twelve to twenty in arboreal monkeys and the gorilla; and fifty or more in baboons and macaques. This small sample may be misleading, but it is clear at present that, in addition to having a larger home range and lower population density, terrestrial monkeys and apes live in larger organized groups.

SEXUAL DIMORPHISM AND DOMINANCE

Field studies of monkeys indicate that dominance behavior, especially of the adult male, is both more frequent and more intense in ground-living monkeys than in other species. This increase in dominance behavior is accompanied by an increase in sexual dimorphism, particularly in those morphological features which equip the adult male for effective fighting—larger body size, heavier temporal muscles, larger canine teeth, etc. (Washburn and Avis, 1958, p. 431). If the apes are compared with respect to sexual dimorphism, there is a clear trend toward increasing sexual dimorphism from the arboreal gibbon, where the sexes are practically indistinguishable, to the chimpanzee, in which the male is appreciably more robust, to the gorilla, where sexual dimorphism is greatest. Only the orangutan is an exception. The trend toward increased sexual dimorphism in terrestrial species of monkeys is also apparent. Some male characteristics, such as the hyoid bone in howler monkeys and the nose of the proboscis monkey, are more pronounced in arboreal species, but morphological adaptations for *fighting* and *defense* are clearly correlated with adaptation to the ground. The various baboons and macaques all illustrate this tendency. Among *Cercopithecus* species sexual dimorphism is procounced in *C. aethiops* and the patas monkey (a species closely related to *Cercopithecus*) and decreases in the more arboreal forms.

The trend toward increased fighting ability in the male of terrestrial species is primarily an adaptation for defense of the group. Zuckerman's

*The most thorough census of macaque troops is that of Southwick et al. (1961a, 1961b). Only the figures for forest and temple troops are included in Table 1 because troops in other habitats were subject to frequent trapping. Counts of 399 troops in all habitat categories revealed an over-all average group size of only 17.6.

account of baboon behavior (1932) would indicate that the acquisition and defense of "harems," and concomitant fighting among the males, places a high premium on aggressiveness and fighting ability in intragroup behavior. Behavioral observations on confined animals can be very misleading, however, and no field study of baboons has found that sexual jealousy or fighting is frequent in free ranging troops (Bolwig, 1959; Hall, 1960, 1961; Washburn and DeVore, 1961a, 1961b). Observations of many baboon troops in close association indicate that intertroop aggression is very rare and that males do not try to defend an area from encroachment by another troop (although baboons were seen trying to keep vervet monkeys away from a fruit tree). The fact that hundreds of hamadryas may gather at one sleeping site would indicate that interindividual tolerance is high in this baboon species as well (Kummer and Kurt, in press). The intergroup fighting of rhesus which Carpenter observed on Cayo Santiago was probably aggravated by unsettled conditions on the island. One index of these conditions is that more infants were killed than were born during his period of study.

Life on the ground exposes a species to far more predators than does life in the trees. Not only are there fewer potential predators in the trees, but also escape is relatively easy. By going beneath the canopy (to escape raptorial birds) or moving across small branches to an adjacent tree (to escape from felines), arboreal species can easily avoid most predators except man. The ultimate safety of all nonhuman primates is in trees, and even the ground-living baboons and macaques will take refuge in trees or on cliffs at the approach of a predator (except man, from whom they escape by running).

Much of the day, however, baboons may be as far as a mile from safe refuge, and on the open plains a troop's only protection is the fighting ability of its adult males. The structure of the baboon troop, particularly when the animals are moving across an open area, surrounds the weaker females and juveniles with adult males. At the approach of a predator, the adult males are quickly interposed between the troop and the source of danger (Washburn and DeVore, 1961b). The structure of a Japanese macaque troop is apparently identical, even though no predators have threatened the Takasakiyama group in many years (Itani, 1954). The ecological basis for sexual dimorphism in baboons has been described elsewhere (DeVore and Washburn, in press). Because only the adult males are morphologically adapted for defense, a baboon troop has twice the reproductive capacity it would have, for the same number of individuals, if males and females were equally large. Adaptation for defense is accompanied by increased agonistic behavior within the troop, but intratroop *fighting* is rare. Stable dominance hierarchies minimize aggression among adults, and male baboons and macaques actively interfere in fights among females and juveniles.

Field studies of other monkeys indicate that intragroup aggressive and agonistic behavior decrease by the degree to which a species is adapted to arboreal life. Among langurs, where sexual dimorphism is less pronounced, females may threaten or attack adult males — behavior that is unparalleled in the baboon-macaques. Langurs are usually in or under trees, where escape is rapid and the need for males to defend the group is much less important. The same argument, with suitable qualifications, holds for the other, more arboreal monkeys. The sexes in the olive colobus, for example, are almost identical in size and form; this species is never seen on the ground (Booth, 1957). This does not imply that the male in other primate species has no protective role. Male colobus, vervets, and howlers have been seen taking direct action against potential predators, and the male in many species is prominent in giving defiant cries and/or alarm calls. With the possible exception of the gibbon, some measure of increased defensive action by adult males is a widespread primate pattern. The evidence does suggest, however, that increased predation pressure on the ground leads to increased morphological specialization in the male with accompanying changes in the behavior of individuals and the social organization of the troop. Although troops of arboreal monkeys may be widely scattered during feeding, a baboon or macaque troop is relatively compact. Some males may live apart from organized groups, either solitarily or in unisexual groups, in less terrestrial species (e.g., langurs), but we discovered no healthy baboons living outside a troop (DeVore and Washburn, in press). Dominant adult males are the focal point for the other troop members in baboons, macaques, and gorillas. When the males eat, the troop eats; when the males move, the others follow. Compared to other monkey species: the baboon-macaques are most dominance-oriented; troop members are more dependent upon adult males and actively seek them out; and the social boundary of the troop is strong.

BEHAVIOR AND TAXONOMY

Species-specific behavior has been valuable in the classification of some vertebrate species, notably the distinct song patterns in birds (e.g., Marler, 1957, 1960; see also Hall, this volume). Spectrographic analyses of primate vocalizations will undoubtedly reveal specific differences in communication patterns, but these are only now being undertaken. Studies of social behavior in monkeys and apes have only begun, and no observations have yet been made in sufficient detail to permit close comparisons. Some general comparisons can be made, however, from studies recently completed.

The African baboons and the Asiatic macaques are very similar in both morphology and general adaptation. Both groups have forty-two chromosomes and their distribution does not overlap, suggesting that they are members of a single radiation of monkeys. Baboons, including drill, mandrill, hamadryas, and savanna forms (but excluding gelada) are probably all species within one genus, *Papio* (DeVore and Washburn, in press). Comparisons between the social behavior of East African baboons and macaques (rhesus and Japanese macaques) have been made elsewhere (DeVore, 1962). At all levels of behavior, from discrete gestures and vocalizations to over-all social structure, baboons and macaques are very much alike. Both groups have an elaborate, and comparable, repertoire of aggressive gestures. In social interactions, the same behavioral sequences occur: an animal who is threatened may redirect the aggression to a third party, or may "enlist the support" of a third party against the aggressor. If support is successfully enlisted, two or more animals then simulaneously threaten the original aggressor. Relations between the adults and the young of both groups are similar.

Play patterns in juvenile groups and the ontogeny of behavior follow the same course. Relationships within the adult dominance hierarchies, and the social structure of the troop are comparable. Some details of gesture and vocalization are certainly distinct, and there is a striking difference, for example, in the form and duration of copulation. Copulation in rhesus monkeys usually involves a series of mounts before ejaculation, as does copulation in the South African baboons ("chacma") studied by Hall (1962), while a single mounting of only a few seconds' duration is typical of East African baboons. Details of gesture and vocalization during copulation also vary between the three groups. On the other hand, most of the behavioral repertoire seems so similar that an infant baboon raised in a macaque troop, or vice versa, would probably have little difficulty in leading the adult life of its adopted group. Behavioral observations clearly confirm the evidence of morphological similarity in this widespread group of monkeys. No other primate, except man, has spread so far with as little morphological change as the baboon-macaque group. With man, these monkeys share the ability to travel long distances, cross water, and live in a wide range of environmental conditions.

On the basis of the present, random studies of monkeys and apes, generalizations regarding trends in behavior must remain speculative, particularly since the majority of the studies are concentrated in the baboon-macaque group. Much more useful statements with regard to the adaptive significance of different behavioral and morphological patterns will be possible when field studies of several species within one genus have been undertaken. The *Cercopithecus* group presents a wide range of ecological adaptations, from swamp-adapted species like *C. talapoin*, through the many forest forms, to the savanna-living *C. aethiops* (see Tappen, 1960). Forms closely related to the *Cercopithecus* group, Allen's swamp monkey, patas, and gelada, would further extend the basis of comparison. A study of patas or gelada monkeys would be particularly useful for cross-generic comparison with baboon and macaque behavior. Although Ullrich has made initial observations, no long term study of an African colobus species is yet available for comparison to Jay's study of Indian langurs. Booth's report (1956, 1957) that olive colobus do not ascend to the upper levels of forests even when these are not occupied by black colobus and red colobus indicate that even brief field observations of the behavior of sympatric primate species would be an immense aid in settling some of the persistent questions in primate taxonomy.

SUMMARY

Field studies of monkeys and apes suggest a close correlation between ecological adaptation and the morphology and behavior of the species. All terrestrial forms occupy a larger home range, and, in monkeys, the geographic distribution of the species increases according to the degree of terrestrial adaptation. Many arboreal species use loud vocalizations in spacing troops; ground-living forms depend more on visual cues. A marked decrease in population density accompanies terrestrial adaptation. Man is part of this continuum, illustrating the extreme of terrestrial adaptation.

Morphological adaptation of the male for defense of the group is more prominent in ground-living species (except man, whose use of tools has removed the selective pressure for this kind of sexual dimorphism), and least prominent in most species that do not come to the ground. The dependence of the other troop members creates a male-focal social organization in terrestrial species. Dominance behavior is much more prominent in terrestrial monkeys, but actual fighting is rare. Terrestrial life and large adult males have not been accompanied by a comparable increase in dominance behavior in the gorilla, however, indicating that defense is more important than intragroup aggression in the development of sexual dimorphism in terrestrial primates. Man's way of life has preserved the division between the male and female roles in adult primate life, but cultural traditions have replaced biological differences in the reinforcement of this distinction.

BIBLIOGRAPHY

Altmann, Stuart 1959. "Field observations on a howling monkey society." *J. Mammal.*, 40(3):317-330.

1962. "A field study of the sociobiology of rhesus monkeys, *Macaca mulatta.*" *Ann. New York Acad. Sci.*, 12(2):338-435.

Bolwig, Niels 1959. "A study of the behaviour of the chacma baboon, *Papio ursinus.*" *Behaviour*, XIV(1-2):136-163.

Booth, A. H. 1956. "The distribution of primates in the Gold Coast." *J. W. Afr. Sci. Ass.*, 2:122.

1957. "Observations on the natural history of the olive colobus monkey, *Procolobus verus* (van Beneden)." *Proc. Zool. Soc. Lond.*, 129:421-431.

Bourlière, François 1956. *Natural history of mammals.* (Rev. ed.). New York: A. A. Knopf.

Burt, W. H. 1943. "Territoriality and home range concepts as applied to mammals." *J. Mammal,* 24:346-352.

1949. "Territoriality." *J. Mammal.,* 30:25-27.

Carpenter, C. R. 1934. "A field study of the behavior and social relations of howling monkeys, *Alouatta palliata."* *Comp. Psychol. Monogr.,* 10(48).

1935. "Behavior of red spider monkeys in Panama." *J. Mammal.,* 16:-171-180.

1940. "A field study in Siam of the behavior and social relations of the gibbon, (*Hylobates lar*)." *Comp. Psychol. Monogr.,* 16:5.

1942. "Sexual behavior of free ranging rhesus monkeys (*Macaca mulatta*)." *J. Comp. Psychol.,* 33:113-162.

1958a. "Territoriality: a review of concepts and problems." See Roe and Simpson, 224-250.

1958b. "Soziologie und Verhalten Frielebendere Nichtmenschlicher Primaten." In *Handbuch der Zoologie,* 8:1-32.

Collias, N. and Charles Southwick 1952. "A field study of population density and social organization in howling monkeys." *Proc. Amer. Phil. Soc.,* 96:143-156.

DeVore, Irven 1962. "The social behavior and organization of baboon troops." Unpublished doctoral dissertation, Univ. of Chicago.

DeVore, Irven and Richard B. Lee In press. "Recent and current field studies of primates." *Folia Primatologia.*

DeVore, Irven and S. L. Washburn In Press. "Baboon ecology and human evolution." In *African Ecology and Human Evolution,* F. C. Howell, ed. Chicago: Aldine.

Haddow, A. J. 1952. "Field and laboratory studies on an African monkey, *Cercopithecus ascanius schmidti* Matchie." *Proc. Zool. Soc. Lond.,* 122, II:297-394.

Hall, K. R. L. 1960. "Social vigilance behaviour of the chacma baboon, *Papio ursinus." Behaviour,* 16(3-4):261-294.

1961. "Feeding habits of the chacma baboon." *Advancement Sci.,* 17(70):559-567.

1962. "The sexual, agonistic and derived social behavior patterns of the wild chacma baboon, *Papio ursinus." Proc. Zool. Soc. Lond.,* 139:283-327.

Imanishi, Kinji 1960. "Social organization of subhuman primates in their natural habitat." *Current Anthropology,* 1(5-6):393-407.

Itani, Junichiro 1954. Japanese monkeys in Takasakiyama. Tokyo, Kobun-sha. (in Japanese).

Jay, Phyllis Unpub. Ms. "The social behavior of the langur monkey."

Kaufmann, John H. 1962. "Ecology and social behavior of the coati, *Nasua narica,* on Barro Colorado Island Panama." *Univ. Calif. Pub. in Zool.,* 60:95-222.

Kortlandt, Adriaan 1962. "Chimpanzees in the wild." *Sci. Amer.*, 206:-128-138.

Kummer, Hans and Fred Kurt In press. "Social units of a free-living population of hamadryas baboons." *Folia Primatologia.*

Marler, Peter 1957. "Specific distinctiveness in the communication signals of birds." *Behaviour*, XI, 1.
1960. "Bird songs and mate selection." *Animal Sounds and Communication, Amer. Institute of Bio. Sci.*, 7.

Murie, Adolph 1944. "The wolves of Mt. McKinley." *Fauna Nat. Parks U.S.*, 5.

Napier, John 1962. "Monkeys and their habitats." *New Scientist* 15:88-92.

Nolte, A. 1955. "Observations of the behavior of free ranging *Macaca radiata* in southern India." *Zeitschrift für Tierpsychologie*, II:77-87.

Pitelka, Frank A. 1959. "Numbers, breeding schedule, and territoriality in pectoral sandpipers in northern Alaska." *The Condor*, 61(4):233-264.

Radcliffe-Brown, A. R. 1930. "Former numbers and distribution of the Australian aborigines." *Official Year Book of the Commonwealth of Australia*, 23:671-696.

Roe, Anne and George Gaylord Simpson (eds.) 1958. *Behavior and evolution.* New Haven: Yale Univ. Press.

Schaller, George 1961. "The orang-utan in Sarawak." *Zoologica, N. Y. Zool. Soc.*, 46:2.

Schaller, George and John T. Emlen, Jr. In press. "The ecology and social behavior of the mountain gorilla with implications for hominid origins." In *African Ecology and Human Evolution*, F. C. Howell, ed. Chicago: Aldine.

Shapera, I. 1930. *The Khoisan peoples of South Africa.* London.

Southwick, Charles H., Mirza Azhar Beg and M. Rafiq Siddiqi 1961a. "A population survey of rhesus monkeys in villages, towns, and temples of northern India." *Ecology*, 42:538-547.
1961b. "A population survey of rhesus monkeys in northern India: II. transportation routes and forests areas." *Ecology*, 42:698-710.

Stark, D. and H. Frick 1958. "Beobachtungen an äthiopischen Primaten." *Zoologische Jahrbücher*, 86:41-70.

Stebler, A. M. 1944. "The status of the wolf in Michigan." *J. Mammal.*, 25:37-43.

Steward, Julian 1936. "The economic and social basis of primitive bands." *Essays in Honor of A. L. Kroeber*, Berkeley: Univ. of California Press.

Tappen, N. C. 1960. "Problems of distribution and adaptation of the African monkeys." *Current Anthropology*, 1:91-120.

Ullrich, von Wolfgang 1961. "Zur Biologie und Soziologie der Colobussaffen." *Der Zoologische Garten*, 25:305-368.

Washburn, S. L. and Virginia Avis 1958. "Evolution of human behavior." See Roe and Simpson, 421-436.

Washburn, S. L. and Irven DeVore 1961a. "The social life of baboons." *Sci. Amer.* 204:6.

Washburn, S. L. and Irven DeVore 1961b. "Social behavior of baboons and early man." *In Social life of early man,* S. L. Washburn, ed. Viking Fund Pub. in Anthropology, 31:91-104.

Wright, Bruce S. 1960. "Predation on big game in East Africa." *J. Wild-life Management,* 24:1-15.

Zuckerman, S. 1932. *Social life of monkeys and apes.* London: Kegan Paul.

18

Evolution of Primate Societies

J. H. CROOK AND J. S. GARTLAN

Research on primate behaviour has tended to concentrate on similarities rather than differences between taxa, thereby allowing authors to make generalizations concerning the evolution of primate and human societies based on the characteristics of single species.[1] Recent developments in field investigation,[2] however, now provide a range of information clearly demonstrating the contrasts that exist even between species the behaviour and ecology of which appeared superficially alike previously. The number of species sampled remains small but enough is known to permit a preliminary examination of the range of social systems uncovered and their functional significance in the habitats concerned. Analysis of function provides some understanding of the selection pressures responsible for adaptive change in social structure, and, given information regarding relevant ecological change in past time, allows the construction of evolutionary hypotheses.

The present approach is based first on the ethological and ecological results of recent field investigations;[3] secondly, on a methodology of evolutionary analysis of social systems used in recent ornithological work;[4,5] and, thirdly, on recent research in the palaeoecology and palaeontology of African primates.[6] Our purpose is to direct attention to patterns of relationships inherent in the data now available and to begin the task of ordering it. The hypothesis put forward gains strength from two aspects not treated here,

Reprinted from *Nature, 210:*1200-1203 (1966) by permission of Macmillan Journals Limited.

Table 1
Adaptive Grades of Primates (see text)

Species, ecological and behavioural characteristics	Grade I	Grade II	Grade III	Grade IV	Grade V
Species	Microcebus sp. Chirogaleus sp. Phaner sp. Daubentonia sp. Lepilemur Galago Aotus trivirgatus	Hapelemur griseus Indri Propithecus sp. Avahi Lemur sp. Callicebus moloch Hylobates sp.	Lemur macaca Alouatta palliata Saimiri sciureus Colobus sp. Cercopithecus ascanius Gorilla	Macaca mulatta, etc. Presbytis entellus Cercopithecus aethiops Papio cynocephalus Pan satyrus	Erythrocebus patas Papio hamadryas Theropithecus gelada
Habitat	Forest	Forest	Forest-Forest fringe	Forest fringe, tree savannah	Grassland or arid savannah
Diet	Mostly insects	Fruit or leaves	Fruit or fruit and leaves. Stems, etc.	Vegetarian-omnivore Occasionally carnivorous in Papio and Pan	Vegetarian-omnivore P. hamadryas occasionally also carnivorous
Diurnal activity	Nocturnal	Crepuscular or diurnal	Diurnal	Diurnal	Diurnal
Size of groups	Usually solitary	Very small groups	Small to occasionally large parties	Medium to large groups. Pan groups inconstant in size	Medium to large groups, variable size in T. gelada and probably P. hamadryas
Reproductive units	Pairs were known	Small family parties based on single male	Multi-male groups	Multi-male groups	One-male groups

Table 1 (continued)

Species, ecological and behavioural characteristics	Grade I	Grade II	Grade III	Grade IV	Grade V
Male motility between groups	—	Probably slight	Yes—where known	Yes in *M. fuscata* and *C. aethiops*, otherwise not observed	Not observed
Sex dimorphism social role differentiation	Slight	Slight	Slight—Size and behavioural dimorphism marked in *Gorilla*, Colour contrasts in *Lemur*	Marked dimorphism and role differentiation in *Papio* and *Macaca*	Marked dimorphism Social role differentiation
Population dispersion	Limited information suggests territories	Territories with display marking, etc.	Territories known in *Aloutta*, *Lemur*. Home ranges in *Gorilla* with some group avoidance probable	Territories with display in *C. aethiops*. Home ranges with avoidance or group combat in others. Extensive group mixing in *Pan*	Home ranges in *E. patas* *P. hamadryas* and *T. gelada* show much congregation in feeding and sleeping. *T. gelada* in poor feeding conditions shows group dispersal

namely cases of parallel adaptive radiation among African, Malagasy and Asian primates and intra-specific adaptations to local differences within habitats of single species. Detailed field investigations will, in due course, test the ideas suggested.

GRADES OF SOCIAL ORGANIZATION

Ornithological investigations have shown that patterns of social organization determining population dispersion are intimately linked to species ecology—the whole interrelationship of behavioural features being co-adapted to certain aspects of the environment forming major selection pressures. In Table 1 we attempt to allocate species recently investigated to a series of "Grades"[7] representing "levels" of adaptation in forest, tree savannah, grassland and arid environments respectively. Anatomical investigations of fossil and living material reveal a progressive adaptive radiation from forest-dwelling insectivorous primates to larger open country animals predominantly vegetarian. It is not surprising, therefore, to find correlated trends in the behavioural data.[8] Table 1 reveals a shift from insectivorous, nocturnal forest animals with markedly solitary habits and a population dispersion based on aggressive contacts through a range of fruit- or leaf-eating forms (diurnal; very small family groups or larger parties showing defensive behaviour of a variety of types in "territorial" encounters with neighbours but, otherwise, little intra-group aggression, sexuality or inter-male competition) to vegetarian browsers of open country normally living in well-structured troops or herds, usually in home ranges and showing much sexuality (often seasonal) and inter-male competition.

The aboreal grades contain several species while fewer are allocated to those of open country. In Table 1 this reflects a large species contingent from Petter's investigations on the lemurs;[9] however, in general, the numbers of forest species do exceed those of open country forms. This is because of the availability of numerous niches in forest where primate populations have specialized in terms of diet, vertical zonation, daily rhythms, etc. The limited ranges of the population units and their confinement to forests means that geographical barriers impose a more extensive speciation than in open country where species show great adaptability in exploiting a considerable range of habitats.

In common with their insectivorous ancestors, the co-adaptations of Grade I primates are clearly related to their nocturnal insect-hunting habits. Two leaf-eating forms are exceptions requiring further explanation. The switch to frugivorous or leaf-eating habits in Grades II and III is linked with diurnal activity and the formation of larger social groups. As with many birds, this development is linked to the change from a diet requiring individual hunting to food sources often locally distributed and at which social responses allow congregation for exploitation in common. "Family bands" of very small size and larger multi-male social units in forest are often markedly "territorial," in that defensive behaviour involving displays and-/or marking in relation to neighbouring groups is shown. While the behaviour may defend a discrete area, this has not often been demonstrated; nevertheless, it does ensure an over-dispersion of population units.

The small size of social units in many forest frugivores is probably related to limiting conditions of food supply occasioned by the relatively stable conditions of tropical rain forests. A non-seasonal climate with a moderately constant availability of various fruits presumably allows increase in numbers to a ceiling imposed by periodic food shortages due to local food crop failures. In this situation the addition of young to the population may be difficult—recruitment necessarily balancing mortality. Breeding appears to be non-seasonal and this may explain the low frequency of copulations observed in some of these species.[10,11] The "territorial" behaviour of forest groups may be interpreted as ensuring an adequate provisioning area for the individuals comprising them. It remains difficult, however, to apply this argument to all leaf-eaters.

In the dry forest, savannah and steppe conditions of Grades IV and V marked climatic seasonality with a harsh period of aridity, and food shortage probably imposes high seasonal mortality rates especially on old or infirm animals. In the rainy season, food for the remainder is superabundant and vigorous seasonal breeding[12] can replenish the population without risks of failures from food shortage in rearing young. It is as yet unclear whether the timing of the breeding seasons confers advantages during pregnancy, lactation or in terms of food supply for young animals; however, in principle, the effects with regard to survival may be the same.

Outside the forest the social units are generally larger and this appears to result from open country conditions of predation and food supply affecting ground dwelling populations. Savannah primates face a number of predators and scattering undoubtedly decreases the chances of individual survival. Groups of *Papio* baboons and *Rhesus* monkeys show a marked cohesion permitted ecologically by local abundance allowing congregation without risk of over-exploitation.

One of the effects of increased group size in open country terrestrial primates is marked competition between males for females and the consequent intra-sexual selection of male characteristics. The open nature of the terrain allows every troop member to be aware of its companion's activities, and the cohesion of most troops makes this doubly sure. The seasonality of mating activities further means that many males are particularly active sexually at the same time, when, in the absence of structural equilibration in the troop, firece competition and fighting would result. Again some Japanese observations[13] suggest that with increased numbers the socionomic sex ratio of *Macaca fuscata* troops increases so that there are fewer females to be shared among the males. This, they observe, may be a factor occasioning the splitting off of small groups from a troop and the exclusion of "all male" parties. Finally there is a possibility that female "open-country" primates may be sexually attractive for longer periods per oestrus cycle than is the case for forest animals.[14] All these factors enhance intra-sexual selection producing sexual dimorphism in size, appearance, and also in behaviour. The marked aggressiveness of male baboons and rhesus monkeys results in differential access to females, and is correlated with the promiscuous behaviour of oestrus females prior to their mating exclusively with the troop despot at about the time of greatest receptivity.

The increased size and aggressive nature of these males are considered to have been pre-adaptive to their role in troop defence. Males disperse around the periphery of moving troops and co-operate against common danger (see, for example, refs. 15 and 16). We do not think size and aggressiveness originated directly as a response to predation (compare with refs. 17 and 18), but the organization of the troop and male co-operative tendencies in defence undoubtedly did so.

Within Grade IV we have included not only typical forest fringe and savannah woodland species such as *Papio cynocephalus* and *Macaca* sp. but also a number of forest forms, which, in secondary habitats provided by human destruction of woodlands, show social organization in some degree similar to these. Characteristic of all the members of this Grade is a marked adaptability revealed in investigations of the species in contrasting habitats. The *Cercopithecus aethiops* population on the rich but small Lolui Island has 'territorial' behaviour not apparent in unlimited but impoverished conditions where the animals have much larger home ranges.[19] Similar, though often less investigated, accounts are available for the langur (*Presbytis entellus*), chimpanzee and baboon. These variations in social systems appear owing to the plasticity of intra-group relations possible in these forms—a plasticity of undoubted survival value in forest-fringe environments.

Grade IV systems seem therefore to be the result of (i) open country conditions of food supply and predation favouring increase in group size, and (ii) intra-sex selection increasing size and aggressiveness of males, thereby producing (iii) marked group structuring and, in some forms, the protective functions of males acting in unison against predators. Changes from Grade IV to Grade V are related to the occupation of habitats in which food supplies are less abundant, and, at least seasonally, more sparsely and infrequently distributed in the environment. In such areas it would be advantageous for groups of Grade IV animals to fragment into smaller, more widely ranging parties to avoid local over-exploitation of food. Indeed there is evidence from *Papio cynocephalus* populations in arid African areas and from *Macaca fuscata* in northern Japan that group size decreases at the limit of the species geographical range. This is probably in part the result of a low population density, their group organization remaining typical of Grade IV. Small groups of this kind are, however, not very efficient population units. The presence of several large males, only functional in mating and playing no part in rearing young, results in the consumption of much food not used in maintaining the species. Furthermore, the role of the male in protection is less efficient in small groups. In these habitats the "one-male groups"[20] of *Papio hamadryas, Erythrocebus patas* and *Theropithecus gelada* are more adaptive in that less food per reproductive unit goes to individuals not involved in rearing young. The exclusive possession of a "harem" probably increases inter-male competition for females still further and the occurrence of "all male" population units indicates a considerable degree of exclusion of potential reproductives from breeding. The intense intra-sexual selection is again doubtless the main factor increasing the dramatic appearance of *P. hamadryas* and *T. gelada* males through the growth of large capes on the backs of these animals[21]—an apparent increase in size not requiring much utilization of energy for its maintenance and providing the animals with a distinctly ferocious appearance enhanced by demeanour and facial expressions.

The differently co-adapted characteristics of the three Group V species suggest clearly the interaction of selection pressures from food shortage, predation and habitat topography in slightly different environments. Thus *T. gelada* in the Ethiopian mountains forms herds in good feeding conditions that split into separately foraging harem groups as conditions deteriorate.[16] Sleeping sites on ever-present cliffs are super-abundant. *P. hamadryas,* although forming "one-male groups," apparently shows less dispersal because of the limited number of sleeping sites available (rock cliffs, outcrops, etc.) in the single area so far investigated in detail.[20] The patas monkey of open savannah grasslands never forms herds, individual "one-male groups" keeping apart.[22] There are no rocky sleeping sites and plenty of predators. At night the animals disperse into separate trees and re-assemble in the morning. The male does not adopt the aggressive defence tactics of *Papio* and *Theropithecus;* rather he plays the part of watchdog, showing alert and diversionary behaviour in relation to predators. He does not exert dominance over the group in the same way as the other two species.

The behavioural features of the Grade V species are thus maintained by three main environmental selection pressures the characteristics and range of lability of which are relatively constant within the three habitats. They contrast markedly with the differing values of these same pressures, maintaining the Grade IV organizations in a richer habitat. Signal flow diagrams relating the environmental input of information and the adaptive response of the population in terms of numerical size, population dispersion and social structure are prepared for each of these types. A generalized diagram to which the characteristics of any one of these systems may be fitted is shown in Fig. 1.

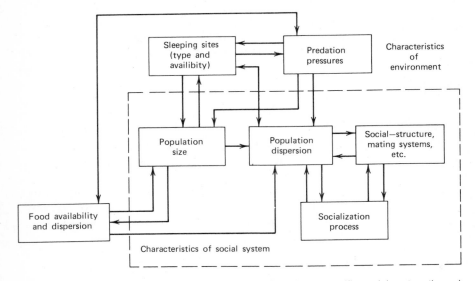

Fig. 1. Diagram showing the role of habitat factors maintaining a specific social system through time in a given environment.

THE GREAT APES AND MAN

Clearly in Table 1 the Pongidae do not fit easily into categories imposed primarily by investigations of cercopithecid primates. Nevertheless, *Hylobates,* in spite of its adoption of brachiation, has much in common with other tree-top frugivores. Similarities between *Gorilla* and *Pan* to other animals in their Grades are much less clear—a function in part doubtless of their pongid status. Reynolds[23] has clarified the adaptive significance of contrasts in social systems between these two genera.

The development by *Pan* of a simple tool-using culture[24,25] suggests a social condition close to that of early hominids, but we do not accept the suggestion[26] that this necessarily implies a regression from a more human condition. Palaeontological investigations suggest a radiation of early hominids from dryopithecine stocks in circumstances similar to those controlling the transitions from Grade IV to V in the cercopithecids. While *Ramapithicus* and *Paranthropus* were herbivores,[27] *Australopithecus* shows evidence from dentition of a more omnivorous food habit including meat eating. Early hominids faced with seasonal food shortage in savannahs probably had initially a social organization not unlike that of the chimpanzee and perhaps formed later either groups guarded by weapon-bearing males or moved in one-male family units—the males playing a part of guardian and watchdog. Possession of weapons would have substituted for gross sexual dimorphism in size. The creatures doubtless retained the ability to congregate in conditions of plenty and in sparsely distributed rocks or tree shelters. These would soon have become bases for hunting raids. Shortage of vegetable food would have accelerated the adoption of weapons for killing animal prey while the trend towards bipedalism, already providing speed in escape from danger, allowed the use of arms in wielding weapons and speed in the chase. The fact that with *Pan* "all-male groups" occur suggests a source for the loyalty and co-operation necessary in the development of group hunting expeditions. Certainly the biomass of animals suitable for hunting was abundantly present in the grassland savannah of Pleistocene Africa,[28] but probably only in *Australopithecus* did an increased reliance on meat lead to major changes in social structure.[29]

SEXUAL SELECTION

The form of a primate social system is a function not only of selection pressures of the environment. As we have seen, sexual selection, especially in terrestrial societies, plays an important part. Wickler[30] has recently stressed this factor as an ultimate determinant of morphological signals functional in the sexual and agonistic behaviour of certain primates. It would seem that the structures concerned have been derived from skin colour changes occurring in certain phases of the oestrus cycle and for which no signal function is known.[31] Wickler suggests that the coloured buttocks normally present in certain species function in protecting the less dominant males from the despots during quarrels in the troops. Weaker or junior males "present" to large animals thereby stimulating a mounting gesture rather than aggression. He considers this to be a case of "intra-specific mimicry," the male buttocks being copies of the feminine original. Again the curiously shaped patch of bare skin on the chest of *Theropithecus gelada* in both sexes, which undergoes cyclical changes including peripheral vesicle formation in the female,[32] has a resemblance in pattern and colour to the perineal skin of the female at oestrus. Wickler considers the chest patches to be substitutes for signals given posteriorly in presentation for mating and he implies, though he does not state, that the chest patches of the males function in a similar way as the coloured buttocks of other male baboons.

Recent investigations (by J. H. C.) in Ethiopia have shown that males do indeed examine the chests of females in oestrus prior to certain copulations, probably those occurring early in an oestrus cycle. Quantitative data show, however, that the colour changes on chest and ischial areas are not in phase. Lactating mothers and non-oestrus females have the reddest behinds. It seems odd that if the red chest colour is attractive to males the perineal areas should be most red at precisely the time when mating would be inappropriate. Long nipples in the mid-line of the chest, assumed by Wickler to be signal imitations of the vulva, appear to arise in the course of suckling. Preparous females have small nipples spaced apart, but, when in oestrus, are nevertheless frequently mated. The role of nipple shape and position in sexual signalling prior to copulation is thus in doubt.

The gelada spends a considerably greater time sitting down than does *Papio doguera* in the same area. A chest signal as an indicator of female sexual condition would therefore be more appropriate than an ischial one. It does not appear essential, however, for the signal to be an exact copy of the ischial design — and indeed we find that it is not. In view of the baby and juvenile geladas' prolonged interest in the nipples — with which they play as well as from which they obtain nourishment — adult concern with this area is readily understandable and any colour changes are likely to develop a significance. The extent to which chest patches really copy the behind remains largely subjective. There is no evidence that the chest patch of the male gelada functions in a manner analogous to Wickler's suggestions for the buttocks in males of other species.

Wickler's hypothesis deserves critical scrutiny as it could have considerable significance in the investigation of human sexual aesthetics. The shape of the human mammary glands is not apparently "essential" for it is not found in the great apes. Possibly they have acquired their function as a sexual releaser in a manner analogous to the evolutionary process suggested for the female gelada's chest. The development could have arisen in co-adaptation with the adoption of the upright bipedal stance in the early Hominidae and the use of the frontal position in mating. However, it is not so much the shape of the human breasts that is the critical sexual releaser as the visual stimulus provided by the nipple and the areolar zone around it.[33] While "intraspecific mimicry" might account for the shape of these organs it seems that normal Darwinian sexual selection (that is, inter-sexual selection) could account for the enhancement of structural features around the nipples. Further research into the significance and evolution of such signals in both nonhuman primates and man will be of considerable interest.

[1] **Washburn, S., and DeVore, I.,** in *Social Life of Early Man,* edit. by Washburn, S. L., 91 (New York, 1961).

[2] **Hall, K. R. L.,** *Symp. Zool. Soc. Lond.,* 14, 265 (1965).

[3] **DeVore, I.,** (ed.), *Primate Behavior* (New York, 1965).

[4] **Crook, J. H.,** *Behaviour Monograph,* 10 (1964).

[5] **Crook, J. H.,** *Symposium Zool. Soc. Lond.,* 14, 181 (1965).

[6] **Howell, F. C., and Bourlière, F.** (eds.), *African Ecology and Human Evolution* (Chicago, 1963).

[7] **Huxley, J. S.,** *Systematics Assoc. Publ.,* 3, 21 (1959).

[8] **Hill, W. C. OSMAN,** *Proc. Roy. Soc. Edin.,* 66, pt. 1 (No. 5), 94 (1955).

[9] **Petter, J. J.,** *Mem. Mus. Hist. Nat. Paris.,* A, 27, (1), 1 (1962).

[10] **Jay, P.,** in *Primate Behavior* (New York, 1965).

[11] **Schaller, G. B.,** in *Primate Behavior* (New York, 1965).

[12] **Lancaster, J. B., and Lee, R. B.,** in *Primate Behavior* (New York, 1965).

[13] **Itani** *et al., Primates,* 4, No. 3 (1963).

[14] **Chance and Mead, S. E. B.** Symp. VII, 395 (1953).

[15] **Hall, K. R. L., and Devore, I.,** in *Primate Behavior* (New York, 1965).

[16] **Crook, J. H.,** *Symp. Zool. Soc. Lond.* (in the press).

[17] DeVore, I., and Washburn, S. L., *African Ecology and Human Evolution* (Chicago, 1963).

[18] Chance, M. R. A., *Primates,* 4, 1 (1963).

[19] Gartlan, J. S., and Brain, C. K. (in the press).

[20] Kummer, H., and Kurt, F., *Folia. Primat.,* 1, 4 (1963).

[21] Jolly, C. J., *Man,* 63, No. 222 (1963).

[22] Hall, K. R. L., *J. Zool.,* 148, 15 (1966).

[23] Reynolds, V., *American Anthropologist,* 67, 3, 691 (1965).

[24] Goodall, J., *Primate Behavior* (New York, 1965).

[25] Koortlandt, A. (personal communication).

[26] Koortlandt, A., and Koij, M, *Symp. Zool. Soć. Lond.,* 10. 61-88 (1963).

[27] Robinson, J. T. *African Ecology and Human Evolution* (Chicago, 1963).

[28] Bourlière, F., *African Ecology and Human Evolution* (Chicago, 1963).

[29] Crook, J. H., in *Symp. "Biology of Co-operation," Eugenics Review* (in the press).

[30] Wickler, W., *Naturwiss.,* 50, 481 (1963).

[31] Rowell, T. E., *J. Reprod. Fertil.,* 6, 193 (1963).

[32] Harrison Matthews, L., *Trans. Zool. Soc. Lond.,* 28, 7, 543 (1956).

[33] Goodhart, C. B., *New Scientist,* 23, 558 (1964).

19

Dimensions of a Comparative Biology of Primate Groups

HANS KUMMER

Abstract. The possible contributions of a comparative study of primate social organization to anthropology are many. Such a study may elucidate the repertory of motivational and organizational raw materials present among primates; it may show us the forms of society that have evolved from this raw material; it may show us the kinds of inhibitions and functional readjustments of phylogenetically old motivations that lead to the types of societies found among primates. These general aims are based upon analysis of specific primate groups. The analysis of a primate group as a functioning system leads to consideration of the anatomy and physiology of groups, the reasons for differences in types of groups, the ontogeny of a group, environmental modifications of types of groups, and the evolution of the adaptive function of groups.

Eight years have now passed since field studies on primate behavior became the active concern of a group of anthropologists in this country, and the broader anthropological audience may rightly ask what this group has done and where they are heading. In asking a zoologist but not an anthropologist to answer such questions, the organizers of this symposium have obviously not wished for a *pro domo* speech but for a broad discussion in terms of our common biological background. This article, therefore, is not a review of facts but an attempt to familiarize the morphologist and physiologist with the special nature of our task in the field.

Let us first ask why we seek our subjects in their native habitats. Every

Reprinted from *American Journal of Physical Anthropology,* 27:357-366. Copyright © 1967 by The Wistar Institute Press and reproduced by permission.

biologist knows that a living system should be grown and studied in as many different environments as possible. No single environment reveals all the phenotypic modifications of which a genotype is capable. Captive animals may develop adaptive behavior patterns that are never observed in their wild conspecifics, and vice versa. It follows that there is nothing basically superior or sacred about studying animals in their natural habitat. Any controversy about a general superiority of either laboratory or field studies is pointless if we agree that our ultimate aim is not a description of one single modification, but an idea of the full genetic potential of our species.* A particular environment is a good or a bad choice only in relation to the questions we ask.

The biologist interested in primate behavior has two reasons for going to the field. The first is his interest in behavior as a means of survival. Behavior can become adaptive in a particular environment either because its genetic substrate was selected by this or a similar environment or because the environment evokes adaptive modifications from the genetic substrate of the present generation. The highest degree of adaptiveness will generally be found in an environment that shaped both the genotypes and the modifications of the genetic substrate of the animals studied. For a given population, this environment is most likely the one in which it presently occurs.

The second reason for field studies is the need to observe *group* behavior. We may without difficulty buy individuals for our laboratory studies, but as yet no dealer sells us an entire group as he found it in the wild. We may have the courage to assemble the system which we are going to study, but before we do so we should learn as much as possible about these fragile structures in their native habitat. For this second reason, the study of social groups may be considered the typical task of field work on primates. The study of social groups will also be the frame of this discussion.

As a first approximation, a social group can be approached as merely another living system, as a form of compound organism which appeared late in phylogeny and in which the metazoan individual is no longer the whole but a part. The organism analogy at one time played its heuristic role in the study of social insects. We shall use it here because it permits us to start out from familiar biological dimensions. We shall ask the questions that every biologist asks when he faces a new kind of organism, namely those concerning the anatomy, the physiology, the ontogeny, the ecology, and the evolution of the "group organism." In answering each of these questions, we shall arrive at a point where the analogy with the organism no longer holds and where the vertebrate group as a form of life reveals its own characteristics. To the biologist working in other fields, these points will best demonstrate the problems of studies on primate societies.

The primate species of the open country of East and South Africa serve

---

*For a discussion of the relationship of field and laboratory work, see Menzel ('67).

as examples throughout this article. Their social organizations show two basic types of groups in various combinations: the multi-male group, in which several adult males live with a number of females and their offspring, and the one-male group, in which females and young are associated with only one male. The multi-male group is typical of the savanna baboons (*Papio anubis, P. cynocephalus, P. ursinus*). In these groups, each male potentially has access to each female (Hall and DeVore, '65). The patas monkeys (*Erythrocebus patas*) of Uganda represent the other extreme; their organizational units are one-male groups which live far apart from each other (Hall, '65). Between these two extremes there are two types of groups. In *Papio hamadryas*, the "desert baboon" of Ethiopia, a number of one-male groups together form a larger association, the "band," which travels and fights as a unit. The band resembles the multi-male group of the savanna baboons except that each female is the exclusive partner of one and only one male (Kummer, '67). A similar two-level organization is found in the gelada baboons (*Theropithecus gelada*) of the Ethiopian highlands. But the gelada one-male groups are more independent of each other; they separate when food is scarce; and, although they join to form large troops, there is no evidence that several one-male groups form a stable well-organized association with the pattern of the hamadryas band (Crook, '66).

GROUP ANATOMY

Like an amoeba, a group constantly changes its shape in space, but certain arrangements of the members of the group occur more frequently than others. Some of them hit the eye immediately. In baboons and macaques, females tend to hold the center of the group with a few dominant males, whereas younger males are more peripheral. This general arrangement, however, can be modified according to the situation. In a troop of geladas walking along the edge of a vertical cliff, the females walk close to the edge, shielded only on the farther side by a belt of large males. When two one-male groups of hamadryas baboons forage in the savanna, their females will form a line between the males. But if the two males start to threaten each other, each set of females swings outward to line up behind its respective male and away from the other group. The obvious advantage of flexible anatomy is that the group can cope with certain situations simply by altering its shape.

The exact quantitative analysis of a group's spacing pattern is rather difficult. My own preliminary attempts have failed because the animals under experimental conditions did not keep one particular arrangement for more than seconds or minutes. Maps of spatial arrangements of hamadryas baboons in the field nevertheless showed that certain simple characteristics survive most changes in formation. For example, the animals of a particular sex-age class are usually surrounded by neighbors of a typical sex-age distribution, and this reveals much about the affinities between the classes.

Although we know so little yet about the spatial dimensions of groups, the question "What is a group?" is in practice answered most often and with reasonable success on the basis of a spatial criterion. In general, a spatial aggregation of primates that travel and rest together and at the same time avoid the proximity of other such aggregations is, at closer inspection, also a functional and reproductive unit.

GROUP PHYSIOLOGY

The members of a group, like the cells of an organism, exchange signals that affect the activity as well as the development of the members that receive them. The effect is a more or less coordinated activity of the group as a whole. From what has been said about the loose spatial structure of the group, it is obvious that group physiology faces a particular task in merely maintaining to group's identity within the population. The individual primate, unlike the metazoan cell, can physically leave the body of the group and survive on his own, at least for some time. In addition, since he is not biochemically earmarked as a stranger, he can enter another group and become its member. Why, then, are certain primate groups as stable in membership as we have found them in the field? What is the "immunological" process by which the stranger is recognized and rejected? Here again, we know next to nothing. One might hypothesize that the members of each group share certain behavioral or morphological traits by which they are recognized, and this may most likely be true in species where the groups respond to each other with "territorial" calls. The alternative is that primates know the members of their group as individuals and recognize a stranger by exclusion, as an unfamiliar individual.

Beyond the necessity of differentiating between strangers and group members, the closed group must have mechanisms of actual inclusion and exclusion if it is to maintain its integrity. Some members of the same species must attract or tolerate each other while avoiding or chasing away others. Groups of savanna baboons avoid each other, and occasional strangers lingering about the group are chased away, although with persistence they may finally enter the group. A patas male spotting another male near his own group will simply chase the stranger away. The integrity of the group is in these cases maintained by the use of distance as an isolating agent. This widespread and relatively primitive technique, however, is obviously replaced by others in species such as the gelada or the hamadryas, where several small groups join to form larger social units without losing their identity. Although the one-male groups of a hamadryas band travel and rest so close together that it can take hours to tell the groups apart, the members of each one-male group mate and groom only among themselves. The spatial proximity of these groups does not destroy their isolation where social interactions are involved. In hamadryas, the mechanisms that maintain this segregation are not the same for females and males. By themselves, the females do not refrain from interacting with other band members, and they actually manage to do so when their male is not watching. Their isolation is imposed on them by the male, who threatens and even bites them when they move too far away from him or when they try to mate or groom with members of other one-male groups. The male hamadryas, in contrast, refrains from mating or grooming with outsiders for reasons not apparent to an external observer. Even a male to whom other animals submit in other contexts will not interact with the females of subordinate males. Of the nature of this inhibition we know nothing except that it is probably limited to the members of the same band: one-male groups artificially transplanted into a distant area are attacked by the local males, and their females are distributed among the resident males without any fighting among the new "owners." Such experimental transplantations may soon teach us more about the mechanisms of exclusion and the permeability of group boundaries.

A closed primate group not only maintains its composition but also travels as a closed body, in a coordinated fashion. How do its members determine the direction of their common travel? In a gorilla group, the members attend to and follow their leader, the single silver-backed male (Schaller, '63). But in a large hamadryas troop ready for departure from a resting place there are several such old males, and the younger males, in addition, reveal clear directional intentions of their own. The ultimate direction is determined by a process in which younger males walk away from the troop in the direction of their choice, thus forming troop "pseudopods" in various directions. The troop, however, does not follow them until an older male accepts one of the proposed pseudopods. At this signal, the younger males in the pseudopod start out in the indicated direction, and the troop follows with the older males in the rear. In this case, the functioning order reveals two male "roles": the role of initiative, typical of younger males, and the role of decision, taken in turn by one of the older males of the troop.

Functioning orders of vertebrate groups are most frequently described by the term "dominance." In its classical definition, a dominance order is the relatively stable sequence in which the animals of an established groups have access to an "incentive," such as a piece of food or a desired social partner. The emphasis that this concept puts on the competitive aspect of group life betrays its origin, the laboratory. In the captive group, competition is indeed a prominent feature of social order, more so, it seems, than in the wild. Field studies show that group function is based on a number of additional orders, which have little to do with sequences of access. Drawing again on the hamadryas data, we find that the number of females belonging to a male (a classical criterion of dominance) is not correlated with his influence on the troop's travel direction: younger males have more females than older males, but it is the older males that determine where the troop will go. The dominance concept, often used so broadly as to be ambiguous

and misleading, must be complemented by other concepts. Among the
social "roles" (*cf.* Bernstein and Sharpe, '66), which may or may not be
correlated with classical dominance, the following may be mentioned:

1. *Leadership* can be provisionally defined as the probability with
    which an animal's movements in space are taken over by others.
    The hamadryas example shows that leadership must not be cor-
    related with classical dominance.
2. The role of *protection* may or may not rest with the dominant ani-
    mal. The spatial arrangement of baboon troops suggests that the
    less dominant males on the troop's periphery may be more active
    in encounters with humans and predators.
3. An animal may be so persistent in asserting his *exclusive access* to a
    particular partner that the latter becomes his social "property."
    The hamadryas male achieves this by interfering with his females'
    interactions with all other males. In a comparable way, baboon
    mothers are more possessive about their infants than are the moth-
    ers of langurs. Logically, this exclusiveness is the extreme expres-
    sion of dominance. However, since even a mother or a male of
    otherwise low dominance status can establish and assert the exclu-
    sive access to a particular partner, the phenomenon of exclusive-
    ness must be distinguished from the dominance concept and be
    investigated as a separate aspect of a group's functioning order.

In short, a dominant primate may be the leader or protector of his
group as well as the exclusive owner of his females, but correlations between
roles cannot be taken for granted. On the other hand, classical dominance
can be correlated with apparently unrelated behavior traits. In a zoo colony
of hamadryas baboons, the four adult females stirred during sleep at night
with frequencies that conformed exactly with their dominance order. The
most dominant female stirred most rarely, the most subordinate most frequently (Kummer, '56). When the dominance order was rearranged, the
frequencies of sleeping movements changes accordingly.

In all these aspects of a group's functioning order, a basic contrast to
the functioning order of an organism is apparent. In a developing metazoan
organism, the parts gradually differentiate as to structure and function, and
at a certain stage of development this differentiation of a part becomes
irreversible. The functional role of a part is then determined for the rest of
the organism's life. In contrast, the only roles in a primate group that are
irrevocably prescribed by morphological and functional differentiation are
those of the two sexes. Within the sexes, the roles of the dominant animal,
of the leader, or of the peripheral defender are not the permanent roles of
clearly differentiated castes. The individual qualities of a dominant animal,
for example, occur on a continuous scale throughout a group, and the actual
role taken by an individual therefore depends on where his qualities place

him in his particular group. The low-ranking male of this year can become the high-ranking male of next year since the relevant qualities change with age; changing alliances in the group add to the lability of role distribution. Even the roles of the two sexes are not exclusively differentiated. One female in a captive hamadryas colony showed complete male herding behavior toward the other females of her group, and the most dominant female of each of our captive gelada groups adopted male roles toward the rest of each group's females. In comparison with the tissues of an organism, the members of a vertebrate group are so similarly equipped that the distribution of roles is never quite stable, and distribution is often achieved with difficulty. In the case of dominance, the ultimate decision may only be reached by overt, aggressive competition. It is this competition for roles among similarly equipped parts in which the functioning order of the vertebrate group most clearly differs from the physiology of organisms. Human history has seen many attempts to abandon this competitive order and violent struggles to reestablish it.

CAUSES OF DIFFERENCES IN GROUP TYPE

In this section we are concerned only with the immediate causes of group organization. The question is not: why did this species evolve its present social organization? Instead, we ask, of what kind are the *immediate* motivational or behavioral causes which, in a given generation, bring about monogamous groups in one species, polygynous groups in another, and multimale groups in still another?

Our first attempt to answer these questions consists of a close look at the social behavior of the individual group member. It has become a tradition, inherited from ethology, that a premilinary field study should include a catalogue of the gestural and vocal signals exchanged by the individuals of the respective species. In studying the signals used among the members of a species we may have hoped to discover some key signals that have a direct effect on the composition of the group. If we had any such hopes, they have failed so far. From the very similar behavioral catalogues of *Papio anubis* and *P. hamadryas* nobody could predict that anubis baboons live in basically promiscuous groups, whereas hamadryas are organized in bands composed of polygynous one-male groups. The behavior that one would expect to be an exclusive hamadryas pattern is the male's particular bite on the nape of the female's neck. This bite brings straying females back to their male and thus is the enforcer of group cohesion. But even this pattern is common to both species. The anubis baboons simply do not use it to herd females. Our captive gelada males also bit their females on the neck, but without relating this action to the female's being too far away. On the other hand, the behavioral repertoires of patas monkeys and geladas are very different, although they both are organized in one-male groups. The field studies to date suggest that the patterns of social *behavior* of a species are related to the taxonomic position of the species. Social *organizations,* however, seem to appear here and there in the order Primates without apparent relationship to taxonomy or to patterns of social behavior. Improbable as it sounds, social behavior as we have so far described it cannot be the cause of differences in social organization. We obviously must separate the two phenomena more clearly in our research than we have done so far. We must also expect to find that apparently similar, convergent organizations in distant species have different causes.

One reason for our failure to explain various types of social organizations is obvious. In the field as in the laboratory, social behavior has almost exclusively been studied on dyads, i.e., on interactions of only two animals. But to study the exclusion of a stranger from a group or the mechanisms responsible for the formation of a monogamous pair, one has to analyze the interactions of at least three animals. If this is done, the causes of differences in type of group may be found to be different affinities among sex-age classes and active interferences of one class with the interactions among two other classes. A comparison of the species forming one-male groups may give us a glimpse of the kind of causal factors we may find. The patas monkeys of Uganda and a population of Hanuman langurs near Dharwar in India (Sugiyama, '66) both live in groups of one adult male and a number of females. Apparently, the adult males of these populations do not tolerate one another in the same group. Surprisingly, however, these populations also include groups of several males without females. This suggests that there is an attraction among the males but that it breaks down in the presence of females. An observation by Sugiyama confirms this hypothesis for the langur. He has sometimes seen an all-male langur group that cooperates in expelling the single male of a one-male group and takes over its females. Immediately after that, however, the new males fight each other until one of them has expelled all the others and becomes the sole leader of the female group. In contrast with these species, the one-male groups of hamadryas and gelada baboons live together in large troops. Here, the attraction among the males seems to outweigh the disruptive effect of the females, but this effect is nevertheless apparent: hamadryas males who have no females often sit close to each other and groom each other, but those who have females keep farther apart and are never seen to groom each other. Our unpublished experiments on captive geladas have shown that the bond between two males regresses to an earlier, more hostile stage as soon as females are added: grooming disappears and aggressive behavior is resumed. Observations of the four species mentioned suggest that there are at least two causal factors determining their social organization: the degree of attraction among adult males and the degree of its reduction in the presence of females. The organizational differences between langurs, patas, geladas, and hamadryas may be an effect of different ratios of these two factors. This interpretation admittedly lacks nothing in crudeness, but it can be experimentally tested, and it seems at least to aim at the kind of phenomena that will ultimately lead to an understanding of the causes of social organizations.

GROUP ONTOGENY

We would certainly understand more about what causes various kinds of groups if we could study their development. But here, unfortunately, the organism analogy seems to fail right away. The ontogeny of an organism is the result of an ordered sequence of gene actions which initiate developmental steps and create temporary states in which the organism is especially sensitive to certain developmental stimuli. In contrast, groups appear to have no distinct life cyle; they do not go through an irreversible ageing process. Potentially the group's capacity to regenerate appears unlimited.

There is, however, a sort of "false exception." A group may be organized around one key animal whose death it cannot survive. In the extreme case, the composition and organization of the group may be so strongly affected by the key individual that the group itself assumes a life cycle, which is the effect of the behavioral ontogeny of that one critical member.

The hamadryas one-male group presents such a case. Subadult hamadryas males have a tendency to "kidnap" and mother infants and young juveniles temporarily. Single infants and juveniles that we released into hamadryas troops were adopted by single, young adult males, who behaved as typical hamadryas mothers. Left to themselves, such young adults will eventually adopt a one-year-old female of their own troop. Thus, the typical one-male group of a young adult male consists of himself and one sexually immature female. There is no overt sexual behavior in these initial groups; instead, the male handles and carries his females as if he were her mother. A number of intermediate stages between this immature form and the groups of old males suggests that the young female eventually becomes the male's sexual consort. Toward the male's early prime, the number of his females increases, for he probably adopts one or two more juvenile females and takes over adult females from older males. During early prime, the male's herding behavior is most intense, and the group's spatial cohesion is strongest. As the male approaches his late prime, however, he becomes more tolerant. His females may now stray away from him as far as 40 meters, and even then the male is unlikely to threaten them. Accordingly, the group of the ageing male decreases in size, but at the same time he becomes more influential in matters of band and troops movements. This influence can still be found in old males who have lost or given up all their females.

This development shows first that certain roles in a primate group may be specifically attributed to certain age classes, even within the time span of adulthood. Secondly, it suggests a hypothesis about the causation of the hamadryas social system: the adult hamadryas female remains a life-long follower of a particular male *because* she was so early in life adopted, mothered, and restricted by a single adult male. Whereas her mother would soon have released her from control, the male who adopted her will keep "maternal" control over her until she is fully adult. Thus, the hamadryas female never develops the independent social habits of female anubis baboons.

A strong "parental" motivation of the males may be only one way by which a species develops a system of one-male groups. Neither Crook ('66) nor Hall ('65) has found evidence of the hamadryas solution in their gelada and patas populations. Since one-male groups appeared independently in one species of each of four genera, it would not be a surprise if they arrived at their similar organizations by different pathways and by exploitation of different behavioral and motivational raw materials. These one-male units would then be typical examples of convergence.

ENVIRONMENTAL MODIFICATIONS OF GROUP STRUCTURE

When Zuckerman ('32) formulated his theory of primate social organization, he appeared to assume that all higher primates were organized in the same basic grouping pattern. Later, Carpenter's ('64) work and the field studies of the last decade suggested that each primate species has its own organization. Only the most recent studies, in which a species was observed in various parts of its distributional range, have shown that there is also a considerable intraspecific variability of social life (e.g., Gartlan and Brain, in press). Unless we assume that these intraspecific variations are genetical, we may interpret them as modifications induced by the particular physical environments. The issue could be settled experimentally by transferring random samples of the same population into different habitats where they could then be studied for at least two generations. The rhesus population introduced by Carpenter and other workers to several islands off Puerto Rico offers research opportunities in this direction, although the value must decrease with every generation that is subjected to the particular selective factors of the new environment before the comparisons are made. Modifications induced by the *social* environment can be more easily studied. It would be sufficient to raise, for example, anubis baboons in a hamadryas troop to explore the capacity of anubis to be modified toward hamadryas group life.

Another opportunity for evaluating the modifying effects of environment, although unsatisfactory, is the study of colonies estabished in zoological gardens and laboratory enclosures. Since these artificial environments differ in nearly all variables from the original habitat of the transferred animals, they serve merely as a test of the resistance of a social organization to a drastic, general change. Furthermore, captive groups are usually composed of individuals that have not grown up together in the same wild group. Instead of growing into an existing group, they have to build one with unfamiliar partners. If the new organization turns out to be different from that of the original population, it will not be possible to attribute the change to any particular environmental factor. But if the original grouping pattern reappears and survives into the next generation, we may assume that it is highly resistant to a wide range of environmental changes. Some species indeed have shown this stability. A colony of geladas, imported from the wild as subadults, formed one-male groups within ten days after being admitted to the 100 by 400 foot enclosure of the Delta Center. Zuckerman ('32) found that the hamadryas colonies of three different zoos were organized in one-male groups (Kummer and Kurt, '65), and the hamadryas baboons of the Sukhumi research station in Russia show the same pattern even after several generations in captivity (Bowden, '66).

Studies of groups under changed environments can, however, produce results that are more interesting than a mere demonstration of organizational stability. The hamadryas colony of the Zurich Zoo regularly displayed a behavior pattern that we never observed in wild hamadryas troops. In this pattern, called "protected threat," an animal enlists the support of a stronger individual against his opponent by a specific set of gestures. Surprisingly, the protected threat is not found in wild hamadryas, but wild anubis baboons (DeVore, '62) and rhesus monkeys (Altmann, '62) show it. This and parallel examples support the assumption (probable already on theoretical grounds) that the genetic potential of a species for social behavior and organization is broader than the overt behavior observed in the wild. Parts of the repertoire of a species must be latent, and among these latent behaviors there may be some that are overt in related species. An investigation into these latent repertoires would certainly add to our understanding of comparative social life in primates.

In interpreting the effects of a change in environment on social structure, one has to consider a possibility that can be disregarded in analogous studies on single organisms, namely the transmission of non-genetical information from one generation to the next by means of "tradition." The work at the Japanese Monkey Center on the spreading of food habits within a group is too well known to be reviewed here. What we must realize, however, is that traditions of group life (i.e., social behavior appearing in younger animals as an enduring response to behavior of their elders) might counteract modifying effects of the physical, non-social environment. Geladas may form one-male groups in captivity because they have been socialized to do so as juveniles in the wild, and their captive-born offspring might be socialized in the same way. Only a partial reduction of the socialization process could reveal the modifications of the physical environment within one generation.

EVOLUTION AND ADAPTIVE FUNCTIONS OF TYPES OF GROUPS

A social organization must have an evolutionary history if it is determined by genes. As we just observed, however, we have as yet no idea of the extent to which a social organization is genetically determined. We know that skeletal and many biochemical characters are under relatively precise genetic control, and we may therefore attribute changes in them during evolution to changes of genotypes. But if we find that two related species differ in their grouping patterns, we cannot decide whether the difference was caused by a change in the constituents of the gene pool or by a modification of the expression of the genotype. It is uncertain whether hamadryas baboon behavior is merely an overt display of an environmental modification which the savanna baboons would show as well if they were subjected to the hamadryas environment or whether the hamadryas organization is the result of an evolutionary change. The latter appears more likely because both the savanna baboons and the hamadryas baboons maintain their type of group in captivity, but the answer must wait for the appropriate experiments.

Regardless of the degree to which modifications and mutations contribute to a particular organization, we should eventually investigate the adaptive functions of the organization in various types of habitat. The question of adaptation has been the foremost concern of the anthropologists studying primates in the field, and therefore I should like to illustrate the kinds of speculations that we may derive from the available preliminary studies.

All three African monkey species that have been found to organize themselves in one-male groups are highly terrestrial, and all live in open habitats where food is sparse and where sleeping trees are scarce or unsafe. Since widely scattered food resources are better exploited by small groups than by large troops, independent one-male groups of the patas type may be an adaptation to the distribution of food supply. In accordance with this hypothesis, Crook ('66) found that geladas live in isolated one-male groups in areas where food is scarce. In richer areas, the one-male groups join to form large troops.

The density and quality of sleeping lairs for the night seem to be the second major factor determining the organization of the open-country species. The best sleeping trees in the habitats of patas monkeys are small and offer little protection against predators. Patas monkeys reduce the risk of predation by dispersal at nightfall: each tree is occupied by only one individual. In a hamadryas habitat, the best sleeping lairs are vertical cliffs. These cliffs, however, are often far apart, and therefore the baboons must form large sleeping parties at night. The hamadryas thus have to cope with two contradictory ecological factors; they form small foraging groups of four to eight animals during the day, but at night they congregate by hundreds on the few available cliffs. No single type of group can satisfy these contrasting needs. Accordingly, the hamadryas has developed a unique multi-level organization, in which small one-male groups, without losing their identity, are organized in large troops or sleeping parties. This interpretation does not explain the intermediate level of hamadryas organization, the band. The similarity of the band to the multi-male group of savanna baboons has already been mentioned. If, as is likely, the hamadrays has evolved from savanna baboons as a specialist of the semidesert, the hamadryas band is probably the surviving homologue of the savanna baboon group. By splitting the original group into small foraging one-male groups for the day and by developing an intergroup tolerance at night, the hamadryas may have arrived at his present three-level stage of organization which permits reversible fusion and fission according to conditions.

CONCLUSION

By subjecting the primate group to the standard questions of biological research, we have tried to point out to the biologist working in other fields some of the special characters of the group organism. The flexibility of a group's anatomy permits the group to assume various shapes, each of which may be adaptive in a special situation. On the other hand, this loose structure leaves the group with the constant task of maintaining its identity, i.e., attracting its members and excluding strangers. The functioning order of the group is further marked by the minimal morphological and functional differentiation of its members. Assumption of some of the roles or functions within the group are therefore subject to competition from a number of similarly equipped individuals. The distribution of roles has only temporary stability, for the established order is for some time recognized rather than challenged by the group members.

The immediate causes of a particular social organization apparently are not related to the repertoire of social signals but probably to the relative strength of affinities among sex-age classes and to the degree to which relationships among certain classes are tolerated or prevented by other animals in the group. The example of hamadryas baboons suggests that phylogenetically old motivations, such as maternal motivations, may assume new functions in group life. Groups in general appear to lack a genetically programmed life cycle except when they are focused on the life cycle of one key individual.

What can a comparative science of primate social organization contribute to anthropology? It may show us the repertoire of motivations and organizational elements present in the order Primates: it may show us what forms of societies have evolved on the basis of this raw material; and it may show us what kinds of inhibitions and functional readjustments of phylogenetically old motivations could lead to the types of societies presently found in nonhuman primates. What, for instance, are the motivational syndromes that can produce monogamous groups or purely male groups? Is primate monogamy, for instance, always based on the same constellation of motivational factors, regardless of whether it appears in the gibbon or in the South American *Callicebus,* or are there several ways in which primates can realize the monogamous pattern? The answers to such questions would establish a body of comparative theory and data against which the specific solutions found by human societies could be evaluated. Which factors of the group-building repertoire does man share with other primates and in which has he specialized? In short, where in the order Primates is man to be placed on the basis of the specific factors underlying social organizations?

A second group of questions concerns the modifiability of social organization. We have seen that a species may in captivity reveal organizational elements that it does not show in the wild but which are found in closely related species under natural conditions. Such observations suggest that a species has a repertoire of social organizations which is only partly realized in any single environment. How large is this latent repertoire? Man may serve as an example. His technical capacities have allowed him to occupy an enormous variety of physical environments, and this variety probably brought out at some epoch and in some place all social modifications of which he is capable. The question of tremendous practical importance is: to how many and to what new kinds of environments can man yet respond with adaptive social organizations without changing his present gene pool? Under what conditions does this capacity of modify fail, and which elements of social behavior and organization are most likley to fail under a particular condition? No nonhuman primate has had man's enormous technical ability to adapt to a large variety of habitats. Each species is more or less confined to one type of habitat, which in turn produces social organizations that vary little within each species. The full genetic potential of such species would only come to the fore if they were experimentally exposed to other environments. The results would not only reveal much more about the evolution of social life than we can presently know, but they could also yield some answers on the critical limits of our own social adaptability.

LITERATURE CITED

Altmann, S. 1962 A field study of the socio-biology of rhesus monkeys, *Macaca mulatta.* Ann. N. Y. Acad. Sci., *102:* 338-345.

Bernstein, I. S., and L. G. Sharpe 1966 Social roles in a rhesus monkey group. Behaviour, *26:* 91-104.

Bowden, D. 1966 Primate behavioral research in the USSR—The Sukhumi medico-biological station. Folia primat., *4:* 346-360.

Carpenter, C. R. 1964 Naturalistic Behavior of Nonhuman Primates. Pennsylvania State University Press, University Park.

Crook, J. 1966 Gelada baboon herd structure and movement. A comparative report. Symp. Zool. Soc. Lond., *18:* 237-258.

DeVore, I. 1962 The social behavior and organization of baboon troops. Ph.D. Thesis, Department of Anthropology, University of Chicago.

Gartlan, J. S., and C. K. Brain In press Ecology and social variability in *Cercopithecus aethiops* and *C. mitis.* In: Primates: Studies in Adaptation and Variability. P. Jay (ed.), Holt, Rinehart and Winston, New York.

Hall, K. R. L. 1965 Behaviour and ecology of the wild patas monkey, *Erythrocebus patas,* in Uganda. J. Zool., *148:* 15-87.

Hall, K. R. L. and I. DeVore 1965 Baboon social behavior. In: Primate Behavior, I. DeVore (ed.), Holt, Rinehart and Winston, New York.

Kummer, H. 1956 Rangkriterien bei Mantelpavianen (Criteria of dominance in hamadryas baboons). Revue Suisse de Zool., *63:* 288-297.

———— 1967 Social organization of hamadryas baboons. Bibl. Primatol. 6. Karger, Basel.

Kummer, H., and F. Kurt 1965 A comparison of social behavior in captive and wild hamadryas baboons. In: The Baboon in Medical Research, H. Vagtborg (ed.), University of Texas Press, Austin.

Menzel, E. W. 1967 Naturalistic and experimental research on primates. Human Development, *10:* 170-186.

Schaller, G. B. 1963 The Mountain Gorilla. University of Chicago Press, Chicago.

Sugiyama, Y. 1966 An artificial social change in a Hanuman langur troop (*Presbytis entellus*). Primates, *7:* 41-72.

Zuckerman, S. 1932 The Social Life of Monkeys and Apes. Kegan Paul, London.

20

Correlates of Ecology and Social Organization Among African Cercopithecines

THOMAS T. STRUHSAKER

INTRODUCTION

The relation between environment and social organization is a problem receiving considerable attention in recent field studies of primate behavior. Notable among the publications contributing to an understanding of this problem are those of DeVore (1963), Hall (1965a, b; 1966), Rowell (1966), Crook and Gartlan (1966) and Chalmers (1968a, b). Unfortunately, most of the reliable information in these studies deal only with savanna or arid-country cercopithecines in Africa. Rowell's (1966) study of baboons living in the gallery forest and savanna of Ishasha, SW. Uganda is a partial exception to this. Chalmers' (1968a, b) is the only comprehensive study on an African rain-forest cercopithecine. Needless to say, the formulation of meaningful hypotheses on the relation between environment and social organization must consider the many species of rain forest cercopithecines which vastly outnumber the non-forest species.

There are a number of factors that can affect the social organization of a species, including a variety of environmental parameters such as habitat, food, predators, competitors, population density, etc. In addition to these immediate or proximate variables, one must also consider the evolutionary history of the species; namely, its phylogeny. To what extent is the social organization of a species a function of its ecology? How do species with

Reprinted and abridged from *Folia Primatologica,* 11:80-118 (1969) with permission from S. Karger AG, Basel.

radically different phylogenies respond to similar environments? How do species with similar or closely related phylogenies respond to different environments?

It is the intent of this article to present data collected between November 1966 and May 1968 on the group size and, to a lesser extent, on group composition of rain-forest cercopithecines living in Cameroon, West Africa. It is further intended to discuss the relevance of these data to the current hypotheses relating ecology and social structure of African cercopithecines and to consider some of the aforementioned questions. A general paucity of detailed information on the ecology and social organization of most of the species mentioned in this article necessitates that the comparisons be made at a very general and gross level. This is unsatisfactory, but will demonstrate those relations that can be explained at a "gross" level and those that require more refined study.

General ecology of species considered in this article

Cercopithecus nictitans and *C. pogonias* are rain-forest species of West and Central Africa, being apparently more successful in mature forest than in younger successional stages and are seen most commonly in the upper heights of the forest. *Cercopithecus erythrotis camerunensia* and *C. cephus* are also monkeys of the West African rain forest, having similar ecologies but differing from the former two in apparently preferring the lower strata and younger or more immature kinds of forest. For most of their range these latter two species are allopatric and seem to be ecological counterparts. *Cercopithecus mona* is found in a variety of forest habitats in West Africa, but is apparently most abundant in mangrove swamps. When in lowland rain forest *mona* can be seen at all heights, but seems to have a propensity for lower levels. *Cercopithecus l'hoesti* occurs in two disjunct populations; one in West Cameroon and one in Eastern Congo Kinshasha and Western Uganda. In Cameroon it seemed most abundant in the montane forests, but did occur in a variety of forests including mature lowland rain forest. This species has a distinct preference for lower levels in the forest and is commonly seen on the ground in contrast to the other aforementioned *Cercopithecus* species. All of the preceding species seem to be opportunistic omnivors. *Cercopithecus aethiops* shares this habit, but differs from the other *Cercopithecus* species in being a savanna and arid-country species and in having a much wider distribution than the others. It occurs in nearly all of the savanna south of the Sahara and is definitely more terrestrial than the other *Cercopithecus* spp. *Erythrocebus patas*, considered to have strong phylogenetic affinities with *Cercopithecus* species, is also a savanna and arid-country species, but seems even more independent of trees than *C. aethiops*. It is also omnivorous, but unlike *C. aethiops* does not occur south of northern Tanzania. Both species of *Cercocebus* are typified as omnivorous, lowland rain-forest species. There is some evidence indicating that *Cercocebus torquatus* is more terrestrial and more common in mangrove and seasonally inundated forests than is *C. albigena* (Jones and Sabater P. 1968). Drills (*Mandrillus leucophaeus*) live in the lowland rain forests of a very restricted geographical region of West-Central Africa. They are most abundant in West Cameroon and their western and northern distribution terminates somewhere in eastern Nigeria, while their southern and eastern limits, although unknown, probably do not extend far beyond the borders of East Cameroon. Drills are omnivors who spend considerable time both on the ground and in the trees. Most populations of *Papio* spp. are living in savanna habitats. Some spend a large proportion of their time in gallery forests and on the edge of more extensive high forests but, as far as is known, always have some contact with savanna. This is in contrast to drills who are strictly a forest species. Most populations of *Papio* spp. can be considered terrestrial omnivors. *Papio* and *Theropithecus gelada* are restricted to the arid high country of Ethiopia, *gelada* apparently preferring higher altitudes than *hamadryas*. Both species are typically terrestrial, but rely heavily on steep cliffs as refuges and sleeping roosts. *Gelada* may be more of a herbivore than *hamdadryas*.

Social group — An operational definition

The term troop or group is widely used in contemporary field studies of primates. However, I have been unable to find an operational definition of social group in any of the recent publications, e.g. DeVore (1965), Altmann (1967). One probable reason for this is that most of the monkey species that have been studied live in relatively open country and their spatial relations are so obvious that a definition seems pedantic. Due to the inherent difficulties of observation in rain forests, these spatial relations are not so evident and one is forced to ask the basic question: what is a social group? Reading the publications on the savanna and arid-country monkeys of Africa gives one a good indication of what the authors mean when they refer to groups or troops. The following features would appear to be the main ones distinguishing monkey groups in these various studies and also represent my concept of a monkey social group. Membership in the same social network would seem to be the major feature distinguishing group members from extra-group conspecifics. Individuals of the same social network, by definition, have the majority (at least 80%) of their non-aggressive social interactions within this social network. A given individual does not necessarily interact with all members of its social network, but is associated with all at least indirectly. For example, if monkey A equally divides its non-aggressive social acts between monkeys B and C, but monkeys B and C don't interact with one another they are, none-the-less, members of the same social group; namely, the group consisting of monkeys A, B, and C. Temporal stability is another key character in describing a social group. Although most field studies of old world monkeys have been for only one or two years, it is generally assumed that group members remain together for longer periods. Subadult and adult males in several species seem more mobile than females and juveniles in changing from one group to another. However, even in most of these cases the males remain with a given group for about five months (Struhsaker 1967d). Seasonal variation in group size and composition has not yet been demonstrated among old world monkeys. However, seasonal changes do occur in the frequency and size of temporary aggregations of two or more social groups. e.g. *gelada* (Crook 1966); *patas* (Struhsaker, pers. observ.). Spatial proximity is often implied in descriptions of intragroup relations, but this seems to have limited value as it is very dependent on the environmental conditions and their effects on the dispersal of the animals and on the observer's ability to measure group dispersal. This factor is discussed more fully in the section on intragroup dispersal. But, one can say that group members occupy the same home range. This also has limited value in defining social group, because in many species the groups have considerable overlap in their home ranges (DeVore and Hall 1965). Distinct social roles seem a key feature among group members, e.g. there are adult males who perform the majority of copulations (Hall and DeVore (1965); Struhsaker (1967)) and adult females who lead the majority of group progressions (Struhsaker 1967d). Synchronized progressions also delineate members of the same group. Such progressions are especially prominent when the monkeys are fleeing from a predator or are moving from one feeding place to another. Group members exhibit different behavior toward non-group conspecifics. This different behavior may be in the form of unique patterns, e.g. intergroup aggressive calls of vervets (Struhsaker 1967a), or may consist of avoidance or a difference in the frequency with which certain behavior patterns are performed, e.g. aggression may be greater among non-group than group conspecifics. Social groups are relatively closed to new members. Exceptions to this have been noted for vervets and baboons, but these cases are relatively infrequent and don't invalidate the concept that social groups are relatively closed. Data presented throughout this paper support the conclusion that the rain-forest monkeys of Cameroon, West Africa live in social groups having these general characteristics and differ only in some details of their social organization from the African savanna and arid-country monkeys previously studied (also see Chalmers 1968a, b).

METHODS

During an 18-month field study in the rain forests of Cameroon, West Africa, attempts were made to count all monkey groups that were encountered. Due to the inherent difficulties of observing monkeys in the thick vegetation of a rain forest combined with the heavy hunting pressure exerted by humans on monkeys in Cameroon, observations and accurate and complete counts of monkey groups were exceedingly difficult to obtain. As a consequence, relatively few reliable counts were made. Often the counts were incomplete, but the amount of noise made by the group permitted what was to be considered a reliable estimate. Sometimes these estimates were confirmed by accurate and complete counts made of the same group at a later time. The most accurate counts were made as the monkeys climbed along the same arboreal route in synchronized progression. During the last year of this study I was accompanied by Mr. Ferdinand Namata, a local African who proved to be extremely perceptive and reliable in his observations. The counts made Mr. Namata were compared with those made by me as a test of observer reliability. A high degree of consistency was attained in our counts and estimates and thus further increased my confidence in their accuracy. In addition to a general survey of many areas and a variety of rain-forest types, one particular area of almost 0.41 square miles was mapped in detail. This area was marked with a grid of trails and numbered aluminium plates were placed at regular intervals along these trails permitting very accurate location of the monkey group. The localization of monkey groups allowed what are considered to be reliable inferences on group stability, i.e., groups of similar size and compositions were often seen at the same specific locality. Every attempt was made to remain with a group or association of monkeys for as long as possible. All social events were described and tabulated.

Compared to the majority of field studies on the behavior and ecology of non-forest African cercopithecines, these data are relatively scant and unsophisticated and may seem unsatisfactory to some readers. However, in the absence of other more reliable and complete data on the species I studied, the data presented in this paper are valuable for our development of theories on the parameters most influential to primate social structure.

RESULTS AND DISCUSSION

Counts made in the major study area that was mapped in detail support the hypothesis that groups of *Cercopithecus nictitans martini* were stable in size and composition. Usually groups of the same size and composition were seen in the same region of this mapped area. These data not only indicate that groups are temporally stable, but also that they are relatively closed. Most of the data for the other species were collected outside of the mapped area and therefore do not necessarily permit this same conclusion. However, prolonged observations of some groups of these other species implied that they too lived in stable social groups. Chalmers [1968a] studied two groups of *Cercocebus albigena* and found that their membership remained relatively constant over an 11-month study. This finding is also consistent with studies on the majority of nonforest cercopithecines. Even in *gelada* and *hamadryas* baboons who form large and unstable aggregations there are small and relatively stable social units (Crook 1966).

SEXUAL DIMORPHISM

DeVore (1963) has advanced the hypothesis that sexual dimorphism, especially regarding body size and canine size, is an adaptation by terrestrial
primates against predation. He states: " . . . morphological adaptations for
fighting and defense are clearly correlated with adaptation to the ground."
And again: "The trend toward increased fighting ability in the male of
terrestrial species is primarily an adaptation for defense of the group." This
hypothesis is not supported by observations on forest cercopithecines.
Drills demonstrate an extreme form of sexual dimorphism. The males are
considerably larger than females in both body (about twice) and canine size
and they have distinct colouring. There is no evidence that adult males play
any role in the defense of the group. Man is the major predator of drills in
Cameroon. When they detected us, the drills either fled immediately or the
adult females and immature drills approached, looked toward, and gave
pant-barks toward us before eventually fleeing. Adult males never did this
and a subadult male only once. On January 2, 1968 while watching a fairly
large group of drills I was given mild threat gestures by a subadult male. He
sat about sixty feet below a cliff edge and below me. While sitting with his
hands on the ground, he jerked his forequarters forward, briefly held this
forward position, and then returned to the upright and original sitting
posture. This jerking was not repeated rapidly and thus differed from the
more rapid, but similar pattern in vervets and baboons. He also jerked his
forequarters sidewards in a lateral movement while retaining his sitting
position. Sometimes he went from a sitting to quadrupedal position in a
rapid and jerky manner and then oriented his longitudinal axis perpendicu-
lar to my line of vision. He then stared toward me. After a few seconds, he
would sit again. This entire sequence was interpreted as a defensive threat
toward me. Usually adult and subadult males were not seen after the group
had detected us and given "alarm" calls. It has been argued that perhaps
the best defense drills can exert against human predators is flight. However,
if this were so, then one would expect to see a similar response to humans
by the adult females and immature drills. As stated earlier, drills spend a
considerable amount of time on the ground. Although not so marked as
among drills, *Cercocebus albigena* is obviously a sexually dimorphic species.
On only one occasion did the larger male demonstrate defensive or protective behavior. This occurred on February 15, 1968 when I was observing a
group of at least twelve or thirteen *C. albigena* who were feeding 60-80 ft
above. Upon seeing me, six fled in one direction and two or three in another.
They were all larger than one-third adult size. Remaining behind in the food
tree were three juveniles (about one-third adult male size) and the only adult
male in the group. For the next ten minutes, after the others fled, the adult
male gave staccato-barks almost continuously, eating fruits intermittently
between barks. The three small juveniles remained close to the adult male

throughout this period. After this ten-minute period, the three juveniles fled in the same direction as did six other *albigena*. The adult male was the last to leave, following the three juveniles. The forest and more arboreal *Cercopithecus* demonstrate as much sexual dimorphism in body and canine size as do the more terrestrial *C. l'hoesti* and the non-forest and more terrestrial *C. aethiops*. The majority of evidence indicates that adult males of forest *Cercopithecus* perform no special role in the defense of the group. Only once did an adult male *C. mona* demonstrate threat or defensive behavior toward me. This occurred on April 18, 1968 when a group of at least six *mona* were encountered. They all fled immediately except for an adult male who remained momentarily, gave low-pitched sneeze calls, and bobbed his head and forequarters toward me. Sometimes his hands left the substrate during the upward portion of the bob. The rate of bobbing seemed more rapid and his body seemed more hunched than a similar pattern in *C. aethiops*, but was clearly a threat. Among the more terrestrial and savanna-dwelling *C. aethiops*, an adult male was seen only once to make threatening lunges toward a potential predator, the Martial eagle (*Polemaetus bellicosus*). The rarity of defensive behavior by adult males does not conform with DeVore's (1963) hypothesis and I am inclined to agree with Crook and Gartlan (1966) in thinking that this sexual dimorphism in body and canine size is not primarily adapted to group defense. It seems reasonable that it is basically a result of sexual selection, either intra and/or inter, and with some species and under certain circumstances has become secondarily adapted to protection of the group, e.g., DeVore's savanna baboons in the Nairobi Park. The basis of this argument is weakened by the relative paucity of positive data and by its heavy dependence on negative data, i.e., few cases of aggression between adult males and defense of the group by adult males have been observed.

GROUP SIZE

What is the relation between group size and the degree to which a species is terrestrial and the degree to which it is a forest dweller? Within the area of detailed study there were five groups of *C. nictitans martini* of whose identity I was quite certain. The average counts of these groups and the mean of the estimates made of these groups (in brackets) are as follows: 7(7.3); 7.2(10.3); 17.0(19.0); 14.2(16.1); and 9.0(9.5). The average (mean) count and estimate of these five groups are 10.9 and 12.3 respectively and are probably not significantly different from the means for all counts of this species. Most of the other forest *Cercopithecus* species lived in heterosexual groups of similar size. The data for *C. cephas* are few and not too reliable. *C. l'hoesti preassi* is an outstanding exception, having a mean group size about half that of the other forest *Cercopithecus* species. It is noteworthy that *C. l'hoesti* is also the most terrestrial of the forest *Cercopithecus* monkeys in this sample. It invariably ran along the ground when fleeing from man and was often seen foraging on the ground. None of the other forest *Cercopithecus* fled on the ground and were very rarely seen there.* This conflicts with DeVore's (1963) conclusion that " . . . a trend toward larger groups in the terrestrial species is apparent . . . ." In contrast to this and consistent with DeVore's (1963) hypothesis, drills (*Mandrillus leucophaeus*), who spent a considerable amount of time feeding, progressing, and fleeing on the ground, have mean group sizes larger than any of the *Cercopithecus* forest monkeys observed, who are primarily arboreal. However, the drills apparently spent more time in the trees than non-forest baboons and yet the group sizes of these species who occupy such radically different habitats do not appear significantly different. This also conflicts with DeVore's hypothesis. Hall (1965b), in his study of *Erythrocebus patas*, was one of the first to demonstrate the inadequacy of DeVore's preceding hypothesis; " . . . patas, which are as terrestrial as the baboons in the same habitat region, have a mean group size of about half that of the Murchison Falls Park baboons. . . . "

The drill data on group size would also appear to contradict the conclusion of Crook and Gartlan (1966) that "Outside the forest the social units are generally larger and this appears to result from open country conditions of predation and food supply affecting ground dwelling populations." A comparison of *C. l'hoesti* group size with *C. aethiops* would support this hypothesis. However, excluding *C. cephus,* the other forest *Cercopithecus* seem to have group sizes of the same size as *C. aethiops* on Lolui Island (Hall and Gartlan 1965) and in Murchison Falls Park (Hall 1965a), but of a smaller size than *C. aethiops* in the Amboseli Reserve (Struhsaker 1967b). The habitat of Lolui Island is a mosaic of moist semi-deciduous forest, fringing forest, thickets, grasslands and swamps, whereas the habitat at Amboseli is best described as semi-arid savanna with very sparse and restricted tree groves in the water hole areas. The vegetation of Murchison Falls Park is rather intermediate to these two extremes and the relation between group size and ecology among *C. aethiops* is worthy of further investigation. Comparison of

---

* It has been argued that because *C. l'hoesti* is more terrestrial than the other forest *Cercopithecus* spp. it may be exposed to greater hunting and trapping pressure by man than the other species and that this might account for its smaller group sizes. If this were so, one would expect to see different sizes of *l'hoesti* groups in areas where different amounts of trapping occur. Not only would it be difficult to measure the effective trapping and hunting pressure being exerted on a given population, but my sample of *l'hoesti* is not adequate for statistical analysis. However, the data on *l'hoesti* group size came from several widely separated areas in West Cameroon and in none of these areas was the impression gained that *l'hoesti* groups differed in size. However, the trapping and hunting pressure varied considerably in these same areas, as evidenced by the number of traps encountered and the number of gun shots heard. These impressions would support the conclusion that, if trapping and hunting of *l'hoesti* have any population effect, they result in a reduction of the number of groups and not a reduction in group size. Furthermore, additional support of this conclusion is given by the fact that other terrestrial forest species such as drills and mangabeys live in large groups with no apparent reduction in group size in areas of different trapping and hunting pressure.

different populations of *Papio anubis* does not support Crook and Gartlan's hypothesis, for the more forest-dwelling baboons of Ishasha (Rowell 1966) lived in groups of the same size or slightly larger than those living in the savanna near Nairobi (DeVore and Hall 1965) and in the arid savanna of Waza, Cameroon. In contrast and in support of Crook and Gartlan are the data on group size on *Papio cynocephalus* in the semi-arid savanna of Amboseli where the mean group size is eighty and significantly larger than the *Papio anubis*, *P. ursinus*, and drill groups. It is of interest that the largest *C. aethiops* groups are also. found at Amboseli.

GROUP COMPOSITION

Comparison of group compositions is hindered by the difficulty of making reliable observations of forest *Cercopithecus*. Adult females could often be distinguished by the presence of pendulant nipples. Sometimes the scrotum of adult males could be seen. The slightly larger size of adult males was most apparent when they were near adult females. Sex determination of immature *Cercopithecus* was not possible under the prevailing observation conditions. Age and sex determination of drills and mangabeys was easier, but never was the complete age-sex composition of a group determined. The vocalizations of adult males were distinct from all other age-sex classes in the forest *Cercopithecus*, drills, and probably, but less certainly, in the mangabeys. This sexual difference in behavior provided indirect clues on group composition.

Among the forest *Cercopithecus* in this sample, never more than one adult male was seen in each heterosexual group. With only four exceptions, there was one source of adult male loud calls in each group. In the four exceptions, two sources of these loud calls were heard once each in *C. nictitans n.* and *C. l'hoesti preussi* and twice in *C. pagonias grayi*. On one occasion two very large and presumably male *C. nictitans martini* were seen running away from a group. This encounter appeared to be a chase. Only one of these, which proved to be an adult male, returned. Solitary monkeys were commonly observed in forest *Cercopithecus*. With one exception these solitaries were adult and when their sex was verified, proved to be adult males. The following data are the frequencies with which solitaries were seen and the number of these that were definitely verified as adult males: *C. nictitans martini* (17, 1A♂); *C. nictitans n.* (5); *C. pogonias p.* (1, A♂); *C. pogonias grayi* (1, not A♂); *C. mona* (2, 1A♂); *C. cephus* (1); *C. erythrotis* (5, 2A♂); and *C. l'hoesti preussi* (1). Many of those listed whose sex was not determined were probably adult males judging from their large size, lack of nipples, and more contrasting pelage coloration that is typical of adult males. These observations suggest that heterosexual groups of forest *Cercopithecus* are characterized by the presence of only one adult male. The other adult and/or subadult males are solitary or very peripheral to those heterosexual groups. This structure would appear to be maintained by the aggressiveness of the adult male in the heterosexual group.

Haddow (1952) collected fifteen solitary *Cercopithecus ascanius* which is also a forest species. These fifteen included: ten adult males, three old adult males, one subadult male, and one adult female.

A different system prevails among drills, *Cercocebus albigena*, and probably, though less certainly, *Cercocebus torquatus*. Even though solitary or extremely peripheral adult and subadult males are moderately common, the heterosexual groups usually contain more than one adult male. Verified solitaries were seen with the following frequency: drills (2 A or SAδ, 3 Aδ, 1 SAδ) and *Cercocebus albigena* (1 SAδ). Adult male drills often give a low-pitched two-phase grunt that carries for a considerable distance and presumably facilitates group cohesion. The number of sources of this call in each group provides indirect, though less reliable, information on the number of adult males present. Such information indicates that drill groups of about twenty or less have only one adult male, whereas those greater than twenty usually have more than one adult male. The few reliable counts would support this. There is no evidence for all-male groups in any of the forest monkeys in this sample.

Apparently no relation exists between the number of adult males in heterosexual groups and the extent to which the species is terrestrial or aboreal among the forest cercopithecines. *C. l'hoesti* is considerably more terrestrial than the other forest *Cercopithecus* and yet they are similar in having only one adult male per group. Drills are more terrestrial than *Cercocebus albigena* and yet both species have several adult males per group. The one relationship which does seem consistent is that solitary adult and subadult males are common among forest cercopithecines but not among non-forest species. Although the ecology of the forest *Cercopithecus* corresponds to Crook and Gartlan's "Grade III" their social structure does not (see selection 18). Drills have a social organization comparable to their "Grade IV," but an ecology between "Grade II" and "Grade III." Revision of their categories seems necessary in this regard.

In contrast to the forest cercopithecines most of the non-forest species have more than one adult male in each heterosexual group. The most notable exception is *Erythrocebus patas* in which one-adult-male heterosexual groups are the rule. No exceptions have yet been observed. It seems relevant to mention that most taxonomists consider *patas* to be closely related to the *Cercopithecus* monkeys. Jolly (1966) has subdivided the subfamily Cercopithecinae into three tribes. One of these, Cercopithecini, consists only of the genera *Cercopithecus* and *Erythrocebus*. Although *gelada* and *hamadryas* have one-adult-male heterosexual units, there are usually several of these units which move together and thus there are in effect several adult males present in the band and herds at any given time. Solitary animals are rare among non-forest cercopithecines. Kummer (1968) has observed only one solitary *hamadryas*; an adult male. Similarly, among baboons of South Africa and *patas* monkeys of Uganda, Hall (1965a) has only observed an occasional isolated adult male. In Cameroon I observed one solitary adult male *patas* and two adult male *patas* who were each very peripheral to heterosexual groups. It seems justifiable to conclude that selection has favored solitary males in forest cercopithecine species and not in non-forest species. Presumably, this is related to lower pressures of predation and greater conceal-ment from predators in the forest. It seems significant that among the non-forest cercopithecines those species having one-adult-male heterosex-ual groups are also the only cercopithecines having all-male groups. Hall (1965b) observed one such group among *patas* monkeys in Uganda. This consisted of " . . . one full-grown adult male, two near-adult males whose fur colouration was similar, but their size was somewhat smaller and one young adult male." I observed three all-male *patas* groups in the Waza Reserve of Cameroon. In 1967 one consisted of three adult males and three subadult males and the other consisted of four subadult males. In 1968 what was presumed to be a third and different group had two adult males and four subadult males. Crook (1966) mentions that all-male groups occur in the herds of *gelada*. Young and subadult male *hamadryas* move on the periphery of the one-adult-male units (Kummer 1968). These all-male groupings and the peripheralness of males support the hypothesis that the solitary mode of existence is selected against in the non-forest cercopithecines.

CONCLUSIONS

The data presented in this article on rain forest primates in Cameroon call for revision of several hypotheses that correlated certain aspects of social organization and ecology. 1) The direct relation between group size and the degree of terrestriality was not supported by these data. 2) The hypothesis that larger groups occur in non-forest habitats was contradicted by some of these data. 3) Several combinations of ecological and social structure varia-bles were described that did not fit into the classification scheme of Crook and Gartlan (1966). This is not to say that ecological factors cannot have an important effect on social organization. Evidence was presented that sup-ports a hypothesis directly relating the occurrence of solitary males with a forest habitat and their absence or rarity in non-forest areas. The fact that all-male groups have been seen only in open, non-forest habitats further supports the idea that the amount of vegetational "cover" can have an effect on social structure. It would, thus, appear that certain ecological factors and certain aspects of social organization can be correlated, but obviously not all. Hall (1965a) has stated " . . . social adaptations shown in the varying patterns of dominance, from the conspicuous adult male hierarchies of baboons and rhesus to the lack of any such demonstrable relationships in other species, can readily be understood only within the ecological frame-work with all its complexity of food-getting, sheltering and resting, avoidance of predators, and the changing seasonal internal pattern of births and mating." This statement has several attributes and in some respects is applicable to many other aspects of social organization. However, in considering the relation between ecology and society it must be emphasized that each species brings a different phylogenetic heritage into a particular ecological scene. Consequently, one must consider not only ecology but also phylogeny in attempting to understand the evolution of primate social organization. The interrelations of these two classes of variables determine the expression of the character, in this case social structure. In some cases, the immediate ecological variables may limit the expression or development of social structure and, with other species and circumstances, variables of phylogeny may be the limiting parameters. For example, heterosexual groups with only one adult male seem to be typical of most *Cercopithecus* spp. and the closely related *Erythrocebus patas.* The only notable exception is *C. aethiops.* They are the only member of the Cercopithecini Tribe that have multi-male heterosexual groups, no solitary males, and no all-male groups. One wonders why they developed the multi-male heterosexual group rather than the all-male group as did *patas* in adapting to savanna life. In contrast, drills, baboons, and *Cercocebus* spp. typically have several adult males in their heterosexual groups regardless of gross ecological differences. *Gelada* and *hamadryas* are rather intermediate between these two extremes. This appears to be a case in which phylogenics are at least as important as ecology, if not more so, in understanding an aspect of the social structure. Eisenberg (1963) in his detailed behavioral study of heteromyid rodents reached a similar conclusion: " . . . it appears that the phylogenetic background has been very important in the expression of the social organization in the family (Heteromyidae)."

When one considers the distribution of African Cercopithecinae species in the various kinds of habitats, two trends become readily apparent. In terms of numbers of species and numbers of individuals the *Ceropithecus* species are more successful in forest habitats than are baboons (*Papio* spp.) who in contrast are more successful in savanna and arid-country than are the *Ceropithecus* species. This would imply that these two genera are and have been exposed to different selective pressures. Consequently, one would expect to find differences between these two groups in behavior having adaptive advantage such as social organization. Apparently, social organization among African Ceropithecinae is a relative inert or stable character.

I would certainly agree that more detailed ecological information is needed before the causal and casual factors of social organization of primates can be fully understood. But I would stress with equal emphasis that a better understanding of phylogenics can also contribute greatly to our comprehension of the evolution of primate social structure.

SUMMARY

New data are presented on the social organization of several species of rain-forest cercopithecines from Cameroon, West Africa. These data are compared with studies on savanna and arid-country cercopithecines and discussed in view of earlier hypotheses that have attempted to relate ecology and social organization. Many observations of the rain-forest species are inconsistent with these earlier hypotheses and call for revision of them. It is concluded that, although some aspects of social organization are clearly related to ecology, others are not and are perhaps best understood from the standpoint of phylogenetic affinities.

REFERENCES

Altmann, S. A. (ed.): Social communication among primates (Univ. of Chicago Press, Chicago 1967).

Chalmers, N. R.: Group composition, ecology and daily activities of free-living mangabeys in Uganda. Folia primat. *8:* 247-262 (1968a). — The social behaviour of free-living mangabeys in Uganda. Folia primat. *8:* 263-281 (1968b).

Crook, J. H.: Gelada baboon herd structure and movement — A comparative report. Symp. zool. Soc. Lond. *18:* 237-258 (1966).

Crook, J. H. and Gartlan. J. S.: Evolution of primate societies, Nature, Lond. *210:* 1200-1203 (1966).

DeVore, I.: A comparison of the ecology and behavior of monkeys and apes. In: Washburn, S. L., Classification and human evolution, pp. 301-319 (Aldine Chicago 1963). — Primate behavior. Field Studies of monkeys and apes (Holt, Rinehart and Winston, New York 1965).

DeVore, I. and Hall, K. R. L.: Baboon ecology. In: DeVore, I., Primate behavior: Field studies of monkeys and apes, pp. 20-52 (Holt, Rinehart and Winston, New York 1965).

Eisenberg, J. F.: The behavior of heteromyid rodents. Univ. of Calif. Publ. Zool. 69 (1963).

Haddow, A. J.: Studies on *Cercopithecus ascanius schmidti.* Proc. zool. Soc. Lond. *122:* 337-373 (1952).

Hall, K. R. L.: Social organization of the old-world monkeys and apes. Symp. zool. Soc. Lond. *14:* 265-289 (1965a). — Behaviour and ecology of the wild patas monkey, *Erythrocebus patas*, in Uganda. J. Zool., Lond. *148:* 15-87 (1965b). — Distribution and adaptations of baboons. Some recent developments in comparative medicine. Symp. zool. Soc. Lond. *17* (1966).

Hall, K. R. L. and DeVore, I.: Baboon social behaviour. In: DeVore, I., Primate behavior: Field studies of monkeys and apes, pp. 53-110 (Holt, Rinehart and Winston, New York 1965).

Hall, K. R. L. and Gartlan, J. R.: Ecology and behaviour of the vervet monkey, *Cercopithecus aethiops*, Lolui Island, Lake Victoria. Proc. zool. Soc. Lond. *145:* 37-56 (1965).

Jolly, C. J.: Introduction to the Cercopithecoidea with notes on their use as laboratory animals. Symp. zool. Soc. Lond, *17:* 427-457 (1966).

Jones, C. and Sabater Pi, J.: Comparative ecology of *Cercocebus albigena* (Gray) and *Cercocebus torquatus* (Kerr) in Rio Muni, West Africa. Folia primat. *9:* 99-113 (1968).

Kummer, H.: Social organization of hamadryas baboons: A field study (Univ. of Chicago Press, Chicago 1968).

Rowell, T. E.: Forest-living baboons in Uganda. J. Zool., Lond. *149:* 344-364 (1966).

Struhsaker, T. T.: Auditory communication among vervet monkeys, (*Cercopithecus aethiops*). In: Altmann, S. A., Social communication among primates, pp. 281-324 (Univ. of Chicago Press, Chicago 1967a). — Social structure among vervet monkeys (*Cercopithecus aethiops*). Behaviour *29:* 2-4 (1967b).—Behavior of vervet monkeys (*Cercopithecus aethiops*). Univ. of Calif. Publ. Zool. 82 (1967d).

21

Problems of Social Structure in Forest Monkeys

F.P.G. ALDRICH-BLAKE

INTRODUCTION

The study of primate social organization began with Carpenter's pioneering investigations of howler monkeys, red spider monkeys, and gibbons in the forests of Central America and the Far East (Carpenter 1934, 1935, 1940). The subject did not develop further until the early 1950s when long term projects were established to study the Japanese macaque *(Macaca fuscata)* (e.g. Itani 1954 and many subsequent papers), and Washburn *et al.* began work on baboons in Africa (i.e. see DeVore and Hall, 1965, Hall and DeVore 1965). The last decade has seen a great resurgence of interest in primate field studies, but research has not been spread evenly throughout the order. Most of the attention has been focused on savanna and open country animals rather than on the more numerous forest species. Some taxa such as the baboon-macaque group and the apes have been investigated fairly thoroughly, while others have been largely neglected.

Until recently the few available studies of forest monkeys had been made mainly on New World species (e.g. Collias and Southwick 1952, Altmann 1959, Mason 1966), while the Old World received scant attention. Of the Asian monkeys, forest populations of langurs and rhesus have been investigated, but to a lesser extent than populations of these same species living under more open conditions. The only study of a forest monkey was for many years

405

Haddow's (1952) paper on the redtail *(Cercopithecus ascanius)*; but this work contains little information on social organization, nor indeed is it intended to be an authoritative statement on this aspect of the animal's biology. Similarly the many papers by Haddow's colleagues at the Virus Research Institute at Entebbe (e.g. Haddow, Smithburn, Mahaffy and Bugher 1947, Lumsden 1951, Buxton 1952) consider aspects of primate behaviour only inasmuch as they are relevant to epidemiology. Booth's (1957) and Ullrich's (1961) studies of the olive colobus *(Procolobus verus)* and the black and white colobus *(Colobus abyssinicus)* are useful but limited in scope. In the last two or three years the situation has improved, with more detailed work on the black and white colobus (Schenkel and Schenkel-Hulliger 1966), Chalmer's study of the black mangabey *(Cerocebus albigena)* (Chalmers 1967, 1968a, b), Gautier-Hion's (1966) study of the talapoin *(Miopithecus talapoin)* my own work on the blue monkey *(Cercopithecus mitis)* (Aldrich-Blake, in preparation) and various other projects still in progress. The original imbalance is thus gradually being corrected.

The reasons for the initial concentration of research on the more terrestrial species are clear. Open country animals are far easier to study than those living in dense vegetation. Once their confidence has been gained they can be followed throughout the day and long periods of concentrated observation are possible. Favourable conditions of observation permit the recognition of individual animals, and hence detailed investigation of the relations between members of a troop. By comparison forest primates are difficult even to see, and even more difficult to follow. Carpenter (1965) gives an eloquent account of the problems likely to be encountered by the would-be student of such species. The return on time, energy, and money expended is correspondingly lower.

In addition, much of the earlier work on primates in the 1950s was carried out by people whose prime interest was in the making of inferences to the social evolution of man. It was thought that animals living in a habitat comparable to that of early man would provide the greatest insight into the problems faced by our simian forbears. Under the circumstances, concentration of research on open country primates was a perfectly reasonable strategy. It has, however, had certain unfortunate consequences.

In the early stages of the development of primatology the great diversity of social organization to be found within the order was not suspected (for example, see Washburn and DeVore 1961). As the scope of field studies increased, it became apparent that there was considerable variation in social structure not only between but within species, and attempts were made to relate such contrasts to differences in habitat (e.g. DeVore 1963, Crook and Gartlan 1966). Early hypotheses as to the determinants of social organization suffered from two great drawbacks. Firstly, there was little detailed information on forest monkeys to go on, and the authors were forced to rely to a large extent on fragmentary and incidental accounts. It will be shown

that such initial impressions and short term studies can be highly misleading. Secondly, most of the reliable accounts of forest monkeys were of New
World species; comparisons between savanna and forest monkeys therefore
contrasted South American monkeys and to a lesser extent Old World
colobinae with Old World cercopithecinae. It was not clear whether contrasts in social organization were dependent wholly on differences in habitat, or in part to differences in genetic constitution and basic behavioural
repertoire. A more reliable picture could have been obtained by comparing
African forest cercopithecines with open country cercopithecines but information on such forest monkeys is only now becoming available. While some
of it fits in well with previous hypotheses, some is at first sight contradictory.

The purposes of this article are two-fold. First, a detailed consideration
of the problems of bias that arise in the collection of data on forest monkeys
will be presented. Second, certain theories of social organization will be
briefly reappraised in the light of recent knowledge of such species. While
problems of bias are mentioned in previous studies (e.g. Rowell 1966,
Chalmers 1967, 1968a, b) there have been few systematic attempts to evaluate their effects. These, indeed, may be more profound than has been
realized in the past. Such a cautionary note seems particularly appropriate
at present when publications on forest species are increasing in frequency.
For this same reason an exhaustive review of the literature on forest monkeys would be out of place here, as it would already be out of date by the
time it appeared in print. I will confine myself, therefore, to a consideration
only of certain features that seem to be of particular interest.

PROBLEMS OF BIAS IN THE COLLECTION OF DATA

The figures on page 408 illustrate the contrast in observational conditions
between forest and more open habitats. The top figure shows a party of
baboons *(Papio anubis)* on grassland at the edge of riverine bush in the
Awash valley in Ethiopia. The thicket in the background is denser than
would be found in most baboon study areas; the grassland represents typical
observational conditions. The animals are clearly visible; most of them
could be sexed without difficulty, and fine details of behaviour could be
recorded. The bottom figure shows a party of blue monkeys *(Cercopithecus
mitis)* in the Budongo forest in Uganda. One adult of indeterminate sex is
grooming another, while a young infant sits beside them. At least three
other monkeys are concealed in the foliage behind. The photo was taken on
the forest edge; inside the forest the visibility is even more restricted.

The poor conditions of observation in forest affect the collection of
data in several distinct ways. Firstly, the time spent in contact with the
monkeys is less than it would be in open country; savanna animals such as
baboons can generally be followed from dawn to dusk, while forest monkeys
may be readily visible for only two or three hours a day. In forest the animals
may be difficult to find, despite their higher population density, and they are

Baboons (*Papio anubis*) in the Awash Valley, Ethiopia.

Blue monkeys (*Cercopithecus mitis*) in Budongo Forest, Uganda.

more likely to be lost if they move away. As the monkeys are in contact with the observer for shorter periods it takes longer for them to become habituated to his presence; baboons in the Awash valley come to tolerate an observer within 20 yards after only three weeks (personal observation), while blue monkeys in Budongo took at least three months to become that tame. It therefore takes very much longer to accumulate a reasonable body of data when working on forest animals, and long periods of continuous observation are rare; one is forced to rely on relatively brief interludes of contact.

Secondly, the poor visibility affects the quality of the data obtained. At any one moment many of the monkeys in the vicinity will be concealed. Typically only about half of the animals known to to be present might be visible, and some of these will be partially obscured by foliage or branches. At the simplest level this means that many observations are fragmentary or incomplete. For example, throughout a behavioural interaction, all the participants can seldom be seen or identified clearly so that the amount of reliable information collected in a given time is further reduced. At a more complex level incompleteness can lead to considerable difficulties of interpretation. The relation between what is seen and what is actually happening is never entirely clear, and initial impressions can be highly misleading. Counts of group size and composition, for example, if taken at face value, can give a totally erroneous picture of social organization. Likewise as Chalmers (1967, 1968a, b) has pointed out, some types of activity and certain sex and age classes will be more readily visible than others. Monkeys that are moving around, playing, or chasing will be more conspicuous than ones that are sitting about doing nothing or indulging in some less vigorous social activity such as mutual grooming. Similarly certain sex and age classes might be more conspicuous than others by virtue of differences in size or behaviour. Behaviour recorded by the observer will therefore be a biased sample of the total behaviour of the animals. Such a bias is probably present to some extent in all field studies, but it is exaggerated in forest work because the screening effect of the environment is superimposed upon and reinforces the natural "filter" of the observer.

There are two problems, then. The quantity of information that can be collected is small, and even such data as can be obtained is suspect in quality. The confusion that may arise is well illustrated by recent research on the blue monkey (Aldrich-Blake, in preparation), and much of this section will be exemplified by references to my work on this species.*

In build and appearance, blue monkeys are fairly typical forest guenons (*Cercopithecus* spp.). They are predominantly grey in colour, with black forelimbs, a black tip to the tail, the crown of the head black, and a pale grey band across the brow. They are found in montane and lowland rain forests

* This research was supported by a grant from the Medical Research Council, London.

in Uganda, the eastern Congo, Kenya west of the rift valley, and the southern Sudan. Related races occur in isolated areas from Ethiopia south to Natal and west to Angola (Tappen 1960). My own work was carried out mainly in the Budongo forest, on the escarpment above Lake Albert in North-west Uganda.

Most monkeys and apes that have been studied to date live in well defined social groups. These vary in their size and composition, and the basic units may sometimes be combined into larger aggregations, e.g. geladas *(Theropithecus gelada)* (Crook 1966), hamadryas baboons *(Papio hamadryas)* (Kummer 1968) but there is seldom any difficulty in establishing the existence of such groups. The only major exception is the chimpanzee *(Pan troglodytes)*. Chimps, at least in two study areas, are found in small parties of varying composition that join together and split up again depending on feeding conditions, the only bond of any permanence being that between a mother and her offspring (Reynolds and Reynolds 1965, Goodall 1965).

In the blue monkey a discrete group structure is not immediately apparent. The monkeys are generally encountered in small parties of four or five animals, but these do not appear to be constant in their composition and they seem continually to coalesce and split up again. Because of the poor conditions of observation very few of the animals in the study area could be recognized individually, but those that could be distinguished clearly were seen sometimes together and sometimes in separate parties up to a quarter of a mile apart. At first sight, therefore, the blue monkey pattern of social organization appears very similar to that of the chimpanzee.

But can this initial impression be taken at its face value? It is possible that the apparent lack of well defined groups might result from the difficulty of seeing the animals clearly. In most parts of the forest the visibility is very restricted; if a group of monkeys was spread over an area greater than the observer's field of vision one would only expect to see a small proportion of them at any one moment. This then is the critical question; is there in fact no definite group structure, or is the apparent lack of discrete groups an artifact of the conditions of observation?

Data on the apparent size and composition of parties was subjected to critical analysis to try to determine the causes of variation. Firstly, it was found that the apparent size of parties increased progressively with time after the initial sighting. Mean size on first sighting was 3.6, but this increased to 6.1 after 15 min of observation, and to 7.7 after 30 min. Thereafter the increase was less marked, but even after an hour's observation more monkeys might appear. These figures are for the total number of monkeys known to be in the vicinity, rather than for those actually visible at any one moment. Some of the increase, particularly the rapid initial rise, is no doubt due to monkeys partially concealed by foliage and branches becoming visible, and some to monkeys moving through the observer's field of vision from further away, though it is difficult to distinguish between these in the field. If the monkeys do move in discrete groups, therefore, the typical "spread" of a group must be greater than the distance one can see in the forest. As visibility is so restricted this conclusion is not in itself particularly surprising, but it does have important consequences. Most periods of observation are of short duration, 30 to 40 min or less, so one will hardly ever have seen all the monkeys in the vicinity before contact is lost. Figures for group size and composition will therefore be incomplete.

This would not be such a serious drawback if it were possible to predict the actual number of monkeys in an area, given that a certain proportion of them had been seen in a short time. If there was little variation in the "growth rate" of parties after the initial sighting, this would be possible, but unluckily there is too much variation in the rate of increase for such an approach to be feasible. An alternative method is to determine the proportion of cases in which the apparent total increases further after remaining constant for a particular length of time. It was found that if the apparent total had remained steady for 15 min the chances of it increasing further were over 80%, and even after numbers had been constant for half an hour there was still a 50% chance of further increase. Numbers would have to remain steady for nearly an hour before one could be confident that few if any more monkeys would materialize. In practice such constancy in numbers was seldom observed, so prediction of actual party size, even during prolonged periods of observation let alone the more typical brief interludes, was not possible.

Various other factors were also found to affect the numbers of monkeys seen. The density of the vegetation, for instance, had an obvious influence on numbers. In those parts of the forest where the canopy was thin and the field of vision correspondingly extensive, parties of monkeys appeared larger than in regions where the luxuriance of the vegetation impeded visibility.

Similarly, the time of day was found to have a slight effect on apparent party size. Parties appeared larger in the morning and evening and smaller in the middle of the day. Careful observation indicated that these contrasts were the result not of coalescence and splitting of smaller units, but of changes in the level of activity. During the heat of the day the monkeys were inactive and retreated into dense foliage, while in the morning and evening they moved around feeding in more open areas, and so were conspicuous.

Much, therefore, of the apparent variability in group size is an artifact of the conditions of observation, but even when the sources of variation described above have been eliminated there remain fluctuations that appear to be a genuine reflection of social organization. During certain periods of from one to two weeks parties appeared consistently larger, up to twice their usual size. These changes seemed to be related to differences in feeding conditions. The diet of blue monkeys in Budongo was a mixture of fruits (in the widest sense of the word), leaves, flowers, and young shoots, from a variety of plants. Sources of food were therefore usually widely scattered. Sometimes, though, a particular area contained a single, concentrated source of food, such as a fruiting fig tree, (*Ficus sp.*), and under these circumstances larger parties were seen. There was also a third type of situation; at certain seasons food was very abundant but still distributed over a wide area. Such a state of affairs was found when common lower canopy trees such as *Celtis* spp. were fruiting; these often grow as large stands rather than the small clumps or isolated specimens characteristic of most forest trees. Under these conditions small parties were again the rule, even though any one tree contained enough fruit to feed a large party of monkeys. It would seem, therefore, that wide dispersion is the normal pattern for the species, but that the monkeys can form larger aggregations if feeding conditions so dictate. Such flexibility in dispersion is of obvious adaptive value in the exploitation of a shifting pattern of food resources. Some of the apparent variation in party size, therefore, is a consequence of the monkeys' social organization, but this does not bring us any nearer to determining what the basic social structure of the species is. Variation in party size with changes in feeding conditions might take place within an overall group structure, or alternatively there might be no distinct groups at all. Initial impressions, during the first few months of the study, favoured the second alternative, but subsequently patterns of behaviour were observed that were difficult to fit in with such a structure.

A typical incident would be as follows. A large party of monkeys would be feeding on some rich food source such as a fruiting fig tree. A second big party would move into the fig tree with it, whereupon a mature male from each party would start to make loud calls and bound conspicuously from branch to branch, and adults from the two parties would threaten and chase one another. This would continue for about five minutes, and one party would then withdraw.

Such behaviour is a complete contrast to the invariably peaceful mingling of smaller parties. When small parties coalesce the only overt behaviour is a few quiet croaking noises as they come into contact with one another. Threat and aggression are virtually never seen outside the circumstances described above, even under the same sort of feeding conditions. It is tempting to conclude that such incidents are examples of territorial behaviour, and that the large parties involved represent distinct groups.

There is further evidence that suggests that discrete groups do exist. If the home ranges of the few animals that could be recognized individually are superimposed on one another, it is found that the ranges of those monkeys that were sometimes seen in the same party occupy almost the same areas, while they overlap only marginally if at all with those of other individuals. Such a pattern would be expected if there were distinct groups with home ranges that overlapped but little. If, on the other hand, the social structure resembled that of the chimpanzee ranges of individual animals might be expected to show all degrees of overlap.

There is on the one hand, then, considerably flexibility in dispersion as revealed by changes in average party size with feeding conditions, and, on the other, evidence strongly suggestive of a division of the population into distinct groups. The only hypothesis capable of accommodating both sets of facts would seem to be that blue monkeys do indeed live in distinct groups, but that the groups do not move as compact, integrated units. Under typical feeding conditions the group would be scattered over a wide area and individual parties might forage independently of one another. Only when food supplies are rich and localized is the whole group found together.

If it is accepted that distinct groups exist, and much supporting detail not mentioned here favours this view, the next step is to determine their size, composition, and home range. With open country species repeated counts over a period of days are fairly consistent, and an accurate picture of group size and composition can be established fairly easily. As we have seen, with forest species such counts are unreliable. Indirect methods have to be used. If it is assumed that there is no extensive interchange of animals between groups, group compositions can be calculated from a combination of counts of parties containing recognizable individuals. Suppose, for example, that a particular recognizable animal was seen on one occasion with a male, two females, and a juvenile, and on another occasion with three females and two infants. One could conclude that the group contained at least one male, three females, a juvenile, and two infants, besides the recognizable individual. From repeated sightings a complete picture of group composition can be built up.

The composition of six groups was so determined. It was found that group size ranged from 12 to 17, with a mean of 14. With one exception, each group contained only one mature male. The ratio of mature and sub-adult males to adult females in groups was about 1 : 2, but the sex ratio in the population as a whole was less discrepant since further mature males appeared to live outside the groups. The number of these solitary males in the area occupied by the six groups could not be established with any certainty, but it was probably about four or five.

Similar indirect methods have to be used to determine the home range of each group. With species such as baboons a troop can generally be followed throughout the day and a picture of the range can be built up by plotting daily movements over several days. In the case of the blue monkey the fragmentation of the group and the difficulties of maintaining contact with the animals over long periods render this approach impracticable, particularly since it is not always clear to which group the party of monkeys being watched belongs. Home ranges have therefore to be deduced from sightings and movements of the recognizable individuals in each group and other monkeys seen associating with them. This method of calculation gives group ranges of from 0.020 to 0.045 sq miles in area, with a mean of 0.031 sq miles. Adjacent groups' ranges overlap marginally; allowing for this overlap the population density is 475 per square mile.

Considering the small size of group ranges and the high population density, territorial encounters as described above are surprisingly infrequent. Only 14 such incidents were observed during the 20 months of the study, and none at all during the first six months. Groups may be kept apart normally by noises. The calls made by mature males during territorial encounters are sometimes heard at other times, and, if one male starts calling, others within a quarter of a mile may answer him. Interchange of calls is in itself no evidence of territoriality, but as the calls are the same as those used during aggressive inter-group encounters it seems reasonable to conclude that they may act as a spacing mechanism. Calls of similar function are made by other forest monkeys such as howlers *(Allouatta palliata)* (Carpenter 1934), and titi monkeys *(Callicebus moloch)* (Mason 1966), and also by gibbons *(Hylobates lar)* (Carpenter 1940, Ellefson 1968). In the blue monkey actual territorial encounters were only seen when there were concentrated food sources in the overlap zone between adjacent groups' ranges and a scarcity of food elsewhere. At other times parties from different groups appeared to avoid one another.

The final picture of blue monkey social organization is thus very different from the impression that would be gained from superficial study. A casual observer walking through the forest at a time when feeding conditions were typical would seldom see more than five or six monkeys together, and any notions of group size and composition based on such sightings would most likely be totally erroneous. Yet it is not unknown for "group counts" based on a single sighting to reach the literature, and, while the authors themselves are usually no doubt aware of the shortcomings of such counts, there is a risk that they may be accorded too great a weight in subsequent citations. The results of a casual encounter become elevated to the "mean group size" for a species, and may be used in comparison with data on other species much of which will rest on a more secure foundation. This is not to say that data derived from small numbers of sightings is entirely without value, but the dangers should be appreciated. If, for instance, a visitor to Budongo happened to hit on a time when fig trees or some other prolific species were fruiting, his impressions of blue monkey grouping tendencies would be entirely different from those of our casual observer above. Even if an observer spent several weeks in intensive study he still might not establish the basic pattern of social organization unless he was lucky enough to see clear cut territorial encounters; indeed after six months of my own study I was almost convinced that blue monkeys had no distinct group structure, but rather a pattern of organization similar to the chimpanzee. Only after nine months was it clear that this view was errone-ous. One wonders what further revisions of opinion would be brought about by an even longer study, say three or four years rather than just 20 months! The moral to be drawn is clear. A reliable picture of the social organization of a forest species can only be obtained by prolonged study; the results of casual observation or short term projects can be positively misleading.

To give a specific example of the difficulties of interpretation that can arise, it was stated in a paper on the redtail monkey *(Cercopithecus ascanius)* (Buxton 1952) that group size varies at different times of day. Large bands were most frequently seen at about 0900 hours and again from 1400 to 1600 hours, while at other times smaller bands predominated. It was concluded that members of small sleeping parties joined to form larger aggregations in the early morning and separated again for the mid-day rest period. Larger bands were again formed for the afternoon feeding period, and these split into small sleeping parties in the evening. It will be recalled that a similar fluctuation in apparent party size, though admittedly less marked, was seen in the blue monkey. With blue monkeys, however, the apparent change in size was caused not by actual splitting and coalescence of parties, but by the difficulty of seeing monkeys during the heat of the day when they retreated into thick foliage. It is at least possible that similar considerations might apply to the redtail. The redtail data was derived not from continuous observation of individual parties but from counts of monkeys, by assistants, at regular intervals. Such figures by themselves do not enable one to decide whether the apparent fluctuations in size are the result either of genuine changes in grouping patterns or of changes in activity level, or, as is perhaps most likely, a combination of the two. Buxton himself inclines to the first alternative; while the times at which large bands are most frequent coincide with peaks of activity in morning and afternoon, a further peak in the evening is not marked by any increase in the apparent size of bands.

The actual pattern of organization is most likely to be obscured in species such as the blue monkey and the redtail in which groups are not compact and well defined; but even in species that do live in compact groups data may be distorted. This is well illustrated in Chalmers' study of the black mangabey *(Cercocebus albigena)* (Chalmers 1967, 1968a, b). Mangabey groups are seldom spread out over more than 50 yards, so there is no difficulty in establishing that well-defined groups exist and their composition can readily be determined after a few weeks observation. Problems arise, however, when social interactions within the group are studied. Chalmers points out that behaviour recorded by the observer will not be a truly representative sample. Some activities, such as rapid movement or fighting, will attract the observer's attention, while others such as sleeping or mutual grooming are liable to be overlooked. Likewise certain sex or age classes may be more conspicuous than others. Figures for the frequency of different types of interaction between the various members of the group will therefore not be accurate, and this may lead to erroneous conclusions about the pattern of organization within the group.

Chalmers attempted to devise methods of counteracting such bias. Censuses were taken at regular intervals recording not only the sex and age class and activity of those monkeys visible but also whether they were in full view or partially concealed by foliage and branches. It was found that the ratio of "exposed" to "partially concealed" monkeys was greater for moving animals than for ones engaged in sedentary activities, 4.2 : 1 as opposed to 2.6 : 1. From this it was calculated that a monkey that was not moving had only 0.89 of the chance of being exposed as one that was moving. When comparing the frequency of different activities a "correction factor" of 0.89 was employed.

Unfortunately, the validity of such a factor can be questioned. The method of calculation takes no account of the monkeys that are not seen at all, and it is these, not the ones that *are* seen even though partly obscured, that distort the results. The factor would only be valid if the ratios of "seen" to "unseen" monkeys for each activity bore the same relation to one another as the ratios of "exposed" to "partially obscured" monkeys among those that *were* seen. The figures quoted above certainly imply that some sedentary monkeys are being overlooked, but they do not tell one how many.

Similarly Chalmers found that the relative frequency of sightings of the various sex and age classes in censuses differed from the proportions that would be expected from their relative numbers in the group. Adult males were seen more often than would be expected if all sex and age classes were equally visible, and adult females, sub-adults, and juveniles less often. Comparisons of the frequency of different patterns of behaviour in the various classes were therefore based on their relative visibility to the observer rather than on the known number of individuals in each class within the group. In this case such a correction is valid, since group composition was known with some confidence. Difficulties would arise, though, were data on social interactions drawn from a sample of animals of indeterminate composition.

These, then, are a few of the problems of bias likely to be encountered during a study of forest monkeys. Some are found, though perhaps to a lesser extent, in all field studies, while others may be unique to the forest situation. With care in devising research procedures some of them can be overcome, but reliable information on forest species will always remain more difficult to collect than comparable data on open country animals.

SOME ASPECTS OF SOCIAL ORGANIZATION IN FOREST MONKEYS

Apparent variations in social structure may be the result of distortions of data or observer bias rather than real differences between the animals being studied. Even so, when due allowance is made for such factors, it is clear that variations in social organization among forest monkeys are as great as those to be found in open country animals. Table 1 gives the "vital statistics" of populations of nine species of monkey. Of these the first five are all inhabitants of tropical rain forest and are exclusively arboreal. The next two are less arboreal in habit and spend more time on the ground, 20 to 40% in the case of the Dhawar langurs and as much as 80% in the Kaukori langurs (Yoshiba 1968). Of the two remaining species baboons are generally regarded as terrestrial, savanna animals, though as Rowell (1966) points out they are highly adaptable and equally at home in the forest. The patas monkey, on the other hand, is found only in open grassland and savanna and represents an extreme adaptation to terrestrial life both anatomically and behaviourally. These nine species, then, could be arranged along a continuum of differing degrees of adaptation to arboreal and terrestrial life, with the rain forest species at one extreme and the patas at the other (cf DeVore 1963). Table 1 is not supposed to be an exhaustive review of the literature but merely to illustrate certain trends.

First, as regards group size, home range, and population density there is a considerable degree of uniformity among the rain forest species. Apart from *Callicebus*, with its widely different social structure the typical group size is generally between 12 and 20, the limits suggested in earlier reviews (e.g. DeVore 1963). Savanna animals such as baboons have larger groups, though the patas is an exception for reasons that will become clear shortly. Likewise home ranges are small, 0.03 to 0.06 sq miles or even less in *Callicebus*, as against ranges several square miles in extent for terrestrial species. The howler monkey range is an apparent exception to the general trend, but range boundaries in this species are not clearly demarcated and while a group might wander over an area of perhaps half a square mile during the course of a year the effective range over shorter periods is probably less than this (Carpenter 1965, Chivers, personal communication). Population densities of forest species are correspondingly higher than those of open country animals. Stricly speaking comparisons between species should be made on a basis of biomass rather than number of individuals in a given area but even so forest species clearly achieve much greater densities. The *Presbytis* populations that live in the more heavily wooded habitats incline rather to the rain forest than the savanna pattern in these features, while the Kaukori langurs with their greater group size, larger home range, and lower population density resemble savanna species such as baboons more closely than populations of their own species from forest areas. Similarly the Ishasha baboons have smaller ranges and a higher population density than savanna populations of the same species; in practice the discrepancy is far greater than the figures imply since the baboons spent 60% of their time in the riverine forest and bush although it constituted only 18% of their range. It seems, therefore, that as Chalmers (1968a) suggests, ". . . an arboreal habitat imposes a certain uniformity on group size, population density and home range size on the taxonomically diverse primates

Table 1
Socio-demographic details for certain monkeys.

Species	Locality	Habitat	Group size
Blue monkey, *Cercopithecus mitis*	Budongo, Uganda	Rain forest	14
Black mangabey, *Cercocebus albigena*	Bujuko, Mabira, Uganda	Rain forest	17
Black and white colobus, *Colobus abyssinicus*	Mt. Meru, Kenya	Rain forest	13
Howler monkey, *Allouatta palliata*	Barro Colorado, Panama	Rain forest	18
Titi monkey, *Callicebus moloch*	Barbascal, Colombia	Rain forest	2–4
Lutong, *Presbytis cristatus*	Kuala Selangor Malaysia	"Parkland"	31
Common langur, *Presbytis entellus*	Dharwar, India	Dry deciduous forest	16
Common langur, *Presbytis entellus*	Orcha, India	Moist deciduous forest	22
Common langur, *Presbytis entellus*	Kaukori, India	Dry scrub and cultivated land	54
Baboon, *Papio anubis*	Nairobi Park, Kenya	Savanna	41
Baboon, *Papio anubis*	Amboseli, Kenya	Savanna	80
Baboon, *Papio anubis*	Ishasha, Uganda	Riverine forest and grassland	45
Patas, *Erythrocebus patas*	Murchison Falls Park, Uganda	Open grassland	15

[a] In *Colubus guereza* Schenkel and Schenkel-Hullinger (1966) report multimale groups with a

Range (sq mile)	Population density (sq mile)	Socionomic sex ratio	Social structure	Source
0.03	475	1:2	One-male groups and solitary males	Aldrich-Blake (in prep.)
0.05	200	1:1–1:1.25	Multimale groups	Chalmers (1967, 1968a, b)
0.06	220	1:1–1:1.7	Multimale groups	Ulrich (1961)[a]
0.5?	136	1:2.5	Multimale groups and a few extra-group males	Carpenter (1965) (1959 census)
0.002	1100	1:1	"Family" groups	Mason (1966).
0.08	375	1:13	One-male groups	Bernstein (1968)
0.07	260	1:6	One-male groups and all-male parties	Yoshiba (1968)
1.5	7–16	1:6	Multimale troops	Yoshiba (1968)
3.0	7	1:3	Multimale troops and a few one-male troops and all-male groups	Yoshiba (1968)
3–15	10	1:2.5	Multimale troops	DeVore and Hall (1965)
3–15	25	1:2.5	Multimale troops	DeVore and Hall (1965)
1.5–2	28+	1:1	Multimale troops	Rowell (1966)
20	1	1:4–1:12	One-male groups and all-male parties	Hall (1965)

dominant adult male whereas Marler (1969) reports one-male groups.

living in that habitat." Blue monkeys, mangabeys, and black and white colobus come from widely different Old World genera and howlers are New World monkeys, yet all are remarkably similar in these features of their social organization.

As Crook (p. 110) points out, group size and dispersion will be influenced by many environmental factors such as the dispersion and availability of food and sleeping sites and predation pressure. Many of these factors will be the same for all forest species; hence similarity in group size and dispersion is to be expected. There is no shortage of suitable sleeping sites in forest, and hence this will not set a lower limit to group size. In contrast Kummer (1968) found that in the open country hamadryas baboon (*Papio hamadryas*), troops were larger in those parts of the range where sleeping sites were scarcer and food more abundant. The only predators of forest monkeys are leopards, eagles, and occasionally man. Leopards can be escaped readily by crossing into another tree along thin branches, and eagles by dropping from the canopy into the lowest layers of vegetation. On the other hand open country animals have no such ready means of escape from predators and must either rely on the fighting potential of the group, as a whole and in particular the adult males (baboons), or else on rapid and silent evasive action (patas). In open country monkeys limited sleeping site availability and potentially high predation pressure will therefore lead to the production of large troops, always provided that feeding conditions permit this. Baboon troops living in relatively rich savanna are consistently larger than typical forest monkey groups. Patas groups, on the other hand, are again small, presumably because their habitat is not sufficiently productive to support a large troop on an area that can readily be covered during a day's ranging. Potential predation pressure is still high, but the patas' slender build and great speed enable it to outrun all but the fastest predators.

In forest monkeys, then, availability of sleeping sites and predation pressure will set few constraints on group size, and the availability and dispersion of food will be the major environmental influence.

Another factor that might be expected to influence group size is the thickness of the vegetation. In open country there are few problems of communication, but, in forest, monkeys will seldom be in direct visual contact if the group is spread over more than a few yards. This might be expected to lead both to increased reliance on vocal rather than visual signals and to a restriction of group size. Problems of communication would militate against the establishment of large, coordinated social units.

The high population density and small range size of forest monkeys is presumably explained by the greater year round productivity of forest compared to savanna or grassland. In addition the ranges of forest monkeys are three-dimensional, while those of open country animals are largely two-dimensional. Admittedly many forest species show a preference for a particular level in the canopy, but their vertical range is nevertheless greater than that of open country species. For example blue monkeys in the Budongo forest were more often seen in the middle layers of the canopy, from 35 to 70 ft up, than at other levels; 55% of sightings were in the middle canopy as against 28% in the upper and 17% in the lower canopy. Likewise the mangabeys studied by Chalmers (1968a) showed a preference for the upper layers of the canopy; 64% of records were at 70 ft and above, though only 52% of the available canopy was at such heights. Not all parts of the canopy are used equally, therefore, but ranges are more three-dimensional than those of savanna species.

When one turns from population size and dispersion to a consideration of the detailed social structure of forest monkeys, the picture is not one of uniformity but of considerable and unexpected diversity. Earlier reviews (e.g. DeVore 1963, Crook and Gartlan 1966) have considered arboreal primates as living either in multimale troops with roughly equal numbers of males and females or more rarely in small "family parties" of a single pair of adults with associated young. Some species such as the black mangabey and the black and white colobus do indeed live in multimale troops. The titi monkey *(Callicebus moloch)* provides an example of the "family" structure, groups being composed of a male, a female, and one or two young (Mason 1966), and a comparable state of affairs is found in the gibbon (Carpenter 1940, Ellefson 1968). Other species, though, such as the blue monkey and the lutong *(Presbytis cristatus)* (Bernstein 1968) have a "one-male group" structure with groups containing several females but only one mature male. Yet another pattern is found in the talapoin *(Miopithecus talapoin)* (Gautier-Hion 1966), a monkey of swamp and riverine forest but living, in the area studied, as a commensal of man. In this species bands numbered 60 to 80 individuals. The determination of details of social structure proved impracticable, but it is recorded that at night the band split into units of a single female with a young infant and the infant of the previous year. Males slept apart from the females and young.

The diversity of social structure among forest monkeys is thus as great as the differences between forest and open country species. In particular, some forest species have the "one-male group" structure formerly considered an adaptation to seasonally arid conditions (Crook and Gartlan 1966). In the remainder of this section close attention will be paid particularly to this type of social organization. The argument for the adaptiveness of one-male groups is as follows. One male can fertilize several females; thus males are in some senses biologically expendable. If a population is subject to periodic food shortages it will be advantageous for the number of males to be reduced, minimizing competition between males and females for scarce food resources. Hence, unless other factors such as defence against predators militate against it, the trend will be towards a reduction in the overall number of males and the separation of some of them from the females and young as individual extra-group males or all male parties. This hypothesis appears plausible; socionomic sex ratios are generally found to be disparate in monkeys living under extreme environmental conditions, and the first species for which the one-male group structure was described, the hamadryas (Kummer 1968), the gelada (Crook 1966) and the patas (Hall 1965), are all inhabitants of severe environments. In the case of the patas, in Uganda at least, individual one-male groups remain separate, while in the gelada and hamadryas they form larger congregations but show flexibility in dispersion with changes in feeding conditions.

For the hypothesis to stand up to critical examination it has to be shown, firstly, that separation of extra-group males from females and young does indeed minimize intra-specific competition between the sexes and, secondly, that under extreme conditions the extra-group males suffer greater mortality than females and young. There is some evidence for the former (Crook 1966), but little as yet for the latter unless disparity in overall sex ratio is taken as proof of selective mortality. Figures for the relative abundance of the two sexes must be interpreted with caution; in species such as the patas extra-group males are very much harder to find than one-male groups so samples may be unrepresentative of the population as a whole. Even if the overall sex ratio is derived from a representative sample a superfluity of females cannot be taken as conclusive proof of differential mortality; in a small population random fluctuations from a 1 : 1 sex ratio at birth can produce quite large discrepancies (Rowell 1966), and if mature males only are considered differences in the rate of maturation of the two sexes will also affect the ratio.

If it is accepted, though, that the one-male group structure is an adaptation to periodic food shortages, why is it found in forest monkeys? Tropical forest is generally thought to provide a constant and abundant supply of food. And yet not only do semi-terrestrial species such as langurs and lutong have one-male groups, but also arboreal monkeys such as the blue monkey from climax rain forest. Indeed it seems that the one-male group pattern may be fairly widespread in such species; Gartlan (personal communication) reports that it is probably found in *Cercopithecus nictitans* in Cameroon, and possibly also in *C. mona* and *C. erythrotis*. Likewise Bernstein (1968) records it not only in *Presbytis cristatus* but also in *P. obscurus* and *P. melalophus* in Malaya. Far from being confined to arid country animals, the one-male group structure is found in a wide variety of habitats.

In some cases such a structure can be explained on the same basis as in open country species. The Dharwar langurs, for example, live in semi-deciduous forest subject to a very severe dry season, and there is some evidence that the population may in any case be artificially high as a result of a recent forest clearance (Yoshiba 1968). During the dry season fruit and leaves are unavailable; the monkeys live primarily on buds and nuts and are sometimes seen to eat the bark off trees (Sugiyama 1964). However ever-green forest does not show such seasonal fluctuations in productivity, and yet some monkeys in this habitat nevertheless have one-male groups.

Crook (elsewhere), suggests that at the high population densities found in forest monkeys, food supplies, though relatively constant, may be limiting. If a population is up against a ceiling imposed by food shortage a one-male group structure might be advantageous. This view is doubtless a simplification; over the forest as a whole productivity may be more or less constant throughout the year, but this is not true of the home ranges of individual monkeys. As Ellefson (1968) points out, the seeming abundance of food in tropical rain forest is partly illusory. Most forest trees, particularly the upper canopy species, grow not in large stands but rather singly or in small clumps. The range of a group of monkeys, being relatively small, may contain only a few of any one tree species. For instance the ranges of four blue monkey groups in Budongo contained between them only five fig trees, two in one group's range, one in each of two other groups, and one in the overlap zone between the first and the remaining group's range. Over the forest as a whole fig trees fruited continuously from September 1966 to March 1967, but the fruit on any one tree would last only about 10 days. Figs were consumed avidly not only by blue monkeys but also by redtails, chimpanzees, various birds, squirrels and fruit bats. Once the fruit started to ripen it was soon stripped; a party of chimps could clear a tree in two or three days. Monkeys would seldom feed in close proximity to chimpanzees; if chimps moved into a fig tree the monkeys moved out, so the effective fruiting period was further reduced. Thus in any one group's range there would be a plentiful supply of food for perhaps a week while the fruit was ripening, and none at all thereafter.

This pattern of fluctuation in resources is found throughout much of the year. Outside the forest there is a marked alternation between wet and dry seasons with associated changes of vegetation, but within the forest seasonality is less clear cut. While there are changes in the vegetation during the course of the year they are often on a different time scale to the climatic cycle and not necessarily related to it. For instance while many trees lose their leaves at the beginning of the dry season others shed them at the beginning of the rains, and some of those species that lose them early in the dry season put out new shoots well before the rains recommence. Likewise trees of some variety or other are fruiting during most months of the year, and while certain species may fruit at the same time each year others are much more erratic in thir pattern. Add to this the limited extent and vegetational heterogeneity of the individual group's range, and it will be clear that feeding conditions fluctuate not from season to season, but rather from week to week. At some times food is abundant, at others it is scarce and the group splits into small parties which forage independently of one another. The same arguments used to account for one-male groups in open country species could therefore be applied to the blue monkey as well.

Social structure is the product of an interaction between ecological factors on the one hand and the basic behavioural repertoire of the population, ultimately dependent on genetic factors but moulded by socialization

processes, on the other. It follows that if species living in the same environment show contrasts in social organization, this may be the result either of
contrasts in the ecological factors operative in the particular niche that each
species occupies or of contrasts in basic behavioural repertoire, or a combination of the two. At present we lack convincing explanations as to why
some forest monkeys have one-male groups and others multimale ones.
Further research seems to require especially careful studies of the differences between the niches occupied by these species.

It is clearly mistaken to regard "forest" as a uniform environment.
Firstly, as Rowell (1966) points out, there is no hard and fast distinction
between forest on the one hand and savanna or grassland on the other; all
types of intermediate are found. Those habitats that are classified as forest
differ greatly in their structure, vegetational composition, and pattern of
seasonality. For instance both the Dharwar langurs and Chalmers' mangabeys lived in "forest," but there the resemblance ends; the langur habitat
is dry, semi-deciduous forest subject to very severe dry season, while the
mangabey habitat is evergreen rain forest. Secondly, even within a particular
type of forest there may be different ecological niches. Evergreen and moist
semi-deciduous forests often support a rich and varied primate fauna. The
Kibale forest in Uganda has eight species of monkeys, including four *Cercopithecus* spp. and two *Colobus* spp., and a comparable diversity is found in many
other forests. Presumably these species occupy contrasting ecological
niches, though in some cases niche separation must be rather fine since
mixed species parties are commonly encountered (Haddow 1952, Gartlan,
personal communication and personal observation).

The ecological relations between the members of such a forest primate
community have yet to be worked out in detail. Nevertheless, during my
study of the blue monkey in Budongo I was able to obtain some information
on the relations between this species and the other primates, redtails, black
and white colobus, baboons, and chimps, found in the same forest. While
no species confined itself exclusively to any one level there was some stratification of species in the canopy, colobus tending to occupy the upper, blue
monkeys the middle, and redtails the lower layers. The relative frequency
of species varied in different parts of the forest; baboons were based on the
forest edge and foraged out into the savanna and a short way into the forest,
redtails were found in young colonizing forest and in secondary growth on
the edge of tracks and clearings, while the remaining species were found
deeper into the forest. There were varying degrees of overlap in diet as
assessed by species composition of the food and parts of the plant eaten;
colobus, for instance, were exclusively leaf eaters while blue monkeys and
redtails were more catholic in their tastes. While these findings must be
considered as preliminary it does seem reasonable to conclude that individual species do occupy contrasting niches. Hence it is not surprising to
find that their social structures also differ.

Full understanding of the critical variables involved must await a more detailed delineation of niches, but contrasts in the pattern of food productivity may prove to be of major importance. Other factors such as predation pressure and the availability of sleeping sites probably vary little between niches. For instance blue monkeys, with their one-male group structure, are largely frugivorous and are subject to periods in which the availability of food is limited. Black and white colobus, on the other hand, are exclusively leaf eaters, and while they prefer young shoots to old leaves the latter are perfectly acceptable. Their food supply would therefore be more constant, and the ecological pressures correspondingly different.

This cannot be the whole story; black mangabeys, for example, have a diet comparable to that of the blue monkey yet live in multimale groups. As Crook (loc. cit) points out, social organization is influenced by many factors other than ecological variables, and the resultant pattern will be a balance between a variety of forces. Those features of forest monkey social organization that show marked similarities are related closely to environmental factors, while details of social structure, which show a greater diversity, are affected by environment only indirectly. Conversely differences in behavioural repertoire and frequency of interactions exert their effect directly upon social structure and only indirectly upon population size and dispersion. The point at which a balance is struck differs from one species to another; hence similarity in some features of organization and diversity in others is only to be expected.

CONCLUSION

Apparent differences in social organization may reflect genuine contrasts between the animals being studied or may be the result of distortions introduced during the collection of data. Forest monkeys present particular problems of observation, and information on these animals is especially prone to unreliability. Short-term studies can be positively misleading; only after prolonged investigation can the pattern of organization be established with any confidence. In spite of these difficulties, sufficient research of forest monkeys has now been done for certain trends to emerge. On the one hand, there is some degree of uniformity in features such as group size and area of home range. Details of social structure, on the other hand, show considerable and unexpected variation. Many of the similarities can be ascribed to the operation of common ecological factors, and some of the differences may be based upon rather more subtle contrasts between ecological niches than have yet been considered. These conclusions are necessarily speculative and a more complete understanding of the critical determinants of social organization must await further research.

REFERENCES
Aldrich-Blake, F. P. G. (in preparation). The ecology and behaviour of the blue monkey, *Cercopithecus mitis stuhlmanni.* Ph.D. thesis, Univ. Bristol.
Altmann, S. A. (1959). Field observations on a howling monkey society. *J. Mammal.* **40,** 317-330.
Bernstein, I. S. (1968). The lutong of Kuala Selangor. *Behaviour* **32,** 1-16.
Booth, A. H. (1957). Observations on the natural history of the olive colobus monkey, *Procolobus verus* (van Benenden). *Proc. zool. Soc. Lond.* **129,** 421-431.
Buxton, A. P. (1952). Observations on the diurnal behaviour of the redtail monkey, *Cercopithecus ascanius schmidti* (Matschie), in a small forest in Uganda. *J. Anim. Ecol.* **21,** 25-58.
Carpenter, C. R. (1934). A field study of the behavior and social relations of howling monkeys. *Comp. Psychol. Monogr.* **10**(48), 1-168.
Carpenter, C R. (1935). Behavior of red spider monkeys in Panama. *J. Mammal.* **24, 346-352.**
Carpenter, C. R. (1940). A field study in Siam of the behavior and social relations of the gibbon, *Hylobates lar. Comp. Psychol. Monogr.* **16**(5), 1-212.
Carpenter, C. R. (1965). The howlers of Barro Colorado Island. *In* "Primate behavior: field studies of monkeys and apes." (I. DeVore, ed.) Holt, Rinehart and Winston. New York.
Chalmer, N. R. (1967). The ecology and ethology of the black mangabey, *Cercocebus albigena.* Ph. D. thesis, Univ. Cambridge.
Chalmer, N. R. (1968a). Group composition, ecology, and daily activities of free living mangabeys in Uganda. *Folia Primat.* **8,** 247-262.
Chalmer, N. R. (1968b). The social behaviour of free living mangabeys in Uganda. *Folia Primat.* **8,** 263-281.
Collias, N. and Southwick, C. (1952). A field study of population density and social organization in howling monkeys. *Proc. Am. phil. Soc.* **96,** 143-156.
Crook, J. H. (1966). Gelada baboon herd structure and movement: a comparative report. *Symp. Zool. Soc. Lond.* **18,** 237-258.
Crook, J. H. and Gartlan, J. S. (1966). Evolution of primate societies. *Nature, Lond.* **210,** 1200-1203.
DeVore, I. (1963). Comparative ecology and behavior of monkeys and apes. *In* "Classification and human evolution." (S. L. Washburn, ed.) Viking Fund Publications in Anthropology No. 37. New York Wenner-Gren Foundation.
DeVore, I. and Hall, K. R. L. (1965). Baboon Ecology. *In* "Primate behavior; field studies of monkeys and apes." (I. DeVore, ed.) Holt, Rinehart and Winston. New York.
Ellefson, J. O. (1968). Territorial behavior in the common white-handed

gibbon, *Hylobates lar*. *In* "Primates: studies in adaptation and variability." (P. Jay, ed.) Holt, Rinehart and Winston. New York.

Gautier-Hion, A. (1966). L'écologie et l'éthologie du talapoin, *Miopithecus talapoin talapoin*. *Biologica Gabonica* **2**(4), 311-329.

Goodall, J. (1965). Chimpanzees of the Gombe Stream Reserve. *In* "Primate behavior: field studies of monkeys and apes." (I. DeVore, ed.) Holt, Rinehart and Winston. New York.

Haddow, A. J. (1952). Field and laboratory studies on an African monkey, *Cercopithecus ascanius schmidti* (Matschie). *Proc. zool. Soc. Lond.* **122,** 297-394.

Haddow, A. J., Smithburn, K. G., Mahaffy, A. F. and Bugher, J. C. (1947). Monkeys in relation to yellow fever in Bwamba County, Uganda. *Trans. R. Soc. Trop. Med. Hyg.* **40,** 677-700.

Hall, K. R. L. (1965). Behaviour and ecology of the wild patas monkey. *J. Zool.* **148,** 15-87.

Hall, K. R. L. and DeVore, I. (1965). Baboon social behavior. *In* "Primates: field studies of monkeys and apes." (I. DeVore, ed.) Holt, Rinehart and Winston. New York.

Itani, J. (1954). Japanese monkeys at Takasakiyama. *In* "Social life of animals in Japan." (K. Imanishi, ed.) Tokyo: Kobunsaya. (In Japanese.)

Kummer, H. (1968). Social organization of hamadryas baboons. *Bibliotheca Primatologica* **6,** 1-189.

Lumsden, W. H. R. (1951). The night resting habits of monkeys in a small area on the edge of the Semliki forest, Uganda. A study in relation to the epidemiology of sylvan yellow fever. *J. Anim. Ecol.* **20,** 11-30.

Marler, P. (1969). *Colobus guereza:* territoriality and group composition. *Science* **163,** 93-95.

Mason, W. A. (1966). Social organization of the South American monkey, *Callicebus moloch:* a preliminary report. *Tulane Stud. Zool.* **13,** 23-28.

Reynolds, V. and Reynolds, F. (1965). Chimpanzees of the Budongo forest. *In* "Primate behavior: field studies of monkeys and apes." (I. DeVore, ed.) Holt, Rinehart and Winston. New York.

Rowell, T. E. (1966). Forest living baboons in Uganda. *J Zool.* **149,** 344-364.

Schenkel, R. and Schenkel-Hulliger, L. (1966). On the sociology of free-ranging Colobus. *In* "Progress in primatology." (Stark *et al.* ed.) G. F. V. Stuttgart.

Sugiyama, Y. (1964). Group composition, population density, and some sociological observations of Hanuman langurs, *Presbytis entellus*. *Primates* **5,** 7-37.

Tappen, N. C. (1960). Problems of distribution and adaptation of the African monkeys. *Current Anthropology* **I,** 91-120.

Ullrich, W. (1961). Zur Biologie und Soziologie der Colobusaffen. *Der Zoologische Garten* **25**(6), 305-368.

Washburn, S. L. and DeVore, I. (1961). Social behavior of baboons and early man. *In* "Social life of early man." (S. L. Washburn, ed.) Viking Fund Publications in Anthropology No. 31 New York. Wenner-gren Foundation.

Yoshiba, K. (1968). Local and intertroop variability in ecology and social behavior of common Indian langurs. *In* "Primates: studies in adaptation and variability." (P. Jay, ed.) Holt, Rinehart and Winston. New York.

22

Energy Relations and Some Basic Properties of Primate Social Organization*

WOODROW W. DENHAM

Recently published field data on non-human primate ecology and social behavior has stimulated much discussion of the control exercised by different environments upon social behavior of populations living within those environments. Here the author attempts to contribute to that discussion by presenting a model of certain relations between food density and distribution patterns and characteristics of predator populations on the one hand, and some basic properties of primate social organization including space and resource allocation, socionomic sex ratios, mating strategies, and the structure of multi-male social groups on the other hand. Some of the better documented primate populations are examined from the vantage point provided by the model, and a new typology of primate societies emerges. Sexual dimorphism and individual mate selection within primate groups is briefly discussed. The article concludes with a statement of its implications for the anthropological study of human social behavior.

In their article on the "Evolution of Primate Societies," Crook and Gartlan (1966) tabulate a large amount of data reported in several field studies of non-human primate ecology and social behavior, detect some interesting patterns in their table, and attempt to account for those patterns by constructing a model in which certain environmental characteristics are treated as independent variables and certain primate behavioral characteristics are

Reproduced by permission of the American Anthropological Association from the *American Anthropologist, 73*(1), 1971.

* I wish to express my appreciation to J. Atkins and G. Orians, both at the University of Washington, for their criticisms of an early draft of this article, and to L. Hiatt and all those who violently but constructively attacked a later draft when it was presented in a seminar at the University of Sydney, Sydney, N.S.W., Australia. Many of the article's virtues are attributable to the efforts of the critics, but all of its faults are my own creations.

treated as dependent variables. Their article is an important and stimulating one; coupled with recent field reports (in, e.g., Jay 1968), it provides ample reason for agreeing with Gartlan's (1968: 103) statement that " . . . social structure in many widespread primate species is largely habitat—rather than species—specific." However, their paper suffers from weaknesses related to the unsystematic presentation of the major argument resulting in some conspicuous holes in that argument, and to the manner in which the authors deal with sexual dimorphism and individual mate selection. Here I present an alternative model which I believe is more systematic, complete, and internally consistent than the one provided by them, rearrange the table published in their paper adding substance to the set of abstractions presented in the first part of the article, use the first two steps as a background against which to view some of the data currently available concerning individual mate selection among non-human primates, then suggest some possible implications of this work for the anthropological study of human social behavior.

Primates invariably engage in multiple energy exchanges. Within a food web, each primate is both a consumer and a producer—he obtains energy from plants or from other animals that are directly dependent upon plants, and he in turn is an energy source for parasites and predators. Given this position in the web, he displays two different sets of responses: (1) those that culminate in the acquisition of energy from sources at lower trophic levels, and (2) those that prevent other animals from using him as an energy source. Energy acquisition may take the forms of hunting, foraging, plant and/or animal domestication, or biochemical synthesis. For convenience, only the first two categories will be dealt with in this paper. Those responses that effectively terminate major energy transfers at the organism in question may assume likewise a variety of forms such as passive concealment, escape, confrontation, and counterpredation. Again for convenience, the last form will be disregarded in this article.

There is a third set of responses in which acquired energy is expended: those associated with reproduction. For all primates, the male and female subsystems of the reproductive system must be linked physically, usually several times, before the development of a new organism can begin. While copulation is the only natural form of linkage between reproductive subsystems that generates offspring, there are several different associations of males and females that can provide the frequency of copulations that will result in the survival of the population. These include the following mating patterns: monogamy, polygyny, polyandry, and promiscuity.

We have, then, three sets of response types that will be displayed by most members of all continuing primate populations: energy acquisition, defensive, and reproductive strategies. All of these energy exhanges occur within an analytically isolable ecosystem which, for our purposes here, can be divided into two parts: (1) the subject population and its individual members, and (2) the physical environment. To this point we have dealt with the former; now let us turn to the latter.

Corresponding to the physiological characteristics of the members of the subject population are certain characteristics of the environment which may vary only within certain limits if the population is to survive there. In considering primates in habitat, those parts of the environment with which we must deal directly are food and predators. Food may be used as a handy though sometimes unreliable index to a multitude of other critical factors that determine the overall energy content of the environment—composition of the atmosphere, temperature, water availability, soil type, etc. Various combinations of these factors produce food supplies of different composi- tions, densities, and temporal and spatial distributions (predictability), all of which necessarily affect the behavior of primate populations that might utilize those resources. For the remainder of this article, I assume that a specific primate population will occur only where food of a kind usable by that species is present. Thus we can control for, or disregard, the "food supply composition" variable. The energy stored in the environment affects other animals as well; for our purposes here, those that are members of predator species. The characteristics of predators of significance to a pri- mate population are size and density of the predator population, and behav- ior patterns of individual predators bearing upon avoidance or defense by the prey.

Relations between characteristics of the environment, the biological structure of the subject organisms, and the behavior of those subjects are indicated in Fig. 1. I acknowledge the existence of feedback relations be- tween behavior and structure, and behavior and environment; however, since those relations are of only secondary importance to me in this article,

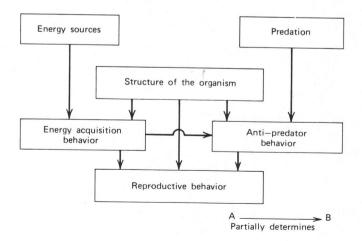

Fig. 1. Flow chart showing relations between environment, structure, and behavior.

I have omitted them from the flow chart. Obviously, the chart should apply to all sexually reproducing organisms. How to get from such an extremely general chart to something applicable to a specific population is the next problem. In the process of solving it, I shall attempt to justify the selection and arrangement of the elements in the chart.

The remainder of this article is based upon the assumption that in energy acquisition, anti-predator behavior, and mating, the most efficient strategy compatible with the structure of the organism is used by members of a primate population in a natural habitat; i.e., that inefficient strategies are selected against by the differential reinforcement of efficient and inefficient responses during the lives of individual organisms, and by the diminished genetic contribution to future generations of those individuals whose behavior is less efficient than that of their conspecifics in the same or similar situations. As used here, efficiency is to be measured exclusively in terms of energy utilization.

Now look at Fig. 2, 3, and 4.* Figure 2 shows the relationship between the density of usable energy, ED, in the environment and the size of an area adequate for maintaining a given organism. As ED decreases, the size of an area required by the organism increases, a relationship that should be self-explanatory. (The exact shape of the curve is of no importance in this or in any of the other figures in this article.) However, within a viable population with a primary sex ratio of 1:1, many males are expendable; i.e., one male is adequate for the impregnation of several females, all of whom can give birth at approximately the same time. Therefore, as ED decreases (Fig. 3), the socionomic sex ratio can shift from 1:1 to progressively fewer actively

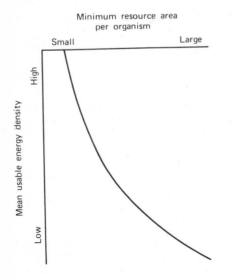

The more abundance the resource the smaller the size of the home range and visa/versa.

Fig. 2. Relation between energy density and size of resource area.

* The relations discussed in conjunction with Fig. 2 to 5 are analyzed in greater detail in, e.g., Orians 1969: 647-668.

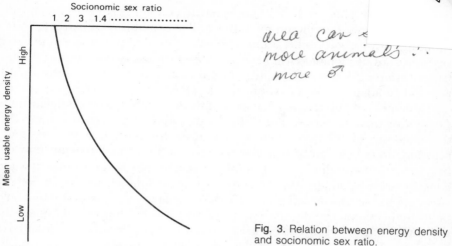

Fig. 3. Relation between energy density and socionomic sex ratio.

mating mature males per female (or, conventionally, more females per male) without impairing the survivability of the population. The superfluous males may completely disappear from the population, or they may simply become sexually inactive while surviving on marginal energy sources. Although this might seem to suggest altruistic behavior on the part of the "extra" males, that is not intended. It is simpler to account for their exclusion or isolation in terms of superior mating skills of other males. (Skill, used in this sense, is a multi-dimensional trait related primarily to the problem of individual mate selection, the topic dealt with in the next section of this article.)

In Fig. 4, energy density is held constant but the degree of predictability of resources is allowed to vary. When food is evenly distributed in the

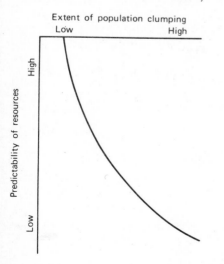

Fig. 4. Relation between energy predictability and population clumping.

environment (highly predictable), it is most economical for the consumers of it to be evenly distributed also. But under conditions of extreme food clumping (low predictability), an evenly distributed population will perish, since most of its members will have no access to food. Consequently, population clumping is the best strategy in this situation, and those individuals who fail to adapt to life in groups will leave few descendants.

Members of populations that are evenly distributed throughout an area can utilize their resources most efficiently if each organism has a clearly delimited set of resources that are adequate for body maintenance and reproduction. This can be had economically, if the animal expends part of his energy in defending a food supply—either in physical encounters with conspecifics who might use the resources in his immediate vicinity, or in signaling to conspecifics the boundaries of the areas already in use by him, or in both. As energy density decreases and space requirements increase correspondingly, resource defense through contact becomes increasingly expensive to the organism, while signaling becomes a relatively more economical defense strategy. The solid line in Fig. 5 represents the energy required for physically defending areas of different sizes, while the dashed line represents one of the consequences of allocating energy for area and resource defense. In the left half of Fig. 5, increases in defensive activity result in gains in energy available for body maintenance and reproduction; in the right half, any increase in defensive behavior would result in a loss of energy for other activities. Since individuals who attempted to defend resource areas at the "large" end of the scale would be at a great reproductive disadvantage or would completely fail to reproduce, the extension of the two lines into that region is purely hypothetical. This, of course, applies

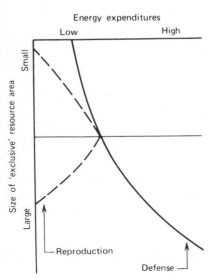

The amount of energy expended should be porpotional to reward.

Fig. 5. Relation between size of resource area and energy expenditure for defending that area (see text for explanation).

primarily to physical defense: signaling, or symbolic, defense can remain inexpensive to the organism even in relatively large areas.

Area and resource (i.e., territorial) defense by individuals is incompatible with population clumping, but the appropriation of specific areas by groups of animals is not precluded by that situation. In this case, where animals are separated already by energy poor regions, physical defense would be called for only rarely, but signaling defense might be employed regularly with varying degrees of success. In other words, physical defense of resource areas by groups might be more effective temporarily and intermittently, but in the long run it would prove to be prohibitively expensive. Hence, we should expect to find relatively imprecise boundaries between groups, with overlap of the areas of different groups. Where extreme resource/population clumping is the case, territorial defense of any kind is unnecessary since groups of conspecifics are adequately spaced by the conditions of energy distribution, i.e., by "no-man's lands."

Fig. 6 summarizes the preceding figures and paragraphs. Each of the four cells is a different type of environment defined in terms of food predictability and density. Henceforth, the environmental types are referred to by the two-letter abbreviation appearing in the top left corner of each cell. In the top right corner of each, there is a schematic representation of the food predictability and density pattern characteristic of the environmental type. Listed in order in each cell are: socionomic sex ratio, population distribution, and mode of area/resource allocation.

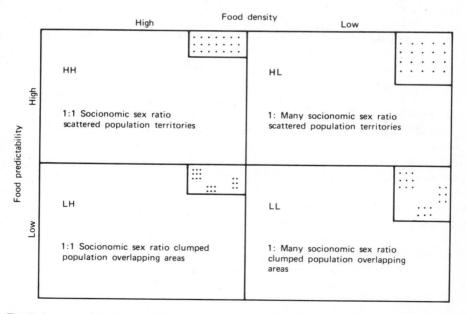

Fig. 6. Demographic characteristics of primate populations under four different conditions of energy availability.

Now that we have examined the organism as an energy consumer, let us look at him as an energy source. The various food acquisition strategies discussed above have their defensive counterparts for all animals that live long enough to reproduce. Although there are several kinds of relationship in which one animal uses another as an energy source, I shall deal only with the predator-prey relationship.

In order for an animal to qualify as a predator against another animal, it must be structurally equipped to overcome the other animal. However, any one predator, even a perfectly equipped one, is a significant threat to a viable prey population only if it frequently encounters members of that population, or if there are several such predators operating in the same area simultaneously. In other words, both the capabilities of the predator and the frequency of his encounters with the prey must be considered in evaluating the effects of predator behavior upon the behavior of prey. Given structurally qualified predators with a population density great enough to make them a threat to a prey population, the prey may use any of the survival reponses listed previously. But the responses he uses must be compatible with his food acquisition strategy, his structural capabilities, and the physical characteristics of the environment in which he lives. Specifically, since the acquisition and storage of energy by an organism is prerequisite to his being a usable energy source, the anti-predator strategy must depend in large part upon the food acquisition strategy. For example, primates that forage in LL environments (Fig. 6) cannot defend against predators by passive concealment—the size of the primate group and the physical characteristics of the environment make a successful passive defense virtually impossible. Similarly, evenly distributed primates in HH environments cannot rely upon the aid of conspecifics in warding off predators, because spacing between conspecifics is too great and territorial boundaries are too firmly established to permit anything other than independent action.

Direct confrontation of predators is a dangerous business, but in some environments it is necessary. Females who participate in those confrontations are less likely to leave descendants than are their female conspecifics who avoid them, so that selection favors females who stay out of interspecific fights. This means, of course, that in environments where confrontation is necessary in contending with predators, either (1) the males fight for themselves and consequently defend the females, or (2) the population is likely to become extinct.* In other words, the female who does not fight when a predator appears is more likely to leave offspring than is one who does fight, but in this environment it is the males that succeed in interspecific confrontations who will father the next generation of their own species. Two other anti-predator strategies, escape and concealment, are superior to confronta-

* A third possibility is the development of sexual polymorphism with males being either "fighters" or "lovers" but not both. That, however, would imply an extremely limited and inflexible behavioral repertoire for each individual and an extremely rigid social order for the group, conditions found among the social insects but not among any species of mammals.

tion when they are compatible with energy acquisition patterns and the physical characteristics of the environment, because both reduce the chance of failure always present in an interspecific fight. For reasons to be given, it is convenient to put these three anti-predator response types into two categories: active (confrontation and escape) and passive (concealment).

Now let us integrate the discussions of food acquisition strategies and anti-predator strategies, and proceed from there to a discussion of reproductive strategies.

Refer again to Fig. 6. Each cell is defined in terms of density and distribution of resources, both of which are treated as two-valued variables. The four combinations are listed in Table 1 in conjunction with names frequently applied to biomes characterized by those conditions. In HH environments, most available energy is stored well above ground level. Progressively lower entries in the table correspond to conditions in which more and more of the energy is stored at or below ground level. Still assuming that the most efficient energy acquisition strategy is used, it is reasonable to expect a high positive correlation between HH and arboreality, LL and terrestriality, and mixed arboreal-terrestrial habitation in the transitional resource zones. This realtionship is expressed by the dashed line in Fig. 7 However, some caution must be used with each of the biome terms. The biome designation is simply a handy label for an extremely complex set of environmental conditions — specific conditions must be dealt with in any analysis. For example, a terrestrial species living in an area marked as "rain forest" on plant distribution maps might experience food availability conditions similar to those typical of low density environments, since little food is available at ground level within rain forests. Similarly, a mixed arboreal-terrestrial population living in a large woods surrounded by arid mountains has energy sources quite different in density and patchiness from those available to fully terrestrial primates living in the open, barren country nearby. Therefore one must focus upon the *relevant* parts of a macroenvironment before attempting to predict the behavioral characteristics of its inhabitants.

Population characteristics can be deduced — given an environmental type and the assumption that the most efficient energy acquisition strategy compatible with the structure of the organism will be employed. But predator defense strategy cannot be arrived at in the same way, since the physical structure of the habitat and the kind and number of predators in

Table 1

Common names for types of environments defined in terms of energy distribution and density

HH: rain forest
LH: mixed savanna (trees + grass)
HL: open savanna (grasslands)
LL: desert or semi-desert

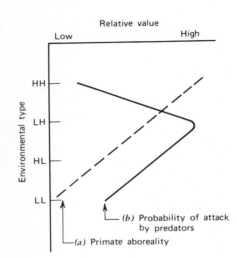

Probability very high in LH, HL decreases in HH, LL.

Fig. 7. Relation between environmental type and (a) degree of primate arboreality, and (b) probability of attack by predators.

that habitat must be taken into consideration. In Fig. 7, the solid line represents a rough estimate of the relative probability of attack upon primates by predators in the four biome types of Table 1. At this time, it is a frankly intuitive rating based upon a general but superficial knowledge of animal life in Subsaharan Africa.*

Given the correlations shown in Fig. 6 and 7, what antipredator strategy is available to primates living in each environmental type? In HH there are few predators, the primates are largely arboreal, and they are evenly distributed in space. Encounters with predators are rare; escape and concealment can be equally economical. This situation, taken in conjunction with earlier comments on the relative merits of the various defensive strategies, suggests that among primates in HH environments there will be no selection for males larger, more powerful, or more aggressive than females, either when the anti-predator strategy is passive or when it is active.

In LH environments, primates tend to form large groups and lead a mixed arboreal-terrestrial life. Encounters with predators would be relatively frequent occurrences, with confrontation and flight into scattered trees being viable anti-predator strategies. General clumping of the primate population makes effective concealment of individuals very difficult to accomplish, but the very fact of population clumping does, in its own right, make the task of locating prey more difficult for a predator. Flight will produce no selection for sexual dimorphism, but confrontation will. However, confrontation with a predator can be experienced either by a lone male, or by most or all males in the group. Lone confrontations select for

---

* Research such as that described by Schaller (1969) and Schaller and Lowther (1969) can do much to eliminate that superficiality. Hopefully the necessary information on predation against primates in various habitats will be available in the near future.

more powerful males, but group confrontations are far less significant in this regard.

In HL, encounters with predators are less common than in LH, and the primates are generally terrestrial due to the rarity of trees. With trees uncommon, escape into them can be a successful response to predators only if the primates can move more rapidly than the predators; i.e., there will be heavy selection for running speed in both males and females. Since the primates are not strongly clumped in this case, concealment is facilitated. But when both escape and concealment fail, confrontation follows. Escape and concealment do not select for sexual dimorphism, nor does confrontation when the population is distributed in individual territories. But when territories contain one male and several females, the necessarily defensive role of males results in heavy selection for sexual dimorphism — only *lone* males confront predators.

In LL, population clumps with 1:many socionomic sex ratios live in areas where there is a relatively low probability of attack by predators. Immediate access to safety by flight is rare during a normal day of foraging, and concealment in the open country is not likely to be successful. This leaves confrontation by males as the most probable response when a predator appears. The population clumps that occur under LL conditions resemble those from LH environments, although each LL clump is relatively more dispersed, thereby decreasing energy demands within narrowly restricted areas. With this increased dispersion, single males rather than groups of them must confront any predator that appears, just as in HL environments. And even more than in the latter case, there is heavy selection for sexual dimorphism.

Table 2 summarizes the relationships between environmental type, anti-predator strategy, and sexual dimorphism. Table 3 is a summary of everything presented to this point.

Now let us look at mating systems — not at their functions but at those things of which they are a function. Of primary importance, of course, is the primate reproductive system. Due to variability of ovulation timing and other characteristics of the female subsystem of no immediate concern to

Table 2

Relative degree of sexual dimorphism for each environmental type and anti-predator strategy

Environmental Types		Anti-Predator Strategy	
		Active	Passive
	HH	no dimorphism	no dimorphism
	LH	moderate dimorphism	—
	HL	marked dimorphism	no dimorphism
	LL	extreme dimorphism	—

Table 3
Summary of figures and tables presented to this point

Energy class	HH	LH	HL	LL
Common name	rain forest	mixed savanna	open savanna	desert
Socionomic sex ratio	1:1	1:1	1:many	1:many
Population distribution	scattered	clumped	scattered	clumped
Area allocation	territory	home range	territory	home range
Sexual dimorphism	none	moderate	none or marked	extreme

us here, several copulations must ordinarily occur for each impregnation. Demographic properties of the population, based upon energy density and predictability, constitute a second set of relevant factors. They alone do not determine which mating strategy will be employed, but they place important constraints upon system development. Further limitations arise from predator pressure and corresponding defensive strategies used against members of those higher tropic levels. By combining these factors with the basic assumption that the chances of survival of individuals and of populations are enhanced by the efficient utilization of resources, we are led to the following predictions concerning the nature of mating systems characteristic of each of the four ideal types of environment.

HIGH PREDICTABILITY ENVIRONMENTS

HH

Primates in this environment will mate either seasonally or permanently depending upon their anti-predator strategy. Even distribution of food makes the most efficient food acquisition strategy that of evenly distributed animals each defending its territory from conspecific encroachment. Since primary sex ratios ordinarily are approximately 1:1, and since the HH environment can support a 1:1 socionomic sex ratio, the 1:1 ratio will persist. But different animals can defend against predators in various ways. A slow moving animal might do best to remain concealed, while a fast moving brachiator might do best to flee at the approach of a predator. (In both cases, we should expect to find sexual monomorphism.) When concealment is best, maximal dispersion is most efficient with both males and females having discrete territories. This pattern necessarily results in temporary matings during short breeding seasons. When escape is the best response to predators, there is the problem of weighing individual territories for both males and females against somewhat larger territories defended by a male for himself and a mate who stays with him permanently. The latter arrangement permits a more efficient division of labor: the male uses his energy in acquiring and defending a resource area, while the female uses hers in reproduction and in caring for the young. Whether matings be temporary or permanent, high food density will make them monogamous.

HL

Here conditions are similar to those in HH environments, except for a changed socionomic sex ratio and an increased frequency with which confrontation is used as a defense against predators. Again matings may be permanent or temporary, but they will be polygynous in either case. For animals that respond actively to the approach of predators, permanent polygynous matings on relatively large territories defended against conspecific encroachment by markedly dimorphic males are to be expected. Animals who defend against predators passively will mate temporarily, they will be sexually monomorphic, and during non-breeding seasons they will be solitary in relatively large territories.

LOW PREDICTABILITY ENVIRONMENTS

In high predictability environments, the only division of labor among adults is along sex lines. Both monogamous and polygynous reproductive/foraging units are isolated from other similar units, and each adult must perform all the roles appropriate to his sex within his unit. However, food clumping permits the development of a more complex and efficient division of labor among the males. Population clumps that occur in conjunction with patchy food distribution patterns have boundaries definable in terms of relative densities of population. So unless they form a circle — one animal deep, for example — some of the members of the group will be in central and some in peripheral locations.

LH

LH environments permit a very high concentration of animals; i.e., most of the members of a group will be close to each other while foraging in the high energy food patches. Territorial defense against conspecifics by individuals is incompatible with this arrangement, and since males rather than females ordinarily confront predators, it is the peripheral males who form a first line of defense for the entire group. These two factors, taken together, indicate that there can be no gains in energy utilization efficiency resulting from the development of permanent matings of any kind. Promiscuous mating of all mature males with all mature females is to be expected.

LL

Due to decreased food density, primate groups in LL environments will be more dispersed than those in LH environments. This, of course, affects defensive strategy. Peripheral males are less effective in providing protection against predators when they surround a widely dispersed group. Consequently, each male is more fully responsible for his own safety in an LL environment. These conditions generate relatively autonomous foraging parties consisting of a powerful male (sexual dimorphism is exaggerated) and several females (1:many socionomic sex ratio) mated polygynously and permanently. Each of the several such units that comprise the larger population group will behave in concert with other units only at those points within the area occupied by the group as a whole where essential but very scarce environmental components occur.

Fig. 8 summarizes the predictions concerning mating systems for each of the four types of environment. It shows the nature of the mating strategy to be expected for a hypothetical population of six adult males and six adult females under the various conditions of food density and predictability.

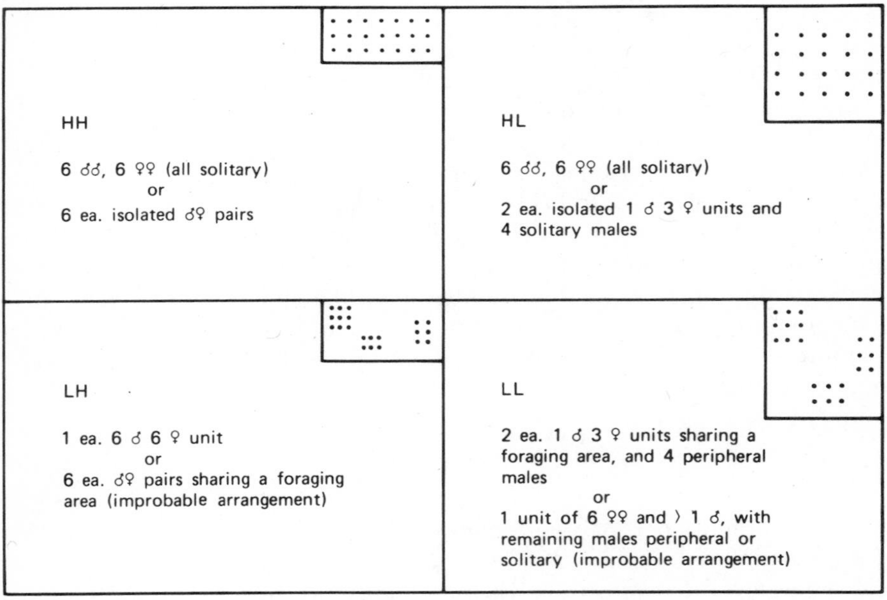

Fig. 8. Predicted social organization for a hypothetical population of six adult females and six adult males in each of the four types of environments.

Offspring are omitted from this figure just as they have been throughout the article.

In nature, energy density and predictability are not two valued variables. Each dimension is a continuum, and the existence of primates in environments that are intermediate between HH and HL, LH and LL, etc., must be dealt with accordingly. Disregarding the factors that might force a population from one type to another, we might predict that during a transition those characteristics of the social organization under consideration here will shift gradually from their original form to that appropriate to the new conditions. From HH to HL, resource defense (territoriality) is retained, but the shift in the socionomic sex ratio leads to polygyny as a replacement for monogamy. From HH to LH, the 1:1 sex ratio is retained, but territories and monogamy are replaced by home ranges and promiscuity. From HL to LL, polygynous mating is retained, but isolated territories disappear. From LH to LL, the communal use of a resource area is retained, but promiscuity gives way to polygyny. From LH to HL, the large population clumps break into smaller units, polygyny replaces promiscuity, and the shared resource area is replaced by isolated territories. Transitions in opposite directions are equally feasible. Because of structural adaptations to exclusively arboreal and exclusively terrestrial habitats, shifts from HH to LL or vice versa by highly specialized species would not occur readily, but given sufficient time during which structural modifications can proceed, any transition can occur. And except for possible problems related to structural specializations, any of these transitions can be either unidirectional (i.e., evolutionary adaptation to a new and stable situation) or cyclical (e.g., periodic changes in behavior controlled by seasonal variations in food density and/or predictability).

Now if we convert Fig. 8 into a three-dimensional space as in Fig. 9, with food predictability and food density as two of the dimensions and antipredator strategy as the third, we notice something that was left implicit in Fig. 8; viz., that this formulation provides for eight possible types of primate social organization, each type characterized by a unique combination of three values. However, we have already established that there can be no successful passive response to predators in low predictability environments. That leaves six types that seem to be possible both logically and ecologically. For example, we should be able to find in the real world primates living in high density, highly predictable environments where there is a 1:1 socionomic sex ratio, with monomorphic individuals living in defended territories scattered relatively evenly throughout the forest. Those animals that defend against predators actively (Fig. 9, HH-Active cell) will mate monogamously and permanently, while those that depend upon concealment (HH-Passive cell) will mate monogamously but temporarily during short breeding seasons. Similar sets of predictions can be made for the other cells. It is interesting to note here that a primate population living in a predator-free environment would have a mating system such as that found among primates living under the same conditions of food density and distribution and using an active anti-predator strategy, but in the predator-free environment there would be no selection for sexual dimorphism.

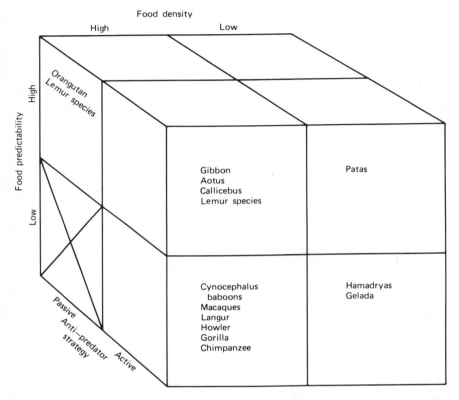

Fig. 9. Categorization of primate species by environmental type and anti-predator strategy. "Typical" populations are entered here; "atypical" populations are omitted.

An examination of the table provided by Crook and Gartlan (1966: 1201) plus data from other sources (Jay 1968; Napier and Napier 1967), indicate that there are indeed several species that fit easily into the HH-Active cell of Fig. 9. The four items entered there include one ape, two monkeys, and several prosimian species. Some are nocturnal and some are diurnal; some are insectivorous and some are frugivorous; but all of them share the characteristics by which the cell is defined. The LH-Active cell contains several promiscuously mating monkeys and apes, and it is probable that several prosimian species can go into this category as soon as more information becomes available. The LL-Active cell contains two species that have been recognized for several years as being distinctively different from those entered in the HH- and LH-Active cells. The HL-Active cell is interesting, because it very easily accommodates one species and several populations of other species that have given primatologists some problems recently. For example, Crook and Gartlan (1966) attempt to squeeze the patas into the same category as the hamadryas and the gelada. Hall (1968: 117) noticed similarities between patas and hamadryas, and between hamadryas and cynocephalus baboons living in "very harsh habitats" — but he detected enough differences between them that he simply left open the question of the relationships between them. Although there are several prosimians and perhaps one ape that fit into the HH-Passive cell, there are not to my knowledge any primates that can go into the HL-Passive cell. Perhaps this is simply an oversight on my part; perhaps it is because of the small size of the sample of primate species/populations that have been adequately studied in their natural habitats; perhaps it is because there are no living primates that fit the requirements for inclusion in that cell. The last situation could be the result of excessive spacing between animals in very large individual territories, or the operation of factors not considered in this paper. Then there are several known primate populations, not included in Fig. 9, which seem to be transitional between LH and HL, and LH and LL (i.e., the species names in Fig. 9 represent "typical" populations of those species; "atypical" populations of all species are to be expected). I know of none that appear to be transitional either from or toward HH environments.

INDIVIDUAL MATE SELECTION

Discussion of this topic has been delayed until now for the following reason: although individual mate selection must occur among all surviving primate populations, it is an activity that can occur only within the context provided by the three survival strategies discussed above. To attempt to make cross-species comparisons of individual mate selection procedures without first attempting to understand energy acquisition and anti-predator defenses and their relations to the general topic of reproductive strategies is to get the "natural" order of things mixed up. Perhaps the following paragraphs will provide some justification for these assertions.

When primates live in large groups in open country and mate temporarily, it is relatively easy to observe their mate selection activities. In those conditions — typical of LH environments — sexual dominance orders among males (i.e., dominance vis-a-vis temporary sexual access to fertile females) and morphological changes associated with changes in estrous phases among females are common occurrences. Although the complementary relationship between these two sets of phenomena has been recognized, relatively little attention has been paid to the logical properties of "dominance order" and "estrous signal," or to their relations to food density and distribution patterns. Sexual dominance, as briefly defined above, is possible only where population clumping and temporary mating occur; i.e., it is logically impossible for an isolated male or the sole male in an isolated and/or stable monogamous or polygynous group to behave dominantly toward other male conspecifics. Likewise, intraspecific competition between females (in estrous) for temporary sexual access to males is impossible among animals mated monogamously (e.g., gibbons), but it is possible where mating is promiscuous (chacmas) or polygynous (patas and hamadryas). This means, of course, that intraspecific competition between members of the same sex for temporary sexual access to members of the opposite sex is *habitat specific.* Neither male nor female competition of this kind can occur in HH environments, both can occur in LH environments, and in both HL and LL environments competition between females can occur but male sexual dominance orders cannot. (I am referring here only to the primates in the Active Defense plane of Fig. 9.)

Now, if male sexual access dominance orders are confined to inhabitants of LH environments, what is the male's role in mate selection in other types of environments? According to Kummer (1968), hamadryas males in LL environments "adopt" young females — with no apparent conflict with other males and with no apparent correlation between age or social position of the male and the likelihood of his adopting a young female. Although the factors that determine which males will mate with which females are not yet identified with certainty, it is clear that the concept of a dominance order is totally inapplicable here.

With regard to individual mate selection (and most other topics), highly predictable environments (HH and HL) present data collection problems not found in patchy environments. In the first place, highly predictable environments may be rain forests in which continuous detailed observations are almost impossible to make because of obstructions to visibility, or they may be open grasslands where locating and closely approaching one's subjects is very difficult. But of even greater importance is the fact that in HH and HL environments, primates live in clearly defined and sometimes quite large territories, and they mate permanently (except HH-Passive). This means that being in just the right place at just the right time to observe mate selection and acquisition as it occurs is much smaller under these conditions than it is if the subjects live in large groups and mate only temporarily, therefore with relative frequency. It is for these reasons that neither Carpenter in his field study of the gibbon (1940) nor Hall in his study of the patas (1965) obtained any information at all on individual mate selection. Nevertheless, mate selection among territorial primates has been equated on occasion with mating patterns among territorial birds; i.e., it is said that the acquisition and defense of a territory by a male primate improves his chance of acquiring a mate. While it is true that possession of a territory in highly predictable environments improves the chance of survival of the occupant(s) of the territory, it is not yet clear just how the possession of a territory operates in the process of primate mate selection and acquisition. However, the postulated relationship between territoriality and mate selection seems reasonable at this time, and pending the publication of additional data of this topic, we will accept "territoriality" as being somehow equivalent to "dominance ordering" and "adoption."*

To summarize then, field research shows that the active role of the male in individual mate selection may take a variety of forms including territoriality, dominance orders of various degrees of rigidity, and adaptation. The first appears in HH and HL environments, the second in LH environments and the third in LL environments.

Throughout the discussion of survival strategies, I treated sexual dimorphism (i.e., males more aggressive and/or powerful than females) as a function of interspecific relations rather than as a function of intraspecific competition between males for sexual access to females. That choice is supported by the scanty comparative data on individual mate selection. A short trip to the local zoo should convince anyone that gibbons, chacmas, patas, and hamadryas can be ranked validly (if not precisely and quantitatively) in that order according to increasing degree of sexual dimorphism (cf. Table 2), but there is no evidence to suggest that they could be ranked in the same order according to increasing degree of competitiveness between males for sexual access to females. If competitiveness were used as the criterion — and measures of competitiveness are probably even less precise and more subjective than measures of degree of sexual dimorphism — the resulting order of those primates might resemble the following: gibbon, patas/hamadryas, chacma. The chimpanzee, which resembles the chacma baboon in terms of dimorphism, appears to rank somewhat lower than the chacma on the competitiveness scale.

A different way in which to classify male roles in mate selection is to consider the duration of the resultant matings. Dominance orders relate to temporary mating, while territoriality and adoption relate to permanent mating (with one exception mentioned below). There are corresponding sets of mate selection activities displayed by females. As indicated earlier, competition between females for temporary sexual access to males is possible only when mating is promiscuous or polygynous. When those situations occur, cyclic changes in fertility are accompanied by major changes in the

---

* The use of the terms "territory" and "territoriality" in connection with individual mate selection has resulted in a bit of terminological confusion. In the discussion of survival strategies, I have used "territory" as the label for the type of area and resource allocation found in highly predictable environments, and I contrasted it with "home range" in patchy environments. Elsewhere in discussions of individual mate selection, others have contrasted "territory" with "dominance order." There is no reason why one word should not do two jobs, but it is most important that a term not be used ambiguously. Perhaps we could reserve the "territory-home range" opposition for discussions of area and resource allocation, and use "territoriality-dominance order-adoption" and so forth in dealing with individual mate selection.

behavior of the female, and sometimes by striking changes in her physical appearance. These changes, in turn, are related to changes in the frequency of copulation between the female displaying them and the male or males perceiving them. Among promiscuously mating primates (in LH environments), where mating and copulation are virtually synonymous, temporary consort relations tend to develop between females at the peak of estrous and males of high rank in a sexual dominance order.

Where permanent polygynous mating occurs, mating and copulation are not synonymous, and we must deal instead with one short-duration phenomenon and one of long duration. Short-duration competition between fertile females for sexual access to a male may take the same forms in a polygynous group as in a promiscuous group, but the long-term phenomenon — the establishment of permanent relations between male and female — has no counterpart among those that mate promiscuously. The long-term phenomenon is similar, however, to permanent pair formation among monogamous populations, where competition between females for temporary access to males is impossible.

In other words, the roles of females in actively selecting a mate are of two kinds, and as with the male roles, they are related to the environmental types presented above. In HH environments, where mating is permanent and monogamous, stable pair formation occurs but short term competition during estrous does not.* Among animals in the HH-Passive cell of Fig. 9, and in LH environments where animals live in large groups and mate promiscuously, the short-term competition occurs but a long-term component is absent. In HL and LL environments, stable polygynous mating occurs, and we find both a short and a long term component; i.e., we find both the establishment of permanent polygynous units, and within those units the sexual competitiveness of females changes in conjunction with changes in fertility.

The primatological literature contains much about morphological and behavioral changes associated with estrous, just as it contains much about

---

* Although short-term behavioral changes in estrous females are likely to occur in conjunction with permanent monogamous mating, short-term morphological changes would appear to be unnecessarily redundant in that situation and therefore less likely to occur. Any signal, even a minimally perceptible one, should be as effective as any other in communicating to the male the required information on the sexual receptivity of the female. But in any event, changes associated with estrous cannot be viewed as competition between females when mating is permanent and monogamous. A further comment on morphological changes in estrous. Crook and Gartlan note that among the gelada, chest patches and anogenital regions are less conspicuous during estrous than at other times, especially during lactation. While it is true, as they say (1966:1203), that this correlation argues against interpretations of red as a color that is sexually attractive to males, it does not imply that color changes fail to function as sexual signals to male geladas. *Any* change in coloration that normally accompanies a change in sexual receptivity can act as a sexual signal; to put it a bit more bluntly, a red behind can mean either "stop" *or* "go" depending upon the learning history of the animal perceiving it.

dominance orders. Also there are many references to the long-term component of the male role in mate selection, appearing in discussions of territoriality and in Kummer's study (1968) of the hamadryas system of adoption of young females by adult males. However, the long-term component of the female role seems to have been neglected. Perhaps this derives from the assumption that females only react passively in the establishment of permanent relations with a male. Certainly Kummer's report on the hamadryas baboon suggests that is a reasonable assumption with regard to the populations that he studied, but it seems less reasonable in connection with populations living in highly predictable environments where territoriality is viewed as the male's role in mate selection. While the hamadryas male may actively adopt a female, then keep her with him by using the "neck bite" response, gibbon and patas males compete with other males for a territory to which a female then moves (or so it would seem at this time). In other words, in the second case it appears to be a female who actively selects a male with a territory, not a territorial male who captures a passive female.

In the list of male roles discussed earlier, one was a short-term component in LH environments where a long-term component is precluded, and the other two were long-term components in HH, HL, and LL environments. No short-term components were given for the last three environments because there seems to be nothing in the behavior or morphology of males in those environments that complements the female's estrous in the way that dominance orders do in LH environments. This situation is consistent with the model for the same reason that we found no short-term competition between females in HH environments: a short-term component for either sex is impossible when there is only one member of that sex within a stable reproductive unit.

Animals classified as HH-Passive in Fig. 9 mate temporarily, but the formation of breeding pairs shows an interesting combination of territoriality — in this case, a *short*-term sharing of a common resource area by animals that ordinarily maintain separate territories — with marked behavior changes indicating sexual readiness of the female. It has the territorial basis of mating in highly predictable environments and the transitory nature of matings in LH environments.

Fig. 10 should be used in conjunction with Fig. 9 even though they differ slightly in form. Each cell in Fig. 10 contains the code for the environmental type in the upper left corner; the remainder of the notations in the cell summarize the contents of the foregoing paragraphs on individual mate selection. Beside each sex are the short-term (S) and the long-term (L) components of the role of the designated sex in the process of individual mate selection and reproductive unit formation. When read together, all of the entries in a cell provide an extremely tentative and generalized description of the mate selection activities characteristic of primate populations entered in the corresponding cell of Fig. 9.

HH-Passive	HL-Passive

Male   S: territoriality
   L: none
 	
Female   S: territoriality;
     behavior changes
   L: none
	(no known populations)

HH-Active	HL-Active

Male   S: none
   L: territoriality
 	
Female   S: behavior changes*
   L: joins male with a
     territory
	Male S: none
	   L: territoriality
	
	Female S: behavior/morphological
	     changes*
	   L: joins male with a
	     territory

LH-Active	LL-Active

Male   S: sexual dominance
    order
   L: none
 	
Female   S: behavior/morphological
     changes*
   L: none
	Male S: none
	   L: adoption
	
	Female S: behavior/morphological
	     changes*
	   L: none

Fig. 10. Characteristics of mate selection activities according to environmental type. Male and female roles are classified according to the duration of the heterosexual relationship based upon them: S = short-term component, L = long-term component.

*Short-term morphological changes in estrous are likely, but certainly not inevitable, among polygynous or promiscuous populations; they are unlikely, but not impossible, among animals that mate monogamously.

CONCLUSIONS

I have here attempted to construct a model that can improve our under-standing of nonhuman primate ecology and social behavior; I have used the model as a basis for reorganizing the table that appeared in the article by Crook and Gartlan; and I have used it as a framework for systematizing some of the data on individual mate selection among non-human primates. I do not claim to have constructed the best model of the relations with which I am concerned. On the contrary, it is put forth as a tentative suggestion. But if this article stimulates further discussion of these topics so that the model can be improved upon or replaced by a superior one, then the effort will have been worthwhile. The most obvious way in which to improve upon it or replace it is to conduct experimental tests of the hypotheses contained in the paper. Four predator-free environments, one corresponding to each of the four "ideal" environmental types schematized in Fig. 6, could be artificially populated with young adults or subadults of a primate species known to be structurally equipped to cope with a wide range of physical conditions. (Macaques and savanna baboons are likely candidates.) Control over food density and distribution patterns within the several square miles of enclosures required for the experiment would present a number of technical difficulties, but conceptually it is rather simple. Controlling for the effects of prior learning among the test populations could be done simply by stocking the enclosures with third or fourth generation breeding colony animals raised in small groups of uniform composition. Theoretically, control over learning history is unnecessary, but with such control the experiment could be expected to yield results sooner than if recently captured animals were used. Behavioral adaptations to the "ideal" environments should occur rapidly, with completely unambiguous results available from second generation inhabitants of the enclosures. And if the experiment were continued through several generations, it could reveal the relative importance of interspecific versus intraspecific competition in generating sexual dimorphism.

Since I consider the stimulus function of this article to be of major importance, I shall conclude with a few brief comments on the relevance of primatological research for socio-cultural anthropology. Although anthropologists are becoming interested in primatological research, the prevailing attitude seems to be that the study of non-human primate social behavior can be of greatest value by providing clues to the evolution of human social behavior. I disagree with that position, and the model presented above contains my reasons for disagreeing. While there may be links between human and non-human primate social behavior, the large variety of social orders among the non-human primates, or even the terrestrial primates alone, makes it obvious that a "unilinear" evolution of human from non-human primate society is not to be found. Research directed toward finding those links rests upon the usually implicit assumption that the social behavior of non-human primates and early man is species specific, an assumption that definitely is not supported by the field data. This being the case, if it were possible to establish beyond doubt that men evolved originally in LH environments in East Africa, what could that tell us about "second generation" men who moved into different habitats throughout

Africa, Europe, and Asia? If the model presented here is valid, knowledge of the social organization of the first generation men would tell us virtually nothing about that of their second generation descendants; whereas information on food distribution and density patterns should tell us much about the social order of a primate population using those resources regardless of when they lived. This applies, of course, to modern non-human primates, early hominids, and modern hunter-gatherers. Thus what I am suggesting here is that primatological research may tell us little about the *evolution* of human societies, but that it may tell us much about the social organization of all primates *including man.* In this case, to attempt to reconstruct the evolutionary sequence(s) connecting non-human with human society may be to chase a red herring.

It is now, and it probably will remain forever, impossible to make meaningful analyses of human hunter-gatherer ecology and social behavior comparable to that attempted in this article for the simple reason that the required data were never collected and most hunter-gatherer societies are now extinct. That situation reflects the well-established though far from universal habit among socio-cultural anthropologists of ignoring one of Charles Darwin's basic insights, that man is an animal. And many have done so while pursuing—if not continuously, at least very frequently—questions of human origins and social evolution. Crude analogies between gibbon territories and Australian Arboriginal tribal countries, or chacma baboon sexual dominance orders and the chain of command within the United States Department of Defense, may be entertaining and suggestive, but until really comparable data on human and non-human societies becomes available, it is most unlikely that those analogies will lead to any important advances in our understanding of human social behavior. Similarly, discussions of differences between human and non-human social behavior can be meaningful, I believe, only after their similarities are reasonably well known.

REFERENCES CITED

Carpenter, C. R. 1940 A field study in Siam of the behavior and social relations of the gibbon, *Hylobates lar.* Comparative Psychology Monographs, 16(5):1-212.

Crook, J. H., and J. S. Gartlan 1966 Evolution of primate societies. Nature (London) 210(5042):1200-1203.

Gartlan, J. S. 1968 Structure and function in primate society. Folia Primatologica, 8(2):89-120.

Hall, K. R. L. 1965 Behavior and ecology of the wild patas monkey, *Erythrocebus patas,* in Uganda. *Reprinted in* Primates: studies in adaptation and variability (1968). P. Jay, ed.

Jay, P., ed. 1968 Primates: studies in adaptation and variability. New York: Holt, Rinehart and Winston.

Kummer, H. 1968 Social organization of hamadryas baboons. Chicago: University of Chicago Press.

Napier, J. R., and P. H. Napier 1967 Handbook of living primates. London: Academic Press.

Orians, G. H. 1969 The study of life: an introduction to biology. Boston: Allyn and Bacon.

Schaller, G. B. 1969 Life with the king of beasts. National Geographic 135:494-519.

Schaller, G. B., and G. R. Lowther 1969 The relevance of carnivore behavior to the study of early homonids. Southwest Journal of Anthropology 25(4):307-341.

23

The Relation between Ecology and Social Structure in Primates

J. F. EISENBERG, N. A. MUCKENHIRN, AND R. RUDRAN

Selected data on primate social systems will be discussed and reinterpreted in this article, with a view toward modifying the existing concepts about the adaptive nature of these social systems (1-3). In light of recent data, we limit the concept of the multi-male troop (2) and call attention to the reproductive group as an organic unit that shows stages of growth and decline which may vary under different environmental circumstances. To this end, we introduce a new category of social structure for the primates—the age-graded-male troop. Before proceeding with the definitions and discussions, we must put the data into a historical framework.

THE EXISTENCE OF "SPECIES TYPICAL" SOCIAL ORGANIZATION

The cornerstones of Carpenter's theories of primate social structure (4) can be summarized as follows: (i) primate troops tend to have more or less exclusive home ranges (5); (ii) the average size of a troop tends to be typical for a given species (the term "apoblastosis" defines the process whereby primate troops divide to restore a species-specific balance in numbers); and (iii) the composition of the troop, with respect to the proportions of sex and age classes, tends to be relatively invariant, regardless of troop size. The ratio of males to females was termed the "socionomic sex ratio." This ratio tended to be species-specific, and, for most species, there were more adult

Reprinted from *Science*, May, 26 1972, Volume 176, pp. 863-874. Copyright © 1972 by the American Association for the Advancement of Science and reproduced by permission.

females than adult males in a given troop. Two corollaries followed: first, there was a strong polygynous trend in most primate species; second, extra adult males were excluded from the troop by some process, to dwell either in a peripheral subgroup of their own or as solitary individuals.

In line with Carpenter's generalizations and some early theories of social evolution, most field workers believed that the behavioral attributes of a species were relatively constant and reflected species-specific adaptations. It was assumed that social structures resulting from the defined patterns of interaction of a given species would manifest themselves as rather predictable entities (1, 2). This assumption still has heuristic value, but the task is much more complex, since, for most primate species, social structure varies with habitat. This has been amply demonstrated for the olive baboon, *Papio anubis* (6, 7), the gray langur, *Presbytis entellus* (8-10), and the vervet monkey, *Cercopithecus aethiops* (11, 12). On the other hand, some species appear to have a standardized form of social organization—for example, the lar gibbon, *Hylobates lar* (13, 14), and the hamadryas baboon, *Papio hamadryas* (15, 16).

The causes of intraspecific variation in social structures are related, in part, to differences in habitat (especially factors of food availability and predation), as well as to differences intrinsic to the troop itself (3); however, once the range in the variation of troop structure is described, a "modal" social organization for a given species can often be discerned, thus facilitating comparisons with other species. It would appear that oné generalization is possible: those species that exhibit a wide range of adaptation to differing habitats often show an equally wide range in social structure [for example, *Papio anubis* (6) and *Presbytis entellus* (9)]. On the other hand, species that show a uniform adaptation to specific kinds of habitats often show a corresponding uniformity in their grouping tendencies (1). In fact, when a group of allopatric species shares the same relatively narrow range of adaptation, then this group begins to exhibit a predictable "adaptive syndrome" with respect to feeding, anti-predator behavior, spacing mechanisms, and social structure. Examples from two separate "syndromes" are (i) the arboreal, leaf-eating monkeys of Africa and Asia [*Presbytis cristatus* (17), *Presbytis johni* (18), *Presbytis senex* (19), and *Colobus guereza* (20)] and (ii) the slow-moving, insectivorous lorisoids of Africa and Asia [*Arctocebus, Perodicticus, Nycticebus,* and *Loris* (21-23)].

GRADES OF SOCIAL STRUCTURE—A REASSESSMENT

It has not been uncommon to find primate societies classified into grades based on supposed increases in social complexity. The implicit suggestion was that higher grades were achieved through evolutionary stages, but it was well recognized that, within and among each major primate taxon, considerable parallel evolution had occurred (1, 2). Even within the morphologi-

conservative Prosimii, social organizations equal in complexity to those of the Cebidae and Cercopithecidae were formed (24, 25).

The so-called solitary-living species are characterized by a minimum amount of direct social interaction with conspecifics of either sex in the same age class. Typically, the "mother family" (that is, an adult female and her dependent offspring) forms the only cohesive social unit that indulges in daily, intimate interaction. Nevertheless, solitary species, whether primates, carnivores, or rodents, have a social life (26), and indirect communication is maintained among adults that have neighboring or overlapping home ranges. The communication channels of solitary species are characterized by olfactory and auditory modalities which maintain spacing except at mating times (1, 22). The terms "solitary," "asocial," and "dispersed" have been objected to (27) because they obscure the fact that a given pair of adults and their sub-adult descendents can share a home range completely or partially, even though they do not indulge either in communal nesting or regular physical contact. In addition, when there are overlapping home ranges, the same adult pair can reproduce in subsequent years; thus, although individuals are dispersed, a family structure and relatively closed breeding unit (28) are maintained. For example, in *Microcebus murinus* and *Galago demidovii* (29), the home ranges of adult females may overlap considerably. A reproductive male's home range includes the home ranges of from one to six females and their juvenile offspring, while extra males live on the periphery of the dominant male's home range, either as solitary individuals or as a noncohesive bachelor group. The spatial distribution of the adults (29) implies a polygynous breeding system (see Fig. 1). Other nocturnal prosimians appear to exhibit a similar spacing system—for example, *Cheirogaleus major, Daubentonia madagascarensis, Loris tardigradus, Perodicticus potto,* and *Lepilemur mustelinus* (22-24, 30) (see Table 1).

The parental family structure is characterized in its extreme form by a bonded pair of adults and their immature descendents. This bonded state occurs rarely within the order Primates [assuming the term "bond" includes only social relationships between specific individuals based on the performance of mutually reinforcing activities, in addition to mating behaviors (31)]. It follows then that grooming, huddling, and other nonsexual behavior engaged in on a daily basis by two individuals of the opposite sex defines the pair bond.

Marmosets of the genera *Saguinus, Cebuella,* and *Callithrix* are examples of the bonded parental family (32, 33). These marmosets exhibit a unique form of parental care, in that the male participates to an extent unparalleled by the other families of Primates (1, 32). The male marmoset typically transports the young from the time they are born, transferring them to the female only for nursing (33).

The gibbons, *Hylobates* and *Symphalangus* (13, 14, 34), as well as some species of the family Cebidae [notably the Titi monkey, *Callicebus moloch*

Table 1

Range of social organization and feeding ecology for selected primate species (see text for descriptions).

Solitary species	Parental family	Minimal adult δ tolerance[a] (uni-male troop)[b]	Intermediate δ tolerance[c] (age-graded-male troop)[b]	Highest δ tolerance[d] (multi-male troop)[b]
A. Insectivore-frugivore	A. Frugivore-insectivore	A. Arboreal folivore	A. Arboreal folivore	A. Arboreal frugivore
Lemuridae *Microcebus murinus*	Callithricidae (Hapalidae) *Saguinus oedipus* *Cebuella pygmaeus*	Cebidae *Alouatta palliata*	Colobinae *Presbytis cristatus* *Presbytis entellus*	Indriidae *Propithecus verreauxi*
Cheirogaleus major	*Callithrix jacchus*	Colobinae *Colobus guereza* *Presbytis senex* *Presbytis johni* *Presbytis entellus*	Cebidae *Alouatta palliata*	Lemuridae *Lemur fulvus*
Daubentoniidae *Daubentonia madagascarensis*			B. Arboreal frugivore	B. Semiterrestrial frugivore-omnivore
Lorisidae *Loris tardigradus* *Perodicticus potto*	Cebidae *Callicebus moloch* *Aotus trivirgatus*	B. Aboreal frugivore	Cebidae *Ateles geoffroyi* *Saimiri sciureus*	Cercopithecidae *Cercopithecus aethiops* *Macaca fuscata* *Macaca mulatta* *Macaca radiata*
B. Folivore	B. Folivore-frugivore	Cebidae *Cebus capucinus*	Cercopithecidae *Miopithecus talapoin*	*Papio cynocephalus* *Papio ursinus*
Lemuridae *Lepilemur mustelinus*	Indriidae *Indri indri*	Cercopithecidae *Cercopithecus mitis* *Cercopithecus campbelli* *Cercocebus albigena*	C. Semiterrestrial frugivore-omnivore	*Papio anubis* *Macaca sinica*

Table 1 (continued)

Solitary species	Parental family	Minimal adult ♂ tolerance[a] (uni-male troop)[b]	Intermediate ♂ tolerance[c] (age-graded-male troop)[b]	Highest ♂ tolerance[d] (multi-male troop)[b]
	Hylobatidae *Hylobates lar* *Symphalangus syndactylus*	C. Semiterrestrial frugivore Cercopithecidae *Erythrocebus patas* *Theropithecus gelada* *Mandrillus leucocephalus* *Papio hamadryas*	Cercopithecidae *Cercopithecus aethiops* *Cercocebus torquatus* *Macaca sinica*	Pongidae *Pan satyrus*
			D. Terrestrial folivore-frugivore	
			Pongidae *Gorilla gorilla*	

[a] Troop with one adult male and strong intolerance to maturing males.
[b] "Troop" refers to the basic social grouping of adult females and their dependent or semidependent offspring.
[c] Troop typically showing age-graded-male series.
[d] Troop with several mature, adult males and age-graded series of males.

(35)], show grouping tendencies similar to those of the marmosets in that a single adult pair and their dependent offspring occupy a given home range, but the male's participation in the care of the young is limited. For example, *Callicebus* and *Aotus* males (36) transport the young to a certain extent and thus appear to be somewhat closer to the marmoset pattern (33), while gibbons and *Indri* males participate little in the rearing of infants. Thus, at least two subvariants of the parental spacing system must be designated: (i) mutual participation by the male and female in the rearing of

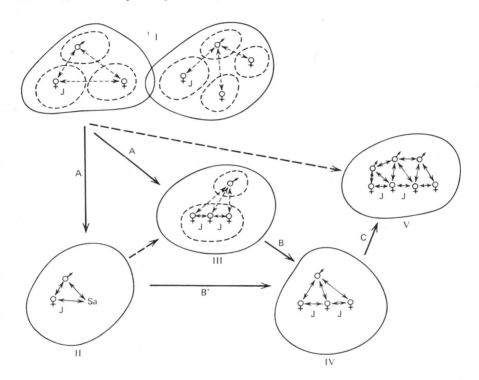

Fig. 1. Space utilization patterns for selected mammals with hypothetical evolutionary pathways. (1) Solitary pattern: adult males and females have separate centers of activity and encounter infrequently. The extended ranges of the males overlap with the home ranges of the females. Only polygynous patterns are presented: *Microcebus murinus* (29). (II) Family group: a bonded pair and their subadult offspring travel as a homogeneous unit in an exclusive home range: *Saguinus oedipus* (*32*). (III) Uni-male group (extended mother family): an adult male is in periodic contact with a cohesive group of adult females and their progeny: *Ateles geoffroyi* (*46*) (IV) Uni-male group: an adult male is in relatively constant contact with a cohesive group of adult females and their progeny; *Papio hamadryas* (*16*). (V) Multi-male group; a cohesive group of several adult males and females with their progeny; *Papio anubis* (*49*). Hypothetical phylogenetic steps in the formation of mobile, cohesive groupings are indicated by arrows. Routes A, B', and C or A', B, and C are the most probable steps in the formation of cohesive, multi-male groups (*I*). The arrows formed from dashed lines are less probable evolutionary pathways (*I*). For simplicity, the symbols *J* (juveniles) and *Sa* (subadult) have been included only occasionally to indicate the presence of immature animals.

offspring and (ii) limited or no direct participation by the male in the rearing of dependent young (see Table 1).

Parental groups have probably evolved by at least two different pathways (Fig. 1). The parental group, a more cohesive form of primate social structure, can easily be derived, in a phylogenetic sense, from the more dispersed systems of some of the prosimians, for example, *Galago demidovii* or *Microcebus murinus* (29).

If only adult female primates were to form affiliations, an extended mother family would exist, given that the exclusion of excess adult males takes place through the aggressive action of the parental adult male. If the parental male were closely bonded to the extended group of mothers and daughters, the result would be a typical uni-male group (Fig. 1). Extragroup males would exist on the periphery of the reproductive units. Recent research on terrestrial African primates (the Patas monkey, *Erythrocebus patas; Papio hamadryas;* and the Gelada baboon, *Theropithecus gelada*) has indicated the existence of troops composed of several females, their dependent offspring, and one sexually mature, adult male (16, 37, 38). In such species, troops can also contain a few maturing males as a normal transitional stage in the life cycle of the troop; however, the young males generally leave the parental troops as subadults and join bachelor bands.

The uni-male troop is more complex in its structure than is the parental group because there is an increased representation of sex and age classes in the troop. Adult females must be tolerant of one another and have affiliation mechanisms to promote cohesiveness (39). The behavior of the adult male toward females can also contribute significantly to cohesiveness, since he may "discipline" the females that stray from his group, thereby keeping the group intact [for example, *Papio hamadryas* (16,40)]. Such herding behavior by the adult male is not typical of all species with a uni-male configuration; it has not been reported for *Cercopithecus campbelli* (41), for example. New uni-male troops can be formed in *Papio hamadryas* when a solitary younger male is able to "capture" a subadult female from a structured breeding unit. Other variations on this theme are discussed by Kummer (16).

One of the more intriguing aspects observed in some primate societies was that several adult males could and did associate continuously with adult females and young. Because of the contrast with other mammalian taxa that have complex social structures in which males are not permanent members of the group, the "multi-male group" came to be thought of as an advanced and almost unique characteristic of higher primates (1, 2, 4). While some species do have a multi-male troop, it is obvious that the concept has been applied too broadly (42). An intermediate form of social organization, between the uni-male and the multi-male structures, should be recognized. This may be termed the age-graded-male troop. Although several males of varying ages coexist in such troops, there are proportionately fewer males in these troops than there are in true multi-male troops (whose sex ratio may approach 1 : 1) (6). The linear male dominance order is based on the age of the males, with no definable subunit of several males in the oldest age bracket. The lack or absence of fully adult males of equivalent age is the characteristic that defines an age-graded-male troop [for example, troops of *Mandrillus leucocephalus, Presbytis entellus,* and *Ateles geoffroyi* (43-47) (see Table 1)].

The age-graded-male troop may be considered a phylogenetic step toward the true multi-male configuration (see Fig. 1), with the former having an intermediate level of male tolerance that allows several young males to mature longer within the troop of their birth than do young males in uni-male troops. Nevertheless, a fundamental tendency toward polygyny and the possibility of the troop's splitting and returning to a uni-male condition remain. Thus, the age-graded-male troop is a variation on the uni-male theme (see Fig. 2). What appears to distinguish species with an age-graded-male troop from species with a strong uni-male tendency is (i) the adult "leader" male exhibits a wider range of tolerance of young males near his own age and (ii) part of the tolerance shown by the dominant male appears to derive from the fact that these species generally have larger troops. In addition, the larger troops have larger home ranges and therefore more possibilities for dispersing into subgroups while foraging. This very tendency toward fractionation can generate new troops by apoblastosis (4).

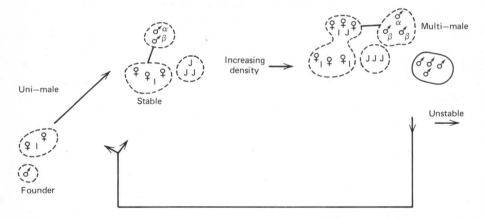

Fig. 2. Hypothetical diagram of troop growth for arboreal primates. The assumption is that a uni-male tendency is the most typical configuration at moderate population densities. Given a founder situation at the left, consider an adult male attached to a cohesive unit of two adult females and their young. The troop grows by recruitment, yielding a subgroup of juveniles and a beta male that is subdominant to the alpha male or father. The two older males form a subgroup in their own right. At greater densities, younger males may form a peripheral subgroup of their own that has no direct contact with the basic subgroup of mothers and their young. The subgroup of adult males may now be augmented slightly to three, with the founding father still dominant. Although the troop now appears to be multi-male, it would be more correct to consider it an age-graded-male group. Splitting of the new unstable troop can lead to the original uni-male configuration (*19, 47, 66*).

Members of the genus *Papio,* particularly the species *ursinus, anubis,* and *cynocephalus,* are adapted to savannas and forage a great deal on the ground, although they retire to rocky places or trees for sleeping at night (48, 49). These species exhibit classic multi-male groups [as do, with some limitations, the semiterrestrial macaques, including *Macaca mulatta, M. fuscata,* and *M. speciosa* (50-52)]; however, to classify other primate species that have more than one male in the troop as multi-male and then to rank these species within the same grade of social structure as these terrestrial macaques and baboons is to oversimplify matters.

The multi-male troop is characterized by an oligarchy of adult males that are roughly equivalent in age (49). These males show affiliation behaviors, and, although they may be ranked in a dominance order, the ranking is not pronounced within the oligarchic subgroup. Cooperation exists among the superdominant males (39), and they actively oppose and dominate younger males in the subadult age class (49). Species exhibiting a multi-male configuration do, of course, have within them males that may be arranged in an age-graded series (53, 54). Furthermore, there may be solitary males that live independently of the troops and that join and leave troops in a semiregular pattern (55, 56).

The multi-male troop was first defined for *Papio anubis,* primarily in the pioneer studies of Washburn and DeVore (48, 49). Baboons were portrayed as living in cohesive troops of 30 to 50 animals; solitary males were not described originally (48), but were subsequently noted for *Papio cynocephalus* (56). The multi-male troop appeared to serve as an antipredator device, since the presence of many adult and subadult males permitted collective attack should the troop be menaced by a terrestrial predator.

Although the sex ratio at birth is almost equal in *Papio* and *Macaca,* the number of reproducing females for each adult male varies from one to three. Any difference in the adult sex ratio is presumed to result from both differential mortality and the differential maturation rates of the sexes. The adult males are organized in a dominance hierarchy, and the alpha male does most of the breeding during the peak of a female's estrous period (49, 54).

Recent studies by Rowell (6) and Altmann and Altmann (56) indicate that in both forest and savanna habitats baboons of the genus *Papio* show neither such strong dimorphism in social roles nor such a disparity in the male-female sex ratio as was originally described (57). In light of more recent evidence, we will probably have to alter our concept of the baboon social life. Nevertheless, the existence of several adult males of an equivalent age serves to crystallize and define the concept of the multi-male troop.

In previous reviews of primate social organizations, the uni-male groupings characteristic of *Erythrocebus patas* (37), *Papio hamadryas,* and *Theropithecus gelada* (16, 38), all three of which are adapted to arid climates, were considered specialized offshoots of the multi-male grouping that supposedly characterized most advanced primates (2, 3). This tendency to characterize most primate species by a multi-male social structure and to set aside those species exhibiting uni-male groups as cases of adaptation to extreme environments was motivated, in part, by a desire to emphasize the uniqueness of primate societies when contrasted with the social groupings of other mammals.

The preoccupation with the concept of the multi-male organization was unfortunate for at least two reasons: (i) male-male tolerance mechanisms were studied to the virtual exclusion of the equally complex spacing and affiliation mechanisms exhibited by females [as an example of such exclusive attention see (58)], and (ii) differences in the roles of males among the various species exhibiting so-called multi-male systems were masked by lumping a broad range of species-typical social organizations under the term "multi-male groups." In support of this last statement, it should be pointed out that the multi-male troops of the lemurid primates, such as *Lemur catta* and *Propithecus verreauxi* (25, 59), are organized around a female matriarchy that differs markedly in discrete social control mechanisms from the multi-male troops of the savanna baboons, *Papio anubis, P. ursinus,* and *P. cynocephalus* (56, 57).

The preoccupation with the "unique," the desire to relate primate behavior to human behavior, and the lack of sound ecological studies are hindrances at this stage of our understanding. What is needed more than ever is a clear appraisal of primate species as mammals exploiting an ecosystem and subject to ecological pressures similar to those acting on other mammalian species (3).

We propose that not only terrestrial primate species adapted to arid climates but also most primate species adapted to forests are characterized by either uni-male troops or age-graded-male troops. The true multi-male system is a less frequently evolved specialization, and the term "multi-male" should be restricted to those species having large troops that include several functionally reproductive adult males, as well as nonreproductive males of different ages.

REGULATION OF TROOP SIZE

One of the key problems in describing a multi-male troop involves the definition of an adult male. In censusing primate troops, most workers use the category "adult male" for males that are sexually mature; however, it is well known that the age of sustained spermatogenesis does not necessarily correlate with the age of "social" maturity. In most medium-sized primates, a male that has become sexually mature at 3 to 5 years of age may not be sociologically mature or physically dominant until 8 or 9 years of age—yet all of these males are lumped together as adults. A similar social maturation sequence has been described in elephants, for which the problems of age-class definition are comparable to those encountered in primates (60, 61).

Long-term studies of *Macaca fuscata* and *M. mulatta* show how roles are established and maintained and how males succeed to leadership (54, 62-64). The rank of young males in later life is, in part, dependent on the status of their mothers (63). Fractionation of a large troop to form two new troops may be accompanied by the deposition of an old leader, the assumption of leadership by a solitary male, or the succession to leadership of second- or third-ranking males upon the removal or loss of an old leader (62, 64). Without recounting the extensive research on this, we wish to make the point that the processes of splitting do not necessarily involve extensive mortality in the infant and juvenile age classes. Furthermore, the new troops thus generated are not essentially uni-male in structure.

Fig. 2 shows the hypothetical growth of a primate troop from a founder situation to eventual instability and breakup because of internal recruitment and crowding that resulted from close neighbors. More data are needed on the natural genesis of troops, since we believe that there is a strong tendency toward a polygynous, uni-male reproductive unit for most species of forest-dwelling primates in the New World and Old World (65).

The case of a Mona monkey, *Cercopithecus campbelli lowei* (66), troop parallels the diagram in Fig. 2. From 1964 to 1968, the troop increased in size because of births to the founder adult male and four adult females. In the years 1968 to 1970, the number of births was curtailed and four young adult males from 3 to 4 years of age emigrated. During this transition period, the tendency toward a uni-male structure was masked and the troop might have been called multi-male.

Ateles geoffroyi and *A. belzebuth* maintain an age-graded-male troop (46, 47, 67). In contrast to *Cercopithecus campbelli lowei,* adult male spider monkeys are not in continuous association with adult females and young. Instead, an *Ateles* "group" is composed of units that contain one or more females and their dependent young and that forage independently in a common home range. The adult male will accompany the female units when females are in estrus and when there exist special positive, dyadic relationships between individuals. The age structure of the males, based on birth intervals, lends itself to the formation of an alpha-beta-gamma dominance hierarchy. Among subadult males, mutual support may be shown during offensive and defensive behavior toward potential predators and intruders, but with increasing age some young males become peripheral.

The range of social organization encountered in *Ateles geoffroyi* in either space or time is given in Fig. 2. At one extreme is the founder situation, or so-called uni-male group. The social organization of this species is generally several females and a semidetached male group that is, in essence. an age-graded group, since there is a male hierarchy based on age and most of the mating during a peak of a given female's estrus is probably accomplished by the older, dominant male. With denser populations, there are larger groups of females and more peripheral males, and there may be fighting among males when associated with female groups (68).

The genesis of two new troops from a single large troop in those species exhibiting the age-graded-male system can occur without a take-over of leadership or fighting among males. For example, division of a large *Presbytis entellus* troop consisting of approximately 30 individuals would begin by a departure from the sleeping tree and a fractionation into two foraging subgroups, each subgroup being under the leadership of a different male. The largest subgroup of females would follow the dominant male, the second subgroup following a subdominant male. At the conclusion of the day's foraging, the animals would return to a common sleeping place. The females themselves would exert a powerful influence on the ultimate composition of the subgroups, depending upon which male they followed. The second stage in troop development would consist of separate foraging patterns and occasional utilization of different sleeping trees. Eventually, the subgroups would be foraging and sleeping independently; thus, two new troops would be created. Examples are further amplified for *Presbytis entellus* by Muckenhirn (*44*).

Long-term population studies on Barro Colorado Island permit some generalizations concerning the grouping tendencies of *Alouatta palliata,* the howler monkey (*69, 70*). Chivers (*71*) has summarized the data for the island population and for one troop (the laboratory group) in particular. Although this species is often cited as exhibiting a multi-male structure, peripheral or extra-group males exist and uni-male troops occur frequently. It would appear that, at low population densities, this species approximates a uni-male structure (*72*), while at higher densities a temporary age-graded-male structure appears. We must reemphasize that, when there exists an age-graded series of related maturing males with one older, dominant male, then the multi-male structure is more apparent than real. In fact, in large troops of *Alouatta palliata* with several mature, adult males, there appears to be curtailed reproduction (*73*). This, then, leads us to consider some of the ecological factors that lead to the division of troops and what forms of social pathology may result in high-density populations.

SOCIAL PATHOLOGY AND DENSITY

For many primate species, the conditions under which splitting and male "take-overs" can occur are, in part, related to the density of, and degree of disturbance in, the population (*19*). Alternatively, fractionation of large groups can occur under ecological conditions (such as dispersed resources) that favor the maintenance of small groups and that may even approximate the founder situation (*12*).

Presbytis senex of Ceylon shows some rather instructive trends in population growth and composition when high-density populations are compared with low-density populations. As Table 2 indicates, *Presbytis senex* tends to live in uni-male reproductive units (*19, 74*) and is found in a wide range of

Table 2

Comparison of the populations of *Presbytis senex* found at Polonnaruwa and Horton Plains (*19*). (Population densities are minimum estimates that may show higher levels in more restricted sample areas.)

Presbytis senex	*Polon-naruwa*	*Horton Plains*
Number of troops studied	33.0	27.0
Total population studied	278.0	229.0
Population density (per square kilometer)	215	92.6
Percent of adults in population	63.8	51.3
Percent of subadults and juveniles in population	14.1	30.7
Percent of adult males in adult population	28.2	36.7
Percent of adult females in adult population	71.8	63.3
Ratios of adult males to adult females in total population	1:2.5	1:1.7
Average number of animals in uni-male troops	8.4	8.9
Average number of animals in predominantly male troops	7.5	7.5
Ratio of adult males to adult females in uni-male troops	1:4.1	1:3.3
Number of infant deaths[a]	19.0	2.0

[a] Based on one complete reproductive period.

habitat types, from a lowland dry zone (Polonnaruwa) to a highland wet zone (Horton Plains). Extra-troop males are organized into groups having an average size of 7.5 individuals. In high-density populations, the uni-male reproductive groups are subject to harassment from the peripheral bachelor groups, which occasionally results in infant mortality and leadership take-overs. Infant mortality is reflected, in part, by the different percentages of subadults and juveniles in low- and high-density populations (see Table 2). In Sugiyama and Mohnot's studies in India, *Presbytis entellus* infants were actually attacked and killed by invading adult males (*10, 45*). In the Ceylon langurs, Rudran did not witness such events directly, but infants and juveniles were found to be injured or missing after a male replacement had occurred. This type of male replacement appears to occur under conditions of high population density (*9*) or in marginal habitats (*45*); in either case, the altered age structure that results from male take-overs curtails population growth to some extent.

The suggestion that the uni-male structure is a response to crowding stress has been made before (*12*). The tendency for captive primate groups to assume a uni-male or "despot" male configuration was often assumed to be a pathological response; however, we believe that, although this uni-male condition may manifest itself at crowding densities, it is erroneous to think of the uni-male structure as being pathological.

SOME CORRELATIONS BETWEEN SOCIAL STRUCTURE AND ECOLOGY

The history of primate evolution has been subject to many reviews (75). Beginning with an insectivore-like form exhibiting certain arboreal adaptations with rather enlarged eyes, the primates underwent an extensive radiation throughout the Paleocene, giving rise to two main branches. One major branch differentiated into the present-day galagos, lorises, and lemurs, while the other differentiated into the Old World monkeys, New World monkeys, tarsiers, pongids, and hominids. The lorises and galagos still persist in Africa and South Asia as nocturnal, forest-adapted forms. These may be considered the most morphologically conservative primates, representing most nearly Paleocene forms.

The island of Madagascar served as a reservoir for the lemuroid primates, and this isolated radiation resulted in the occupancy of feeding niches that replicate, in part continental niches (76). Deriving from the second major radiation, the neotropical monkeys (Hapalidae and Cebidae) began their adaptations in the Oligocene in isolation from the Old World monkey radiation. Thus, the Madagascan, neotropical, and Palaeotropical radiations can be compared to elucidate convergences.

Judging from the habits of the living primates and the structure of fossil forms, the early primates were nocturnal and arboreal and subsisted on an omnivorous diet, including fruits, small invertebrates, and perhaps small vertebrates. The early radiation took place in a tropical forest habitat, and, in general, present-day primates remain tropical in their distribution, with the greatest diversity of species occurring in the rain forests of West Africa, Indo-Malaysia, South America, and Madagascar. The tropical rain forest, then, is the habitat in which the most complex problems of primate evolution are to be found. Those primate species that have extended their ranges into seasonally arid areas characterized by thorn scrub or savanna are often confined to areas surrounding riverine forests.

The forests were retained by primates as their primary environment. The acquisition of terrestrial habits is recent in primate history and occurred in some species of the now extinct giant lemurs, Arceolemurinae (77), from Madagascar, as well as in the cercopithecine genera *Papio, Mandrillus, Erythrocebus, Theropithecus, Macaca,* and *Cercocebus* in the Old World tropics. Terrestrial adaptation within the Pongidae are exemplified by *Pan* and *Gorilla.* Only in South America has a truly terrestrial form not evolved.

Fig. 3 compares four geographic areas with respect to the kinds of niches occupied by their respective primate genera. We have classified the genera by activity cycle, height of feeding, diet, and relative size. Many species weighing less than 2 kilograms are nocturnal and show a pronounced tendency to feed on high-energy food resources such as insects and fruit. All small primates are arboreal, and, with the exception of the

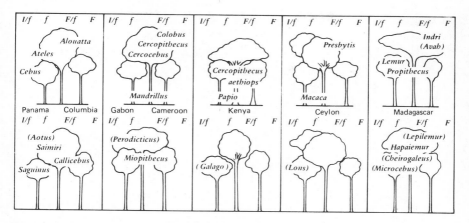

Fig. 3. Ecological equivalence for selected primate species. Horizontal scaling: *I/f*, insectivore-frugivore; *f*, frugivore; *F/f*; folivore-frugivore; *F*, folivore. Since the feeding categories are not absolute, only relative trophic preferences are indicated. The midpoint of the name should lie at the modal feeding classification. Vertical scaling; relative feeding height is illustrated by the position of the name. Lowest position: semiterrestrial; middle position: arboreal, second growth; highest position: upper canopy feeder. Names in parentheses indicate a rhythm of nocturnal activity. The upper series are all medium to large primates (adult weight, greater than 5 kilograms); the lower series are all small primates generally (adult weight, less than 2 kilograms).

neotropical forms and the West African *Miopithecus,* are derived from the morphologically conservative lorisoid or lemuroid stocks. The larger primate species are almost all diurnal, with the exception of the Madagascan genus *Avahi.* Only among the larger primates do we find several genera exhibiting semiterrestrial adaptation; these genera include *Mandrillus, Papio, Pan, Gorilla,* and *Macaca.*

The large diurnal primates (the Lemuridae, Indriidae, Cebidae, and Cercopithecidae) display parallel evolution with respect to their feeding strategies; there have been trends away from dependency on energy-rich invertebrates and fruit toward the more readily available cellulose found in leaves. Gut modifications associated with the change in diet can include a larger caecum, as in *Alouatta, Indri, Avahi* (*78*), and *Lepilemur* (*79*), or a chambered stomach and bacterial symbionts, as in *Colobus* and *Presbytis* (*80*). The latter modifications are convergent with the physiological and morphological adaptations evolved by terrestrial ungulates. Those primate species that can utilize cellulose in leaves are referred to as folivores and are distinguished from frugivores, which cannot utilize cellulose (*81*). Of course, these basic categories do intergrade, since folivores may supplement their leaf diet with fruit. Conversely, frugivorous species may utilize leaves for the simple sugar in their parenchyma cells (*81*).

Not surprisingly, the arboreal folivores are the most numerous of the

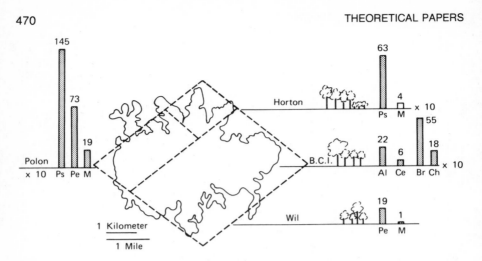

Fig. 4. Primate biomass comparisons for Ceylon and Panama. To compare the sizes of survey areas, the Ceylon study areas are superimposed on an outline map of Barro Colorado Island (B.C.I.). The dashed perimeter outlines the extent of the Wilpattu (Wil) survey area; for comparison, the Horton Plains (Horton) and Polonnaruwa (Polon) study areas are superimposed, and the latter is indicated by shading. To aid scaling on a single diagram, the biomasses for Polonnaruwa, Horton, and B.C.I. are displayed at 1/10 the actual values in kilograms per square kilometer. (Ps, *Presbytis senex;* Pe, *Presbytis entellus;* M, *Macaca sinica;* Al, *Alouatta palliata;* Ce, *Cebus capucinus;* Br, *Bradypus infuscatus;* Ch, *Choloepus hoffmani.*)

larger forest mammals, sometimes accounting for 30 to 40 percent of the arboreal mammalian biomass (see Fig. 4, Tables 3 and 4). Only in the Neotropics does the primate biomass rank second to that of another arboreal mammalian order — the edentate grazer of the treetops, represented by the three-toed sloth *Bradypus* (*82*) (see Fig. 4).

Table 3
Some comparisons of home range and group sizes. [Average home range and group size will vary widely from one study area to another. We present only species groups from the same areas for the best relative comparison. (B.C.I. indicates Barro Colorado Island.)]

Species	Locality	Average home range (km²)	Group size (average)	Feeding class
Alouatta villosa (71)	B.C.I., Panama	< 0.08	14.6	Folivore-frugivore
Ateles geoffroyi (46,87)	B.C.I., Panama	0.60	12.0	Frugivore
Cebus capucinus (91)	B.C.I., Panama	0.85	15.0	Frugivore
Presbytis senex (19,74)	Ceylon, Polonnaruwa	0.06	8.4	Folivore
Presbytis entellus (88,96)	Ceylon, Polonnaruwa	0.14	~ 20.0	Folivore
Macaca sinica (53)	Ceylon, Polonnaruwa	0.15	24.1	Frugivore
Presbytis entellus (44)	Ceylon, Wilpattu	> 1.00	~ 20.0	Folivore
Macaca sinica (53)	Ceylon, Wilpattu	> 1.00	20.0	Frugivore

Table 4

Comparisons of arboreal biomasses. [Data for Ghana from Collins (*108*); data for Barro Colorado Island are preliminary estimates and represent the minimum; data for Ceylon are minimum estimates that may show higher levels in restricted sample areas.]

Species	Biomass (kg/km^2)	Estimated arboreal biomass[a] $(\%)$
Barro Colorado Island		
Alouatta villosa	220.0	22
Cebus capucinus	60.0	6
Bradypus infuscata	550.0	54
Choloepus hoffmani	180.0	18
Ghana		
Colobus (three species)	55.08	79
Cercopithecus (two species)	5.3	8
Cercocebus (one species)	2.6	4
Polonnaruwa, Ceylon		
Presbytis senex	1450.0	61
Presbytis entellus	730.0	31
Macaca sinica	190.0	8
Wilpattu, Ceylon		
Presbytis entellus	19.0	< 95
Macaca sinica	< 1.0	> 5
Horton Plains, Ceylon		
Presbytis senex	630.0	94
Macaca sinica	< 40.0	6

[a] Percentage is based on the known mammalian biomass totals but not the total arboreal mammalian biomass, which may be one-sixth again as high.

SOCIAL STRUCTURE OF ARBOREAL PRIMATES

The smaller, nocturnal, insectivorous prosimians of Asia, Africa, and Madagascar seem to exhibit the same form of social organization. They are solitary or organized into dispersed family groups (*22-24*). In part, their solitary habits may result from their being nocturnal, since the coordination of groups would be difficult. The solitary state may also correlate with their insectivorous habits, since foraging patterns may demand a solitary technique (*22*). *Lepilemur* is a folivore, however, and retains the solitary pattern, which suggests that the rythm of nocturnal activity restricts group formation and co-ordination. The neotropical night monkey *Aotus* is derived from a diurnal form, and nocturnality is a secondary adaptation. *Aotus* exhibits a parental family structure and is thus partly an exception to the rule established for the prosimians. This may indicate the retention of a phylogenetically old parental system that is still compatible with the rythm of nocturnal activity (*36*).

The small diurnal primates concentrated in the New World (*Callicebus, Saguinus, Callithrix,* and *Saimiri*) are insectivore-frugivores and exhibit two grouping tendencies. *Callicebus, Saguinus,* and *Callithrix* live in parental families, while *Saimiri* lives in age-graded-male troops. The parental structure of *Callicebus* and *Saguinus* may be the retention of a phylogenetically conservative trait.

The only small diurnal primate of the Old World is *Miopithecus talapoin*. It shows strong convergences in social behavior and ecology with *Saimiri sciureus* of the Neotropics (*83, 84*). Both species live in large troops (20 to 100 individuals) that divide into subgroups based on age and reproductive classes (*83-85*). Although these species are superficially multi-male, adult males show little affiliation with each other or with females except during the breeding season, and young males may form a peripheral group. Furthermore, the form of the social structure is influenced by the breeding season, which leads to dominance among males and the formation of temporary uni-male breeding units (*86*); hence, the classic multi-male structure is not matched. Table 1 indicates our interpretation of the form of social structure for the smaller species of primates and the correlations with their feeding habits.

Given the two broadly defined feeding niches, folivores and frugivores, the following generalizations can be made about the larger arboreal primates: (i) given comparable habitats, the frugivores have larger home ranges and move more widely during their daily activities than do folivores of an equivalent size class [for example, compare *Ateles* with *Alouatta* (*87*), and *Presbytis senex* with *Macaca sinica* (*88*)] (see Table 3); (ii) both trophic types tend toward either a uni-male structure or an age-graded-male system (see Table 1), with folivores especially tending toward a uni-male organization; and (iii) many folivores [*Colobus, Alouatta,* and *Presbytis* (*19, 20, 71*)], but only some of the frugivores [*Symphalangus* and *Hylobates* (*14, 34*)], employ troop or individual vocalizations in maintaining spacing between adjacent troops.

Comparisons among the African, Asian, and South American forest-dwelling species show the marked tendencies toward a uni-male reproductive unit (see Table 1). In an extensive survey, Struhsaker lists almost all species of West African rain forest frugivores of the genus *Cercopithecus* as typically exhibiting a uni-male reproductive unit (*65*). *Cercopithecus mitis*, a frugivore studied in Uganda (*89*), and *Colobus guereza*, a folivore (*20*), show similar modes. Examples of uni-male reproductive units found among the strongly arboreal folivores of Asia include *Presbytis cristatus, P. johni,* and *P. senex (17-19)*. Although some troops of *P. cristautus* may, under certain circumstances, contain more than one male (*90*), this reflects an age-graded-male system. In Central America, the folivore *Alouatta* may exhibit a uni-male system under conditions of low population density, but at high densities several males may be included in an age-graded-male troop. The frugivorous, white-throated capuchin, *Cebus capucinus,* appears to exhibit a uni-male reproductive system (*91*), while *Ateles geoffroyi* tends toward an age-graded-male system (*92*).

The arboreal folivores may be characterized by their vocalizations, which usually take the form of dawn choruses. These announcement calls may also be produced at various times during the day, since they appear to occur prior to progressions and are also triggered by exogenous factors. The role of these calls is definitely related to intraspecific spacing. This function has been demonstrated for *Colobus* (*20*) and *Alouatta* (*71*) and is undoubtedly being served by the calling of *Presbytis* (*19*). The frugivorous gibbons and the siamang exhibit comparable chorusing behavior (*14, 93*). No investigator has specifically attempted to correlate possible seasonal increases in the frequency of chorusing behavior with breeding peaks for any of the chorusing species. McClure, however, offers observations on *Symphalangus syndactylus* and *Hylobates lar* in Malaysia that suggest such a correlation may exist (*94*).

SOCIAL STRUCTURE OF SEMITERRESTRIAL PRIMATES

In general, semiterrestrial species tend to live in variable habitats and are frugivores or varied feeders. *Presbytis entellus* and *Gorilla gorilla* are the only folivorous primates that do forage extensively on the ground. Semiterrestrial species also tend to live in larger groups, compared to arboreal forms of a similar body size, and tend to form age-graded-male troops (see Tables 1 and 3).

All but one species of *Presbytis* and of *Cercopithecus* fit into the two forest niches (arboreal folivore and frugivore) described above. The exceptions are *Presbytis entellus,* which is terrestrially adapted and occupies riverine forests, seasonally dry forests, and seasonally arid scrub (*8*), and *Cercopithecus aethiops,* which occupies gallery forests and savanna areas (*11*). Both species show a variable troop structure, with *Cercopithecus aethiops* tending to exhibit an age-graded-male to multimale configuration (*11, 12*), and *Presbytis entellus* showing an age-graded-male organization in Ceylon and in some parts of India. *Presbytis entellus,* however, may also show a strictly unimale pattern when crowded into remnant patches of roadside forest in India (*8, 10, 44, 45, 95, 96*).

The tendency toward a larger group size and an increased number of males in a group is well illustrated in the forest-adapted mangabeys (*Cercocebus*) when a species that forages on the ground (*C. torquatus*) is compared to a conspecific of similar size that is primarily arboreal (*C. albigena*). The more terrestrial *C. torquatus* forms larger troops with more than one adult male per troop. The sympatric species *C. albigena* forms smaller troops and characteristically has a uni-male group stucture (*97*); however, Chalmers (*98*) presents data that suggest a multi-male group for *C. albigena* in Uganda.

The tendency toward larger group sizes than those of strictly arboreal forms and an accompanying age-graded-male structure may also hold for semiterrestrial macaques and *Mandrillus*. *Macaca sinica* troops, for example, have a distinct age gradation in the so called adult male class, as well as extra-group males (*53*). Recent research by Gartlan (*43*) suggests that the drill (*Mandrillus leucophaeus*), which forages on the ground in dense forests, has retained the uni-male to age-graded-male structure.

Papio anubis, P. cynocephalus, P. ursinus, Macaca mulatta, and *M. fuscata* forage primarily on the ground (*7, 50, 56*). They exhibit a multi-male grouping pattern, which seems to be effective as an antipredator mechanism (*7, 49*) (see Table 1).

Terrestrial species of the genera *Theropithecus* and *Erythrocebus,* as well as *Papio hamadryas,* present other problems. *Theropithecus gelada* and *Papio hamadryas* are adapted to extremely arid environments and break up into foraging parties that are uni-male in composition; hence, the selective advantage of group attacks on terrestrial predators is lost. Crook and Gartlan (*2, 3*) have suggested that this pattern promotes feeding efficiency by reducing the competition for food that would occur if a large number of males accompanied the females and young; hence, these species have approximated a uni-male grouping at the expense of group offensive behavior toward terrestrial predators. Such foraging units reassemble at selected sleeping sites in the evening.

The mode of antipredator behavior is extremely important to a full understanding of the selective advantage of either a multi-male or uni-male group. The terrestrial *Erythrocebus patas,* for example, lives sympatrically with *Papio* but has retained the uni-male grouping tendency, with the male keeping watch while the females forage. *Erythrocebus* relies mainly on speed, distraction displays by the adult male, and dispersed hiding to avoid predation; thus, its antipredator behavior is not and never was built around a mobbing response (*37*). Since *Erythrocebus* is derived from a *Cercopithecus*-like form, it may have retained the uni-male structure in its new habitat and have undergone selection for greater speed instead of evolving a multi-male structure, as is the case in some species of the genus *Papio* (*65*).

The chimpanzee (*Pan troglodytes*) occupies a range of habitats, from rain forest to forest-grassland. Troops seem to have a loose grouping pattern (*99*), although cohesive troop behavior may be shown during long marches in savanna areas (*100*). These frugivores break up into foraging units in which strong cohesion exists mainly in the immediate mother family, although patterns of long-term affiliation are displayed when the independently foraging subunits come together (99, 101).

Gorilla gorilla may be considered a semiterrestrial folivore that has a cohesive group structure. The males appear to be age-graded, with one old, silver-back male as the leader (102).

ROLES, GROUP FUNCTIONS, AND SOCIAL STRUCTURE—SOME SELECTIVE ADVANTAGES

Crook (3) has evaluated the inter-relations between ecology and social structure in primates. We offer here some comments on critical issues that deserve further research. The fundamental questions are (i) Why do some primate species exhibit a multi-male troop composition? and (ii) Why do adult female primates find it advantageous to form extended mother families? In an evolutionary sense, the number of males in a given troop will depend on what advantage the males are to the reproducing females (103).

It would appear, in comparing the various species of higher primates, that there is a strong trend toward polygyny. Certainly a given male increases his individual fitness by distributing his genes among the greatest possible number of females. However, many advantages can accrue to a dominant male through the presence and actions of subordinates (16, 39). A given male's dependence on other males for support in enhancing his survival value will determine, in part, the short-term advantage of having many males in a troop. Nevertheless, if fewer males can do the same task better than many males, then all things will tend to favor a polygynous mating system. Even in a multi-male troop, such as has been reported for various species of *Papio* and *Macaca,* it may well be that the descendents of the alpha male would be more predominant within the breeding unit than would those of subordinate males, since subordinate males engage in less sexual activity than do alpha males (49, 54).

In most species of primates, the role of the male involves little parental care. There are, however, exceptions within the families Cebidae and Hapalidae, as well as in some species of the genus *Macaca.* The functions of the adult male, either alone or in company with other males, seem to be (i) to maintain spacing with respect to neighboring troops of the same species, (ii) to reduce competition within the group by driving out younger males, and (iii) to enforce some degree of protection against predators. These behaviors may involve vigilance, mobbing, pursuit-invitation displays, or outright attack.

A fourth aspect of the adult male's activities, which has received only sporadic attention, is his role of providing leadership. By initiating and maintaining movement in a certain direction, the male is influential in promoting cohesion of the troop and serves as a focus for the troop's movements (44, 102). The dominant adult male is the one most likely to initiate a following response from the females, hence his leadership role is dependent upon the active participation of females and attendant juveniles in his movements (44).

In contrast to the roles of adult males, the roles of adult females are dominated by infant care. During the early phases of the infant's development, the mother is responsible for protecting it; when the infant enters the juvenile stages, the mother serves as a focus for its socialization. Indeed, the juvenile's status is, in part, predetermined by its mother's status within the female hierarchy (63).

To some extent, each female old enough to reproduce is in competition with males and other females of the same age class for food, sleeping space, and so on. Nevertheless, certain advantages (for example, increased feeding efficiency and antipredator behavior) may be inherent in a group of several females traveling together and maintaining a close liaison. The extent to which individual females benefit from group rearing of their progeny and the extent to which their own chances of survival are increased by associating with other females will determine the size to which the troop can profitably increase.

Ultimately, the size of the troop is, in part, a compromise between competition among members for resources that are in short supply and the advantage of having many members in locating resources that are scattered and available for restricted periods (96). The advantage of feeding in a group is often overlooked by field workers, although feeding calls, which promote aggregation in fruit trees, are a well-known phenomenon for some species.

We speculate that, although primary folivores such as *Colobus guereza* and *Presbytis senex* eat considerable quantities of fruits, their feeding strategy is not predicated on a daily need of finding ripening fruit trees within their home range. Small, cohesive, uni-male social units are permitted within this strategy. However, in frugivores such as *Ateles geoffroyi* and *Pan satyrus* (46, 101), the best feeding strategy involves breaking up the troop into small, independently foraging units that spread out to locate fruit trees within their home range and then "announce" the location of feeding spots. In *Saimiri*, foraging for small, dispersed canopy insects leads to the formation of subgroups that forage at different rates (84). The interaction between the distribution of primary resources (food and water) in a specific habitat and the population density in that habitat can profoundly affect social structure (12).

In rain forests, sympatric species often form mixed feeding groups that move together and show very little overt competition. The phenomenon has been described for Central America and for Gabon, West Africa (104). Such species associations should be distinguished from instances in which an occasional male of one species may associate with a troop of another species [for example, a *Saimiri* male in a *Callicebus* group (35) and two *Macaca sinica* males in a *Presbytis entellus* group (*44, 95*)]. The exact significance of either the casual or the frequent mixed-species group has not been ascertained, although in both types predator avoidance or feeding efficiency, or both, may be increased for all members. The whole problem requires attention (*105*).

We have referred to antipredator behavior throughout our consideration of selective advantages for the various social systems. Yet no single aspect of primate field studies has less supportive data than the generalizations concerning the survival value of the various presumed antipredator mechanisms. The primate troop can and does exhibit antipredator behavior. Emphasis on different patterns (including vigilance, alarm calls, distraction displays, mobbing, and attack) may vary between species and between age classes and sex classes within a species (*44*).

Only a few individuals in a troop need spend a great percentage of their time in vigilance for all members of the group to be benefited; typically subadult and adult males perform this role. Visual scanning from high positions may serve to locate intruding conspecific males as well as predators (*37, 49*), and troop members may be warned of the presence of a predator by alarm calls. The manner in which monkeys respond to a potential predator depends partly upon their own size and mobility and partly upon the relative size and position of the predator.

The jumping and vocalizations of adult males are common alarm behaviors, and are described for such diverse primate species as terrestrial *Erythrocebus patas* (*37*) and arboreal *Presbytis senex* (*19, 74*). It is generally considered that such behavior distracts the predator while the females encumbered by young scatter and hide (*44*). Protective males typically position themselves between the group and a terrestrial predator.

Mobbing, another alarm behavior, requires the presence of a group. A predator may be harried through vocalizations, group defecation and urination (*46, 69*), and, in large species, branches that may fall from the weight of leaping adult males (*Presbytis*) or be purposely broken off and dropped (*Alouatta, Cebus*) (*46, 69, 91*).

If a predator should surprise a troop at short range, individual flight responses may scatter troop members in various directions, thus confusing a predator. This confers some selective advantage on group life, even though no altruistic tendencies can be detected.

Unlike the smaller primates, larger species attack outright, and some (for example, gorilla and chimpanzee) are more than a match for even the largest predators. Baboons exhibit a pronounced sexual dimorphism, and, since the baboons of the genus *Papio* typically forage in large multi-male groups, the larger males can form an effective attack unit and displace a predator. The responses of baboons and chimpanzees to leopards have been studied (*106*) and have been found to include the use of actual weapons (sticks and rocks) (*107*).

GENERAL CONCLUSIONS

When the major radiations in the Old World, New World, and Madagascar are compared with each other, no doubt some of the differences in social structure seen in those species adapted to similar ecological niches will be found to result from phylogenetic differences; that is, the social structures of the ancestral forms have been carried forward in the adaptive radiation of these species. For example, the consistent tendency in lemuroid primates such as *Propithecus* and *Lemur* to show multi-male groups with more males than females, a dominance of females over males, and the segregation of troops into all-male and all-female subgroups does not exist in any known continental species (*25, 59*). The South American radiation has produced a trend toward male participation in parental care and the formation of pair bonds between given males and females.

Although we can generalize about the selective advantages of primate social structures, we must remember that the history of the population under study, its particular adaptation to local environmental conditions, and the idiosyncratic nature of its dyadic relations (which have been ontogenetically established within the particular group) can result in a great deal of variability in social structure, even within the same species when it occurs in widely differing habitats. Hence, in making generalizations about social structure, we must remember that we are talking about behavioral modes or behavioral medians.

SUMMARY

It has been the custom for ethologists to divide mammalian societies into grades. Each ascending level of complexity denotes an increase in the complexity of interaction patterns among the members of the group. The multi-male group traditionally represented a high level of social organization, as well as the higher primate norm, but it was defined from early studies on terrestrial primates. What we have tried to show is that the uni-male system occurs in a wide variety of primate species in both the cercopithecoid and ceboid radiations. Furthermore, we have attempted to illustrate that multi-male systems are more apparent than real and that many should be considered age-graded-male systems.

The three proposed classes (uni-male, age-graded-male, and multi-male) of social structure (above the level of the parental family) are gradations and represent an increased complexity based on an increased tolerance among adults at the maximum "sociological" age level. There are only a few species for which the data are sufficient to place them in a class. The multi-male system is apparently a specialized form of social grouping that represents a particular adaptation to terrestrial foraging by intermediate-sized primates. It is readily derived from an age-graded-male system. The multi-male system does not differ profoundly from an age-graded-male system, but the former does allow for increased affiliation and cooperation among adult males.

The uni-male system or the age-graded-male system is favored in arboreal species, both frugivores and folivores. The structure of a species' social organization is more predictable for diurnal, leaf-eating forms than it is for the frugivores. We can correlate an arboreal, diurnal, leaf-eating niche with a species having a social structure that tends toward a uni-male system with a small home range and the employment of chorusing behavior to effect spacing. Such species tend to be sedentary and are typified by *Alouatta, Colobus,* and *Presbytis.* It would appear, then, that similar predation pressures (semiarboreal felids probably being most important) and similar foraging problems have forced similar behavioral solutions upon these species.

One should be wary of generalizations concerning the form of social structure for any given species, since social structure may vary with habitat. Similar variations have been noted in response to problems of density and habitat disturbance. Parallel trends can be noted in both the ceboid and cercopithecoid lines of evolution. More nearly accurate correlations will be possible only when we have more data concerning feeding efficiency and anti-predator mechanisms for a wide variety of species, each studied within a range of habitats.

REFERENCES AND NOTES

[1] J. F. Eisenberg, *Handb. Zool.* 8 (No. 39), 1 (1966).

[2] J. H. Crook and J. S. Gartlan, *Nature* 210, 1200 (1966).

[3] J. H. Crook, in *Social Behaviour in Birds and Mammals,* J. H. Crook, Ed. (Academic Press, London, 1970), pp. 103-168.

[4] C. R. Carpenter, *Trans, N.Y. Acad. Sci.* 4, 248 (1942); *Hum. Biol.* 26, 269 (1954).

[5] In Carpenter's original formulation, he suggested that territorial behavior toward conspecific troops was shown. We now, know that such exclusive patterns can be effective without overt fighting, and the generalization can be modified to state that all primate troops appear to have exclusive rights to certain areas of their home range at certain critical times of the year, even though they may share parts of their home range with neighboring troops.

[6] T. E. Rowell, *J. Zool. London* 149, 344 (1966).

[7] K. R. L. Hall and I. DeVore, in *Primate Behavior,* I. DeVore, Ed. (Holt, Rinehart & Winston, New York, 1965), pp. 53-110.

[8] P. Jay, in *ibid.,* pp. 197-249; S Ripley, in *Social Communication among Primates,* S. A. Altmann, Ed. (Univ. of Chicago Press, Chicago, 1967), pp. 237-253.

[9] K. Yoshiba, in *Primates,* P. Jay, Ed. (Holt, Rinehart & Winston, New York, 1968), pp. 217-242.

[10] Y. Sugiyama, in *Social Communication among Primates,* S. A. Altmann, Ed. (Univ. of Chicago Press, Chicago, 1967), pp. 221-236.

[11] T. T. Struhsaker, *Ecology* 48, 891 (1967); *Univ. Calif. Publ. Zool.* 82, 1 (1967).

[12] **J. S. Gartlan and C. K. Brain,** in *Primates,* P. Jay, Ed. (Holt, Rinehart & Winston, New York, 1968), pp. 253-292.

[13] **C. R. Carpenter,** *Comp. Psychol. Monogr.* **16,** 1 (1940).

[14] **T. Ellefson,** in *Primates,* P. Jay, Ed. (Holt, Rinehart & Winston, New York, 1968), pp. 180-200.

[15] **H. Kummer,** *Z. Psychol.,* No. 33 (1957).

[16] _____ *Social Organization of Hamadryas Baboons: A Field Study* (Univ. of Chicago Press, Chicago, 1968).

[17] **I. Bernstein,** *Behaviour* **32,** 2 (1968).

[18] **F. E. Poirier,** in *Primate Behavior,* L. A. Rosenblum, Ed. (Academic Press, New York, 1970), vol. 1, pp. 251-383.

[19] **R. Rudran,** thesis, University of Ceylon, Colombo (1970).

[20] **P. Marler,** *Science* **163,** 93 (1969).

[21] **G. H. Manley,** *Symp. Zool. Soc. London* **15,** 493 (1967).

[22] **P. Charles-Dominique,** *Biol. Gabonica* **7,** 121 (1971); *ibid.* **2,** 347 (1966).

[23] **J-J. Petter and C. M. Hladik,** *Mammalia* **34,** 394 (1970).

[24] **J-J. Petter,** *Mem. Mus. Nat. Hist. Natur. Paris Ser. A Zool.* **27,** 1 (1962).

[25] **A. Jolly,** *Lemur Behavior: A Madagascar Field Study* (Univ. of Chicago Press, Chicago, 1966).

[26] **P. Leyhausen,** *Symp. Zool. Soc. London* **14,** 249 (1965).

[27] **P. K. Anderson,** *ibid.* **26,** 299 (1970).

[28] The objection to the term "solitary" is valid if and only if the term is taken literally. If due consideration is given to the existence of indirect communication and the possibility of the maintenance of reproductive continuity among related members of such dispersed population units (*1*), then we can maintain the older term; if not, then we may well substitute the more cumbersome term, "dispersed, noncohesive family group."

[29] **P. Charles-Dominique,** *Fortschr. Verhaltens-forschung* **9,** 7 (1972); R. D. Martin, *ibid.,* p. 43.

[30] **A. Petter-Rousseaux,** *Mammalia* **26,** 1 (1962); P. Charles-Dominique and C. M. Hladik, *Terre Vie,* part 1(1971), p. 3.

[31] **H. Fischer,** *Z. Tierpsychol.* **22,** 247 (1965).

[32] **N. A. Muckenhirn,** thesis, University of Maryland (1967).

[33] **G. Epple,** *Folia Primatol.* **7,** 37 (1967); M. Moynihan, *Smithson. Contrib. Zool.* **28,** 1 (1970).

[34] **M. Kawabe,** *Primates* **11,** 285 (1970).

[35] **W. A Mason,** in *Primates,* P. Jay, Ed. (Holt, Rinehart & Winston, New York, 1968), pp. 200-216.

[36] **M. Moynihan,** *Smithson. Misc. Collect.* **146,** 1 (1964).

[37] **K. R. L. Hall,** *J. Zool. London* **148,** 15 (1965).

[38] **J. H. Crook,** in *Play, Exploration and Territory in Mammals,* P. A. Jewell and C. Loizos, Eds. (Zoological Society of London, London, 1966), pp. 237-258.

39 ———, in *Man and Beast: Comparative Social Behavior,* J. F. Eisenberg and W. Dillon, Eds. (Smithsonian Institution, Washington, D.C., 1971), pp. 235-260.

40 **H. Kummer,** W. Goetz, W. Angst, in *Old World Monkeys,* J. R. Napier and P. H. Napier, Eds. (Academic Press, New York, 1970), pp. 351-364.

41 The uni-male system may be maintained ontogenetically from a founding pair through the combined processes of internal recruitment by birth and antagonism exercised by the founding adult male toward his sons. The maintenance of such uni-male groupings is paralleled in the Carnivora [D. G. Kleiman and J. F.Eisenberg, "Comparisons of canid and felid social systems from an evolutionary perspective" (paper presented at the First International Symposium on World Felidae, Laguna Hills, Calif., 1971)].

42 In recognition of this fact, Crook has distinguished two grades of "multi-male" social structures (2, 3), but he has not elaborated on his reasons for doing so or on the criteria for separation.

43 **J. S. Gartlan,** in *Old World Monkeys,* J. R. Napier and P. Napier, Eds. (Academic Press, New York, 1970), pp. 445-480.

44 **N. Muckenhirn,** thesis, University of Maryland (1972).

45 **S. M. Mohnot,** *Mammalia* **35,** 175 (1971).

46 **J. F. Eisenberg and R. E. Kuehn,** *Smithson. Misc. Collect.* **151,** 1 (1966).

47 **L. Klein and D. Klein,** *Int. Zoo Yearb.* **11,** 175 (1971); J. F. Eisenberg, unpublished data.

48 **I. DeVore and S. L. Washburn,** in *African Ecology and Human Evolution,* F. C. Howell and F. Bourlière, Eds. (Aldine, Chicago, 1963), pp. 335-367.

49 **I. DeVore and K. R. L.** Hall, in *Primate Behavior,* I. DeVore, Ed. (Holt, Rinehart & Winston, New York, 1965), pp. 20-52.

50 **C. H. Southwick, M. A. Beg, M. R. Siddiqi,** *idib.,* pp. 111-159.

51 **K. Imanishi,** *Curr. Anthropol.* **1,** 393 (1960).

52 **M. Bertrand,** *Bibl. Primatol.* **11,** 1 (1969).

53 **W. Dittus,** unpublished data (1968-1971).

54 **J. H. Kaufmann,** *Ecology* **46,** 500 (1965).

55 **T. Nishida,** *Primates* **7,** 141 (1966).

56 **S. Altmann and J. Altmann,** *Baboon Ecology: African Field Research* (Univ. of Chicago Press, Chicago, 1970).

57 **K. R. L. Hall,** *Proc. Zool. Soc. London* **139,** 283 (1962); I. DeVore, in *Classification and Human Evolution,* S. L. Washburn, Ed. (Viking Fund Publications in Anthropology, No. 37, Wenner-Gren Foundation, New York, 1963), pp. 301-319.

58 **L. Tiger,** *Men in Groups* (Random House, New York, 1968).

59 **N. Bolwig,** *Mem. Inst. Rech. Sci. Madagascar Ser. A Biol. Anim.* **14,** 205 (1960).

[60] **H. Hendrik,** in *Dikdik und Elefanten* (Piper, München, 1970) pp. 70-77; G. M. McKay, thesis, University of Maryland (1971).

[61] **J. F. Eisenberg, G. M. McKay, M. R. Jainudeen,** *Behavior* **38,** 193 (1971).

[62] **C. B. Koford,** in *Primate Behavior,* C. Southwick, Ed. (Van Nostrand, New York, 1963), pp. 136-152.

[63] **D. S. Sade,** *J. Phys. Anthropol.* **23,** 1 (1965).

[64] **M. Kawai,** *Primates* **2,** 181 (1960); Y. Furuya, *ibid.,* p. 149.

[65] **T. T. Struhsaker,** *Folia Primatol.* **11,** 80 (1969).

[66] **F. Bourlière, C. Hunkeler, M. Bertrand,** in *Old World Monkeys,* J. R. Napier and P. H. Napier, Eds. (Academic Press, New York, 1970), pp. 297-350.

[67] **L. L. Klein,** "Ecological correlates of social grouping in Colombian spider monkeys," paper presented at the Animal Behavior Society Meeting, Logan, Utah, 1971.

[68] **C. R. Carpenter,** *J. Mammal.* **16,** 171 (1935).

[69] ――――, *Comp. Psychol. Monogr.* **10,** 1 (1934).

[70] ――――, in *Primate Behavior,* I. DeVore, Ed. (Holt, Rinehart & Winston, New York, 1965), pp. 250-291.

[71] **D. J. Chivers,** *Folia Primatol.* **10,** 48 (1969).

[72] **N. E. Collias and C. H. Southwick,** *Proc. Amer. Phil. Soc.* **96,** 143 (1952).

[73] **J. B. Calhoun,** in *Physiological Mammalogy,* W. V. Mayer and R. G. Van Gelder, Eds. (Academic Press, New York, 1963), vol. 1, p. 91.

[74] **G. Manley,** in preparation.

[75] **W. E. LeGros-Clark,** *The Antecedents of Man* (Quadrangle, Chicago, 1960).

[76] **J. F. Eisenberg and E. Gould,** *Smithson. Contrib. Zool.* **27,** 1 (1970).

[77] **A. Walker,** in *Pleistocene Extinctions,* P. S. Martin and H. E. Wright, Eds. (Yale Univ. Press, New Haven, Conn., 1967), p. 425.

[78] **C. M. Hladik,** *Mammalia* **31,** 120 (1967).

[79] ――――, P. Charles-Dominique, P. Valdebouze, J. Delort Laval, J. Flanzy, *C. R. Hebd. Seances Acad. Sci. Paris* **272,** 3191 (1971).

[80] **T. Bauchop and R. W. Martucci,** *Science* **161,** 698 (1968).

[81] **C. M. Hladik and A. Hladik,** *Biol. Gabonica* **3,** 43 (1967); *Terre Vie,* part 1 (1969), p. 27.

[82] **G. G. Montgomery and M. E. Sunquist,** in preparation.

[83] **A. Gautier-Hion,** *Folia Primatol.* **12,** 116 (1970).

[84] **R. W. Thorington,** Jr., in *The Squirrel Monkey,* L. Rosenblum and R. W. Cooper, Eds. (Academic Press, New York, 1968), pp. 69-87.

[85] **F. V. DuMond,** in *ibid.,* pp. 88-146.

[86] **J. D. Baldwin,** *Folia Primatol.* **9,** 281 (1968).

[87] **A. Richard,** *ibid.* **12,** 241 (1970).

[88] **C. M. Hladik and A. Hladik,** *Terre Vie,* in press.

[89] **F. P. G. Aldrich-Blake,** in *Social Behaviour in Birds and Mammals,* J. H. Crook, Ed. (Academic Press, London, 1970), pp. 79-102.

[90] **Z. Y. Furuya,** *Primates* **3,** 41 (1961-62).

[91] **J. R. Oppenheimer,** thesis, University of Illinois (1968).

[92] The frugivorous species do not show such uniform trends in social structure. As with the folivores, there is tendency toward a uni-male situation, but at least two kinds of group organization can be discerned: (i) the cohesive uni-male band, which is generally quite small, 20 members (for example, *Cebus capucinus* and *Cercopithecus mitis*), or (ii) the less cohesive, extended group, in which mother families and age-graded-male units forage independently (for example, *Ateles geoffroyi*), reassemble only from time to time at preferred resting loci, and so on. Although all of these species have long-distance calls, the calls are generally not given in a chorusing fashion and their relation to intraspecific spacing is not well understood.

[93] **H. E. McClure,** *Primates* **5,** 39 (1964). It shoud be noted that, although the gibbons are classically considered frugivores (*13*), they do feed on leaves at certain seasons of the year (*14*). It may be that further research will lead to their being classified as folivore-frugivores. If such is the case, their spacing system is more easily comprehensible. See also D. J. Chivers, *Malayan Nature J.* **24,** 78 (1971).

[94] In *Symphalangus,* the peak number of morning calling sessions heard in June was several times greater than the peak number heard between September and January, the months of minimal calling. The only observation of a newborn was recorded in February. While such evidence is obviously not sufficient to establish a birth peak for this species, it suggests that some breeding did occur in June. Likewise, 2.5 times more calls of *Hylobates* were heard in June than in November and February, the months of minimal calling.

[95] **J. F. Eisenberg and M. Lockhart,** *Smithson. Contrib. Zool.* **101,** 1 (1972).

[96] **S. Ripley,** in *Old World Monkeys,* J. Napier and P. Napier, Eds. (Academic Press, New York, 1970), pp. 481-512.

[97] **C. Jones and J. Sabater Pi,** *Folia Primatol.* **9,** 99 (1968).

[98] **N. R. Chalmers,** *ibid.* **8,** 247 (1968).

[99] **J. Goodall van Lawick,** *Anim. Behav. Monogr.* **1,** 165 (1968).

[100] **K. Izawa,** *Primates* **11,** 1 (1970).

[101] **V. Reynolds and F. Reynolds,** in *Primate Behavior,* I. DeVore, Ed. (Holt, Rinehart & Winston, New York, 1965), pp. 368-424.

[102] **G. B. Schaller,** *The Mountain Gorilla* (Univ. of Chicago Press, Chicago, 1963).

[103] For a theoretical discussion, see J. E. Downhower and K. Armitage, *Amer. Natur.* **105,** 355 (1971).

[104] **R. W. Thorington, Jr.,** in *Progress in Primatology,* D. Starck, R. Schneider, H.-J. Kuhn, Eds. (International Primatological Society, Frankfurt, 1967), pp. 180-184; T. T. Struhsaker, in *Old World Monkeys,* J. Napier and P. Napier, Eds. (Academic Press, New York, 1970), pp. 365-444; J. P. Gautier and A. Gautier-Hion, *Terre Vie,* part 2 (1969), p. 164.

[105] Interspecific antagonisms can occur, however, even between those species that form mixed feeding associations, since the propensity to show interspecific aggression apparently varies with respect to the history between the two populations under study and the distribution of resources in the habitat under consideration. In a broad sense, the density of primate populations, either arboreal or semiterrestrial, may be severely limited by the presence of competitors. It should not be assumed, however, that the most severe competition typically comes from other species of primates. Struhsaker (*11*) has reported that the major food competitor of *Cercopithecus aethiops* in Kenya is the elephant. This competition results from the fact that, during the dry season, the elephants push over the major species of trees that provide food and refuge for the monkeys.

[106] **A. Kortlandt,** in *Progress in Primatology,* D. Starck, R. Schneider, H.-J. Kuhn, Eds. (International Primatological Society, Frankfurt, 1967), pp. 208-224; ———and M. Koij, *Symp. Zool. Soc. London* **10,** 61 (1963).

[107] Although the members of the group that attack the leopard may act individually, the effect of several adult males charging can be quite intimidating to a potential predator and thereby be a distinct selective advantage for a multi-male social group. Such antipredator maneuvers are generally in the province of males, and those species that tend to show responses of this kind generally exhibit a great dimorphism, with the males being much larger than the females; however, dimorphism is less pronounced in the chimpanzee than in the baboon.

[108] **W. B. Collins,** as quoted by F. Bourlière, in *African Ecology and Human Evolution,* F. C. Howell and F. Bourlière, Eds. (Aldine, Chicago, 1963), pp. 43-54.

[109] This article is an expanded version of a talk presented by J.F.E. at the Animal Behavior Society Meeting, Logan, Utah, 1971, and is an outgrowth of fieldwork that has been in progress since 1964. Our primary effort was in Ceylon from 1968 to 1970 (*95*), but studies were also conducted in Central America and Madagascar (*32, 46, 76*). In 1967, J.F.E. and Suzanne Ripley formed a research group that studied the Ceylon primates from the standpoints of ecology and behavior. Some eight investigators participated in the project from 1968 to 1971. The adaptive patterns that have emerged will be or are being published by individual investigators (*19, 23, 44, 53, 88, 96*), and this article does not attempt to synthesize all interpretations. Research was supported, in part, by National Science Foundation grant GB-3545, awarded to J.F.E.; National Institute of Mental Health grant RolMH15673-01; research grant 686 from the National Geographic Society; and Smithsonian Institution Foreign Currency Program grant SFC-7004, awarded to J.F.E. and Suzanne Ripley. The authors wish to thank D. G. Kleiman for a critical reading of the manuscript.

24

Survival, Mating and Rearing Strategies in the Evolution of Primate Social Structure

J. D. GOSS-CUSTARD, R. I. M. DUNBAR, AND F. P. G. ALDRICH-BLAKE

Abstract. This article discusses the bearing of variations in certain aspects of social structure on the survival and reproductive success of individuals. Possible factors limiting the female's litter size and frequency are discussed. The consequences of the male's ability to sire and look after several litters at once and the effects of the resultant sexual selection on social structure are considered. The relations between environmental variables, particularly resource distribution and predation, and aspects of social structure are discussed and their bearing on reproductive success and individual survival evaluated. The conclusions reached are illustrated by reference to selected species, and the types of information required for a fuller evaluation of the ideas presented are listed.

Key Words social structure, evolution, sexual selection, reproductive success

INTRODUCTION

Primate social structures vary greatly in their form. Differences between populations in the behaviour of individuals give rise to contrasts in the

Reprinted from *Folia Primatologica, 17*:1-19 (1972) with permission from S. Karger AG, Basel.

following fundamental features of social organisation: (1) individuals may be solitary or form groups of sizes varying from 2 to 100 or more; (2) the permanency of groupings ranges from temporary aggregations at localised resources to long-term, stable relatively closed social units; (3) in groups, the numbers of mature males and the socionomic sex ratio may vary; (4) individuals or groups may occupy undefended and usually overlapping home ranges, or defended and hence exclusive territories.

While the adaptive significance of these features has been much discussed (DeVore, 1963; Crook and Gartlan, 1966; Kummer, 1968; Crook, 1970), previous authors have considered them largely in terms of the *survival* of individuals or groups. In this article, we concentrate more on the bearing that variations in social structure have on the *reproductive success* of individuals. Consequently, we examine the possible effects not only of ecological selection pressures but also of sexual selection on each of the features of social organisation outlined above.

Selection favours characteristics that maximise an individual's contribution to the gene pool of succeeding generations (Williams, 1966). The evolution of adaptations allowing individuals to raise the greatest number of offspring to breeding age would thus be expected. Aspects of the biology of animals other than primates have been exhaustively discussed in this light. The extensive literature on clutch size in nidicolous birds, for example, supports the notion that the average number of eggs laid is that which yields the maximum number of fledged young (Lack, 1966): laying too few eggs means that the food-collecting capacity of the parents is not fully utilised, while laying too many eggs results in each chick receiving too little food for proper growth. The same kind of argument has been applied successfully to other aspects of avian reproductive biology, such as various features of social organisation (Crook, 1965; Lack, 1968). Until now, however, this approach has seldom been applied explicitly to the analysis of the adaptive significance of primate social organisation.

There are two means by which the number of offspring reaching maturity may be increased. Firstly, the number of births could be increased. Secondly, a greater proportion of those born might be reared to maturity by devoting more time and energy to looking after the slowly developing young. These two methods may to some extent be incompatible since the production of more young may mean that less time and energy can be devoted to caring for each of them. Hence the successful system may be a compromise between the needs of these two contradictory requirements. In the interests of clarity, however, we will examine the two components separately. In the first section, we consider social tendencies and other features which may increase the number of live births per individual, while in the second, adaptations promoting the successful rearing of the young are examined. Obviously both means of increasing the number of offspring produced require healthy adults, so some social adaptations for individual survival also are considered in the second section. The approach we have adopted is as follows. Firstly, the possible adaptive significance each feature may have for mating, rearing and individual survival is examined. Secondly, an attempt is made, illustrated by reference to some typical patterns of social structure, to discuss their combined effects; in nature it is the whole integrated range of adaptations shown by an individual that will determine its success.

NUMBER OF BIRTHS

The number of births achieved may be increased by increasing litter-size, by increasing the frequency with which litters are produced, and in the case of males by siring several litters at the same time. The limit to litter-size and frequency is likely to be set by the amount of time and energy a parent can devote to the young. Males and females differ in the type of care they can give. Food for the infant prior to weaning, for instance, can only be provided by the female. While in principle both sexes would seem equally capable of grooming and carrying the young, in the majority of species these tasks also are performed by the female. Usually the male contributes only indirectly, for example by deterring predators or defending a territory. Since these are activities which can be performed for several young at the same time, a male may be able to look after several litters at once.

This is in contrast to many species of monogamous birds where both parents are capable of feeding the young. In these cases, the female can probably rear far more young with the help of the male than she could on her own. Monogamy may likewise enable the male to raise more young than he would if he spread his time and energy over several simultaneous broods (Crook, 1965; Lack, 1968). The release of male primates from direct care of the young appears to have had a profound influence on the evolution of primate social systems, as indeed it has in many mammals and birds (Orians, 1969). Before the consequences of this are discussed, however, it seems worthwhile to suggest possible factors that may limit the basic litter-size and frequency.

Number of Offspring per Female

Most primate species usually have one young at a time, although some regularly have up to three. Two possible reasons for the small litter-size are suggested, although present data do not enable either to be categorically rejected.

One possibility is that, on average, females can collect at a particular stage of the reproductive cycle, such as gestation or lactation, enough energy or particular nutrients for only one to three healthy young at a time. As a result, tendencies towards small litter-size would be selected for, and the reproductive physiology of the species modified accordingly. It is becoming increasingly clear that the apparent superabundance of food for many herbivorous mammals is misleading, and that their breeding success may be related to the availability of a suitable *quality* of food (Klein, 1970; Ellis, 1970). In primates likewise, the typical quality and abundance of the food supply, in conjunction with the foraging capacities of the mother, may have determined whether one, two or three is the most successful litter-size. It should be noted that food need not necessarily be limiting throughout the growth period; indeed, it seems unlikely that this would often be the case. It may need only a short period of feeding difficulty at some stage to restrict litter-size. If food limits litter-size, the frequency with which litters are produced may then depend on how long it takes the mother to recover from looking after one and build up reserves in preparation for the next.

The second possibility is that the parent is able to carry only one young at a time, particularly at the stage in growth when the infant is large but still dependent on the mother. Few individual primates carry more than one young for sustained periods. Apart from *Callithrix* and *Saguinus,* litters of two or three are kept in a nest for most of the time (e.g. *Cheirogaleus*). In semi-solitary species which do not use nests, a litter-size of one is the general rule (e.g. *Pongo, Perodicticus* and *Lepilemur*). Litter frequency likewise may perhaps be influenced by the difficulty of carrying one infant inside the body and one outside simultaneously, although this in itself seems unconvincing as a prime determinant of litter frequency.

Number of Young per Male

While the capacity of the female to increase the number of young she produces may be limited by one or more of the above factors, the male is generally not so restricted. A male may increase the number of offspring he sires simply by fertilising as many females as he can. If the sex ratio is more or less even, however, one male can only acquire several females if another fails to obtain any at all. Polygamy would, therefore, give rise to competition between males for the available females. Hence characters that enhance an individual's success in sexual competition would be expected to evolve.

The nature of sexual selection and its consequences for sexual dimorphism and mating patterns have been discussed by a number of authors (Huxley, 1938; Fisher, 1958; Maynard Smith, 1958; Orians, 1969) since Darwin's (1871) original formulation. It is now generally accepted that sexual selection may operate in two ways. Firstly, intra-sexual selection favours the evolution of structural and behavioural characteristics that enhance a male's chances of competing successfully with other males. This would have given rise to features useful in fighting. Secondly, certain features, apparently nonfunctional in combat, are thought to have arisen by inter-sexual selection, in which the female's preference for certain charac-

teristics determines the evolution of male behaviour and physical adorn-
ments.

Intra-sexual selection in primates seems to have favoured increased
male size and changes in pelage colour and length (Struhsaker, 1969;
Crook, 1970). Inter-sexual selection, on the other hand, does not at first
sight seem to have given rise to any additional features comparable to the
spectacular courtship and plumage of certain birds. Since the female
devotes so much time and energy to rearing an infant, however, and each
litter may represent a sizeable proportion of her total reproductive effort,
it would be to her advantage to be selective in her choice of male (Maynard
Smith, 1958; Orians, 1969). Females would benefit from mating with males
that firstly, sired the most viable young, and secondly, provided the most
effective indirect care for the young during rearing. It would also be to the
advantage of a female to select a male with characters that gave him success
in inter-male competition. Such success might not only indicate a general
fitness that her offspring might inherit, but perhaps also make it more likely
that her own male offspring in their turn would fertilise many females and
thus contribute disproportionately to the gene pool (McClaren, 1967). Secondary sexual characteristics such as the capes of male *Papio hamadryas* might
hence owe their origin as much to inter- as intra-sexual selection (Jolly,
1963). The extent of female choice is as yet uncertain (Saayman, 1970), and
if such choice exists, the criteria on which it is based and their relation to
the male characters enumerated above are unknown.

While these ideas on the influence of sexual selection are widely accepted, the possible influence of such selection on the evolution of the
various features of social structure enumerated earlier have seldom been
examined. Differences in the duration of male/female bonds, in group structure and in patterns of dispersion could influence profoundly the number
of females a male can fertilise. Hence it seems important to examine the
possibility that some aspects of social structure evolved through intra-sexual
selection; i.e. they have resulted from male behaviour which enables the
animals concerned to compete most successfully for females.

It can be argued, for instance, that intra-sexual selection may be implicated in determining the kinds of mating bonds that develop between
males and females within a group. To fertilise a female, a male must copulate with her in that part of the oestrous cycle in which she is likely to ovulate.
While males of many species are undoubtedly capable of detecting gross
changes in the female's sexual condition (Michael and Keverne, 1968; Saayman, 1970), they may lack the ability precisely to predict the optimum
moment for copulation. Under such circumstances, it would be advantageous to the male to associate closely with a receptive female over a period
of a few days. By so doing, he would not only ensure that he copulated with
her during the optimum period for achieving fertilisation, but might also be
able to prevent other males fertilising her instead. Such short-term "consort

relationships" are found in the multi-male troops of many baboon and macaque species. Males would presumably differ in their ability to form and maintain the exclusivity of such relationships. One would expect differences between the males in a multi-male troop in the number of fertilisations they achieve. Altmann (1962) suggests on theoretical grounds that if access to females is controlled by a rigid linear hierarchy, males will be sharply demarcated into those which achieve several fertilisations and those which achieve few or none. Even though this view of rigid hierarchical access to females is an oversimplification, the observations of Kaufman (1965) on *Macaca mulatta* lend support to Altmann's thesis; males do indeed differ substantially in the number of copulations at peak oestrous that they achieve.

At first sight it might seem reasonable to argue that sexual selection would favour males which both increased the duration of the consort relationships and formed such associations with more than one female at once; i.e. took a harem. Such males would seem, thereby, to ensure for themselves at least a proportion of the limited supply of females available. Although many species have birth peaks, few have a restricted breeding season (Lancaster and Lee, 1965). If females were to come into oestrous throughout much of the year, permanent maintenance of harems might then be an advantage.

There are a number of difficulties, however, which may render the suggestion implausible. Although it is possible that a male might increase his chances of fertilising a female by forming a bond with her earlier in the oestrous cycle and perhaps even permanently, this is far from certain. Moreover, taking a harem will not necessarily maximise the number of fertilisations a male achieves during his lifetime; staying with a harem when none of its members are receptive could debar him from association with other females that are in oestrous. In *Papio hamadryas*, for instance, oestrous cycles of females within each one-male group are synchronised and there are long periods when none are receptive (Kummer, 1968). It seems that a male has two alternative strategies; to form a permanent bond with a small number of females or attempt to establish a short-term relationship with several in turn. At present, we cannot assess which of these is the more competitive.

While sexual selection *per se* may not provide a sufficient explanation for the origin of harems, such selection might militate in their favour under certain ecological conditions. If conditions are such as to favour the population being dispersed in small coherent parties it might be both advantageous and, because of the small group size, practicable for a male to ensure that all the mature individuals in the group to which he belongs are females. The presence of other mature males in the group would mean that the maximum possible number of females was not being realised, and would also increase the risk of the females being fertilised by another male. Hence in this situation, sexual selection might favour males which prevented other mature males from joining the group.

Sexual selection may likewise be implicated in the evolution of spatial relations between harems. A male's hegemony over his harem may be further ensured if spatial separation between his females and other males is such as to eliminate effectively all risk of clandestine copulations, i.e. if territories are formed. A territorial one-male group system might be particularly advantageous if the cover is of such a density that the male cannot watch his females the whole time, or if feeding conditions are such as sometimes to promote wide scattering.

Sexual selection might give rise to territoriality under other circumstances. In situations in which individuals of the two sexes forage in separate but overlapping home ranges, e.g. *Galago demidovii* and *Microcebus murinus* (Charles-Dominique and Martin, 1970), a male which defended his range might thereby secure more exclusive access to the females within it. In effect, this would mean that the male would be defending a portion of an evenly dispersed resource, but whether or not this strategy would actually increase the number of fertilisations achieved is difficult to determine. A male might be unable to defend the whole of his potential home range; the benefit of exclusive access to a few females must, therefore, be balanced against the opportunity to compete, sometimes unsuccessfully, for a larger number.

To conclude this section, it seems that under certain circumstances sexual selection might indeed favour male tendencies which would result in a harem group structure within small groups, and in spatial separation by territorial activity of both individuals and groups. On the other hand, it is clear that certain ecological conditions may be necessary to promote the particular dispersion pattern within which sexual selection may have the effects we have postulated. Hence only a broader discussion, which includes consideration of the possible influence on social structure of factors affecting adult survival and rearing success, can provide the perspective from which the influence of sexual selection can realistically be viewed.

REARING OF THE YOUNG AND ADULT SURVIVAL

Long-lasting groups, as opposed to short-term aggregations, may be adaptive in a number of ways, not only by improving adult survival but also by ensuring a more successful rearing of the young. As Dr. J. E. Ellis has pointed out to us, the young primate's lengthy period of dependence on the mother means that the growth periods of successive litters overlap, with the result that females may be caring at any one time for several young at different stages of growth. The care provided by the mother may not only include such direct activities as carrying, grooming and feeding, but also benefits derived from her greater experience of the local environment. The young animal's chances of survival may be considerably increased if this knowledge enables it, for example, to locate new sources of food or particularly safe sleeping sites sooner or more certainly than it would be able to do on its own.

While the latter factor would favour permanent groups of females and young, it does not in itself provide an adequate explanation of the presence of mature males in such groups. One would expect experience and knowledge of the habitat to depend primarily on age rather than sex. Hence the adaptive significance of the male remaining with the females and young needs to be sought elsewhere.

A male which merely fertilised any females he could, and thereafter had no association with them, might in the long run actually leave relatively few offspring even if he fertilised many females. Unaccompanied females and young may be particularly vulnerable to predation, for instance, or at a competitive disadvantage when food or sleeping sites are scare. As a result, a male might improve his females' and youngs' chances of survival, and hence his own reproductive success, if he stayed with them. The advantage to the females and young of associating with one or more mature males might thus be the benefits obtained from the males' superior abilities in competing for limited resources or with any protection from predation that their presence provides.

In the remainder of this section, possible ways in which long-lasting groups may both facilitate rearing of the young and enhance adult survival will be examined in detail.

Of the possible agents which could favour a particular kind of group structure or a particular pattern of spatial relations between groups, the abundance and dispersion of food and the kind and severity of predation appear most significant. Accident and disease, on the other hand, would seem unlikely to favour one social system rather than another. The possible influences of food and predation will be examined in turn.

Food Dispersion and Abundance

The abundance and quality of the food of a population may vary considerably both in time and in space. Differences in the food supply between wet and dry seasons are an obvious example of seasonal variability, while the patches of food provided by fruiting trees in forests illustrate spatial variability. Furthermore, the continually changing pattern of such fruiting trees in forests shows how the food's spatial distribution may itself also vary throughout the year (Aldrich-Blake, 1970). It may be an advantage to an animal living in variable conditions to forage in a limited home range as opposed to being nomadic. A detailed knowledge of the locality will not only familiarise the animal with the position of the safest routes and refuges but may also amongst other things, enable it to find the best feeding sites at different times of year (Altmann and Altmann, 1970). This might not only save it from starvation when food is scarce but also promote the efficient

collection of food at all times, and hence allow mor͏
devoted to other essential activities.

If food supplies fluctuate, there may be ce͏.
scarce and competition between animals severe. In ͏
food supply may be more or less constant but the population ͏.
to a point at which competition occurs most of the time. In eithe͏.
male that secures himself a food supply and prevents his females and you͏.
coming into competition with other animals would be at an advantage. This
would be achieved by his preventing other animals taking food which has
been found (direct competition) or would later be found (indirect competition) by other members of the group.

For competition to be minimised, some degree of spatial separation
seems necessary. While it is possible that the harem leader in a one-male
group could prevent other animals from supplanting his females and young
at food sources and hence eliminate direct competition, he would have to
keep other animals out of the area altogether to prevent indirect competition.

Spatial separation could result either from other animals leaving of
their own accord, or by their being excluded. For example, it might be to
the advantage of males without females to remain near harems if by so doing
they increased their chances of acquiring females themselves. Under certain
conditions, however, it might benefit them to leave the females and forage
independently. For instance, Crook (1970) suggests that male geladas are
less timid than are females and young, and therefore, forage further from
the safety of cliffs. By doing so, they may benefit from exploiting different
and less used sources of food. However, it seems that separation would be
achieved in this way only seldom. It is not only the "surplus" males that may
compete with a harem but other harems as well, and these may share the
same preferred foraging areas. Hence, if indirect competition is to be minimised, one-male groups must behave in such a way as to prevent not only
surplus males but also other harems feeding in the same area, i.e. form
territories.

There will be a limit, however, to the amount of time and energy that
a male can devote to defending a territory, and this may determine the area
that can economically be defended. Only when all requirements can be
found within a defendable area is competition likely to give rise to territoriality (for this argument applied to birds, see Brown, 1964). Thus there must
be enough food, and in addition safe refuges and water resources, for the
group to survive, at least in pessimal times. In the case of food, for example,
a uniformly distributed supply which is dependable throughout the year
might allow territories to form. Similarly, a temporally and spatially
clumped supply, such as fruiting trees, would allow territoriality so long as
there was always at least one clump available within a suitable area. In
contrast, some groups may have to forage over a large home range as, for

le, when food is clumped but the clumps are widely separated or
.i food is sparse. In other cases, a local population may be forced to
ggregate" at a few safe sleeping sites, e.g. *Papio hamadryas* (Kummer,
968), or water holes, e.g. *Papio cynocephalus* in Amboseli (Altmann and
Altmann, 1970), *Erythrocebus patas* in Cameroon (Struhsaker and Gartlan,
1970). In such circumstances forming a territory would be either uneco-
nomic or simply impossible.

Presumably, the available resources within a territory would set a limit
to the number of animals that it could support. This factor may of itself
account for the size of harems in territorial species. Ellefson (1968) puts
forward a similar argument to account for group size in *Hylobates lar*. He
suggests that the area that a male can defend is capable of supporting only
two adults and their young.

Predation

Membership of a group, irrespective of its structure, might reduce in various
ways the risk of an animal being taken by a predator. For instance, several
animals may be more likely to detect an approaching predator than is one.
Again, certain predators, even after they have located a group, may be
"confused" by having a number of animals from which to select a victim and
hence be less likely to kill.

It is also possible that predation might favour a particular kind of group
structure. Thus it has been suggested that multi-male groups may provide
protection in open country habitats where predators are numerous and the
primates cannot easily escape by climbing trees (DeVore, 1963; Crook and
Gartlan, 1966). Here the greater predation risk for both young and adults
that would exist in one-male groups may have favoured the development of
inter-male tolerance and hence multi-male groups. While there are few
examples in the literature of male baboons combining to beat off predators,
the observations of DeVore and Washburn (1963) and Altmann and Altmann (1970) suggest that the presence of several mature males may act as
a deterrent, at least against smaller carnivores. That patas one-male groups
are viable in savanna does not invalidate this argument, since their speed
and crypticity provide them with alternative means of evading predators
[Hall, 1965).

In some species having multi-male troops the small size of mature
males might render them less effective as a deterrent to predators. On the
other hand, males may be more alert to potential danger since they are not
involved in direct caring for the young. Thus, the presence of several males
may enhance group awareness. Gartlan (1968), for instance, has shown that
adult male *Cercopithecus aethiops* spend more time than other age-sex classes
engaged in "look-out" behaviour. It is also possible that in the same way the

single mature male in one-male groups may provide the young with added safety from predators (Hall, 1965).

DISCUSSION

The major conclusions reached so far may be summarised as follows. While females are restricted in their reproductive capacities, males are capable of siring several litters from different females at the same time. This may give rise to competition between males for females, and hence to intra-sexual selection. A case can be made for males increasing their share of females by forming permanent harems, and in some circumstances, by being territorial. Parents may rear greater numbers of young to maturity by maintaining long-lasting associations with them, and in the case of the males, by reducing competition for scarce resources between their young and other animals. The presence of several mature males in such long-term groups may be related to the kind and severity of predation. The size of groups and the spatial relationships that exist between them may depend on the abundance and pattern of resource distribution, particularly on that of the food supply.

These suggestions will now be discussed in relation to the social structures of various species in an attempt to evaluate the respective influences of the different factors upon them. The examples chosen have been drawn from a broad spectrum of primate societies but no attempt is made to classify such societies into discrete categories.

Many prosimians such as *Microcebus murinus* and *Galago demidovii* do not form long-lasting heterosexual groups. During the day the nests used for sleeping generally contain parties of females or single males; at night both sexes forage solitarily (Charles-Dominique and Martin, 1970). Home ranges of the heavier males overlap but little; lighter males occupy smaller and less exclusive ranges. This pattern of home ranges suggests that the heavier males are territorial (Martin, personal commun.). The ranges of such males overlap with those of several females.

Crook and Gartlan (1966) suggest that the nature of the food supply in nocturnal solitary prosimians militates against group foraging. The larger males' exclusive home ranges may secure them and their offspring a reliable food supply, and may also increase the males' exclusivity of mating with the females within their ranges. However, there are difficulties with both interpretations. If breeding is seasonal, as is the case in *Microcebus* but not in *Galago* (Martin, personal commun.), it might not be necessary for the males to hold exclusive ranges throughout the year to secure females during only part of it. On the other hand, it could be argued that a male would increase his chances of fertilising females by establishing such a range well before the beginning of the breeding season and, perhaps, permanently. Likewise it does not necessarily follow that a male's range would provide food solely for his own females and young. Some of the females foraging within his range may have been fertilised by other males. Clearly a male would not benefit by securing a food supply for the offspring of his neighbour. Hence the determinants of dispersion in these species remain uncertain.

In other cases, however, territoriality may be more clearly related to the provision of an adequate supply of food or some other resource. The forest-living blue monkey *Cercopithecus mitis,* for instance, lives in groups typically made up of a single mature male with associated females and young (Aldrich-Blake, 1970). Such groups are territorial, with only marginal overlap between home ranges. Solitary males tend to be seen in these overlap zones rather than within the group territories. Indirect competition for food both between the various groups and perhaps between groups and solitary males may hence be largely eliminated. Thus a territorial one-male group can be viewed as an adaptation which firstly enhances adult survival, and secondly enables a male to sire and rear several litters at once and a female to increase the success with which she rears her young.

In contrast, the one-male groups of many species are not territorial. In some cases, as for instance that of *Erythrocebus patas* in Uganda (Hall, 1965), the area over which the animals must range is so large that its defence would be impracticable. In other cases, the need for local populations to aggregate upon limited resources such as water holes, e.g. *E. patas* in Cameroon (Struhsaker and Gartlan, 1970), concentrations of food, e.g. *Theropithecus gelada* (Crook, 1966), or sleeping sites, e.g. *P. hamadryas* (Kummer, 1968) precludes territoriality.

In such cases, the ways in which the single male may give benefit to his females and young, and thereby himself, are not clear. Whereas in the territorial species the male may minimise indirect competition between his females and young and other animals, this argument is not applicable to non-territorial groups. While it is possible that the harem leader in such groups may discourage direct competition for food, the food items may be so small or handled so rapidly that supplanting attacks would be unprofitable. Moreover, in gelada herds females and young are often so widely separated from their males that the latter would be unable to prevent such interference. Alternatively, it could be argued that the male serves in some way to minimise the risk of predation, whether by aggressive defence or other means. Hall (1965), for instance, attributes to the male patas the role of watcher and performer of diversionary displays. As regards aggressive defence against predators, Struhsaker and Gartlan (1970) describe males in herds of patas in Cameroon combining to chase off jackals, but whether individual males would be able to do so at times when the harems are moving separately, is not known. Defence against predators, however, is a less satisfactory explanation in areas where predation pressure is said to be minimal.

In the absence of any convincing grounds for believing that the males' presence in these groups enhances adult survival or the success with which the young are reared, it may be necessary to invoke possible mating advantages. We have already argued that sexual selection might favour males that took harems when ecological conditions are such as to necessitate the population being dispersed in small, coherent parties. In all three of the species under consideration, herds have at times to split up, albeit for varying periods, into smaller units.

While the continued association of males with the females bearing their young can often be explained in terms of rearing and mating strategies, that between possibly non-reproducing mature males and the rest of the group is at first sight less accountable. Such a situation may pertain in the multi-male groups characteristic of many *Macaca* and *Papio* populations. If "subordinate"* males achieve some fertilisations, however few, then their continued association with the rest of the group poses no problem. If, on the other hand, this is not the case, why should such males remain within the group? By remaining near the females they may increase their chance of achieving fertilisations in the future, both by their increased familiarity to the females and by being on the spot should they become available. Moreover, should a male have little success in one group he may be able to move to another (Rowell, 1969; Altmann and Altmann, 1970), wherein he may in time be more successful. In addition, by remaining with the group a male may benefit from its accumulated experience of the local habitat. Any such benefits must be balanced against possible advantages of becoming solitary or joining an all-male party.

We must now ask what, if anything, the "dominant" males, females and young gain from the presence of the subordinate males. By tolerating such males within the group, dominant males run the risk of the number of fertilisations they achieve being reduced, or their young suffering through increased competition for food. It could be that the time and energy needed to eject other males would be too great to justify any putative gain in reproductive success. Alternatively, dominant males may actually benefit from the continued presence of other males through any added protection against predators they may provide.

If protection against predators, whether through active defence or increased group awareness, has been a major factor in the evolution of multi-male troops, one would expect some correlation between the occurrence of this type of social structure and the kind and severity of predation. The presence of multi-male troops in open areas largely free of predators — many *M. mulatta* populations, for example (Southwick *et al.*, 1965) — does not accord with this expectation, although it is likely that predators have only recently been eliminated from such areas. More problematically some forest monkeys, of similar size and living in the same area, and hence presumably subject to the same kind of predation, show a wide range of

*In this article "dominant" and "subordinate" are used only as a convenient means of referring to a male's success in competing with others for females, irrespective of whether such success is correlated with other criteria of rank.

social structure. Thus *C. mitis* in Uganda has one-male groups (Aldrich-Blake, 1970), while the black mangabey, *Cercocebus albigena*, in the same region has multi-male troops (Chalmers, 1968).

A complication here is the indeterminate effect of phylogenetic heritage on social structure, and hence on correlations between social structure and environmental factors. As Struhsaker (1969) points out," . . . each species brings a different phylogenetic heritage into a particular ecological scene." Hence the black mangabey may owe its multi-male group structure as much to the close relationship between *Cercocebus* and the baboon-macaque group, of which multi-male troops are characteristic, as to present ecological conditions.

Of course, phylogenetic characteristics must in the long-term themselves have been moulded by ecological factors. The issue, therefore, is not so much whether phylogenetic heritage or present ecological factors play the greater part in determining social structure, but the rate at which such structure can change in response to changing ecological conditions. It appears that different taxa vary greatly in this respect; all *Macaca* spp. so far studied have multi-male troops irrespective of habitat, while some colobines show extensive variation even within species; e.g. *Presbytis entellus* (Yoshiba, 1968), *Colobus guereza* (Schenkel and Schenkel-Hulliger, 1967; Marler, 1969).

It will be clear that much of this article has been speculative. Speculation can only be justified if it identifies issues and generates ideas useful for further research. Thus our purpose has been not so much to produce a general theory of primate social structure, as to reappraise existing information in the hope that critical gaps in our knowledge will thereby be revealed. We will conclude, therefore, by drawing attention to certain topics that may merit further study.

Firstly, more information is needed on many aspects of reproductive behaviour. For instance, further studies comparable to those of Kaufman (1965) on relative mating success of males would be desirable, with the important proviso that they should be long-term. While present studies indicate that males do indeed differ in this respect at any one time, it is important to know whether such differences exist when males' entire reproductive lifespans are considered. It could be, for instance, that todays' subordinates become tomorrow's dominants, and hence that short-term contrasts are cancelled out over a longer period. Such information is required both for further multi-male group populations and also for populations having one-male groups. It is known, for instance, that harem leaders in *Presbytis entellus* may be ejected by males from all-male groups (Sugiyama, 1965). Thus while harem leaders clearly achieve more fertilisations over a short period, one cannot assume that this is necessarily the case in the long run. Only when such information is available can one assess the intensity of sexual selection in the various types of society.

Secondly, it is not as yet clear to what extent female choice affects the numbers of fertilisations a male achieves, nor, if such choice exists, on what basis it is made. On theoretical grounds, one might expect females to be selective, since they have such an investment of time and energy in each offspring (Maynard Smith, 1958). Such choice would contribute to the intensity of sexual selection in the society: indeed, it could play an important role in determining the sex ratio within breeding units, if females form bonds with only certain males as, for example, with those within a particular age range.

Turning now to ecological factors, virtually nothing is known about the relative importance of different causes of mortality. In particular, it is not known whether food is ever so sparse or of such poor quality as to influence adult survival and reproductive success. Even if food is at times in short supply, it is not certain whether aspects of social structure have the effect of minimising direct or indirect competition for this or other resources. It has been suggested, for example, that the division of gelada populations into one-male and all-male groups may reduce competition between non-reproducing males and other animals (Crook and Gartlan, 1966). However, there is as yet no conclusive evidence that all-male groups forage in different areas to one-male groups at time of food shortage, and hence that indirect competition is avoided. Where one-male groups are territorial, as in *C. mitis* and *P. entellus,* the exclusivity of range use so attained ensures that competition between adjacent one-male groups is minimised, but it is not yet certain to what extent, if at all, indirect competition between one-male groups and extragroup males is avoided.

As regards predation, the evidence for the frequent statement that the several males in multi-male troops serve actively to defend the group against predators or even to deter attacks is equivocal (DeVore and Washburn, 1963; Rowell, 1969; Altmann and Altmann, 1970). It must be stressed, however, that active defence or deterrence is not the only means whereby such males might reduce the risk of predation. We have already noted Gartlan's (1968) observation that male *C. aethiops* spend more time "looking out" than other age-sex classes; comparable measures on other species would be valuable.

Finally, if ecological factors affect social structures in the ways postulated, one would expect differential mortality between different sections of a population. Solitary males, for instance, might be expected to suffer greater mortality than individuals living in groups. Evidence on this point is again lacking.

These, then, appear to us the major topics on which further information is desirable. Only when such data are available can the validity of the ideas here presented be evaluated.

SUMMARY

The number of offspring reaching maturity may be increased firstly by greater numbers being born, and secondly by a larger proportion of those born being reared to maturity. Females are limited in their capacity to increase the number of births, but males are capable of siring several litters at once. The resultant sexual selection may under certain ecological conditions give rise to a harem group structure, and also to territoriality between groups or individuals. Rearing of young may be facilitated by the formation of long-term groups in which the young can benefit from the accumulated experience of the adults. Males may increase the chances of survival of their females and young by ensuring for them an adequate supply of resources, and perhaps by minimising the risk of predation. The optimum mating, rearing and survival strategies, hence, vary with ecological conditions.

ACKNOWLEDGEMENTS
We are grateful to the many people who provided stimulating and critical discussion of earlier drafts of this manuscript, in particular Mr. J. M. Deag, Dr. J. E. Ellis, Dr. J. S. Gartlan and Mr. I. E. Inglis. Especial thanks are due to Dr. J. H. Crook, discussions with whom led us to develop the ideas presented in this paper.

REFERENCES
Aldrich-Blake, F. P. G.: Problems of social structure in forest monkeys; in Crook Social behaviour in birds and mammals, pp. 79-101 (Academic Press, London 1970).

Altmann, S. A.: The social behaviour of anthropoid primates. An analysis of some recent concepts; in Bliss Roots of behaviour, pp. 277-285 (Harper & Row, New York 1962).

Altmann, S. A. and Altmann, J.: Baboon ecology. African field research (Chicago University Press, Chicago 1970).

Brown, J. L.: The evolution of diversity in avian territorial systems. Wilson Bull. *76:* 160-169 (1964).

Chalmers, N. R.: Group composition, ecology and daily activities of free-living mangabeys in Uganda. Folia primat. *8:* 247-262 (1968).

Charles-Dominique, P. and Martin, R. D.: Evolution of lorises and lemurs. Nature, Lond. *227:* 257-260 (1970).

Crook, J. H.: The adaptive significance of avian social organizations. Symp. zool. Soc., Lond. *14:* 181-218 (1965).

Crook, J. H.: Gelada baboon herd structure and movement. A comparative report. Symp. zool. Soc., Lond. *18:* 237-258 (1966).

Crook, J. H.: The socio-ecology of primates; in Crook Social behaviour in birds and mammals, pp. 103-166 (Academic Press, London 1970).

Crook, J. H. and Gartlan, J. S.: Evolution of primate societies. Nature, Lond. *210:* 1200-1203 (1966).

Darwin, C.: The descent of man and selection in relation to sex (Murray, London 1871).

DeVore, I.: Comparative ecology and behaviour of monkeys and apes; in Washburn Classification and human evolution, pp. 301-319 (Wenner Gren, New York 1963).

DeVore, I. and Washburn, S. L.: Baboon ecology and human evolution; in Howell and Bourlière African ecology and human evolution, pp. 335-367 (Wenner Gren, New York 1963).

Ellefson, J. O.: Territorial behaviour in the common white-handed gibbon, *Hylobates lar* Linn; in Jay Primates. Studies in adaptation and variability, pp. 180-199 (Holt, Rinehart & Winston, New York 1968).

Ellis, J. E.: A computer analysis of fawn survival in the pronghorn antelope; Ph.D. thesis Davis, Calif. (1970).

Fisher, R. A.: The genetic theory of natural selection (Dover, New York 1958).

Gartlan, J. S.: Structure and function in primate society. Folia primat. *8:* 89-120 (1968).

Hall, K. R. L.: Behaviour and ecology of the wild patas monkey, *Erythrocebus patas,* in Uganda. J. Zool., Lond. *148:* 15-87 (1965).

Huxley, J. S.: The present standing of the theory of sexual selection; in De Beer Evolution; essays presented to Prof. E. S. Goodrich, pp. 11-42 (Oxford University Press, Oxford 1938).

Jolly, C. J.: A suggested case of evolution by sexual selection in primates. Man *63:* 178-179 (1963).

Kaufman, J. H.: A three-year-study of mating behaviour in a free-ranging band of rhesus monkeys. Ecology *46:* 500-512 (1965).

Klein, D. R.: Food selection by North American deer and their response to over-utilization of preferred plant species; in Watson Animal populations in relation to their food resources, pp. 25-44 (Blackwell, Oxford 1970).

Kummer, H.: Social organisation of hamadryas baboons (Chicago University Press, Chicago 1968).

Lack, D.: Population studies of birds (Oxford University Press, Oxford 1966).

Lack, D.: Ecological adaptations for breeding in birds (Methuen, London 1968).

Lancaster, J. B. and Lee, R. B.: The annual reproductive cycle in monkeys and apes; in DeVore Primate behaviour. Field studies of monkeys and apes, pp. 486-513 (Holt, Rinehart & Winston, New York 1965).

Marler, P.: *Colobus guereza.* Territoriality and group composition. Science *163:* 93-95 (1969).

Maynard Smith, J.: Sexual selection; in Barnett A century of Darwin, pp. 231-244 (Heinemann, London 1958).

McClaren, I. A.: Seals and group selection. Ecology *48:* 104-110 (1967).

Michael, R. P. and Keverne, E. B.: Pheromones in the communication of sexual status in primates. Nature, Lond. *218:* 746-749 (1968).

Orians, G. H.: On the evolution of mating systems in birds and mammals. Amer. Naturalist *103:* 589-603 (1969).

Rowell, T. E.: Long-term changes in a population of Ugandan baboons. Folia primat. *11:* 241-254 (1969).

Saayman, G. S.: The menstrual cycle and sexual behaviour in a troop of free-ranging chacma baboons (*Papio ursinus*). Folia primat. *12:* 81-110 (1970).

Schenkel, R. and Schenkel-Hulliger, L.: On the sociology of free-ranging colobus; in Starck, Schneider and Kuhn Progress in primatology, pp. 185-194 (Fischer, Stuttgart 1967).

Southwick, C. H.; Beg, M. H., and Siddiqi, M. R.: Rhesus monkeys in North India; in DeVore Primate behaviour. Field studies of monkeys and apes, pp. 111-159 (Holt, Rinehart & Winston, New York 1965).

Struhsaker, T. T.: Correlates of ecology and social organization among African cercopithecines. Folia primat. *11:* 80-118 (1969).

Struhsaker, T. T. and Gartlan, J. S.: Observations on the behaviour and ecology of the patas monkey, *Erythrocebus patas*, in the Waza reserve, Cameroon. J. Zool., Lond. *161:* 49-63 (1970).

Sugiyama, Y.: On the social change of hanuman langurs, *Presbytis entellus*, in their natural condition. Primates *6:* 381-418 (1965).

Williams, G. C.: Adaptation and natural selection. A critique of some current evolutionary thought (Princeton University Press, Princeton 1966).

Yoshiba, K.: Local and intertroop variability in ecology and social behaviour of common Indian langurs; in Jay Primates. Studies in adaptation and variability, pp. 217-242 (Holt, Rinehart & Winston, New York 1968).

25
Primate Social Organisation and Ecology

T. H. CLUTTON-BROCK

Attempts to relate interspecific differences in social organisation among primates to gross differences in habitat or diet type have been largely unsuccessful. This is probably partly because distantly related species have adapted to similar ecological situations in different ways and partly because much finer ecological differences are important.

The accumulation of primate field studies in the course of the last two decades has emphasised the extent to which social organisation varies between species. Marked inter-population differences have been observed in a number of cases,[1] but most species possess characteristic modal patterns of social organisation. Members of some species are solitary[2] for much of their time, while others live in family groups[3,4] and others in semi-stable groups of fifty or more animals.[5,6] Groups may include approximately equal numbers of males and females[7] or, as is more usual, there may be a preponderance of females.[8,9] In a number of species, groups regularly split up into parties which forage separately[10,11] while in others all members of the group stay together.[6,8] The average size of the area used by groups differs from <0.01 to >50 km^2.[12]

At present, little is known about the adaptive significance of these differences. Several reviews have attempted to relate them to gross differences in diet or habitat type.[12-18] The first and perhaps the most influential of these was produced by Crook and Gartlan[13] in 1966. Primate species

Reprinted from *Nature*, 250:539-542 (1974) by permission of Macmillan Journals Limited.

were divided into five "grades" according to the nature of their diets, the kind of habitat they occupied and the pattern of social organisation which they displayed. The paper emphasises the similarity of social organisation in species sharing the same diet and habitat type and is widely quoted as evidence that social organisation and ecology are closely related. But when it is viewed in the light of current knowledge, differences between groupings are less impressive than differences within them. For example, *Saimiri sciureus*, *Colobus* spp., and *Gorilla* are all placed in the same grade, despite the gross differences in diet, foraging behaviour and social organisation which exist between them.[19-26] In addition, the paper illustrates a fundamental problem in classifying primate social systems. Since different aspects of social organisation are not well correlated across species, categories defined by a single criterion will include social systems which differ widely in other ways.

Jolly[12] avoids this problem by considering different aspects of social organisation separately. She groups species into six ecological divisions ("nocturnal," "arboreal leaf-eaters," "arboreal omnivores," "semi-terrestrial leaf-eaters," "semi-terrestrial omnivores" and "arid country species"), and compares troop size, range and territory size, population density, intergroup behaviour and day-range size separately between ecological groups. Again, differences within most groupings are more striking than differences between them. For example, between species allocated the "arboreal leaf-eaters" category, average troop size differs from <10 to >50 and average range size from <0.02 to >1 km^2, while other categories show even more variability.

To criticise these reviews in the light of current knowledge is not to discount their value. The paper by Crook and Gartlan in particular has stimulated much interest and a considerable body of research. There are, however, theoretical reasons why social organisation should not be expected to be closely correlated with gross ecological variation.

PHYLOGENY, ECOLOGY AND ADAPTATION

One probable reason for the absence of close correlation is that different species tend to react to similar environmental pressures in different ways. When a novel adaptation evolves, its form will be partly determined by the various environmental factors through which selection is operating and partly by the species' phylogenetic inheritance.[27,28] Consequently, distantly related species are likely to evolve different traits with similar functions. For example, anatomical specialisations permitting the digestion of foliage have evolved in several primate families. In some this has been achieved by the development of a large caecum (for example, *Alouatta*, *Indri*),[29] in others by chambered stomachs with bacterial symbionts (for example, *Colobus*, *Presbytis*).[30] Similarly, many interspecific differences in social organisation may well prove to represent different methods of overcoming the same ecological problems.

Another reason is that the wrong kind of ecological variation has been assumed to be important. Within habitat categories as broad as "forest," "woodland" and "savanna," and dietetic categories such as "insectivore," "frugivore," "folivore" and "omnivore" there is room for vast ecological differences. Food supplies may be static or mobile, sparse or dense, heavily clumped or evenly dispersed, reasonably stable throughout the year or extremely variable.[17] Intensity of predation may vary in the same way. If ecological differences at this level affect social organisation (and, by analogy, the plentiful literature on birds suggests that this will be found to be the case[31]) one should expect social organisation to vary widely within habitat and diet types.

A recent study of two species of colobus monkeys in East Africa[32] provides an example of the level at which differences in social organisation may be related to ecology. Both the red colobus (*Colobus badius*) and the black and white colobus (*C. guereza*) are widely distributed across tropical Africa.[33] The two species are sympatric throughout much of their range,

A male black and white colobus watches an intruder.

though the black and white colobus is found both in wet and dry forest while the red colobus is less commonly found in dry forest.[25,34] Although both species are largely folivorous, arboreal and forest-dwelling,[35] their characteristic patterns of social organisation differ widely. Red colobus live in large, multi-male troops of 40 or more animals which occupy extensive ranges of around 1 km^2 in size (see Fig. 1). Their different calls intergrade.[36] Oestrous females show pronounced swelling of the perineal region. Infants have a black natal coat and are apparently handled only by the mother during the first months of life. In contrast, black and white colobus live in small troops of five to 10 animals which often contain only one adult male.[24-25] The troops occupy defended territories usually <0.2 km^2 in size and inter-troop relationships are normally hostile.[24] Males give a booming roar which may serve to space groups[17] while intragroup vocalisations inter-

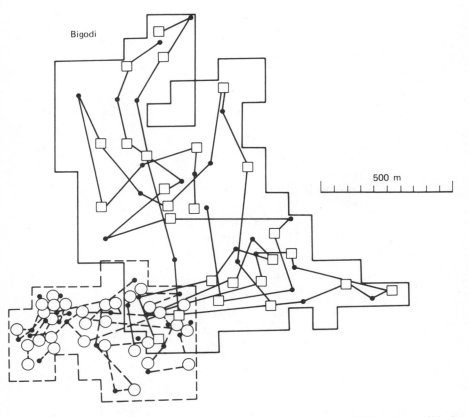

Bigodi

500 m

Fig. 1. Movements of a troop of red colobus (squares: September 6 to 31, 1970) and one of black and white colobus (circles: August 29 to September 31, 1970) about their ranges in Kibale Forest Reserve, Uganda. Squares and circles shows the troops' night-resting positions. Points show the positions of the troops at 12 noon on each day.

✗ Important for research paper

grade as in the red colobus. Females show no obvious oestrous swellings. Infants have a white natal coat and are handled by females other than their mothers from the day of birth.[38]

Between August, 1969, and June, 1970, I observed one troop of red colobus in the Gombe National Park for 9 months.[32] Afterwards, for approximately a month each, I watched one troop of red colobus and one of black and white colobus in each of two areas in Kibale Forest, Uganda. I measured the amount of time which the animals spent feeding on different foods by recording, at quarter-hourly intervals, the foods that all visible animals were eating.[39] In each area, I also measured the relative abundance of the different tree species.

The feeding behaviour of all three troops of red colobus was very similar. The animals ate the flowers, fruit, shoots and leaves of a variety of tree species. They were extremely selective in their choice of food, regularly choosing certain parts of particular tree species. Most tree species were not evenly distributed through the forest, with the result that the animals fed on different foods in different parts of their ranges. In addition, the availability of food on most species varied seasonally and this was reflected in seasonal changes in the animals' diet. When shoots, flowers and fruit were less abundant, the animals fed to a greater extent on mature leaves, though in no month did they spend more than 60% of their feeding time eating mature leaves. In all months of the year at Gombe, and in both study areas at Kibale, the animals fed on a wide variety of food species. To measure the variability of their diet, in each month I ranked the animals' foods on the amount of time spent feeding on them. At Gombe the proportion of time spent feeding on the top-ranking food species varied from 13-42% between months, that on the second ranking species from 11-21%, while the amount of time spent feeding on the top five varied from 51-80%. Figures for the two Uganda troops were similar. In one study area the red colobus spent 11% of their time feeding on the top-ranking food species, 10% on the second and 46% on the top five. In the other they spent 16% of their time on the top species, 14% on the second and 50% on the top five.

The diet of the black and white colobus troops differed in two important ways from that of the red colobus. Both troops fed almost exclusively on two tree species (*Celtis durandii* Engl. and *Markhamia platycalyx* (Bak.) Sprague). In one area *Celtis durandii* accounted for 71% of all feeding records and *Markhamia platcalyx* for 19%. In the other, *Celtis durandii* accounted for 88% and *Markhamia platycalyx* for 5%. These species were utilised to a lesser extent by the red colobus troops in both areas and the difference in feeding behaviour was not a product of reduction in the availability of alternative foods to the black and white colobus.[39] Second, during the first weeks of the Uganda study, when few *Celtis* trees had yet come into flower or shoot, the black and white colobus fed largely on the mature leaves of *Celtis*. At the same time, the red colobus were feeding on the shoots, flowers and fruit of a variety of other food species, but were not observed to feed on mature *Celtis* leaves. Subsequently, many of the *Celtis* trees came into flower and shoot, and the black and white colobus switched to feeding on these parts.

These differences in feeding behaviour may be closely related to the differences in distribution and social organisation between the two colobus species. Black and white colobus may be able to exist on a diet of mature leaves when only these are available. This would allow the species to colonise areas of dry forest where strongly seasonal rainfall produces a high degree of production synchrony across different tree species (so that shoots, flowers and fruit are only available at certain times of year). Second, it would allow them to exist on the products of a small number of tree species throughout the year, since these would provide acceptable food in all seasons. Consequently, a small area of forest would be able to support animals throughout the year. In this situation, small troop size might permit the animals to minimise range size and thus to increase their ability to defend their food supply efficiently.[17,40] It might also allow them to minimise the distance which they would have to travel each day to collect food (both because the total demands of the troop would be less and because they would be able to utilise food sources too small for larger troops).

Female red colobus feeding on flowers of *Combretum molle*.

In contrast, red colobus may need to maintain a high proportion of shoots, flowers and fruit in their diet throughout the year. Consequently, the species may be limited to wetter forests where some tree species carry these parts in all seasons. Since different tree species will carry acceptable foods at different times, the animals would need to maintain access to a wide variety of food species. As the distributions of most tree species are heavily clumped, a large size range would be necessary in order to provide sufficient supplies of acceptable food in all months of the year. A square kilometre of forest can support a considerable population of red colobus. Theoretically, the animals could either aggregate into a single group or split up into several small groups with extensively overlapping ranges. In the red colobus, minimisation of group size may be less advantageous because the range size necessary to support even a small group throughout the year would be too large to be efficiently defended. Aggregation may have a number of advantages: the presence of other feeding animals in the immediate area may show individuals where to find food,[41] while the troop may provide a reservoir of knowledge about the distribution of food and of predators in the past.[42] It also may allow the animals to develop a pattern of regular use of the different parts of the range which would maintain leaf growth at an acceptable stage and maximise the reaction of the tree species to being cropped.[43] Finally, large troop size may enhance the animals' ability to detect and defend themselves against predators. Certainly chimpanzees at Gombe were regular predators of red colobus (R. Wrangham, personal communication) and several instances of cooperative defence by the colobus were observed.

The origins of several other differences between the two species are probably linked with these differences in social organisation. The contrasting pelage of the black and white colobus and the males' roars may help to demarcate troops' territories (J. F. Oates, quoted in ref. 44). Differences between the two species' reactions to intruders are well adapted to the difference in group size. When disturbed, black and white colobus tend to move to thick cover and remain silent while red colobus escape noisily and males may attack potential predators, including chimpanzees. Other differences, such as the absence of obvious sexual swellings in the black and white colobus and of infant-swapping in the red are less easily explained.

This reconstruction is clearly speculative. Few colobus troops were observed and, except at Gombe, they were watched only at one time of year. Current studies of the two species by T. T. Struhsaker and J. F. Oates may throw further light on the problem. But large group and range size is associated with clumped, unstable food supplies in several other animal groups. Studies of *Presbytis entellus* and *P. senex* in Ceylon show that *P. entellus*, which lives in larger troops in larger ranges than *P. senex* also feeds more on fruit and less on mature foliage and utilises a wider range of foods. Among East African ungulates, most of the plains-dwelling, grazing species, whose food supplies tend to be clumped and relatively unpredictable, aggregate in large herds which range widely (for example, most of the *Alcelaphinae*). In contrast, most browsing species, whose food supplies tend to be more uniformly scattered and more stable, occupy small ranges and live either in small groups, in pairs or alone (for example, dik-diks, duiker and most of the *Neotraginae*).[46] With some exceptions, the same association between diet and social organization is found among European and Asiatic deer. Finally, a high proportion of graminivorous bird species feed in flocks, whereas relatively few insectivorous species do so.[31] The diet of the former is commonly assumed to be more heavily clumped than that of the latter. [47] Within the insectivores, the aerial feeders (for example, swallows, martins and swifts) whose food supplies tend to be more clumped and less predictable, almost all feed in flocks.[31]

The colobus study provides one example of the kind of ecological differences which may be associated with interspecific differences in social organisation. In other primate species, different factors are likely to be important. But if many of them are as subtle as those probably involved in this case, considerable variation is to be expected within gross diet and habitat categories. This is not to say that high level ecological variation has no effect on social organization. Indeed, the examples quoted above suggest that in some animal groups there is a relatively close correlation between some aspects of social organisation and certain types of diet. Whether associations at this level can be identified will depend both on the extent of variation within ecological categories and on the number of species which can be compared. Lack,[31] working with a vast array of bird species, was able to demonstrate correlations between social organisation and gross dietetic variation despite the presence of considerable variation within ecological categories. In primates, however, this approach is less likely to be useful both because there are relatively few species and because there is evidently wide variation within gross ecological groupings.

I thank the following for help, advice and criticism: Professor R. J. Andrew, J. Brooke, Dr J. S. Gartlan, Professor R. A. Hinde, Dr A. Jolly, M. Kavanagh, Dr H. Kruuk, Dr J. van Lawick-Goodall, Dr M. J. Simpson, Dr P. J. B. Slater, C. Packer and R. Wrangham, The field work was carried out while I was a member of the Subdepartment of Animal Behaviour at Cambridge and was financed by a Royal Society Leverhulme Scholarship and a Science Research Council Studentship. I am grateful to the Tanzanian National Parks and to the Uganda Forestry Department for permission to work in the Gombe National Park and in the Kibale Forest Reserve.

REFERENCES

[1] **Jay, P. C.,** in *Primates: Studies in Adaptation and Variability* (edit. by Jay, P. C.), (Holt, Rinehart and Winston, New York, 1968).

[2] **Charles-Dominique, P., and Martin, R. D.,** *Advances in Ethology*, **9** (Paul Parey, Berlin, 1972).

[3] **Ellefson, J. O.,** in *Primates: Studies in Adaptation and Variability* (edit. by Jay, P. C.), (Holt, Rinehart and Winston, New York, 1968).

[4] **Mason, W. A.,** in *Primates: Studies in Adaptation and Variability* (edit. by Jay, P. C.), (Holt, Rinehart and Winston, New York, 1968).

[5] **Gautier-Hion, A.,** *Folia primat.*, **12**, 116-141 (1970).

[6] **Altmann, S. A., and Altmann, J.,** *Baboon Ecology* (University of Chicago Press, Chicago, 1970).

[7] **Jolly, A.,** *Lemur Behaviour* (University of Chicago Press, Chicago, 1966).

[8] **Poirier, F. E.,** *Folia primat.*, **10**, 20-47 (1969).

[9] **Hall, K. R. L.,** *J. Zool.*, **148**, 15-87 (1965).

[10] **Eisenberg, J. F., and Kuehn, R. E.,** *Smithsonian Misc. Coll.*, **151**, 1-63 (1966).

[11] **Aldrich-Blake, F. P. G.,** Thesis, University of Bristol (1970).

[12] **Jolly, A.,** *The Evolution of Primate Behaviour* (Macmillan, New York, 1972).

[13] **Crook, J. H., and Gartlan, J. S.,** *Nature*, **210**, 1200-1203 (1966).

[14] **Crook, J. H.,** in *Social Behaviour in Birds and Mammals* (edit. by Crook, J. H.), (Academic Press, London, 1970).

[15] **Crook, J. H.,** *Anim. Behav.*, **18**, 197-209 (1970).

[16] **Denham, W. W.,** *Am. Anthrop.*, **73**, 77-95 (1971).

[17] **Crook, J. H.,** in *Sexual Selection and the Descent of Man* (edit. by Campbell, B. G.), (Aldine, Chicago, 1972).

[18] **Eisenberg, J. F., Muckenkirn, N. A., and Rudran, R.,** *Science*, **176**, 863-874 (1972).

[19] **Schaller, G. B.,** *The Mountain Gorilla* (Chicago University Press, Chicago, 1963).

[20] **Thorington, R. W.,** jun., in *Progress in Primatology* (edit. by Starck, D., Schneider, R., and Kuhn, H-J.), (Fischer-Verlag, Stuttgart, 1967).

[21] **Baldwin, J. D.,** *Folia primat.*, **9**, 281-314 (1968).

[22] **Baldwin, J. D.,** *Folia primat.*, **11**, 35-79 (1969).

[23] **Balwin, J. D., and Baldwin, J.,** *Folia primat.*, **18**, 161-184 (1972).

[24] **Marler, P.,** *Science*, **163**, 93-95 (1969).

[25] **Kingston, T. J.,** *E. Afr. Wildl. J.*, **9**, 172-175 (1971).

[26] **Schenkel, R., and Schenkel-Hulliger, L.,** in *Progress in Primatology* (edit. by Starck, D., Schneider, R., and Kuhn, H-J.), (Fischer-Verlag, Stuttgart, 1967).

[27] **Chalmers, N. R.,** in *Comparative Ecology and Behaviour of Primates* (edit. by Michael, R. P., and Crook, J. H.), (Academic Press, London, 1973).

[28] **Struhsaker, T. T.,** *Folia primat.*, **11**, 80-118 (1969).

[29] **Hladik, C. M.,** *Mammalia*, **31**, 120-147 (1967).

[30] **Hollihn, von K. W.,** *Z. Saugetierk*, **36**, 65-95 (1969).

[31] **Lack, D.,** *Ecological Adaptations for Breeding in Birds* (Methuen, London, 1968).

[32] **Clutton-Brock, T. H.,** Thesis, University of Cambridge (1972).

[33] **Rahm, U. H.,** in *Old World Monkeys* (edit. by Napier, J. R., and Napier, P. H.), (Academic Press, New York, 1970).

[34] **Kingdon, J.,** *East African Mammals,* **1** (Academic Press, London, 1971).

[35] **Clutton-Brock, T. H.,** *Folia primat.,* **19,** 368 (1971).

[36] **Marler, P.,** *Folia primat.,* **13,** 81-91 (1970).

[37] **Marler, P.,** *Behaviour,* **42,** 175-197 (1972).

[38] **Wooldridge, F. L.,** *Anim. Behav.,* **19,** 481-485 (1971).

[39] **Clutton-Brock, T. H.,** *Folia primat.* (in the press).

[40] **Goss-Custard, J. D., Dunbar, R. I. M., and Aldrich-Blake, F. P. G.,** *Folia primat.,* **17,** 1-19 (1972).

[41] **Krebs, J. R., MacRoberts, M. H., and Cullen, J. M.,** *Ibis,* **114,** 507-530 (1972).

[42] **Kummer, H.,** *Primate Societies: Group Techniques of Ecological Adaptation* (Aldine, Chicago, 1971).

[43] **Vesey-Fitzgerald, D. R.,** *E. Afr. Wildl. J.,* **7,** 131-145 (1969).

[44] **Marler, P.,** *Behaviour,* **42,** 175-197 (1972).

[45] **Hladik, C. M., Hladik, A.,** *La Terre et la Vie,* **2,** 149-215 (1972).

[46] **Jarman, P. J.,** *Behaviour* (in the press).

[47] **Emlen, J. M.,** *Ecology: An Evolutionary Approach* (Addison, Wesley, Mass, 1973).

26

Ecology, Diet, and Social Patterning in Old and New World Primates

C. M. HLADIK

Specialization in primate species appears in various ecological niches. Eaters of repugnant insects (*Perodicticus, Arctocebus, Loris*) and tough leaf eaters (*Lepilemur, Indri, Presbytis, Colobus*) are the most clear-cut examples. Among the many primate species that have differentiated in seventy-five million years, man is one of the less differentiated species from the ecological point of view. Consequently man is relatively more adaptable.

Some aspects of social structure are related to adaptation to ecological conditions. To demonstrate this, I will use the Ceylon primate population as an example that was studied in detail* and about which we obtained the most accurate quantitative ecological data. Comparisons with similar ecological studies undertaken in Madagascar† and in America (Barro Colorado)‡ as well as in Africa (Gaboon)§ and consideration of numerous

Reprinted by permission of Mouton Publishers from: "Ecology, Diet and Social Patterning in Old and New World Primates" by C. M. Hladik in *Socioecology and Psychology of Primates* (R. H. Tuttle, ed.). Pages 3-35. World Anthropology. Mouton, The Hague. 1975.

I am indebted to Dr. Russell Tuttle for revising the English translation of this manuscript.

* Smithsonian Biological Program in Ceylon, under the direction of Dr. J. F. Eisenberg.

† Special concerns of Foreign Affairs (Paris), on the occasion of the International Conference of Tananarive, organized by Dr. J. J. Petter.

‡ Smithsonian Tropical Research Institute, under the direction of Dr. M. H. Moynihan.

§ C. N. R. S. (France) Laboratory of Primatology and Equatorial Ecology, under the direction of Dr. A. Brosset.

data collected by other investigators will allow us to see how far we can generalize a theory concerning the way the social patterning of the whole primate order is interdependent with environmental conditions.

ECOLOGICAL NICHES, FOOD AVAILABILITY AND ADAPTATIONS OF PRIMATE GROUPS AT THE POLONNARUWA FIELD STATION, SRI LANKA (CEYLON)

The Polonnaruwa field station is described in Hladik and Hladik (1972). We will mention here only some of its most important features. It is a semi-deciduous forest in the "dry zone" of Ceylon, where the annual rainfall is 1,700 millimeters. The very marked dry season is an important climatic factor considering that at the same latitude some moist evergreen forests do not have more rainfall.

The undergrowth of this forest has been cleared over a large area by archaeologists to bring to light ruins of a city of the twelfth century. In this area, which is very similar to the nearby forest (the canopy is intact except for vines), it was possible to observe in great detail the different species of primates, and to obtain good quantitative estimates of their natural diet (with control in the nearby forest). The whole area was surveyed and mapped, and we calculated the annual production of the different available foodstuffs. Using collecting baskets along a transection, a comparative estimate of the production in the primitive forest was made as a control.

Our field of detailed study covered 54.5 hectares, with a total length of two kilometers. Using an aerial photograph, all trees were mapped after ground identification. The canopy of each one of them was measured. Production was estimated from average measurements.

There are forty-seven species of trees. Though their spatial distribution is not uniform, it is interesting to consider it in relation to the use of the land and the foodstuffs by the different primate species.

Four primate species share the resources: two leaf monkeys of the same genus, *Presbytis senex* and *P. entellus,* one macaque, *Macaca sinica,* which chiefly uses the fruit resources, and a small nocturnal prosimian, *Loris tardigradus,* that can eat many sorts of repugnant insects.

Except for primates, mammalian fauna is fairly scarce at Polonnaruwa. There are no longer big predators (*Panthera*). Only the small nocturnal carnivores characteristic of the Ceylon dry zone (Eisenberg and McKay 1970), some rodents (two species of squirrels are often seen among the monkey groups), ruminants (*Axis axis*) and Tragulidae (*Tragulus meminna*) inhabit the area. Some cattle from a neighboring farm entered the undergrowth to feed on grass or shrubs but they did not interfere with the activity and feeding of the primates.

Presbytis senex

P. senex is a very unobtrusive species forming small groups of five to six individuals that always stay in the canopy. *P. senex* are not easy to observe.

They hide their faces, move around a tree trunk or branch, and quickly flee. Our field station at Polonnaruwa is unique in that we were able to quantify diets by direct counting because of the clearance of undergrowth.

The yearly average of the diet of *P. senex* includes 60 percent leaves and shoots, 12 percent flowers and buds, and 28 percent fruits. One-half of this food is produced by only two tree species: *Adina cordifolia* and *Schleichera oleosa* from which fruits and flowers are eaten as well as shoots and mature leaves. The bulk of the food is obtained from no more than a dozen tree species; nevertheless *Presbytis senex* may eat small parts of many other trees.

The groups stayed in small territories that were three to four hectares wide (Rudran 1970, 1973; Manley f.c.). Shapes of the territories are shown in Fig. 1. After mapping all trees at the field station, we were able to calculate the quantity of food available through the year for each of the groups (for details and methods of calculation, see Hladik and Hladik 1972).

Table 1 indicates the total food available per year for one monkey in each of the groups shown in Fig. 1.

If we consider the spatial distribution of the food trees on the same scale as the home ranges, it seems fairly homogenous. The annual crop is about four tons of foliage (including an average of two tons of *Adina*) and about a hundred kilograms of *Adina* flower for a monkey whose annual needs are about 400 kilograms. Of course, the annual crop exceeds the annual need; but considering that the trees cannot stay very long if they are totally defoliated at regular intervals, the ratio of food requirement to food available which is actually 1 : 10 can be considered as a maximum permissible for leaf eaters using a large amount of young shoots. In fact, for one of the groups (H on Figure. 1) the total annual crop was less than two tons. We noted that some trees in their territory were dying because the leaves and shoots were eaten too frequently. These animals lived in that restricted area because the male of the neighboring group (G in Fig. 1) was very aggressive. Conversely, group G had more food available in its home range than others.

Table 1

Total food in kilograms (fresh weight) available per year for one monkey in each group in Fig. 1

| | Adina cordifolia | | Schleichera oleosa | | Other trees |
	leaves	flowers	shoots	fruits	leaves
A	1,300	60	240	65	1,100
B	3,500	160	380	105	2,200
C	2,100	100	420	115	2,200
D	4,700	215	350	100	2,000
E	3,500	160	510	160	2,000
F	1,000	50	200	55	2,800
G	6,500	300	300	85	5,800
H	600	30	130	35	1,000

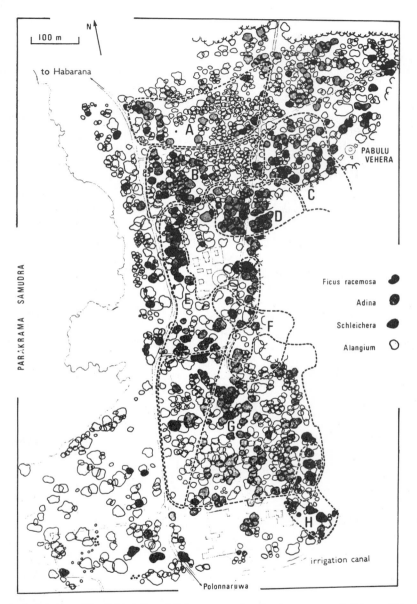

Fig. 1. Aerial view of the field station at Polonnaruwa showing the canopy of all trees. This open forest is located on the shore of an artificial lake (Parakrama Samudra). The limits of the home ranges of *P. senex* are shown by dotted lines, after the data of Manley and Rudran. We note on this map the homogenous distribution of the main food species of *P. senex*: *Adina cordifolia* and *Schleichera oleosa*.

Thus behavior and social structure strongly affect spatial distribution of the species and consequently the habitat use and ecology.

Social structure (see Rudran 1970, 1973) allows division into small groups that use small portions of the field area (Fig. 1). The dispersion of the tree species used by *P. senex* is homogenous enough in those small portions to allow a regular division. On the other hand, many other vegetal species are not so evenly spread, and they are important in food production for other primate species (see below).

Social units of *P. senex* are "one-male groups" that include a few females and their offspring. Some males are necessarily excluded from these basic groups and may stay solitary or form "all-male groups." In these units, individuals are less powerful; they are chased away by the leaders of the "one-male groups" and they live in places neglected by others and which are obviously not the best food-producing areas.

Direct fighting between two groups was not often observed. Contact at the borders of the territories was infrequent. The animals do not have to move very much to find abundant food. Powerful calls emitted by the males early at dawn can be considered as territorial calls. These calls come from the different groups responding to each other. In this manner all of them learn about the location of neighboring groups as soon as the period of activity begins.

Similar types of morning calls are known in other species of leaf-eating primates.

Other species of leaf monkey possess many additional features described for *P. senex*, viz. small groups (with a fairly similar social structure) that move only short distances in a small territory, the source of food always being a few common vegetal species distributed homogenously.

Presbytis entellus

It is noteworthy to compare the gray langur, *P. entellus*, of the Polonnaruwa field station because of noticeable interspecific differences in ecology and social structures.

These monkeys form rather large groups (twenty to thirty individuals), and they are very conspicuous because they come down to the ground and stay there for long periods of time without fear of man.

The home ranges of the different groups of *P. entellus* (Fig. 2) were mapped mainly after Ripley's data. They are wider than those of the *P. senex* groups; nevertheless the biomass of *P. entellus* is remarkably similar (ten to fifteen kilograms per hectare) due to the large size of the groups. The total biomass of folivorous mammals is very high (twenty-five kilograms per hec-

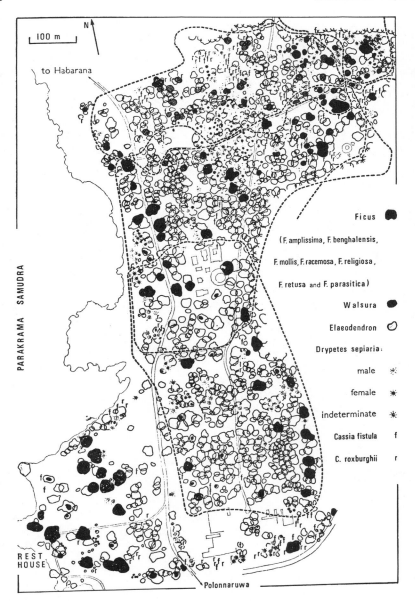

Fig. 2. Map of Polonnaruwa showing the limits (in dotted lines) of the territories of three groups of *Presbytis entellus,* as for *P. senex* on Fig. 1. Land use differs depending upon the scattering of food species in the field.

tare), probably reaching the maximum allowed by the available vegetation, and the maintenance of two rather similar primate species in ecological niches can be explained only by subtle differences in diet and habitat use.

The diet of the gray langur (considered as an average for a one-year cycle) includes 48 percent foliage, 7 percent flowers, and 45 percent fruits. Thus, this last item is eaten in larger quantities by *P. entellus* than *P. senex*. *P. entellus* uses foliage mainly in the form of tender shoots and flushings while *P. senex* includes more mature leaves.

P. entellus also eats some different food plants from those chosen by *P. senex*. But the main difference is in the larger number of food plants included in the diet of the gray langur. The main crops are: *Walsura piscidia*, *Drypetes sepiaria*, and the various *Ficus* forming the bulk (as *Adina* and *Schleichera* do in *P. senex*).

The food plants used by *P. entellus* are also more widely scattered in the field as is shown in Fig. 2. Nevertheless, if we consider their spatial distribution in the home range, we see that it is fairly even. That is to say, in the territories of each group there is approximately the same number of each one of these food-producing tree species. If the territories were not so large (and the groups were smaller, yielding an identical population density) these tree species would not be so evenly distributed. Certain ones would be missing in certain territories and over-abundant in others.

For each individual living in the different territories shown in Fig. 2, there is an annual mean production calculated as shown in Table 2. An important food-producing tree, *Drypetes sepiaria*, which is concentrated in local spots on the general map is, nevertheless, equally distributed between the north and central groups. There are one female (producing fruits) and one and one-half male trees per individual (the data about the south group are not complete because the trees were counted after the fruiting season, but their distribution must be generally the same).

There are not too many *Ficus* in the home range of the north group. This group often tried to extend into the territory of the central group

Table 2
Annual mean production of trees for each individual living in the different territories shown in Fig. 2

	Walsura shoots	*Drypetes sepiaria* shoots	*Drypetes sepiaria* fruits	*Ficus* spp. fruits	*Schleichera* fruits
North group	40	180	23.5	62.5	30
Central group	20	135	20	262	115
South group	(few)	> 50	> 10	197	40

causing frequent battles between groups when the animals were feeding in *Ficus* near the boundaries.

Presbytis entellus feeds on a fairly large number of species. In Polonnaruwa, 90 percent of its food comes from twenty-three woody species. This total includes the twelve species used differently by the other *Presbytis* species. The gray langur uses other species for staple food, uses more fruits, and among leaves takes more shoots and tender leaves (27 percent of the total amount of food).

This greater selectivity in the diet is correlated with a slightly different general behavior and activity pattern. The group has to move more frequently and farther in the home range to find the different resources which are more widely dispersed. Some tree species have very short productive periods (for example the different *Ficus* species as well as the trees from which only young shoots are eaten). This is correlated with a social tradition in the group which permits using these different sources of food during seasonal cycles.

Nevertheless, the social structure of *Presbytis entellus* is quite loose. Jay (1965) characterized groups of langurs as peaceful and relaxed with very little intragroup aggression and no marked hierarchy. Adult males (mostly one dominant among adult males) protect the territory of their group against neighboring groups by chasing other males without really fighting (Ripley 1967). The results of such intertroop encounters in part determine the territorial boundaries.

In India, Sugiyama et al. (1965), observed smaller troops of *P. entellus* with only one adult male. Like *P. senex* "all-male groups" of *P. entellus* also occur.

This similarity in social structure of both species of *Presbytis* was observed at a lower level in our Polonnaruwa field station. In fact, *P. entellus* were organized into large homogenous groups only during the dry season, and the larger groups lived in the most arid places: Jay (1965) found groups of 120 members in South India, and Ripley (1970) observed groups of sixty in Yala (southeast of Ceylon). The structure of the large groups is most effective as an anti-predator adaptation when the animals are on the ground. Many animals can watch and give alarm calls when a predator approaches, and females and young when fleeing can be protected by the more aggressive males who stay behind the group. Because, in the most arid zones, the gray langurs utilize water from water holes, the groups travel long distances on the ground (Beck and Tuttle 1972). In Polonnaruwa, they also seek water during the dry season and spend long periods on the ground to feed on the dried fruits of *Drypetes* and on the leaves of *Mimosa pudica*.

Conversely, during the monsoon season, the langur way of life is obviously more arboreal. The diet includes more foliage and more shoots (fruits are not so abundant) and the animals have to spend more time feeding in the canopy. At this time the groups often split into smaller units, and their social behavior tends to become more similar to that of *P. senex*. During the rainy season, *P. entellus* seems to be more disturbed by human observers, probably because it does not feel the reassuring presence of a large group. Like *P. senex* it uses a "spacing call" (morning whoop) that is not commonly heard during the dry season.

Thus the differences in social structure characterizing the two Ceylonese species of *Presbytis* are related to differences in ecology and habitat use and vary with these factors.

Macaca sinica

The toque macaque (*Macaca sinica*) is ecologically very different from the two species of *Presbytis*. The social structure of its population differs markedly as well. I will now refer to certain observations that Dittus made at the Polonnaruwa field station (see Dittus, this volume). He followed several groups of macaques in large sections of the Polonnaruwa forest and observed variations in population structures over several years. Thus his primatological information is of exceptional value.

I will only cite some aspects of the ecology of the toque macaque in order to clarify the most important ways in which it differs from other primates.

The diet of *Macaca sinica* is comprised chiefly of fruits (77 percent), with some green vegetal growth and flowers (14 percent) and a maximum of 5 percent fungae and 4 percent insects and small prey. The search for animal prey is the principal activity of the macaque. Because it is not abundant, the animals can get only very small quantities (according to Dittus the proportion can be even lower than our estimates). The fruit-eating primates must complement their diet with protein from animal prey and young vegetal shoots. They use many different species to compensate for the deficiencies in essential animo acids of the individual vegetal species (Hladik et al. 1971; 1973). Hence the toque macaque uses certain leaves and shoots like *Strychnos potatorum* and *Randium dumetorum*.

The fruits eaten come from many species widely dispersed in the field. The banyan, *Ficus benghalensis*, gives a fruit (which is eaten in large amounts) very rich in lipids, as do *Schleichera oleosa, Grewia polygama, Glenniea unijuga*, and *Drypetes separia*.

We can only make a general comparison of the distribution of these food-producing species (Fig. 3) with what was accurately calculated for the leaf monkeys. *Ficus benghalensis* is more scattered in the field than any of the species used by the leaf monkeys and so are the other food trees utilized by the macaque. But their territories are very large. According to Dittus' observations, the two groups shown on our map (covering two kilometers) moved in a home range covering about twice the area described.

Despite the scattering of the main food trees, each group has very

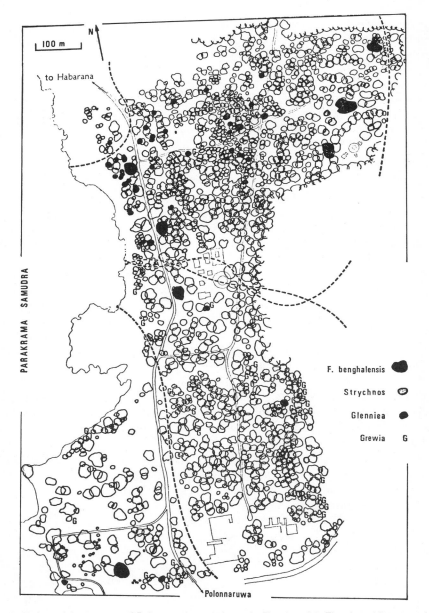

Fig. 3. A view of the canopy of Polonnarywa, as shown in Fig. 1 and 2. The dotted lines are the limits of the home ranges of two groups of macaques. The vegetal food species shown as examples are very scattered.

similar resources available because of the large size of its home range. In this perspective the dispersal of the important tree species seems to be "homogenous." But vegetal production is not an important factor, because it is very much greater than what is the need — two tons of pulp of the fruit of *Schleichera oleosa* are available to the group located in the north on Fig. 3 and the other species produce annually about one hundred times more than what is eaten (this also means that there is no competition with *Presbytis entellus* for fruits). In contrast, looking for prey means that the macaque must move in a very large home range, over irregular pathways to avoid destroying all of the game. The need for this is indicated by the fact that they actually obtain only a very small proportion of their food in animal proteins (compared with other primates of similar weight such as *Cebus*, which feeds on fruits, insects, and a few leaves).

The two groups of macaques which we studied included only twelve and thirteen individuals. Their territories were thirty and forty-five hectares, respectively. (Actually, we must consider their "area of land use" as equivalent to twenty-five and thirty hectares because there were partial territorial superpositions.) The biomass is two and one-half kilograms per hectare. This is the highest figure for primates at this trophic level.

The intragroup social organization is a linear hierarchy evident among males but not so marked for females. This organization allows a close integration of individuals, each with a precise role to play, and greater general efficiency when foraging or detecting predators because most of the time is spent on the ground.

This social model is common to many primate species with similar ecological background. The diet including small prey has the same effect on habitat use. It necessitates long journeys, sometimes on the ground, over a very wide territory.

Loris tardigradus

The slender loris, *Loris tardigradus,* also uses a portion of the animal resources of the Polonnaruwa forest.

We cannot compare social structures among such nocturnal Prosimiae to what we saw in the group of higher primates described above. Nevertheless, it is useful to examine a few examples to try to understand how these primates with complex social structures may have evolved.

The slender loris comes from a branch that might be very close to the most primitive primate ancestor. Its exclusively insectivorous diet is composed of small prey found along branches and vines on which the loris moves very slowly in order to detect by smelling (Petter and Hladik 1970). It is likely that the persistence of this primitive form is due to its nocturnal cycle of activity, slow movements which make it unobtrusive and very spe-

cialized diet, which consists of repugnant insects, myriapods, and ants —
which are neglected by other mammals.

We generally observed single lorises. The species is designated "solitary"; but this does not mean that there are no social contacts between the
individuals of a population.

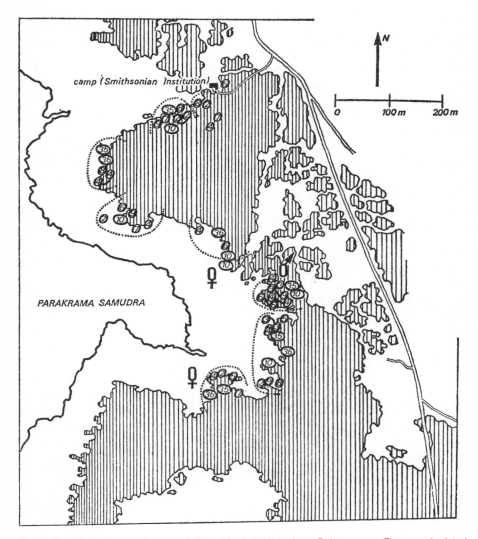

Fig. 4. Some home ranges in a population of *Loris tardigradus* at Polonnaruwa. The area depicted
here is to the south of that shown in Figs. 1, 2, and 3. Each circled number indicates an observation
in December, 1969. Home ranges are shown by dotted lines.

We located the individual home ranges in a population at Polonaruwa (Fig. 4) during a brief field study with Petter, on the edge of a secondary growth forest. The area used by one adult male or female was about one hectare. Calculation of the corresponding biomass gives a fairly high weight for a secondary or tertiary consumer, 0.25 kilograms per hectare.

Social bonds among scattered individuals of this population probably correspond to what was described by Charles-Dominique (1971, 1972) for African Lorisidae: that is, one male individual territory is partially superimposed on one or several individual female territories. This type of primitive social structure is illustrated by the example of *Lepilemur* described below.

ECOLOGICAL NICHES AND SOCIAL STRUCTURES OF MALAGASY LEMURS

Along the banks of the Mandrare River, in the south of Madagascar, is a gallery forest very similar to the Polonnaruwa forest. Five species of lemurs occur in this gallery forest. *Lepilemur leucopus, Cheirogaleus medius* and *Microcebus murinus* are small nocturnal species. *Propithecus verreauxi* and *Lemur catta* are larger diurnal species.

There are similarities between the social organizations of these lemur populations and what was observed, in similar conditions, among the more evolved monkeys of Ceylon. From these similarities we can get an idea about the evolution of social groups of primates.

Lepilemur *and Other Nocturnal Forms*

The most primitive social organization was observed in the sportive lemur (Charles-Dominique and Hladik 1971).

Individual territories observed in the field are shown in Fig. 5. This type of social patterning corresponds to what was described for many other "solitary" primitive mammals. Each female protects its territory against other females but will share it for one or two years with one or several daughters. The male has a wider territory extended over one or several female territories and protects it against other males. As a result of this organization some extra males live outside of the population nucleus (peripheral males).

The most important part of the nocturnal activity of these animals is a motionless watching at the border of the territory. At the beginning of the night, *Lepilemur* makes some specific calls. By these sounds it clearly shows the neighboring animals what its location is at the moment. It is likely that the few animals living nearby are able to identify the sportive lemurs individually by their calls (differing slightly from one animal to another).

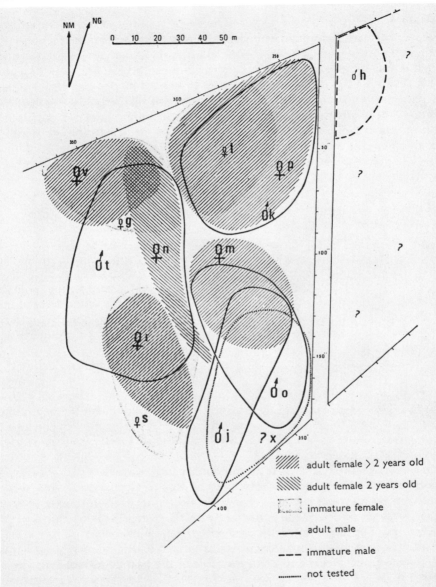

Fig. 5. The individual territories in a population of *Lepilemur leucopus*, in southern Madagascar. Territories of adult males cover territories of one or several females. The adult females are tolerant toward their daughters (territory in grey).

The food of *Lepilemur* is very abundant in all parts of the territory, consisting chiefly of common foliage and flowers. This explains why the animals do not move much and why they spend most of their time watching competitors at the territorial boundaries.

This activity of territorial defense is complementary to the feeding activity. We demonstrated that during a short period the individual territory contains only 1.6 times the minimum amount of necessary food. Thus the surface area of the territories has been selected by this periodic limiting factor. As in the examples previously shown, the population density is precisely adjusted, through social and behavioral patterns, to the maximum available resources.

In the gallery forest, the biomass of *Lepilemur* reaches five kilograms per hectare (810 animals per square kilometer). We must consider, in the ecological system, a necessary balance between folivorous nocturnal lemurs and the larger diurnal forms that are also partly folivorous. Thus, the total size of folivorous biomass in this forest of Madagascar is almost similar to what we observed in the same type of forest at Polonnaruwa.

The *Loris* in Ceylon has a biomass twenty times lower than the biomass of *Lepilemur*. This difference is due to the differences in trophic level of the two species. The *Loris* only uses animal food. It is a secondary or third consumer in the trophic chain. This type of food is ten to one hundred times less abundant than vegetal food.

Yet social structures are very similar* in *Loris* and *Lepilemur* despite the very important differences in their ecologies. For both species, it seems most probable that this organization does not differ very much from that of the primates at the beginning of their evolution. Each of them, at the beginning of the process of differentiation, has become adapted to a highly specialized ecological niche. And so, from the beginning of the Tertiary era they maintained their morphological form as well as their social traditions. Like *Tamiasciurus* (Smith 1968) their population densities are adjusted to food availability only by the system of individual territories. But their social structures are already more sophisticated. The males have permanent contacts with one or several females for whom they "control" territories.

Other nocturnal lemurs, for example the mouse lemur, *Microcebus* (Martin 1972), have similar social structures but with a higher level of gregariousness: several animals might gather in a common nest or hollow tree trunk for the daily rest. From an ecological point of view, they are less specialized than the animals described above, eating insects, fruits, gums, and foliage. We do not have precise information about all the different species of *Microcebus* but it is certain that they are transitional forms between *Loris* and *Lepilemur* which possess a primitive type of social structure.

---

*The exact social structure of the *Loris* population is still to be studied.

Diurnal Lemurs Forming Structured Groups

The ecology and type of habitat use of the sifaka (*Propithecus verreauxi*) and the ring-tailed lemur (*Lemur catta*) is so reminiscent of the two species of *Presbytis* in Ceylon that we may consider them as an experimental model. Two groups of primates evolved separately in Ceylon and in Madagascar. Similar environmental conditions gave rise to partly similar social structures. Thus the social forms found in Madagascar must not differ much from the organization in the first groups of true monkeys.

Groups of sifakas are small, barely mobile units in a small territory. For instance, one group of three sifakas lived in one hectare (Petter 1962). Sifakas feed on leaves and fruits but their diet is less folivorous than the diet of *Presbytis senex* in Ceylon. The social units (Petter et al. i.p.) are "family groups" (in the usual Western meaning of the "family"), including only one couple and their offspring. The sifakas are concentrated a little higher in the gallery forest along the Mandrare River (Jolly 1966). In larger groups (four to six) there is more tolerance among males and among females (Sussman and Richard 1974). Indris (*Indri indri*) are more similar to *Presbytis senex,* with a highly folivorous diet, but they live only in the moist forest of the eastern coast of Madagascar. They also form social units with only one couple (Petter i.p.). This can be considered as the most primitive type of group.

Groups of ring-tailed lemurs are larger, including twenty to thirty individuals (Jolly 1966). This recalls the groups of *Presbytis entellus* of Ceylon. Like these monkeys, ring-tailed lemurs eat leaves and fruits from many species. They often move on the ground over a fairly large territory. There is a hierarchy, with ranks of dominance among males and females. This linear hierarchy is an important factor in group adaptation to environmental conditions. It allows the presence of several males without loss of their aggressiveness, and thus better protection against predators for the whole group when it is moving on the ground, which is an essential aspect of searching out certain types of food distributed over the territory.

Sussman (1974) compared the *Lemur catta* with *Lemur fulvus* in Malagasy forests where both species occur. *Lemur fulvus* has a less sophisticated diet (three vegetal species yielded 80 percent of the food) and eats more leaves than *L. catta.* Comparing the diet and social organization of *Lemur fulvus* to those of *Presbytis senex* in Polonaruwa, it is remarkable how similar they are. *Lemur fulvus* live in small groups without much hierarchy and remain in very narrow territories (Sussman 1974). The biomass may reach twenty-five kilograms per hectare like *P. senex.*

In these few examples of Malagasy lemurs, we note the great possibilities for variation of social structures but mostly for the diurnal species. The "solitary" mode of the nocturnal species is really a prototype of matriarchal society because the young are associated with their mother and share her territory during at least a short period. In the family groups, for instance among Indriidae, the male remains in permanent contact with one female. According to the environmental conditions, there may be some tolerance among adult males or females in wider groups. In more gregarious species of lemurs we observe instances where there is more tolerance among males or among females.

An intermediate social form between the "family group" and the multi-male group might be found in *Hapalemur griseus* (Petter et al. i.p.). These are small diurnal lemurs that live on the bamboos in swamp areas. They are tolerant enough to form small groups adapted to permanent guarding against dangerous predators like *Galidia.* The evolution of contact calls and alarm calls should also be considered as a criterion of adaptation to a hostile environment.

Identical variations in social patterning are found in lemurs and monkeys. In most cases closely similar ecological adaptations require the same type of social structure. Thus we can assume that the actual structures ("family groups") in the less specialized species of lemurs are the best representation of what was the intermediate social condition in the diverse phylogenetic radiations of the Old and New World monkeys.

ECOLOGY AND EVOLUTION OF THE SOCIAL ORGANIZATION IN THE NEW WORLD MONKEYS

The common ancestors of the primates were geographically separated at an early stage and, in the New World, they evolved towards many forms of the Platyrrhini. To exemplify these forms, I will refer to our observations between 1966 and 1968 on Barro Colorado Island (Panama). Determination of types and quantity of food available for each species will give an idea of its way of adaptation to the environment (Hladik and Hladik 1969).

Barro Colorado is a tropical rain forest. Most of the primate species live in this type of forest. Unlike areas discussed earlier it is not possible to get as precise quantitative data in rain forests. Ecological niches in a rain forest are generally more specialized because there are more animal and vegetal species.

On Barro Colorado Island there are still areas with young vegetation but most of it is mature forest. The dry season is marked by intensive sun radiation (no mist or clouds as in equatorial zones), and many trees defoliate at this time. It is semi-deciduous moist forest differing from the evergreen forest of the equatorial zone.

Five species of primates share the resources of this forest with specializations as in the examples of Ceylon and Madagascar. If we consider the ecological similarities and differences, the social patterning is found to be modeled in a series of similar types.

Alouatta palliata

The howler monkey, *Alouatta palliata* is well known through the ethoecological studies of Carpenter. The groups include ten to twenty individuals. The howlers always stay in the upper canopy. They are the most folivorous among New World monkeys. But their diet is not markedly similar to the diet of *Presbytis* in Ceylon. The diet of *Alouatta* consists of 60 percent fruits and only 40 percent leaves. The ecological niche of the true folivores, in Barro Colorado, is filled by two species of sloth (*Bradypus* and *Choloepus*) which constitute the most important part of the biomass (Eisenberg et al. 1972).

With a group of ten howlers using fifteen to twenty hectares of forest, the biomass comes to a maximum of four kilograms per hectare. The howler consumes a fairly small number a vegetal species, especially several kinds of *Ficus* from which it can eat immature fruits as well as leaves. The groups do not move much because the food is easy to find in most parts of the territory. The well known call is equivalent to the morning call of other leaf-eating primates. It evidences presence in their territory without moving.

The distribution of the vegetal species used as food (Fig. 6) is not as homogenous as it is in Polonnaruwa. This explains the need for a larger territory (as for the frugivores) to include enough of each kind of food tree. Consequently for the same density of the population groups will be larger than the "one-male group."

The concept of the "age-graded male troop," introduced by Eisenberg, Muckenhirn, and Rudran (1972) fits this case. According to these authors, howlers have the "one-male group" structure when the population density is low. When the density increases, several mature males may coexist in the same group. This situation corresponds to what was shown for the frugivorous-folivorous lemurs of Madagascar with the difference that, among lemurs, females are generally the less tolerant and the system is closer to a matriarchy.

In howler groups, cohesion is due to the tolerance between males. Nevertheless they might be chased away by the dominant male when they reach sexual maturity. According to population density they may become peripheral males or solitary.

Ateles geoffroyi

The red spider monkey, *Ateles geoffroyi*, was reintroduced on Barro Colorado by Moynihan. The nucleus of the population (six individuals in 1967) had a home range of forty hectares that is supposed to be adjusted to available food. Their biomass was about one kilogram per hectare.

Ateles is typically a frugivorous monkey. Its diet includes 80 percent fruits, is complemented by shoots and young leaves (20 percent) and a very small quantity of animal prey (less than 1 percent). Many species of fruits are used. So, ecologically, *Ateles* differs from *Alouatta* as much as *P. senex* differs from *P. entellus*. But the two Panamanian primate species, both frugivorous, are in a higher trophic level (with a lower biomass).

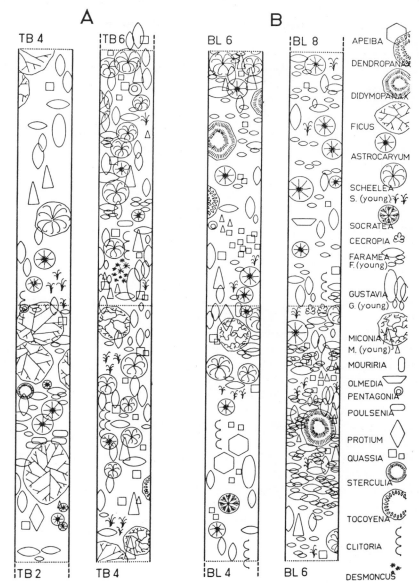

Fig. 6. Distribution of some food species used by the primates of Barro Colorado, mapped on two transactions, twenty meters wide, crossing the home range of the different groups (see comments in the text).

A great tolerance between individuals allows the formation of large troops of spider monkeys that can split into small units. These subgroups are organized around the females and their offspring (Eisenberg and Kuehn 1966) but the general shape of the social organization is the "age-graded male group."

Cebus capucinus

The white-throated capuchins (*Cebus capucinus*) are organized in more structured groups over wide territories. One group whose feeding behavior we observed was made up of fourteen *Cebus* moving over a territory of ninety hectares (Oppenheimer 1968). As with the macaque, the chief activity of the capuchin is foraging for small prey (insects, spiders, and other invertebrates) found in hollow trunks or dead leaves (mostly on the ground, in the litter). This predatory behavior also involves long irregularly patterned walks over a large home range. Furthermore, the animals have to spread out when moving and keep contact by specific calls.

The diet of the capuchin includes 65 percent fruits, 20 percent animal prey, and 15 percent shoots, leaves, and stems. The maximum number of prey they can get in the forest litter is only a small fraction of the total amount they need. Therefore they are forced to be frugivores. We note that they get many more insects than the macaques in Ceylon do, and correlatively, their biomass is much smaller (0.5 kilograms per hectare). Thus they do not need a higher quantity of animal food per hectare.

If fruits are the main food of the capuchin, they are, nevertheless, selected in a particular way in relation to their chemical composition (Hladik et al. 1971). *Cebus* will select many fruits rich in lipids. For instance, among different palm trees it will take the fruits of *Scheelea zonensis*. (Conversely, the spider monkey will choose *Astrocaryum standleyanum* which gives more glucids.) In Fig. 6, we note the abundance and rather even distribution of the *Astrocaryum* palm tree compared to the more scattered *Scheelea* palm tree. Again there is a correlation with the size of the territory, which is larger for the primate using the more scattered food-producing species. As among Ceylon primates this demonstrates the way the social structure adapts to an homogeneous distribution of available food.

There is a linear hierarchy among female and male capuchins. But males of the age-class of four to five years are generally chased away (Oppenheimer 1968) showing a tendency towards the social type of "age-graded male group" (Eisenberg et al. 1972). Integration of individuals in a group of capuchins is accomplished through social exchanges like allogrooming. Grooming is more frequent in this species because it is necessary for animals that spend much time foraging in litter and dead wood to clean their fur. Some types of ritualized behavior such as the aggressive display and various ritualized responses permit the avoidance of true fights, and yet preserve a high level of aggressiveness in a group which is exposed to predators.

Saguinus geoffroyi

The feeding behavior of the rufus-naped tamarin, *Saguinus geoffroyi,* does not differ very much from that of the capuchin, although the diet is much more insectivorus. The tamarin is smaller than the capuchin and thus it needs less food. In the same environment it captures about the same number of small prey, so the relative proportion is higher. Its average diet is 60 percent fruits, 30 percent insects, and 10 percent green parts of vegetation.

The tamarin is becoming rare on Barro Colorado Island, probably because it is specifically adapted to the edge of the forest and to the secondary growth. It lives in thick vegetation and rarely comes down to the ground. Social units are groups of six to nine individuals (Moynihan 1968) including one or several pairs of adults and their offspring.

So, among the Callithricidae (marmosets and tamarins) which are the most primitive radiations of the New World primates, we already find a sophisticated social pattern. But in most of the genera, the groups are of the "parental type" (Eisenberg et al. 1972), which corresponds to Petter's "family type." In such groups, an adult pair lives with permanent bonds, reinforced by social exchanges like grooming and bodily contacts.

Aotus trivirgatus

The night monkey, *Aotus trivirgatus,* is a unique example of a true monkey adapted to a nocturnal cycle of activity. On Barro Colorado Island it lives either alone or in a small primitive group including at most one adult pair and one or two infants (Moynihan 1964). *Aotus* is among the least gregarious species of primates: there is no ritualized display as in the social species, which serves to divert fighting between individuals.

We have no precise ecological data about the night monkey, thus it is difficult to make comparisons with other species. It does not seem very specialized in its diet. It may be like some transitional forms of lemurs.

ECOLOGICAL COMPARISON: PANAMA VS. CEYLON

I will now draw a parallel between the precise ecological data on the Ceylon primates and those just described for the species of Barro Colorado. In Fig. 7b I show a division into more specialized ecological niches in the rain forest. In the dry forest of Ceylon, the primate species are less specialized in their diet. Hence all of them cover the whole available food supply (from the insectivores to the more specialized folivores). In the rain forest of Barro Colorado they have to share the resources with other mammalian species.

So, in Fig. 7b, they are gathered in a narrower space according to their biomass which depends directly on the trophic level, i.e. the type of diet. Similar comparisons could be made with the primate species in Madagascar or in continental Africa but the ecological data are not yet accurate enough.

For a given species, the groups are more or less important not only in relation to their biomass and available food but also according to particular conditions of the environment (for example scattering of vegetal food species). The cohesion of large groups is due to the emergence of ritualized displays that make possible tolerance between the adults, male or female.

It is not possible to say whether ethological evolution of the social structures came from particular pressures of ecological conditions or if the adaptation to the environment followed these socioecological changes. What is certain is that correlation between those two conditions allows the survival of the different species.

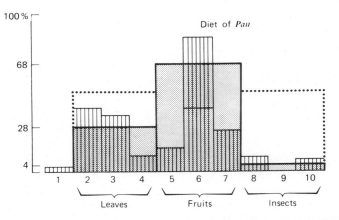

Fig. 7a. Example of the diet of the chimpanzee (*Pan troglodytes troglodytes*). This diagram is made with the same symbols as those used for the diets of other primates in Fig. 7*b*.

Each of the grey parts of the three squares is proportional to the corresponding type of food ingested in a year:

Leaves (28 percent), in the left square;
Fruits (68 percent), in the central square;
Insects (4 percent) in the right square.

A more detailed account of the types of food ingested and their daily variations is expressed by the hachured columns as follows:

    1. Minerals (earth) occasionally ingested.
    2. Barks and twigs   ⎫ Leaves
    3. Leaves, buds and pith   ⎬ and
    4. Flower buds and gums ⎭ twigs
    5. Immature fruits   ⎫
    6. Sweet fruits.   ⎬ Fruits
    7. Seeds and arils   ⎭
    8. Ants, termites, and small arthropods (*Orthoptera*) ⎫ Insects
    9. Large arthropods (*Orthoptera*)   ⎬ and other
  10. Eggs, fledgings, and other large prey   ⎭ prey

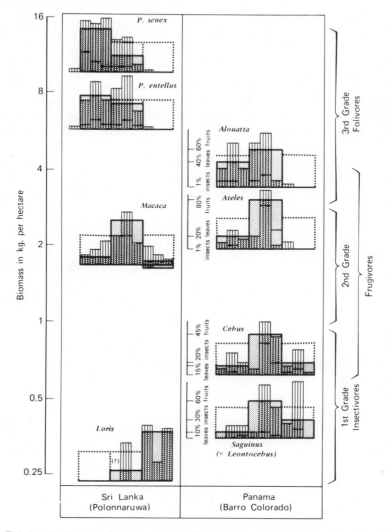

Fig. 7b. Relationship between the type of diet and the biomasses of primates in two different field stations.

Each primate species is characterized by its "dietogram" symbolized as in Fig. 7a. Its level on the present diagram corresponds to its biomass.

It is clear that the different types of diet are directly correlated with the biomass, thus affecting social patterning.

SOCIAL LIFE AND ECOLOGY IN ANTHROPOID APES
A parallel evolution took place in the forests of the Old World, and there are many examples of convergences in socioecological adaptations rather

similar to those shown in our examples (Struhsaker 1969). But there is no place where ecological data have been collected in much detail.

We studied *Pan troglodytes troglodytes* in the forest of Gaboon for one year (Hladik 1973).

The chimpanzee is a true frugivore which complements its diet with green plants and a few insects or other invertebrates (68 percent fruits, 28 percent leaves and bark, 4 percent insects; see Fig. 7a). Occasionally it eats small game animals. This particular predatory behavior is very interesting to consider in itself (Goodall 1965; Teleki 1973) but it has no true consequences for nutritional needs because the average quantity of meat eaten is very low. Furthermore this type of food is used only by adult males and a few females; the juveniles, whose protein needs are the highest, have no access to the meat.

This hunting and meat-eating behavior could be compared with what is known about primitive human tribes of hunter-gatherers. For them, meat-eating is a pleasure as well as an opportunity to organize festivities with social exchanges. But precise ecological data show that their diet is almost entirely vegetarian (80 percent fruits and roots collected by women; Woodburn 1968). In human populations, the meat of the game is a significant source of proteins, but it is not a compulsory need considering that some kinds of insects are included in the food gathered by women (e.g. caterpillars are richer in proteins and lipids than meat).

In the diet of the chimpanzee the proportion of animal food as well as the fruits and other vegetal parts may vary from one place to another. This is not surprising if we consider that the geographical range of *Pan troglodytes* crosses more than half of the African continent, with three subspecies living in many different environments. Goodall described termite "fishing" behavior by East African chimpanzees. They use a long twig as a "tool" to catch the insects. In several hours of such activity a chimpanzee may obtain a large amount of animal protein and fat. Chimpanzees in the rain forest of Gaboon find most of their animal protein in several species of ants' nests. They spend a large part of their time looking for these prey in the manner of gatherers more than hunters. Thus they get a regular supply through gathering. Hunting per contra is an activity of high risk and low return (Lee 1968). In one day we saw 120 grams of insects eaten by chimpanzees (up to 6 percent of the total food amount) gathering the small nests of *Macromiscoides aculeatus,* one of which contains only 0.2 grams of ants and larvae. Other ants eaten in large quantities are *Oecophylla longinoda, Polyrhachis militaris,* and *Paltothyreus tarsatus.* The last species is "fished" with a stick as was described for termites.

To have a regular supply of large quantities of insects, chimpanzees need very wide territories. We have no good data for Gaboon about the size of this area. In the rain forest of East Africa, the densities observed by Reynolds (1965) correspond to a biomass of 1.5 kilograms per hectare, which is somewhat lower than the biomass of the macaque in Polonnaruwa which also feeds on fruits and a small proportion of insects.

Social traditions in local populations of chimpanzees make the use of special techniques possible. An example is the way *Panda oleosa* or *Coula edulis* nuts are opened (Rahm 1971). The hard shells are cracked with a branch of hard wood or a heavy stone used as a hammer. This technique is ignored by the chimpanzees living in the rain forest of Gaboon in spite of the local abundance of *Coula* and *Panda.* But the variety in the types of food eaten and recognized by the chimpanzees is impressive; we collected 210 specimens of food in the primary forest of Ipassa (Gaboon).

The social unit does not appear to have great cohesion. Sometimes small temporary groups of chimpanzees split or rearrange themselves some- times in a few hours. It seems that subtle, permanent bonds exist on the larger scale of a "regional population" (Sugiyama 1968). Some fifty in- dividuals are acquainted with each other, maintaining "friendly" interrela- tions. This type of big "group" lives in a home range much larger than any other territory of frugivorous primates.

The gorilla (*Gorilla gorilla*), studied by Schaller (1963) in a montane forest, forms smaller social units (six to seventeen individuals) that can fit in the category of "age-graded male groups." One male is dominant and the others become more or less peripheral. The gorilla is a leaves-and-fruit eater and there is some similarity with the social patterning of folivorous primates. But territories are among the largest ever seen for primates (twenty-five square kilometers for one group of ten adults). This could be explained by the large number of species included in the diet. More than 100 vegetal food species were collected by Schaller. Only the chimpanzee, being more frugivorous, has a wider range (see Suzuki, this volume).

More classical correlations are found in two species of Hylobatidae, in Malaya. The gibbon, *Hylobates lar,* is predominantly a frugivore while the siamang, *Symphalangus syndactylus,* eats more leaves and shoots of many vines, including many species of *Ficus* (Chivers 1972). Both species live in small familial groups maintaining distances by specific calls. But the more folivorous siamang has a territorial area which is only half the size of the territory of the gibbon.

CONCLUSION

We started this comparative study with a few examples in which the knowl- edge of precise ecological data allowed comparison of several sympatric primate species. But we were only demonstrating the correlations between the types of social patterning and the resources and habitat use in an sup- posedly stabilized environment.

Long term mechanisms for regulating population densities were not examined. The most rapid fluctuations may depend on the groups' social structure. In the case of increasing population density, certain species have slight differences in group structure. The males are more aggressive and may kill most of the new-born infants. Eisenberg et al. (1972) showed that this way of limiting the population is not "social pathology" at all because in this case, the "one-male group" is the structure best adapted to the environment. Wynne-Edwards (1962) showed other general mechanisms regulating the population according to available resources but many of them have yet to be verified by ecological observations of primates.

Social patterning is modeled by environmental conditions but it also comes after a long evolution of each systematic group: so we find now different expressions of it. This conclusion is shared by Struhsaker (1969) and Eisenberg et al. (1972). More general laws (Crook and Gartlan 1966) only concern a few precise predators by having a large, structured group with a linear hierarchy. The most sophisticated unit with many males is the baboon troop (DeVore and Hall 1965). Among forest-living species, the more terrestrial primates are also the larger groups with ranks of dominance among males. But the occurrence of large, polyspecific groups among arboreal primates of the rain forest (Gautier and Gautier 1969) cannot be explained by the same ecological factor.

Group structure is more directly affected by the way the whole population uses food resources. Similar examples are clear for all leaf-eating primates: the biomass is high; they are organized in small units ("family" type or "one-male group"); they have narrow territories; and all of them use territorial calls.

In other cases, similarities are not so clear. But some authors speak about folivorous or frugivorous forms or just about arboreal primates without any precise ecological definition. Many field studies concerning the ethology, structure, and composition of primate groups have been carefully carried out, but very few ecological and quantitative data are available, especially on the diet. That is why we quote only a few examples in this text. The last example mentioned (the comparison between the gibbon and the siamang) only refers to the actual time these two species spend feeding on fruits or leaves and not really to the weight eaten. Among Chivers' projects is the collection of complementary quantitative data. In a few years the ecology will be clearly known. The same could be said of the gorilla and of many other primate species. The exact proportions of the different components of the diet must be known at least to elucidate what the impact of the species is on the environment and how it is integrated with it.

With such precise quantitative data we can hope to separate into finer classes the ecological types of primates as follows:

1. The first grade: from typical types of insectivores (*Loris, Arctocebus*) with a gradation towards diets including fruits and insects (*Saguinus, Saimiri*) in large proportion (see Fig. 7b).
2. The second grade: diets combining insects or other invertebrates

and green parts of vegetals, in fairly small proportions, with a main complement of fruits and/or seeds (*Cebus, Cercopithecus, Macaca, Papio, Pan,* etc.).

3. The third grade: from the frugivores-folivores obtaining their proteins mostly from the leaves (*Ateles, Gorilla, Alouatta*) toward the more specialized folivores-frugivores with a complex stomach (*Presbytis, Colobus*) or special caecum (*Indri, Lepilemur*).

Among the forms of insect-eating primates are the prosimians of continental Africa. For these last species, the ethoecological studies of Charles-Dominique (1971, 1972) showed the very exact specializations in the diet of five species living in the rain forest. Charles-Dominique also remarked on the importance of the vegetal gums in the diet of certain species (*Euoticus elegantulus, Perodicticus potto*). This last type of feeding which we are actually investigating might be very different from the frugivore type and more similar to the leaf-eating specialization shown in our third grade.

The classical term *omnivore* has generally applied to the first or second grade of primate ecological types, including all transitional types. Now it should be avoided, because it is used with different definitions by different authors.

In all these ecological categories, there may be variations in the social patterns of the internal structures of the social units but they are most closely correlated with the mode of sharing local resources.

REFERENCES

Beck, B. B., R. Tuttle 1972 "The behavior of gray langurs at a Ceylonese waterhole," in *The functional and evolutionary biology of primates.* Edited by R. Tuttle, 351-377. Chicago: Aldine-Atherton.

Charles-Dominique, P. 1971 Eco-éthologie des Prosimiens du Gabon. *Biologia Gabonica* 7(2): 121-228.

1972 Comportement et écologie des Prosimiens nocturnes. Ecologie et vie sociale de *Galago demidovii. Zeitschrift für Tierpsychologie,* supplement 9:7-41.

Charles-Dominique, P., C. M. Hladik 1971 Le *Lepilemur* du sud de Madagascar: Écologie, alimentation et vie sociale. *La Terre et la Vie* 25:3-66.

Chivers, D. J. 1972 "The siamang and the gibbon in the Malay peninsula," in *Gibbon and siamang.* Edited by D. M. Rumbaugh, 103-135. Basel: S. Karger.

Crook, J. H., J. S. Gartlan 1966 Evolution of primate societies. *Nature* 210:1200-1203.

DeVore, I., K. R. L. Hall 1965 "Baboon ecology," in *Primate behavior.* Edited by I. DeVore, 20-52. New York: Holt, Rinehart and Winston.

Esenberg, J. F., R. E. Kuehn 1966 *The behavior of* Ateles geoffroyi *and related species.* Smithsonian Miscellaneous Collections 151(8):1-163.

Eisenberg, J. F., G. M. Mc Kay 1970 An annotated checklist of the recent mammals of Ceylon with keys to the species. *The Ceylon Journal of Science, Biological Sciences* 8(2):69-99.

Eisenberg, J. F., N. A. Muckenhirn, R. Rudran 1972 The relation between ecology and social structures in primates. *Science* 176:863-874.

Gautier, J. P., A. Gautier-Hion 1969 Les associations polyspécifiques chez les *Cercopithecidae* du Gabon. *La Terre et la Vie* 23:164-201.

Goodall, J. 1965 "Chimpanzees of the Gombe Stream Reserve," *in Primate behavior.* Edited by I. DeVore, 425-473. New York: Holt, Rinehart and Winston.

Hladik, A., C. M. Hladik 1969 Rapports trophiques entre végétation et primates dans la forêt de Barro-Colorado (Panama). *La Terre et la Vie* 23:25-117.

Hladik, C. M. 1973 Alimentation et activité d'un groupe de chimpanzés réintroduits en forêt gabonaise. *La Terre et la Vie* 27:343-413.

Hladik, C. M., A. Hladik 1972 Disponibilités alimentaires et domaines vitaux des primates à Ceylan. *La Terre et la Vie* 26:149-215.

Hladik, C. M., A. Hladik, J. Bousset, P. Valdebouze, G. Viroben, J. Delort-Laval 1971 Le régime alimentaire des Primates de l'île de Barro-Colorado (Panama): Résultats des analyses quantitatives. *Folia Primatologica* 16:85-122.

Jay, P. 1965 "The common langur of North India," in *Primate behavior.* Edited by I. DeVore, 197-249. New York: Holt, Rinehart and Winston.

Jolly, A. 1966 *Lemur behavior. Madagascar field studies.* Chicago: The University of Chicago Press.

Lee, R. B. 1968 "What hunters do for a living, or, how to make out on scarce resources," in *Man the hunter.* Edited by R. B. Lee and I. DeVore, 30-48. Chicago: Aldine.

Manley, G. f.c. "Aspects of the ecology of *Presbytis senex.*"

Martin, R. D. 1972 "A preliminary field-study of the Lesser Mouse Lemur," in *Behavior and ecology of nocturnal Prosimians. Zeitschrift für Tierpsychologie* (supplement) 9:43-89.

Moynihan, M. H. 1964 *Some behavior patterns of Platyrrhine monkeys. I. The Night Monkey* (Aotus trivirgatus). Smithsonian Miscellaneous Collections 146(5):1-84.

1968 *Some behavior patterns of Platyrrhine monkeys. II.* Saguinus geoffroyi *and some other Tamarins.* Smithsonian Miscellaneous Collections.

Oppenheimer, J. R. 1968 "Behavior and ecology of the white-faced monkey *Cebus capucinus,* on Barro Colorado Island, Canal Zone." Unpublished doctoral dissertation, University of Illinois, Urbana, Illinois.

Petter, J. J. 1962 Recherches sur l'écologie et l'éthologie des Lémuriens malgaches. *Mémoires du Muséum National d'Histoire Naturelle,* série A, 27(1):1-146.

i.p. "A study of population density and home range of *Indri indri* in Madagascar," in *Prosimian biology.* Edited by R. Martin, G. A. Doyle, and A. C. Walker, 75-108. London: Duckworth.

Petter, J. J., C. M. Hladik 1970 Observations sur le domaine vital et la densité de population de *Loris tardigradus* dans les forêts de Ceylan. *Mammalia* 34(3): 394-409.

Petter, J. J., Y. Rumpler, R. Albignac i.p. "Les lémuriens de Madagascar," in *Faune de Madagascar.* Paris: O.R.S.T.O.M. and C.N.R.S.

Poirier, F. E. 1970 "The Nilgiri Langur (*Presbytis johnii*) of South India," in *Primate behavior; developments in field and laboratory research.* Edited by L. A. Rosenblum, 251-383. New York and London: Academic Press.

Rahm, U. 1971 L'emploi d'outils par les Chimpanzés de l'Ouest de la Côte d'Ivoire. *La Terre et la Vie* 25:506-509.

Reynolds, V., F. Reynolds 1965 "Chimpanzees of the Budongo Forest," in *Primate behavior.* Edited by I. DeVore, 368-424. New York: Holt, Rinehart and Winston.

Ripley, S. 1967 "Intertroop encounters among Ceylon gray langurs (*Presbytis entellus*)," in *Social communication among primates.* Edited by S.A. Altmann, 237-253. Chicago: University of Chicago Press.

1970 "Leaves and leaf monkeys: The social organization of foraging in gray langurs (*Presbytis entellus thersistes*), in *Old World monkeys.* Edited by J. R. Napier and P. H. Napier, 481-509. New York and London: Academic Press.

Rudran, R. 1970 "Aspects of ecology of two subspecies of purple-faced langurs (*Presbytis senex*)." Unpublished thesis, University of Ceylon, Colombo.

1973 Adult males replacement in one-male troops of purple-faced langurs (*Presbytis senex senex*) and its effect on population structure. *Folia Primatologica* 19:166-192.

Schaller, G. B. 1963 *The mountain gorilla: ecology and behavior.* Chicago and London: University of Chicago Press.

Smith, C. C. 1968 *The adaptive nature of social organization in the genus of tree squirrel* Tamiasciurus. Ecological Monographs 38:31-63.

Sugiyama, Y. 1968 Social organization of chimpanzees in the Budongo Forest, Uganda. *Primates* 9:225-258.

Sugiyama, Y., K. Yoshiba, M. D. Pathasarathy 1965 Home range, mating season, male group and inter-troop relations in Hanuman langurs (*Presbytis entellus*). *Primates* 6:73-106.

Sussman, R. W. 1974 "Ecological distinctions in sympatric species of lemur," in *Prosimian biology.* Edited by R. Martin, G. A. Doyle, and A. C. Walker, 75-108. London: Duckworth.

Sussman, R. W., A. Richard 1974 "The role of aggession among diurnal prosimians," in *Aggression, territoriality and xenophobia in the primates.* Edited by R. L. Holloway, 49-76. New York: Academic Press.

Struhsaker, T. T. 1969 Correlates of ecology and social organization among African cercopithecines. *Folia Primatologica* 11:80-118.

Teleki, G. 1973 The omnivorous chimpanzee. *Scientific American* 228(1): 32-42.

Woodburn, J. 1968 "An introduction to Hadza ecology," in *Man the hunter.* Edited by R. B. Lee and I. DeVore, 49-55. Chicago: Aldine.

Wynne-Edwards, V. C. 1962 *Animal dispersion in relation to social behaviour.* Edinburgh and London: Oliver and Boyd.

27

MAMMALIAN SOCIAL SYSTEMS: STRUCTURE AND FUNCTION

J. H. CROOK, J. E. ELLIS, AND J. D. GOSS-CUSTARD

Abstract. Mammalian societies are complex socio-ecological systems controlled by the interactions of numerous internal constraints and external factors. We present a simple model describing these systems functionally in terms of the adaptive behavioural strategies for resource exploitation, predation avoidance and mating and rearing of young to maturity shown by the individuals that comprise them. The relations between species-specific limitations on the range of potential individual social behaviour and the environmental variables to which the system is responsive are analysed and hypotheses from correlational and analytical field studies examined. We advocate the continued development of sophisticated systems-analytical approaches to societal analysis taking into account the contrasting informational provenance of factors of different types. This is preferred to either an over-emphasis on environmental determination or excessively formalised neo-darwinian modelling based on assumptions from genetics and selection theory alone.

A major objective in the study of mammalian societies is to develop an understanding of the way in which, through evolution, species characteristics and environmental forces have interacted to shape the structure and dynamics of these diverse and complex systems. Such an understanding should make it possible to state how a society operates and to predict how changes in intrinsic (species) or extrinsic (environmental) characteristics may affect social organizations and the relations between individuals.

Reprinted from *Animal Behaviour*, *24*:261-274 (1976) by permission of Baillière Tindall Publishers.

Mammalian societies are complex systems. influenced and modified by the interactions of numerous external forces and internal constraints. In order to understand them, we need to work toward the development of social systems models which will incorporate this inherent complexity and yet produce realistic simulations of both the structure at one time and dynamic change through time. In working toward this goal, the following procedures need to be undertaken, although not necessarily in the sequence suggested here:

1. Identifying: (a) The principal social system variables which between them describe the structure of a social system (e.g. group size and dispersion, inter-individual relations) (Fig. 1). (b) The external environmental variables which produce changes in social structure and dynamics (e.g. dispersion of food; predators). (c) The species parameters which affect the flexibility of social structure and dynamics (e.g. mobility and body size may determine ranging capacity).

2. An assessment of how, in relation to the environment, the system expresses the life-support and reproductive strategies of the individuals comprising it.

3. The development of conceptual models suggesting how, with respect to these functions, the various environmental 'and species characteristics interact to determine the described structure and underlying dynamics of the society.

4. The proposal and testing of hypotheses derived from these conceptual models.

5. The conversion of the conceptual models (incorporating the sur-

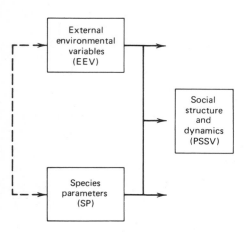

Fig. 1. External environmental variables (EEV) interact with species parameters (SP, e.g. morphological and physiological characteristics) to determine social structure (measured as the principal social system variables (PSSV) and social dynamics (changes in PSSV over time)). The dotted arrow takes note of the fact that EEV's also affect SP but on a slower (evolutionary) time scale than the effects on PSSV's which may change within the life-span of an individual, through learning.

viving hypotheses) into mathematical models which may then be used as tools for predicting the social effects of changes in environmental variables or species parameters.

The initial steps (1-3) form the major topics of this article. The first sections identify what appear to be the principal variables in each of the three 'boxes' shown in Fig. 1. Later sections are concerned with interactions among these variables and are related to a classification of mammal social systems. In doing this, we hope to set out a framework for future studies of mammalian social systems, and through some simple models, to generate tentative hypotheses about the adaptive significance of mammalian societies.

FUNCTIONAL SUB-SYSTEMS IN A SOCIETY

Individuals of a mammalian species neither disperse nor relate to one another randomly. Rather they are found in characteristic patterns of population dispersion, grouping and ranging and form relationships varying in number, complexity and duration. These patterns we refer to as social structure. Structure is both maintained and changed by processes of interindividual behaviour; these we refer to as social dynamics (Crook & Gross-Custard 1972). Both dispersion patterns and relating are mediated by individual behaviours comprising a communication system.

We argue that particular social structures arise because they provide an optimal context within which the individuals comprising them carry out vital functions. These, on a minimal count, include (i) resource expolitation, (ii) predator avoidance, (iii) mating and rearing. Behavioural adaptations maximizing an individual's reproductive success, survival and the survival of offspring to maturity we describe here as "strategies." These include strategies comprising the behavioural determinants of the social structures with which we are concerned. We analyse social structure here in terms of subsystems within which individuals achieve the main vital functions. If desired, resource-exploitation can itself be sub-divided into, for example, food and non-food resources, although for our present purposes it adds little to do so.

These four functions may all be carried out within the same social structure, as in the gibbon's (*Hylobates lar*) monogamous, single-male family group. Often, however, different activities occur within different sub-groupings within a society. In hamadryas baboons (*Papio hamadryas*) for example, rearing of the young is primarily the job of a single individual, the mother. Mating takes place within a mating group or harem of one to five females dominated by a single male. Predator-avoidance is a coordinated function of the whole troop. Resource-exploitation may be accomplished by a lone individual, by a single harem group, or within the context of an extremely large troop, depending upon the density and dispersion of resources (Kummer 1968).

The ways in which the vital functions are satisfied in different species or populations of the same species comprise an organization of strategies that interact in various ways. In some species the strategic solutions occur in spatially or temporally separated social units. In others they are convoluted into complex systems based on a compromise (Hinde & Tinbergen 1958) arising from conflicting needs and which represent the optimal strategy for individuals within the system as a whole. The total system amounts to the adaptive "grade" of a species population in relation to its habitat (Crook 1964).

The division of a society into functional sub-systems is useful analytically in that it helps to focus attention on environmental variables and species parameters which are likely to be particularly important in the accomplishment of a particular function, but which may have little influence on the accomplishment of others. For instance, one species parameter, the maturation rate of the young, provides critical constraints on the structure of the mating and rearing sub-systems, but may be less critical in determining the structure of the resource-exploitation group. It must be remem-

Table 1

Matrix of Some Important External Environmental Variables (EEV), Species Parameters (SP), and Principal Social System Variables (PSSV), by Functional Sub-systems

	Resource-exploitation and predation avoidance sub-systems	Mating sub-system	Rearing sub-system
EEV	Resource density, resource distribution, density of predators	Resource distribution	Resource seasonality
SP	Mobility potential, susceptibility to predation	Role of male in rearing	Role of male in rearing, maturation rate of young
PSSV	Group size and stability, cover utilization, range exclusivity	Duration of male-female bond, no. females mated-male	Duration of the male-female bond, duration of the female—offspring bond

bered, however, that these sub-systems do not exist in isolation but are fully integrated and that individuals usually operate within several sub-systems simultaneously (Ellis, in preparation (a)). What appear to be the main parameters or variables in each are described now and are summarized in Table 1.

PRINCIPAL SOCIAL SYSTEM VARIABLES

Mating Sub-systems

Each sub-system can be usefully described at this stage by quantifying a relatively small number of principal social system variables. For instance, a mating group can be described by: (1) enumerating the number of females with which each male consorts during a breeding season (if mating is seasonal) and (2) noting the duration of the male-female bond. Although other variables will also require consideration in due course, four mating systems can be defined on this two-variable basis: (a) Monogamous pairs: single female per male; lengthy male-female bonds. (b) Monogamous mating with brief pairing; single female per male; short-term bonds. (c) Simultaneous polygamy: more than one female per male; lengthy male-female bonds. (d) Serial polygamy: more than one female per male; sequential short-term bonds. Number of females and males in group forming the potential mating pool.

Similar descriptions can be constructed for each of the other functional sub-groupings utilizing identified variables which form foci of attention in current literature. We have no doubt that future studies in detail will identify further or more subtle variables and extend our account with an appropriate increase in sophistication.

Rearing Sub-systems

Rearing sub-systems can be described by: (1) relating whether or not the male remains with the female while she rears the young (i.e. the duration of the male-female bond) and (2) determining if the young remain with the female past weaning and if so, for how long (i.e. the duration of the female-young bond). Rearing sub-systems arising from variations in these two principal social variables are shown in Table 2 as follows:

Table 2
Classification of Mammalian Social Systems

Sociotype	Rearing strategy	Mating strategy	Grouping and dispersion strategy	Examples
	System variables: 1. Male's presence in rearing group. 2. Duration of male's presence. 3. Duration of mother-infant (M—I) bond.	System variables: 1. Number of female consorts per male. 2. Duration of male-female bond.	System variables: 1. Group size. 2. Group stability. 3. Refuge utilization. 4. Range exclusivity.	
Ia	Male absent from rearing group.	Promiscuous matings.	Single M—I unit, males and juveniles also range individually. Refuge (cache or nest-based) system, exclusive ranges.	Tree squirrels. Gophers.
Ib	Same as Ia. Longer M—I bond.	Same as Ia.	Similar to Ia except that ranges are not exclusive.	Bush-babies.
Ic	Same as Ia. M—I bond persists past weaning.	Promiscuous matings, brief copulatory meetings. Males may enter female herds or females may visit mating territory of male.	M—I units join together in large herds. Males may form separate herds or remain solitary. No refuges. M—I herds free-ranging. Solitary males may have exclusive ranges.	Elephants. Kob.

Table 2 (continued)

Sociotype	Rearing strategy	Mating strategy	Grouping and dispersion strategy	Examples
IIa	Males present with rearing group, but only in loose association, not bonded	Promiscuous; brief copulatory meetings within-in groups.	M—I units, males, and independent juveniles remain together in groups. Non-exclusive ranges.	Some bats. Badgers.
IIb	Same as IIa.	Same as IIa.	Same as IIa, except non-refuge based, free-ranging	Bison, Wildebeest.
IIIa	Male present in rearing group.	Harem; several females per male, membership varies through the season. Some male-female bonds may be re-established in successive years but not necessarily so.	Individual harem groups during breeding season, may gather into colonies—break up after breeding season. Refuge based. Non-exclusive foraging range.	Seals.
IIIb	Same as IIIa.	Same as IIIa	Exclusive rearing and breeding sites; units grouped into colonies, each unit with an exclusive foraging range. May be maintained outside breeding season.	Rabbits, Prairie dogs

Table 2 (continued)

Sociotype	Rearing strategy	Mating strategy	Grouping and dispersion strategy	Examples
IVa	Male remains with rearing (M—I) group, usually for more than one season. M—I bond lasts from 1 to several seasons.	Mated pairs.	Refuge-based groups may include offspring from successive seasons. Ranges may be exclusive or non-exclusive.	Beavers.
IVb	Same as IVa.	Same as IVa.	Same as IVa except non-refuge based.	Marmosets, Gibbon, Titi.
IVc	Same as IVa.	Same as IVa but looser relations between mates.	Several "family" groups may join together, otherwise, same as IVa, and IVb.	Wolves, hunting dogs Hyaenas.
IVd	Same as IVa.	Harem groups of long duration.	"Excess" males may form independent groups or maybe integrated within rearing-mating groups. "Herding" of basic harem groups may occur. May or may not be refuge-based. May or may not maintain exclusive ranges.	Horse, Vicuña, Patas monkey, Blue monkey, Gelada baboon Hamadryas baboon.
IVe	Same as IVa.	Multimale groups with several females of long duration.	Same as IVd except all males are integrated male—female groups; no "excess" males. Herding of basic units not known to occur.	Papio baboons, other than hamadryas, Vervet monkeys, macaques.

Males do not assist:

(a)   Males provide no assistance during rearing and do not remain with females, young leave mother at weaning (i.e. Ia).

(b)   Same as (Ia) except that young remain with mother until subsequent young are born.

(c)   Same as (Ia) except that young remain with the dam long past weaning and after the birth of subsequent young.

Males do assist:

(a)   Males remain with the females during rearing and provide some assistance in the form of feeding or protection for the dam and young, the young mature rapidly and leave the family group at weaning. Male assistance is most frequently associated with slow maturation of the young.

(b)   Same as above except that young leave prior to birth of subsequent offspring. This would be the same as IIa if several litters are born annually; however, it differs from it where only a single litter is produced annually and young overwinter with adults (beavers).

(c)   Males remain with females during rearing (i.e. III-IV in Table): young remain with the pair-bonded parents long past weaning, dependence usually decreasing slowly with age. Adult-young bond is frequently maintained after the birth of subsequent offspring. In some rare cases the attendant young may carry subsequent offspring and thus participate in care (marmoset).

Resource-exploitation and Predator-avoidance Sub-system

The structure and strategy of resource-exploitation sub-systems can be described in terms of: (1) group size, (2) group stability, (3) refuge or cover utilization, and (4) range exclusivity. However, all but the last of these may also express predator-avoidance. Thus, except in specific instances where there is evidence for assuming that these social variables are determined either by resource-exploitation strategies or by predator-avoidance strategies, they are currently most economically considered as the products of both and discussed as grouping and dispersion strategies.

(a) Group size and stability. The potential variation in group size ranges from one to the size of the population deme. Group stability may range from considerable mobility of membership through steady or periodic changes in size or composition over time to complete group stability.

(b) Refuge or cover utilization. One method of classifying cover utilization may be based on the frequency of use. Many mammals utilize cover on a continuous or daily basis. Others use dens or cover only when caring for young, while many with precocious young never use cover or refuges. There are then three refuge-utilization categories: (1) continuous refuge-users, where the refuge also contains the required resources for forage, (2) tempo-

rary refuge-users where foraging may (also) occur elsewhere, and (3) free-ranging types without consistent use of protective refuges.

(c) Range exclusivity. This term is used to avoid the connotations inherent in "territoriality." Thus, it is possible to say that an individual's range is used either exclusively by that individual, or a cohesive group of individuals, (e.g. vicuna), or the range is non-exclusive and other individuals (or other groups) also have access to the resources within the range (e.g. Barbary macaques). If a temporal component is included there are: (i) non-exclusive ranges, (ii) temporarily-exclusive ranges, and (iii) permanently-exclusive ranges.

If the principal system variables describing a social structure are identified and if the functions of the system are defined, the next appropriate step is to attempt to determine the relations between the external environmental variables and the constraining species parameters which give the society its structure (Fig. 1)

SPECIES CHARACTERISTICS

Comparative analysis of social structures often suggest a phyletic ordering. For example, among mammals the simple system of isolated females living in more or less exclusive yet contiguous ranges, themselves distributed within the larger ranges of equally exclusive males is frequent among forms, such as the Insectivores, commonly considered to be primitive. The more elaborate organizations of monkey troops or wildebeeste herds appear derived from such simple beginnings. Although the direction of change is not always easily examined or determined, wherever a range of social systems is found within a taxon an adaptive radiation in relation to ecology may be suspected (Eisenberg 1966; Eisenberg, Muckinhirn & Rudran 1972; Crook 1970; Jarman 1974). Changes in characteristics during cladogenetic divergence may allow new options in social organization to be realized. For example, animals whose size limits mobility may not be able to congregate into large groups since it may not be possible for them to forage over the large distances that may be required once such large groups form. An increase in body size however releases the species from such a constraint. In this sense new "options" in social evolution become possible. Likewise structures maintained through the practice of elaborate social skills are unlikely to be found among mammals which possess relatively primitive brains.

We argue here that both gross morphological features such as body size and more subtle features such as brain neurology related to learning capacity and social flexibility may often be more resistant to evolutionary change than many characteristics of behaviour and relationship. This is because it seems likely that stabilizing selection on a genetically determined trait will tend to make it conservative and more resistant to change under selection pressure than characteristics controlled by information of more diverse provenance. The rates of evolutionary change probably differ for different classes of character. We suspect, with Struhsaker (1969), that traits with long phylogenetic history in a species are likely to act as constraints on changes within systems of which they are a part. Furthermore, in cases of societal convergence the behaviour mediating social organization is likely to show differences providing indications of phylogenetic sources.

In the organization of functions such as rearing or resource-exploitation a number of species morphological and physiological characteristics may limit the kinds of strategies and social structures that might be developed. For instance, an important constraint operating on the form of a rearing strategy may be the maturation rate of the young. We suspect that maturation rate is likely to be a more conservative trait than other characteristics related to it. Thus among species where maturation is slow the duration of the mother-infant bond is, of necessity, lengthy. However, we admit that in the examination of behavioural correlations the attribution of dependency to a variable is often difficult since it is rarely clear which characteristic is the more resistant to change. For example, where feeding and caring for the young so occupy the mother as to make them both susceptible to predators, or where the female alone is unable to provide sufficient food for her offspring, the male may remain with his mate or mates to assist in rearing. While the adaptiveness of the correlation seems clear, the causal relation between such traits (as to which may necessitate or facilitate the other) has rarely been resolved in thorough investigations.

Admitting these difficulties, which arise in part from the current state of knowledge, we would none the less like to propose a preliminary list of some species parameters which appear to act as constraints on the kinds of social structure that a species may adopt. These are (1) The duration of the nutritional dependency of young upon adults (i.e. the maturation rates, as an extension of this the duration of psychosocial dependency in socially complex mammals such as carnivores and primates may also need consideration. (2) The susceptibility of the species to predation. (3) The mobility potential of the species in terms of its capacity for ranging in given habitats. (4) The feeding and foraging behaviour of the species (e.g. digging and dental equipment and use, etc.).

These species characteristics are intermediate parameters in that they are the consequences of specific combinations of more basic characteristics. For example susceptibility to predation depends upon basic species characteristics such as size, speed, and weaponry. Ultimately, these basic characteristics must be considered as the crucial species parameters; however, for the present purposes, the intermediate characteristics are a useful preliminary categorization for the development of our argument.

EXTERNAL (ENVIRONMENTAL) VARIABLES
The constraints imposed by the characteristics of the species on the range of social "options" open to it interact with a number of environmental variables to determine the actual social structure. These environmental

variables include: (1) Resource density (resources including all forms of food, water, cover, etc.); (2) the temporal distribution of resources; (3) the spatial distribution of resources, and (4) the density, distribution and mode of hunting of relevant predators in the environment.

Among recent papers Hamilton & Watt (1970) have discussed in particular the effect on resource exploitation of the adoption of a regular refuge. Crook (1972) analyses the possible relations between the social organizations of terrestrial primates and spatial and temporal fluctuations in the dispersion of patchily distributed food. Jarman (1974) relates different types of grazing and browsing requirements to ungulate social systems. Kleiman & Eisenberg (1973) discuss the socio-ecology of carnivores and Ward & Zahavi (1973) show how refuging combined with gregarious behaviour and a "dispersal system" could operate in information transfer, facilitating effective exploitation of limited resources by birds. Clutton-Brock (1974) and Richard (1974), studying colobine monkeys and Sifakas respectively, both show how a detailed analysis of the distribution of dietary items allows the formulation of hypotheses at a finer level than that available to Crook & Gartlan in 1966. They also show how such refinement is related to the need for careful empirical studies in the formulation and testing of sophisticated models.

Up to this point we have suggested that our three sub-systems operate within the social system and have proposed first a list of some principal social system variables by which the society can be described and second a list of species and environmental variables which may be important determining factors. We now propose to relate these three sets of variables into a classification scheme of mammal social systems which, of necessity, involves assumptions about what appear to be the important relationships determining the overall form. The scheme (Table 2) is derived from an extensive survey of information available on mammals. Our scheme uses the differences in rearing strategies as a prime criterion for distinguishing sociotypes. This is because this sub-system seems particularly conservative, less flexible and least directly responsive to environmental fluctuation. This is probably because of the high level of innate characteristics involved in its determination. We discuss each functional strategy in turn.

Rearing Strategies

Male's presence in the rearing group. The survey of mammal social systems reveals the great importance of the rearing conditions in determining the kind of system that evolves. The basic rearing unit is that of mother-infant (M-I). Since only the mother can provide the milk essential for the young mammal's early survival and growth, this bond is universal in mammalian societies. We may postulate that the male is acceptable to the female and will stay with this basic unit through gestation and rearing only if his presence significantly enhances the survival of their offspring. This seems likely, since merely to mate the male probably needs to stay with the female for only a short time. Hence, we can assume that in general the male staying with his mate and progeny is a rearing strategy; the case that this is a mating strategy being much weaker (Gross-Custard, Dunbar & Aldrich-Blake 1972).

There are a number of ways in which the presence of the male may aid the survival of his young, the particular form of this advantage depending on species characteristics and ecology. Being relatively free from directly caring for the young, he can devote time and energy to protecting resources within an exclusive feeding range and preventing other animals taking the food his own young may find, actually providing food for the offspring (canids), looking out for danger, and in some cases, taking necessary defensive steps. There seems to be little reason why the male's familiarity with an area should exceed that of the female, so this is not necessarily a factor promoting male presence in the rearing group. In one-male units of the hamadryas baboon, however, it is possible that the males pay greater attention to topographical changes than do their dependents so that the life experience of old males may be of especial value to a group. In this species males appear to determine the routes taken in foraging more than in the gelada baboon where older females play an important role (Kummer 1968; Dunbar 1974). In this latter species the prime significance of the male to the reproductive unit appears to be in relation to reducing the frequency of aggression between females and in preventing a high frequency of incursions from non-reproductive males. In addition, he probably performs functions in protecting the young from possible predators.

Within multi-male groups of animals such as baboons where males do not play a role in bringing food for the young, they may none the less be important in child care. While it remains uncertain how far the hierarchy of the males regulate differential access to females in oestrus, it seems certain that the collective policing of the group maintains its stability in the face of recurrent quarrels. Furthermore, the males show collective care of young irrespective of paternity in relation to both fights within the group and to predation risks. Collective caring has presumably arisen due to the close genealogical relationship of the males concerned, and the operation of kin selection. The actual occurrence of large multi-male groups must be considered in relation to ecology and resource distribution (see below). The association of several adult males in a hierarchy none the less expresses an important adaptation ensuring care during rearing through collective male operations. This type of association is therefore to an important degree concerned with the rearing strategies of the group members. The agonistic buffering that occurs within Barbary Macaques whereby relation between males are regulated, and antagonistic encounters within the group probably reduced, may be an important extension of this trait. (See Crook 1975; Deag & Cook 1971).

Duration of the M-I bond. The time period during which the young are nutritionally dependent is likely to determine the length of the particular mother-infant bonds, as well as the male-female bonds if these exist. The bonds may be severed relatively early if the young become independent at weaning or later if this does not happen until the birth of the next young. In some cases, bonds may last over several seasons if the maturation rate of the young is very slow or if the dependency periods of young of several females overlap one another in time. We may assume that the adult-young bond will be maintained until the presence of the young significantly inhibits the survival of the adult or subsequent offspring or until the young male becomes a reproductive rival. (See Trivers 1972).

We have felt that the most illuminating way to classify rearing strategies is to specify four classes (Table 2) based on (i) whether or not the male stays with his females and young, (ii) the duration of his presence, and (iii) the length of the female-young bond.

In type I the male does not stay with his female and young, but the mother-infant bond may be short or long-lived. In type II the males and females probably stay together but no particular bond is involved; rather, all individuals consort in the same group. We cannot in this case rule out benefits for the young accruing from the male's presence, but no specific relationships appear involved. Hence, we use the term "association" to describe this state of affairs.

Male-female bonds lasting for one breeding season are classified in type III and in all cases each male has more than one female.

In type IV, the male-female bond lasts for more than one season and appears to evolve where the period of dependency of the young exceeds one season. The ratio of females to males varies greatly in this class which includes single male-single female pairs, stable one-male reproductive groups (i.e. "harems") and the more-or-less stable multimale groups found in many primates (Crook & Gartlan 1966; Crook 1972). This type includes species whose male-female relationships within their stable groups are the most long-lasting and it is perhaps the most behaviourally complex of all the four rearing groups outlined in the table.

Mating Strategies

If males were completely free from rearing their young, we might generally expect each to pursue a promiscuous strategy since this would maximize the number of females that a male fertilizes. Males that behaved this way would contribute disproportionately to the gene-pool of succeeding generations. Strong intra- and inter-sexual selection would then be expected to operate in this vital and highly competitive activity (e.g. Darwin 1871; Fisher 1929; Huxley 1938a, b; Smith 1962; Goss-Custard et al. 1972; Crook 1972, 1975; Trivers 1972).

The form that the mating sub-system takes varies considerably and probably relates to the species characteristics and ecology. The important species characteristic seems to be whether or not the males contribute to rearing the young. In type I, for example, where the male does not help rear the young and where immobile young in a nest or a food cache limits the freedom of movement of the females, males may take a territory which overlaps as many female ranges as possible (Goss-Custard et al. 1972; Crook, in press). Here the defence of an area may ensure the male more exclusive access to the females contained therein. In these cases, the males that are unable to obtain a territory within the area most utilized by females are excluded and live in peripheral areas with reduced access to females (Charles-Dominique & Martin 1972). In another case, males may sometimes take territories that have a continuous throughput of wandering females, as in the hartebeest (Gosling 1974). Here the territory may ensure its owner undisturbed access to as many females as possible.

In the case of the Uganda kob (Buechner 1961) the males come together and display in locally concentrated territories of small extent resembling the arenas of lekking birds such as ruff or sage-grouse (Hogan-Warburg 1966; Wiley 1973). These "leks" are situated in areas used by the females for foraging.

In herding type II species, no direct male help occurs and inter- and intra-sexual selection is likely to be intense. Males are likely to differ markedly in the numbers of females they fertilize and certainly, very striking weapons and displays have also evolved in some of these animals.

In type III the males stay with their females and young in general, we think, to contribute to their survival, although the ways in which this is achieved may not always be obvious. An alternative case could be made that since the male cannot predict when a female will come into oestrus, the male's best strategy might be to collect every available female for himself. However, as pointed out by Goss-Custard et al. (1972) there are many problems with this view, so that interpreting the long-term bond between males and females as a mating strategy is not convincing.

In situations where the male can contribute to the survival of several young from several females we might expect long-lasting harems to develop, as they have done commonly in the class III and IV species. In these cases, inter- and intra-sexual selection is again likely to be strong, since by mating with more females, a male can contribute disproportionately to the genepool of succeeding generations.

Grouping and Dispersion strategies

Resources and their effects on groupings

(i) *Refuge Utilization.* Anything which forces the animals to forage out radially from some more-or-less fixed point or refuge seems to have a great influence on social structure (Hamilton & Watt 1970). The reasons for having such a fixed site are numerous and depend on the animal's ecology and characteristics. Undoubtedly, all animals benefit from accumulating knowledge of the local area so that staying in a suitable place, once discovered, may be an advantage (Kummer 1971), but returning regularly to the same site imposes important constraints. Thus, Smight (1968) proposes that tree squirrels live solitarily because individuals place their food in a cache. Were pairs or larger groups to evolve, much greater areas would have to be searched for food with a consequent increase in the amount of energy expended in foraging.

In other cases, immobile young may require a nest, or a safe resting place may impose a locus on daily activity. In some cases, a locus of this kind might evolve as a means by which an individual may discover the whereabouts of undiscovered food from the experience of its conspecifics. This "information centre" idea has been discussed in relation to birds (Ward & Zahavi 1973) and may also apply to wide-ranging socially refuging mammal species such as bats that utilize a patchy and fluctuating food supply.

In cases such as this the primary function of social refuging may lie in information exchange concerned with reducing energy expenditure in the daily search for food in foraging. The protective site that is actually used may then evolve secondarily as a necessary adaptation to increased predation that results from the attraction of predators to a large assemblage of potential prey.

Selection favouring congregational behaviour for information exchange (whether at a protective site necessary for sleep or brood care or periodically at non-protected sites, e.g. water holes, etc.) is likely to arise when resources are heterogeneous and dispersed patchily both spatially and temporally. Congregation at sleeping sites generates a "dispersal system" (Crook 1953) centred upon the refuge itself. In general this results in a heavily exploited zone nearest the locus and a gradient in the intensity of resource-exploitation that diminishes with distance. One might expect that on the average an individual or group foraging afar may be as able to replenish its energy expenditure of foraging as an individual or group foraging near the locus, finding less food but expending less energy in travelling there to collect it. Seasonal fluctuations in the overall abundance of nourishment may, however, periodically impose so great a range on individuals that energy expenditure exceeds the returns. Such a system must then break and an alternative strategy such as nomadism may arise (see Hamilton & Watt 1970; Crook 1972).

(ii) *Range Exclusivity.* Brown (1964) introduced an important idea concerning the circumstances in which exclusive use of areas will occur. He proposed that, assuming competition for some limited resources exists, it will be an advantage to defend an area only if an adequate supply of that resource (usually food) is found in an area small enough to defend economically. This may occur under a number of circumstances of food density and distribution. For instance, a food supply that is relatively dense yet highly clumped may allow exclusive ranges to evolve if suitable food patches always occur close enough together for the individual (or group) to have at least one available throughout a period long enough to make defence of its range worthwhile. Similarly, relatively evenly dispersed food supplies may be defendable in the same way. On the other hand, clumped food sources that are widely spaced and irregular in occurrence relative to the species' "cruising" range may be simply undefendable as, indeed, may very sparse but more evenly dispersed supplies. The outcome seems likely to depend on the relationship between the overall density and dispersion of the resources on the one hand and the ranging abilities of the animal in question on the other. In general, defendability is likely to correlate with more-or-less "evenly-dispersed" resources whereas more-or-less spatially and temporally "heterogeneous" supplies are unlikely to be defended. The terms "homogeneous" and "heterogeneous" are used in this article although we recognize that they don't fully express the condition under which range exclusivity and overlap respectively are likely to evolve.

In free-ranging species we may assume that there will be a selective advantage in an individual restricting its activities to a limited home range because of the advantages accruing from acquaintance with an area (e.g. sites of resources and places particularly dangerous from predators). Since the animals have to locate food and water continually, individuals utilizing heterogeneous resources have to range over an area large enough to contain available resources, at least in some regions, at all times. This might necessitate huge ranges in some highly mobile species that do not have a refuge-based foraging system. Since many individuals may be utilizing the same area, the possibility for large associations (e.g. herds of bison) exists if there is a selective pressure, such as predation or a need for more efficient food utilization (Ellis, in preparation (b)), which favours it.

Hence we assume that, where possible, animals maintain exclusive ranges to maintain a competed resource for themselves and that non-exclusive ranges result from the necessity of sharing resources that are uneconomical or impossible to defend. If there are major differences in the distribution of resources over time (seasonally) or in the distribution of two or more resources in space relative to the "cruising range" of the species, range exclusivity is unlikely unless the resources are effectively homogeneous for periods of time sufficient for mating and rearing. In the latter case, seasonal exclusivity may be expected. So, in general, the major subdivision of range use into exclusive and non-exclusive ranges seems to be related to the nature and distribution of the resources, as it does in birds (Crook 1965).

(iii) *Group size and stability.* Contrasts in resource distribution illustrate the way in which species ecology may influence the size and stability of a social group. A male may be able to improve the chances of survival of his young if he defends an exclusive area in which some competed resource is thereby maintained for just his offspring and their mother(s). The resource density in the area that a male can defend economically may then determine the size of the group. In some cases, the male may be able to defend a range which provides sufficient resources for just one female and young (e.g. gibbons, Ellefson 1968), while in others it may be possible for him to provide for several (e.g. blue monkey, Aldrich-Blake 1970). In contrast, a patchily distributed resource where large individual ranges are necessary may result in the common use of a range by several mating groups which together form large herds.

If we assume that group stability depends on resource stability (Estes 1969) and that the number of animals in a particular area depends largely on the density of resources in that area, four broad categories of group size and stability can be identified. If resource distribution changes with time or varies spatially, unstable groups are likely. If the resource density is high, large, unstable groups (e.g. gelada baboon, Crook 1966) are likely; if not, small ones may evolve (e.g. chimpanzee, Reynolds & Reynolds 1965). With temporal and spatial homogeneity of resources, stable groups of any composition may evolve. Again, group size is likely to correlate with resource density.

Predation and its effect on grouping

(i) *Refuge Use.* The extent and kind of predation in conjunction with the possibility of adopting antipredator strategies other than through social means is likely to have a considerable influence on social systems. It can be assumed that animals particularly susceptible to predation will use cover continuously, or at least sleep in some form of refuge, but those which can readily avoid predation by fleeing or fighting will not seek refuge. The species parameter assumed to determine refuge use is the susceptibility of adults and young to predation. Hence, if the adults require cover to avoid predation, continuous or frequent cover and perhaps refuge use would be expected, as occurs in many small, ground living animals. If only the young require cover, temporary refuging during rearing would be expected. If neither adults nor young require cover, complete free-ranging becomes possible.

(ii) *Group composition and size.* One reason for a male staying with his females and young is that his presence may increase their safety. Whether or not this is likely will depend on basic characteristics of the animal in relation to the hunting characteristics of the predators. For instance, highly mobile patas monkeys in relatively treeless savannah pursued by fast-running predators may benefit from the male adopting the role of look-out (Hall 1965). However, small animals that rely on a nocturnal existence under dense cover may not benefit in this way. In these cases, the presence of the male would not add anything to the survival chances of the female and young, which are probably hidden in a camouflaged nest anyway.

The possibility of active defence may also favour the presence of males, e.g. chimpanzees, baboons. In other cases, large associations of the herd kind may be effective in avoiding predation in situations in which large numbers of animals range widely over huge areas (e.g. bison). In these cases, large associations in which males do not directly help only their own young, may be more effective than smaller, more vulnerable, family or harem groups. The ways by which temporal and spatial grouping may reduce predation risk are discussed for example by Hamilton (1971), Goss-Custard et al. 1972), Crook (1972), Vine (1971), Lazarus (1972), and with reference to the formation of polyspecific associations, by Gartlan & Struhsaker (1972). In general the presence of conspecifics provides both a type of cover reducing the probability of an individual's capture and also increased collective vigilance. Synchronization of births has the same effect of survival of individual young.

TOWARDS SOCIO-ECOLOGICAL MODELS BASED ON DETAILED FIELD STUDY

The environmental conditions determining shifts from one-male to multi-male reproductive groups in primates have been discussed in formulation of Crook's (1972) simple models which attempt to account for the socio-ecological relations underlying these types of grouping. Recent studies illustrate (i) the number and subtlety of the factors which need to be considered in field work if advances in theory are to be made, (ii) the complex nature of the interaction between such factors and their variability between species and populations of the same species, (iii) the refined nature of the effects of ecological contrasts on social behaviour and (iv) the fact that, rather than simple correlations of traits, we are now comparing systems.

In a specific case, Clutton-Brock's (1974) comparison of two African *Colobus* species indicates the subtle habitat contrasts that possibly underlie societal differences. Black and white colobus seem able to survive on a diet of mature leaves when only these are available. They can thus live in dry forest areas with marked seasonality and are able to survive on the products of a small number of species providing an acceptable diet throughout the year. Relatively small areas of forest could support a considerable number of animals. The small troop size of this species may permit it to minimize ranging and thus to increase the defensibility of the range through some form of territorial behaviour. The reduction of adult male reproductives to one, may express the extreme sexual competition within such small groups and, furthermore, have adaptive consequences in making a high proportion of the food available to females and young (Crook 1972).

The red colobus, by contrast, evidently requires a high proportion of shoots, flowers and fruit throughout the year. The species is found in wetter forests where suitable food becomes available on different trees sporadically in space and time. A large range is required to provide sufficient supplies of acceptable food all the year. This range size is unlikely to allow easy territorial defence so that larger wide ranging groups are preferred. Furthermore these may increase protection from predators. The presence of several male reproductives will tend to favour group defence, as well as providing additional strategic functions (see ablove). The colobus story will however, need further extension since the sub-species of red colobus living in the Tana river valley is now known to have one-male groups living in small territories. Clive Marsh (personal communication) believes that these also are explicable in terms of contrasting patterns of dietary availability.

Richard (1974) in her *Propithecus* study in Madagascar compares arid and humid forest populations of the same lemur species. Her findings and her explanatory hypotheses resemble those of Clutton-Brock strikingly in form. Her arid country animals showed range exclusivity and group territorial behaviour while her forest animals showed considerable overlap in group ranging. She suggests that the latter is associated with little periodic variation in abundance while in the former there is a greater seasonal fluctuation in overall abundance even though items are similarly dispersed. In this case a strategy of range reservation, even when food is not lacking, provides adequate resources later when stocks are low. In the first case the dispersion of dietary items necessitates range overlap and, while groups tend to avoid one another, range exclusion has not developed.

Richard adds, however, that she attributes some variance between her groups to a randomizing effect of contrasting traditions of local behaviour and remains unsure how far the observed contrasts might not be accounted for in this way. She was unable to correlate important components of behavioural variance with the ecological variables she examined. Under what conditions such behavioural randomization would obscure valid socio-ecological correlations or generate societal structures unrelated to underlying ecology remains to be determined. It seems unlikely, however, that traditions of societal functioning will prove any more independent of ecology than did the apparently "arbitrary" cases of ritualization in bird displays studied in the 1930s. However, the complexity of the interacting determinants in these societies may render clear explanation difficult to obtain.

DISCUSSION

Early attempts to relate interspecific differences in social organization in birds and mammals to ecology were initially successful because the choice of level at which socio-ecological variance was examined and the categories chosen for comparison were appropriate. Furthermore, the patterns of relationship were often repeated at different taxonomic levels. With more advanced and socially complex mammals, experience has shown that the levels of comparison and the choice of categories need to be more finely defined than was at first possible. With an increase in information available, differences within categories become close to those between categories, thus requiring that boundaries between categories and their number be redefined, so that finer resolution may be obtained. Current studies do not represent a break from earlier comparative work using grosser categories but rather their continuation. Furthermore, now that effective socio-ecological modelling has become possible, comparisons are made more between systems than between sets of statically conceived socio-ecological grades. The method of broad correlation between habitat and society remains the only approach initially available when knowledge of a group of animals is sparse.

Parallel evolution of taxonomically distinct stocks necessarily involves convergent modification of characteristics initially different. The degree of apparent convergence becomes a function of the way in which changes in phylo-genetically separated traits produce likeness. For example, similarities between gelada and hamadryas baboon social structures are mediated by social processes which a detailed analysis shows to be very different in the two cases and doubtless dependent on contrasting phylogenetic inheritance. However, the parallelism at the gross level may remain meaningful enough, the interest simply shifting to the way in which convergent modification has occurred. A focus dwelling only on the details of interaction would obscure the interest of the convergence. Thus gross similarities in structure and ecology may be based on those more subtle interspecific differences in organization and dynamics that may represent, as Clutton-Brock (1974) argues, different methods of overcoming the same or similar ecological problems.

In this article we have treated broad problems of socio-ecology in the Class Mammalia as a whole. Clearly the knowledge available varies enormously from order to order, yet, in spite of the approximations this imposes, repeatable patterns emerge that demand analysis and explanation. It seems possible that socio-ecological systems analysis will provide important clarification of these issues. Even so, in those species in which intra-specific variation between populations is dependent on acquired and traditional behaviours we need to be aware that processes of socio-cultural change may obscure the ecological base. Again, here, only the more sophisticated treatment of the variables involved will identify the extent to which ecology is relegated to a less-directive role in societal evolution.

Naturally, it has not been our intention to review all the existing hypotheses that could relate mammalian social structure to ecology and species characteristics. Rather, we have presented a few generalizations and a classification scheme that we hope will have heuristic value in organizing our knowledge of these systems a little more extensively than has been attempted so far. It is of necessity somewhat abbreviated, but the major social-system types appear to be represented.

We have set forth a number of assumptions about how certain environmental variables and species parameters interact to determine social variables. Consequently, social structures have been generated and related to social functions. We stress that these relationships should be considered as working hypotheses to guide further study. It is beyond the scope of this article to expand on these hypotheses, although one of these has been dealt with elsewhere (Ellis, in preparation). It is obvious that these extremely simple models and working hypotheses will require considerable elaboration based on new quantitative field study before they will be suitable for conversion to mathematical models of mammalian societies. None the less, the work of Hamilton & Watt (1970) and Vine (1971) indicates ways in which this may be attempted, and the recent field studies quoted show the type of data analysis and inference to which such models must be anchored. We feel that the study of mammalian social organizations has progressed to the point where the impact of environmental variables and species parameters on the formation of social structures may be investigated empirically. Progress seems more likely to come from such a research orientation than from a continued emphasis on the more general studies of social organization which have brought us to this point.

ACKNOWLEDGMENTS

Manuscript preparation supported in part by National Science Foundation Grant GB-41233X to the Grassland Biome, U.S. International Biological Program, for "Analysis of Structure, Function, and Utilization of Grassland Ecosystems." We are grateful to Robert Hinde and Martin Daly for critical evaluations of the text in preparation.

REFERENCES

Aldrich-Blake, F. P. G. (1970). The ecology and behaviour of the blue monkey *Cercopithecus mitis stuhlmani*. Ph.D. Dissertation, Bristol, England: Bristol University Library.

Brown, J. L. (1964). The evolution of diversity in avian territorial systems. *Wilson Bull.*, **76,** 160-169.

Buechner, H. K. (1961). Territorial behaviour in Uganda Kob. *Science, N. Y.*, **133,** 698-699.

Charles-Dominique, P. & Martin, R. D. (1972). Behaviour and ecology of nocturnal prosimians. Field Studies in Gabon and Madagascar. *Fortschr. Verhaltensforsch Zugleich Z. Tierpsychol. Beih.* Heft 9. (*Adv. Ethol.* Supplements to *J. Comp. Ethol.* Issue 9) Berlin, West Germany: Paul Parey. 89 pp.

Clutton-Brock, T. H. (1974). Primate social organisation and ecology. *Nature, Lond.*, **250,** 539-542.

Crook, J. H. (1953). An observational study of the gulls of Southampton water. *Br. Birds,* **46,** 385-397.

Crook, J. H. (1964). The evolution of social organisation and visual communication in weaver birds (Ploceinae). *Behaviour Monograph,* **10;** Leiden: Brill.

Crook J. H. (1965). The adaptive significance of avian social organisations. *Symp. Zool. Soc. Lond.,* **14,** 181-218.

Crook J. H. (1966). Gelada baboon herd structure and movement: comparative report. *Symp. Zool. Soc. Lond.,* **18,** 237-258.

Crook J. H. (1970). The socio-ecology of primates. In: *Social Behaviour in Birds and Mammals* (Ed. by J. H. Crook). London: Academic Press.

Crook J. H. (1972). Sexual selection, dimorphism, and social organisation in the primates. In: *Sexual Selection and the Descent of Man* (Ed. by B. G. Campbell). Chicago: Aldine.

Crook J. H. (in press). On the integration of gender strategies in mammalian social systems. In: *The Memorial Symposium for D. Lehrmann* (Ed. by J. Rosenblatt). Rutgers University.

Crook J. H. & Gartlan, S. S. (1966). Evolution of primate societies. *Nature, Lond.,* **210,** 1200-1203.

Crook J. H. & Goss-Custard, J. D. (1972). Social ethology. *Ann. Rev. Psychol.,* **23,** 277-312.

Darwin, C. (1871). *The Descent of Man and Selection in Relation to Sex.* London: John Murray.

Deag, J. M. & Crook, J. H. (1971). Social behaviour and "agonistic buffering" in the wild Barbary macaque. *Macaca sylvana* L. *Folia Primatol.,* **15,** 183-200.

Dunbar, R. (1974). Ph.D. Thesis. Bristol University Library. In press. *Bibliotheca Primatologica.*

Ellefson, J. O. (1968). Territorial behaviour in the common white-handed gibbon, *Hylobates lar* Linn. In: *Primates: Studies in Adaptation and Variability* (Ed. by P. Jay), pp. 180-199. New York: Holt, Rhinehart & Winston.

Ellis, J. E. (In preparation) (a). Systems in primate societies.

Ellis, J. E. (In preparation) (b). The social implications of resource distribution.

Eisenberg, J. F. (1966). The social organisation of mammals. *Handb. Zool.,* **10,** 1-92.

Eisenberg, J. F., Muckinhirn, N. A. & Rudran, R. (1972). The relation between ecology and social structure in primates. *Science, N. Y.,* **176,** 863-874.

Estes, R. D. (1969). Territorial behaviour of the wildebeest (*Connochaetes taurinus* Burchell, 1823). *Z. Tierpsychol.,* **26,** 284-370.

Fisher, R. A. (1929). *The Genetical Theory of Natural Selection.* Oxford: Clarendon Press.

Gartlan, J. S. & Struhsaker, T. (1972). Polyspecific associations and niche separation of rain-forest anthropoids in Cameroon, W. Africa. *J. Zool.*, **168**, 221-265.

Gosling, L. M. (1974). The social ethology of Coke's hartebeest. *Alcelaphus busclaphus cokei.* Gunther. In: *The Behaviour of Ungulates and its Relation to Management.* IUCN.

Goss-Custard, J. D., Dunbar, R. I. M. & Aldrich-Blake, F.P G. (1972). Survival, mating, and rearing strategies in the evolution of primate social structure. *Folia Primatol.*, **17**, 1-19.

Hall, H. R. L. (1965). Behaviour and ecology of the wild patas monkey *Erythrocebus patas* in Uganda. *J. Zool.*, **148**, 15-87.

Hamilton, W. D. (1971). Geometry for the selfish herd. *J Theor. Biol.*, **31**, 295-311.

Hamilton, W. J., III & Watt, K. E. F. (1970). Refuging. *Ann. Rev. Ecol. Syst.*, **1**, 263-286.

Hinde, R. A. & Tinbergen, N. (1958). The comparative study of species-specific behavior. In: *Behavior and Evolution* (Ed. by A. Roe & G. G. Simpson). New Haven, Conn.: Yale University Press.

Hogan-Warburg, A. J. (1966). Social behavior of the ruff, *Philomachus pugnax* L. *Ardea*, **54**, 109-229.

Huxley, J. S. (1938a). The present standing of the theory of sexual selection. In: *Evolution: Essays on Aspects of Evolutionary Biology* (Ed. by G. R. DeBeer). Oxford: The Clarendon Press.

Huxley, J. S. (1938b). Darwin's theory on sexual selection and the data subsumed by it in the light of recent research. *Am. Nat.*, **72**, 416-433.

Jarman, P. J. (1974). The social organisation of antelope in relation to their ecology. *Behaviour*, **48**, 215-267.

Kleiman. D. & Eisenberg, J. F. (1973). Comparisons of canid and felid social systems from an evolutionary perspective. *Anim. Behav.*, **21**, 637-659.

Kummer, H. (1968). Social organization of hamadryas baboons. A field study. *Bibl. Primatol.*, No. 6. Basel: Karger.

Kummer, H. (1971). *Primate Societies.* Chicago: Aldine-Atherton.

Lazarus, J. (1972). Natural selection and the functions of flocking in birds: A reply to Murton. *Ibis*, **114**, 556-558.

Reynolds, V. & Reynolds, F. (1965). Chimpanzees of the Budongo forest. In: *Primate Behavior* (Ed. by I. DeVore), pp. 368-424. New York: Holt, Rinehart & Winston.

Richard, A. (1974). Intra-specific variations in the social organization and ecology of *Propithecus verreauxl. Folia. Primatol.*, **22**, 178-207.

Smith, C. C. (1968). The adaptive nature of social organization in the genus of tree squirrels *Tamiasciurus. Ecol. Monogr.*, **38**, 31-63.

Smith J. M. (1962). Disruptive selection, polymorphism, and sympatric speciation. *Nature, Lond.*, **195**, 60.

Struhsaker, T. T. (1969). Correlates of ecology and social organisation among African cercopithecines. *Folia Primatol.,* **11,** 86-118.

Trivers. R. (1972). Parental investment and sexual selection. In: *Sexual Selection and the Descent of Man* (Ed. by B. Campbell). Chicago: aldine.

Vine, I. (1971). The risk of visual detection and pursuit by a predator and the selective advantage of flocking behaviour. *J. Theor. Biol.,* **30,** 405-422.

Ward, P. & Zahavi, A. (1973). Importance of certain assemblages of birds as information centers for food finding. *Ibis,* **115,** 517-534.

Wiley, R. H. (1973). Territoriality and non-random mating in sage grouse *Centrocercus urophasianus. Anim. Behav. Monogr.*

28

Nectar-Feeding by Prosimians and its Evolutionary and Ecological Implications

ROBERT W. SUSSMAN

INTRODUCTION

During July and August of 1973 in Ampijoroa, northwestern Madagascar, I and two other participants in this article session (McGeorge and Tattersall) observed *Lemur mongoz* (= *L.m. mongoz*) behaving in an unexpected manner. In this forest, the animals were nocturnal (although previously reported to be diurnal like other species of the genus *Lemur*), but even more unexpected was their feeding behaviour. Approximately 80% of the observed feeding behaviour of *L. mongoz* during this study was on the nectar-producing parts of four plants and 80% of this was on the nectar of the kapok tree (*Ceiba pentandra*) (see Tattersall and Sussman, 1975). In June, 1974, I returned to Ampijoroa and found *L. mongoz* still behaving in this unusual (at least for a primate) manner. In this article, I will discuss the implications of this behaviour as it relates to inter-taxa interactions and to the ecological concepts of foraging guilds and co-evolution.

CEIBA PENTANDRA: A BAT-ADAPTED TREE

In Asia, Africa, and North and South America, the kapok tree is pollinated by plant-visiting bats and is adapted to attract these pollinating agents. The plant attracts bats with distinctive, easily accessible flowers and nutritious nectar. The bats then aid the plant in reproduction.

Reprinted with permission from *Recent Advances in Primatology*, Vol. 2. EVOLUTION by D.J. Chivers and C.M. Hladik, Editors, in press. Copyright by Academic Press Inc. (London) Ltd.

The kapok tree has coevolved with bats in a number of ways (Baker and Harris, 1959; Baker, 1965). The plants lose their leaves in the dry season before flowering occurs and leaflessness is prolonged until after the distribution of seeds. This not only allows free dispersal of seeds but also allows access to the flowers by flying animals larger than insects. The majority of flowers are produced, in dense clusters, on the terminal branches of the trees in the canopy. Flowering time is regular, with flowers opening about half an hour after dusk and remaining open during the night. Flowers produce a strong scent. The following day the corolla and stamens fall. Each flower produces five stamens which stand well above the corolla, and the pollen grains are sticky due to the secretion of oil drops in the anthers. Nectar is secreted at the base of the flower before it opens and disturbance of a cluster of flowers by wind or by visitors produces a rain of nectar when the flowers are open. Analyses of the nutritional content of the nectar of *Ceiba pentandra* indicate that it contains some sugars and a relatively large quantity of amino acids, with seven different amino acids represented (Baker and Baker, pers. comm.; see Sussman and Tattersall, 1977). Kapok trees are self-compatible as well as cross-pollinating.

There is only one plant-visiting bat in northwestern Madagascar, *Pteropus rufus*. This bat, however, is often destructive to *Ceiba pentandra* flowers and has not been considered a legitimate pollinator of this species. In this area of Madagascar, *Lemur mongoz* may fill a niche usually occupied by plant-visiting bats. It may serve as the major pollinator of the kapok tree and possibly other bat-adapted plants. As Tattersall (in press) reports, however, the behaviour of *L. mongoz* is variable. In some areas *L. mongoz* is diurnal and, I suspect, has a very different diet.

NON-FLYING MAMMAL POLLINATORS

Rodents and marsupials

Whereas bats and birds are common vertebrate pollinators in the tropics, there are very few reports of pollination by non-flying mammals. A number of Australian marsupials and one Australian rat, *Rattus fiscipes*, are known to visit flowers for nectar and/or pollen (Rourke and Wiens, 1977) and the honey possum (*Tarsipes spencerae*) is a very specialized nectar feeder (Glauert, 1958; Vose, 1973). In Ceylon, the tree squirrel, *Funambulus palmarum* is known to feed on the nectar of *Grevillea* and possibly *Ceiba* (Phillips, 1935; Walker, 1968). The interactions between these mammals and the plants they visit, or the indigenous species of birds and bats, have not been studied.

In Hawaii, introduced species of rats are known pollinators of *Freycinetia arborea*. In Asia, *Freycinetia* is pollinated by bats (van der Pijl, 1956) and shows many characteristics of a bat-pollinated tree (Faegri and van der Pijl, 1971). However, in Hawaii there are no plant-visiting bats and before the rats were introduced into Hawaii, *F. arborea* developed some ornithophilous traits and was probably pollinated by birds. It is likely that the retention of bat associated characters pre-adapted *F. arborea* to pollination by rats, which are highly aggressive and out-competed birds as pollinators (Rourke and Wiens, 1977).

Primates

The role of primates as potential pollinators has not been extensively studied, although a number of observations of flower-visiting by primates have been mentioned in the literature. Many primates eat blossoms or parts of blossoms but, in most cases, this is a very small part of their diet and the effect on the flower is destructive. Some of the nocturnal prosimians, however, have been seen to feed on nectar without destroying the flowers. In these primate species, nectar may provide important sources of nutrients and the plant may benefit from the interaction.

In addition to *Lemur mongoz*, *Microcebus murinus* were observed at Ampijoroa licking nectar from the kapok tree without destroying the flowers. In southern Madagascar, Martin (1972) observed the mouse lemur feeding on flowers of *Vaccinium emirnense* and of *Uapaca* trees. In western Madagascar, *Phanar furcifer* have been observed spending long periods of time licking cluster of flowers on the finest terminal branches of *Crateva greveana* trees. They also spend considerable time near the buds on the terminal branches of baobabs (*Adansonia*) (Petter et al., 1971). Outside of Madagascar, both of these species of plant are normally bat-pollinated. At Morondava, in western Madagascar, nectar from flowers of an as yet unidentified species of Caesalpiniacae is the main food of *Cheirogaleus medius* during the beginning of the rainy season (Hladik, pers. comm.).

In the arid regions of southern Madagascar, during the driest part of the year, *Lepilemur* feeds mainly on the flowers of two endemic species of *Alluaudia*: *A. ascendens* and *A. procera* (Charles-Dominique and Hladik, 1971). Charles-Dominique and Hladik remark that it is the successive flowering of the two species that permits *Lepilemur* to survive the severest portion of the dry season. However, *Lepilemur* is a predator upon the flowers and these plants are normally insect pollinated (Hladik, pers. comm.).

The only description of a Lorisidae acting as a possible pollinating agent is that of Coe and Isaac (1965). They observed *Galago crassicaudatus* in East Africa visiting baobab trees and feeding on newly-opened flowers. The animals moved from flower to flower burying their faces within the flowers and around the sepals. The bushbabies lick the flowers, causing only superficial damage to them.

From these few accounts, it seems likely that nectar is an important potential source of nutrition for nocturnal primates. As more detailed observations of feeding behaviour are made on these animals, more examples of nectar feeding may come to light. However, in most of the cases of nectar feeding described, the primates are visiting plants that are normally bat-pollinated.

In areas where plant-visiting bats are rare or absent, the primates may be partially filling a void niche and the plants are able to take advantage of a potential pollinator. The adaptations of the plant to attract bats are sufficient to attract nocturnal primates, although I think that primates are not as adept at exploiting these resources as bats, nor are they ideal pollinating agents. It seems unlikely, but not impossible, that coevolution between plants and primates-as-pollinators has occurred. Much as *Daubentonia* fills in for woodpeckers in Madagascar (Cartmill, 1974), many of the nocturnal prosimians may be occupying (at least to some extent) the plant-visiting bat niche where bats are rare or absent.

EVOLUTIONARY AND ECOLOGICAL IMPLICATIONS

The role of primates in pollination ecology has brought to mind a number of related questions and problems which, I believe, have been relatively ignored in the primate literature. The first of these is the interaction between species of the major vertebrate taxa using the canopy of the tropical forest (e.g. birds, bats, squirrels, and primates). These interactions probably are among the major factors shaping present ecological adaptations and were extremely important in the evolution of the particular adaptive trends of these taxa. Related to this is the problem of defining major varieties of feeding and foraging adaptations of the primates and how these adaptive groups are apportioned in different tropical plant communities. Finally, to what extent is coevolution occurring between primates and the plants that they ulitize? For the remainder of this article, I would like to discuss these issues. The discussion will, because of time and length restrictions, be brief and, mostly, speculative. But my hope is, in the spirit of scientific inquiry, to throw out some ideas for the purpose of further thought and discussion and, perhaps more ambitiously, to stimulate further research in these areas.

Those vertebrate taxa that have the most species represented in the tropical forest canopy and that are most likely to be utilizing resources similar to those used by primates are bats, birds, and tree squirrels. All bats are nocturnal and most birds are diurnal. Although most tropical forest-dwelling mammals are nocturnal, most primates and all tree squirrels, except flying squirrels, are diurnal (see Charles-Dominique, 1975). Bats and birds avoid competition for many of their resources (especially insects and other prey, and nectar) simply by these differences in activity pattern. It seems likely that bats and nocturnal prosimians have been major or potential competitors since the Eocene. The success of bats in the tropics and the relative lack of success of prosimians could very well be related. Existing nocturnal prosimians may use resources not normally exploited by bats, including gums and insect larva, as well as crawling insects, and catch much of their prey in the dense vegetation of the bush and canopy where it is difficult for bats to navigate. I believe more detailed investigations into bat and nocturnal prosimian interactions are warranted (and I wonder what *Aotus* is doing).

Besides basic differences in cycles of activity, there are patterns of resource partitioning and habitat selection within and between major taxonomic groups. Animals with parallel or convergent roles in similarly structured plant communities often share foraging and dietary adaptations, as well as certain general morphological adaptations. Groups of animals filling similar roles have been called "guilds."* In each foraging guild are animals with adaptations that allow them to search, ingest, and digest certain types of available food items and that often share certain locomotor, sensory, dental, enzymic and gastrointestinal adaptations. Once foraging guilds are defined in primates, it may be possible then to attempt to relate various aspects of morphology, locomotion and individual and social behaviour to these dietary and foraging patterns. We expect to find some of these factors more similar and others more variable between members of similar guilds. A study of the relationships between similar guilds across taxonomic boundaries may also raise some interesting ecological and evolutionary questions.

Most research in this area has been done on birds (Root, 1967; Orians, 1969; Snow, 1971; Morton, 1973; Pearson, 1975; McKey, 1975) with a few papers on bats (Tamsitt, 1967; McNab, 1971; Fleming et al., 1972; Heithaus et al., 1975). In Table 1, I list nine foraging guilds of tropical forest birds (from McKey, 1975, and Pearson, 1975). The majority of bird species inhabiting the tropical forest canopy are either exclusively insectivorous or partially frugivorous (usually feeding the young on insects and utilizing fruits heavily as adults). A very small percentage feed entirely on fruit (McKey, 1975; Pearson, 1975). The morphology, size, and behaviour of birds utilizing these resources is strongly determined by the particular type of foliage structure in which they typically forage. Furthermore, the prey species exploited by specific guilds have coevolved mechanisms to minimize predation or, in the case of plant species, to maximize efficiency of seed dispersal (Smythe, 1970; Snow, 1971; Morton, 1973; McKey, 1975). For example, fruits used by partially frugivorous birds usually attract a number of opportunistic dispersal agents, contain a great mass of small seeds, are sources of water and carbohydrates, and have short, displaced fruiting seasons. Fruits

---

* "*A guild* is defined as a group of species that exploit the same class of environmental resources in a similar way. This term groups together species, without regard to taxonomic position, that overlap significantly in their niche requirements. The guild has a position comparable in the classification of exploitation patterns to the genus in phylogenetic schemes. . . . One advantage of the guild concept is that it focuses attention on all sympatric species involved in a competitive interation, regardless of their taxonomic relationship." (Root, 1967, p. 335).

attracting specialized frugivorous birds are typically large, have one large seed, supply the birds with most of their lipids and proteins, as well as carbohydrates, and have fruiting seasons spread over a long period (McKey, 1975). Species of trees utilizing the former seed dispersal strategy are most often found in the understory, in open vegetation areas or on the forest edge; those with the latter strategy are usually associated with the climax plant association in the canopy and emergent layers of forest (Snow, 1971; McKey, 1975). The birds utilizing these two types of fruits make up two frugivorous guilds differing in size, morphology, physiology, and behaviour.

Bats, squirrels, and primates also utilize these fruits and are seed dispersal agents, but specific information on fruit usage and seed dispersal in these animals is not extensive (however, see van der Pijl, 1957; Hladik and Hladik, 1967, 1969; Heithaus et al., 1975; and Smith, 1975). Fruit-eating guilds of birds, bats, squirrels and primates probably overlap in various ways. Pearson (1975), for example, found the proportion of fruit-eating birds, in two South American forests, to be inversely correlated with monkey population size, excluding the mainly leaf-eating species, *Alouatta seniculus*. He states that: "In general, it appears that a certain biomass of monkeys will offset a certain biomass of birds even though no effect can be detected at the level of species richness" (p. 464).

It is likely that leaf-eating is one of the major means by which diurnal primates have avoided extensive resource overlap and competition with birds and arboreal rodents. Besides a few specialized rodents, marsupials and edentates, primates are the only major group of vertebrates that feed on leaves in the forest canopy. Although some folivory may be beneficial to plants (Oppenheimer and Lang, 1969), in most cases it is detrimental (Ehrlich and Raven, 1964; Freeland and Janzen, 1974; Levin, 1976). Just as coevolution has occurred between plants, their pollinators and seed dispersal agents, mechanisms for protection from leaf predators and the strategies

Table 1
Foraging Guilds of Tropical Birds

Mainly insect foraging
Glean insects
Sally (both bird and prey on wing)
Snatch (bird on wing, prey not)
Peck and probe
Glean and sally
Glean and snatch
Ant follower
Mainly fruit foraging
"Opportunistic" frugivore
"Specialized" frugivore

of folivory by mammals are interrelated (Janzen, 1970; Freeland and Janzen, 1974; Westoby, 1974; Atsatt and O'Dowd, 1976). Hence, among primates we may expect to find more than one pattern of folivory. For example, some primates are specialized leaf-eaters and have developed specialized gastrointestinal mechanisms for detoxifying plant secondary compounds (for example *Lepilemur,* Hladik et al., 1971, and many colobine monkeys). These species should use large amounts of several related toxic foods that are present in a year-round supply (Freeland and Janzen, 1974). Other, less specialized species of primate may adapt to a folivorous diet by using a particular foraging strategy such as: consuming a variety of plant foods at any one time, treating new foods with caution, ingesting small amounts at first encounter, sampling foods continuously, and feeding on leaves that contain small amounts of secondary compounds (see, for example, Glander, 1975, on *Alouatta seniculus*).

Extensive comparison and categorization of groups of species of primates that exploit the same class of resources in similar ways has not been done. However, recent research is pointing to some very interesting patterns of resource partitioning and utilization, especially in arboreal primates (see, for example, Clutton-Brock, 1975; Hladik, 1975; Struhsaker and Oates, 1975; Sussman, 1977). I believe that further research into primate foraging guilds, inter-taxa interactions, and coevolution between primates and the plants they utilize will lead to a better understanding of primate morphology, behaviour and evolution.

ACKNOWLEDGEMENTS

This research was supported by National Science Foundation Research Grant No. BG - 41109 and Biomedical Research Support Grant RR - 07054 from the Biomedical Research Support Program, Division of Research Resources, National Institute of Health. Funds to attend the Conference were from a Washington University Summer Research Grant. I would like to thank the following people for their comments on various drafts of this article: Linda Barnes, Marc Bekoff, J. Buettner-Janusch, Matt Cartmill, Jeremy Dahl, Bill D'Arcy, Steve Easley, Annette and Marcel Hladik, R. D. Martin, John McArdle, Stephen Molnar, Peter Raven, Bill Sawyer, Brian Suarez, Ian Tattersall and Steve Ward. I, of course, am responsible for the omissions and shortcomings.

Atsatt, P.R. and O'Dowd, D.J. (1976). *Science, 193,* 24-29.
Baker, H.G. (1965). *In* "Ecology and Economic Development in Tropical Africa," (D. Brokensha, ed.), pp. 185-216, Inst. Internat. Studs., Univ. California, Research Series 9, Berkeley.
Baker, H.G. and Harris, B.J. (1959). *Jl. W. Afr. Sci. Ass., 5,* 1-9.
Cartmill, M. (1974). *In* "Prosimian Biology," (R.D. Martin, G.A. Doyle, A.C. Walker, eds), pp. 655-670, Duckworth, London.
Charles-Dominique, P. (1975). *In* "Phylogeny of the Primates," (W.P. Luckett and F.S. Szalay, eds), pp. 69-88, Plenum Press, New York.
Charles-Dominique, P. and Hladik, C.M. (1971). *Terre et Vie, 25,* 3-66.

Clutton-Brock, T.H. (1975). *Folia primatol., 23,* 165-207.

Coe, H.J. and Isaac, F.M. (1965). *E. Afr. Wildlife, J., 3,* 123-124.

Ehrlich, P.R. and Raven, P.H. (1964). *Evolution, 18,* 586-608.

Faegri, K. and van der Pijl, L. (1971). "The Principles of Pollination Ecology," 2nd. ed., Pergamon Press, Oxford.

Fleming, T.H., Hooper, E.T., and Wilson, D.E. (1972). *Ecology, 53,* 555-569.

Freeland, W.J. and Janzen, D.H. (1974). *Am. Nat., 108,* 269-289.

Glander, K.E. (1975). *In* "Socioecology and Psychology of Primates," (R.H. Tuttle, ed.), pp. 37-57, Mouton, The Hague.

Glauert, L. (1958). *Austr. Mus. Mag., 12,* 284-286.

Heithaus, E.R., Fleming, T.H., and Opler, P.A. (1975). *Ecology, 56,* 841-854.

Hladik, C.M. (1975). *In* "Socioecology and Psychology of Primates," (R.H. Tuttle, ed.), pp. 3-35, Mouton, The Hague.

Hladik, C.M., Charles-Dominique, P., Valdebouze, P., Delort-Laval, J., and Flanzy, J. (1971). *C. r. hebd. Seanc. Acad. Sci., Paris, 272,* 3191-3194.

Hladik, C.M. and Hladik, A. (1967). *Biol. Gabon, 3,* 43-58.

Hladik, A. and Hladik, C.M. (1969). *Terre et Vie, 1,* 25-117.

Janzen, D.H. (1970). *Am. Nat., 104,* 501-528.

Levin, D.A. (1976). *Am. Nat., 110,* 261-284.

Martin, R.D. (1972). *Z. Tierpsychol.,* Beiheft *9,* 43-89.

McKey, D. (1975). *In* "Coevolution of Animals and Plants," (L.E. Gilbert and P.H. Raven, eds), pp. 159-191, Univ. of Texas Press, Austin.

McNab, B.K. (1971). *Ecology, 52,* 352-358.

Morton, E.S. (1973). *Am. Nat., 107,* 8-22.

Oppenheimer, J.R. and Lang, G.E. (1969). *Science, 165,* 187-188.

Orians, G.H. (1969). *Ecology, 50,* 783-801.

Pearson, D.L. (1975). *The Condor, 77,* 453-466.

Petter, J-J., Schilling, A., and Pariente, G. (1971). *Terre et Vie, 25,* 287-327.

Phillips, W.W.A. (1935). "Manual of Mammals of Ceylon," Publ. Ceylon J. Sci., Dulau and Co., Ltd., London.

Root, R.B. (1967). *Ecol. Monogr., 37,* 317-350.

Rourke, J. and Wiens, D. (1977). *Ann. Missouri Bot Garden, 64,* 1-17.

Smith, C.C. (1975). *In* "Coevolution of Animals and Plants," (L.E. Gilbert and P.H. Raven, eds), pp. 53-77, Univ. of Texas Press, Austin.

Smythe, N. (1970). *Am. Nat.,* 104, 25-35.

Snow, D.W. (1971). *Ibis, 113,* 194-202.

Struhaaker, T.T. and Oates, J.F. (1975). *In* "Socioecology and Psychology of Primates," (R.H. Tuttle, ed.), pp. 103-123, Mouton, The Hague.

Sussman, R.W. (1977). *In* "Primate Feeding Behaviour," (T.H. Clutton-Brock, ed.), pp. 1-36 Academic Press, London.

Sussman, R.W. and Tattersall, I. (1976). *Folia primatol., 26,* 270-283.

Tamsitt, J.R. (1967). *Nature, 13,* 784-786.

Tattersall, I. (in press). *In* Recent Advances in Primatology, Vol. 2. Evolution (D.J. Chivers and C.M. Hladik), Academic Press, London.

Tattersall, I. and Sussman, R.W. (1975). *Anthrop. Pap. Am. Mus. nat. Hist.,* *52,* 193-216.

van der Pijl, L. (1956). *Acta bot. neerl.,* *5,* 135-144.

van der Pijl, L. (1957). *Acta bot. neerl.,* *6,* 291-315.

Vose, H. (1973). *J. Mammal.,* *54,* 245-247.

Walker, E.P. (1968). "Mammals of the World," 2nd ed., Johns Hopkins Press, Baltimore.

Westoby, M. (1974). *Am. Nat., 108,* 290-304.

AUTHOR INDEX

SUBJECT INDEX

585